Electromagnetic Field Theory

Yaduvir Singh

Department of Electrical and Instrumentation Engineering
Thapar University, Patiala, Punjab

Delhi • Chennai • Chandigarh

Associate Acquisitions Editor: Sameer Gupta
Production Editor: Jennifer Sargunar
Composition: Mukesh Technologies Pvt. Ltd., Pondicherry
Printer: Pushp Print Services

ISBN 978-81-317-6061-1

First Impression

Published by Dorling Kindersley (India) Pvt. Ltd, licensees of Pearson Education in South Asia.

Head Office: 7th Floor, Knowledge Boulevard, A-8(A), Sector 62, Noida 201 309, UP, India.
Registered Office: 11 Community Centre, Panchsheel Park, New Delhi 110 017, India.

Dedicated to my parents

Dr Ganga Sagar and Smt. Vidya

About the Author

Yaduvir Singh is currently working as Associate Professor in the Department of Electrical and Instrumentation Engineering at Thapar University, Patiala, Punjab, India. Dr Singh obtained his Bachelor of Engineering degree from Dayalbagh Educational Institute (Deemed University), Agra, in 1991, his Master of Engineering degree from Motilal Nehru National Institute of Technology, University of Allahabad, Allahabad, in 1993, his Master of Business Management in Marketing Management from Newport University, the USA, in 1997 and his Doctor of Philosophy from Thapar University, Patiala, in 2004.

Dr Singh was a teaching faculty at the Institute of Engineering and Technology, Lucknow; N. E. Regional Institute of Science and Technology, Itanagar; G. B. Pant Engineering College, Pauri Garhwal and Harcourt Butler Technological Institute, Kanpur, before joining Thapar University as Assistant Professor in 2000. He later became Associate Professor in 2006. With more than 157 research publications in various conferences and journals, he has over eighteen years of teaching and research experience at UG, PG and Ph.D. levels of engineering and technology. He has guided students in more than 55 projects, 57 master-level theses/dissertations and several Ph.Ds. Dr Singh has already authored six bestselling technical books in the areas of electrical, electronics and soft computing. He is a member of several professional bodies and societies.

Dr Singh's research interests include areas of electrical and electronic systems, intelligent systems design, soft computing, automated control systems, artificially intelligent systems, industrial electronics, robotics, system modelling and identification, graph theory and design of instrumentation systems.

Contents

Preface

Books like friends, should be few and well-chosen.

—*Samuel Johnson*

Electromagnetic Field Theory is a fundamental subject of great research, academic and industrial importance and with a large number of applications. This subject has lead to the emergence of various state-of-the-art technologies in engineering areas. My teaching experience and interactions with the students of UG, PG and Ph.D. in Engineering and Technology of this course at IET, Lucknow; NERIST, Itanagar; GBPEC, Pauri; HBTI, Kanpur and Thapar University, Patiala, for nineteen odd years provided the incentive and impetus to write this book.

Electromagnetic Field Theory is designed to serve as a text book for physics and engineering students at the advanced undergraduate level or beginning graduate level of all major technological universities in India and abroad. This book provides an integrated treatment of all aspects of electromagnetic field theory like electrostatics and magnetostatics and demonstrates how these can be unified in one theory, classical electrodynamics and Lorentz's microscopic formulation of Maxwell's equations. The concepts on electromagnetic field theory are well treated in this text book, which provides an axiomatic foundation on this subject matter. Emphasis has been placed on modern electromagnetic concepts while programming tools used like MATLAB programs for some of the fundamental operations are presented as an appendix. It retains all the classical and modern concepts. A detailed analysis of important topics such as Smith Charts, Transmission lines and wave guides, some exposure to radiation and antennae, coordinate systems, vector calculus, inductors, capacitors, magnetic materials, electromagnetic waves and so on add to the utility of this book making it extremely useful for students.

Pedagogical features of the book include clearly labelled illustrations interspersed throughout the text to aid clear understanding, a large number of solved examples that help in clarifying concepts and solved problems and exercises at the end of each chapter that test the amount and quality of knowledge grasped by students. These features make the book an invaluable asset for students while preparing for semester examinations, GATE and various UPSC, PSU and private-sector tests and examinations. The student friendly approach used while developing this text book makes it a comprehensible and complete text book.

Contents and Structure

The book has been divided into 11 chapters. The topics covered in each chapter are described as given below:

Chapter 1 reviews the basics of vector algebra control and delves into unit vector, addition and multiplication of vectors, dot and cross products, scalar triple product, vector triple product and vector components.

Chapter 2 is devoted to the study of coordinate systems, viz. Cartesian coordinate, cylindrical coordinate and spherical coordinate systems. The chapter also delves into various transformations from one coordinate system into another.

Chapter 3 discusses vector calculus. Concepts of line and surface and volume integrals have been discussed. Further, gradient of scalar field, divergence of scalar fields, divergence theorems and its proof have been given. Concepts of curl of vector, Stoke's Theorem, various Laplacian operators and harmonic field have also been discussed.

Chapter 4 deals with conductors, insulators, resistance and Ohm's Law, basic components like capacitors and their various types, inductors and inductances of long solenoid, toroid, coaxial cable, solution for perfect dielectric conditions, nature of magnetic materials, permeability, magnetic susceptibility, magnetic d dipole and its classification.

Chapter 5 focuses on electrostatics. The topics covered in the chapter include Coulomb's Law and field intensity, electric dipole, electric field intensity on axial line of electric dipole, field intensity on equatorial line of electric dipole, electric field due to continuous charge distributions, electric flux density, electric potential, Gauss' Law and its applications, electrostatic energy or energy density. Most importantly, the Continuity Equation, Poisson's and Laplace's Equation and Boundary Condition at the interface of two mediums have been well discussed.

Chapter 6 provides magnetostatics and discusses the concepts of magnetic forces, magnetic torque and moment, Faraday's Laws of electromagnetic induction, Lenz's Law, Biot–Savart Law and its applications, Ampere's circuit and its applications, magnetic flux density, magnetic scalar and vector potential, magnetic boundary conditions, etc. in detail.

Chapter 7 deals with the most important aspect of electromagnetic field theory, i.e. Maxwell's first, second, third and fourth equations. Further, chapters include a summary of Maxwell's equation for time-varying fields and Maxwell's equation for sinusoidal time-varying field.

Chapter 8 delves into electromagnetic waves, wave equation, wave propagation in lossy ,dielectric medium, good conductor, good dielectrics and free space. The topics covered in this chapter also include skin depth, Poynting's Vector, Poynting's Theorem, reflection of a plane wave at normal and oblique incidence and parallel and perpendicular polarization.

Chapter 9 describes transmission line and its types. The Smith Chart and its applications have also been given. Further, this chapter comprises microstrip line, transmission line equation, lossless and lossy transmission lines, distortionless transmission line, input impedance, reflection coefficient and standing wave ratio and transmission line impedance.

Chapter 10 is devoted to wave guides. Topics covered in this chapter include the rectangular wave guide, various modes like transverse magnetic (TM) modes and transverse electric (TE) modes.

Chapter 11 is the last chapter and focuses on the application of electromagnetic field theory like antennas. This chapter includes topics on radiation and antennas. In addition, basic antenna parameters, arrays, retarded potentials, short dipole antenna, radiation resistance and broadside array have been dealt with.

Acknowledgements

During the course of writing this book, many people have supported me in countless different ways. I express my deepest sense of gratitude to all of them, whether mentioned here or not.

It had been most refreshing to be associated with the dynamic Pearson Education team. I greatly acknowledge the subject help rendered to me by my guru Dr P. S. Bimbhra from Thapar University, Patiala. I greatly acknowledge the motivation, professional help and support rendered to me by my buddies, my young and old friends from IIT Roorkee; IIT Delhi; IIT Kanpur; IIT Kharagpur; IIT Bombay; IIT Rajasthan; DEI Agra; NIT Allahabad; NIT Bhopal; BIET Jhansi; HBTI Kanpur; IET Lucknow; G. B. Pant University, Pant Nagar; DMRC; RVNL; CSIR; DRDO; HAL; Infosys and Samsung India.

This book is a tribute to my late grandparents, Late Sri Ganga Prasad, Late Smt. Rama Devi and Late Sri Chandrika Prasad. I am grateful to my grandmother Smt. S. Kumari, parents Dr Ganga Sagar and Smt. Vidya, my brother Dr Onkar Singh and his family members—Prateek, Sneha and Parveen—for their blessings, affection and inspiration. I am indebted to my maternal uncles Dr H. Ram and Mr Hans R. and their family members—Atul, Ekta, Pravin, Naveen, Aruna and Urmila—for their love and support. I am beholden to all my paternal uncles and their families and my in-laws. I place on record the technical help given to me by my PG students, Mr Amit Agrawal, Mr Sachin Agrahari and Mr Subhranshu Padhee.

I would also like to acknowledge the unfailing support of my entire family, especially my parents, Dr Ganga Sagar and Smt. Vidya; my wife, Prachi; and my sons, Vibhav Baba and Vinit Baba, who have encouraged me at every step of this process.

Yaduvir Singh

Introduction to Vector Algebra

1

1.1 INTRODUCTION

Vector analysis is a mathematical tool. Electromagnetic (EM) concepts are conveniently expressed and explained by vector analysis. A "scalar" is a quantity having only magnitude. It is described simply by a number. For example, physical quantities like mass, temperature, time, density, electrostatic potential, length, volume, distance, population, entropy, etc. are scalar quantities.

A "vector" is a quantity having both magnitude and direction. For example, force, velocity, electric field intensity, displacement, acceleration, etc. are vector quantities.

To differentiate between vector quantities and scalar quantities, we show a vector by an alphabet with an arrow over it. For example, $\vec{P}$ and $\vec{Q}$. Vector $\vec{P}$ is graphically represented by a line. The length of a line is equal to the magnitude of $\vec{P}$. It is denoted by $|\vec{P}|$ with an arrowhead at the end of the line. The arrow points towards the direction of $\vec{P}$. Four directions, east, west, north and south, are shown in Figure 1.1. For example, as also shown in Figure 1.1, vectors $\vec{P}$ and $\vec{Q}$ are directed in the northeast and north directions, respectively.

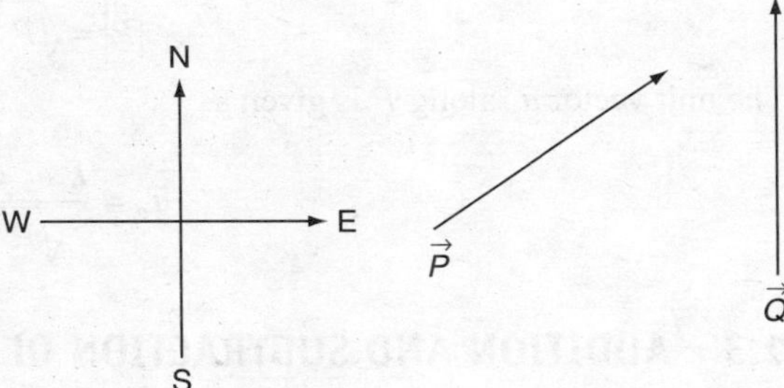

Figure 1.1 *Four directions and directed vectors*

1.2 UNIT VECTOR

It is a vector $\vec{P}$ that has both magnitude and direction. A unit vector has the same direction as that of the main vector. However, its magnitude is unity. A unit vector can be written in various forms as 1_p, i_p, a_p or u_p. A unit vector is given as the ratio of vector itself to its magnitude. It is given as

$$\vec{a}_P = \frac{\vec{P}}{|\vec{P}|}$$

where $|\vec{a}_P| = 1$ (unit magnitude). Thus, $\vec{P}$ is given as

$$\vec{P} = |\vec{P}|\vec{a}_P$$

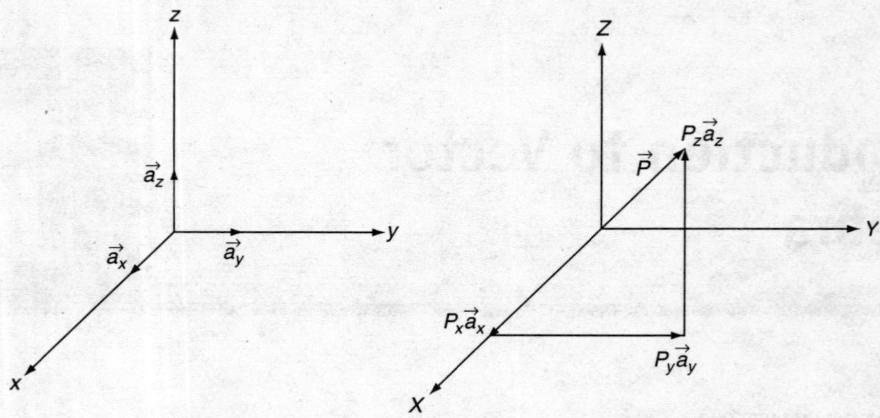

Figure 1.2 *Vector $\vec{P}$ in Cartesian coordinates*

In the expression above, $\vec{P}$ is specified in terms of its magnitude $|\vec{P}|$ and its direction $\vec{a}_P$ given by its unit vector.

The coordinates of a vector $\vec{P}$ in Cartesian are represented by (P_x, P_y, P_z), or alternatively, $P_x\vec{a}_x + P_y\vec{a}_y + P_z\vec{a}_z$. P_x, P_y and P_z are the components of $\vec{P}$ in the x, y and z directions. Alternatively, $\vec{a}_x$, $\vec{a}_y$ and $\vec{a}_z$ are unit vectors in the x, y and z directions, respectively. This representation is shown in Figure 1.2.

The magnitude of this vector $\vec{P}$ is given as

$$|\vec{P}| = \sqrt{P_x^2 + P_y^2 + P_z^2}$$

The unit vector $\vec{a}_p$ along $\vec{P}$ is given as

$$\vec{a}_p = \frac{P_x\vec{a}_x + P_y\vec{a}_y + P_z\vec{a}_z}{\sqrt{P_x^2 + P_y^2 + P_z^2}}$$

1.3 ADDITION AND SUBTRACTION OF VECTORS

When the sum (addition) of two or more than two vectors is done, the resultant is called a "vector sum." Two vectors can be added by using the Law of Parallelogram. It is shown in Figure 1.3. Two vectors to be added are represented by the sides of a parallelogram. Their addition or sum is represented by the diagonal (bigger vector).

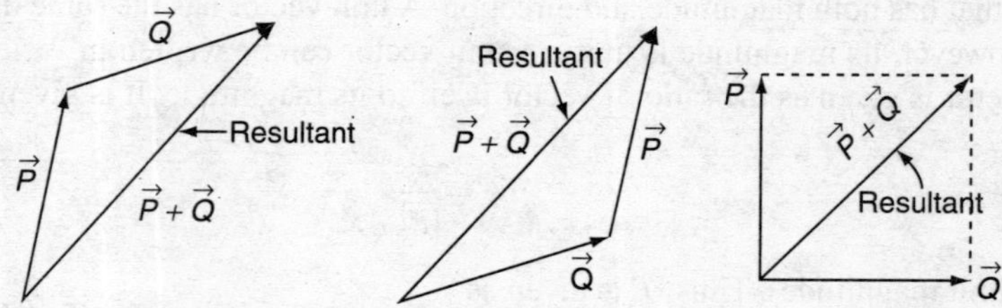

Figure 1.3 *Law of Parallelogram for the addition of a vector*

Addition of two or more than two vectors follows the Commutative Law. It is given as $\vec{P}+\vec{Q}=\vec{Q}+\vec{P}$. Also, the addition of two or more than two vectors follows the Associative Law. It is given as $\vec{P}+(\vec{Q}+\vec{R})=(\vec{P}+\vec{Q})+\vec{R}$. Subtraction of two or more than two vectors can be considered a special case of addition like $\vec{P}-\vec{Q}=\vec{P}+(-\vec{Q})$. It is shown in Figure 1.4.

The resultant vector $\vec{P}-\vec{Q}$ is shown in Figure 1.4. The resultant vector $\vec{P}-\vec{Q}$ is the diagonal of the parallelogram. It lies between the tips of $\vec{P}$ and $\vec{Q}$ vectors.

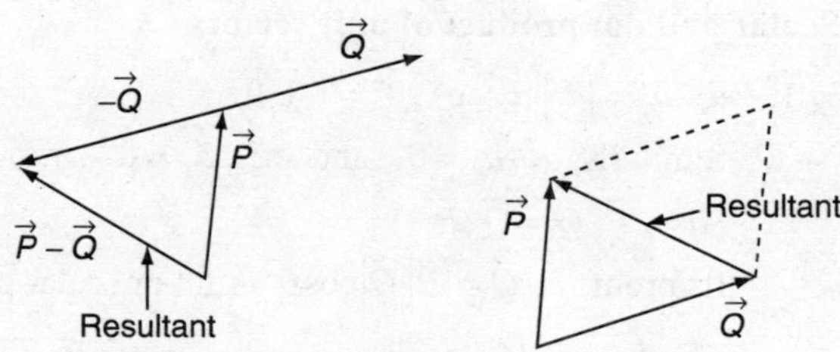

Figure 1.4 *Subtraction of two vectors*

1.4 MULTIPLICATION OF TWO VECTORS

If a vector is multiplied by a positive scalar quantity, the vector magnitude only changes, whereas the vector direction remains the same, as earlier. If the scalar is a negative quantity, then the direction of the vector will become reverse (opposite).

Multiplication of two vectors $\vec{P}$ and $\vec{Q}$ is either a vector or a scalar. It depends on how the multiplication has been carried out. Two ways of multiplication of two vectors are

(a) Scalar (or dot) product
(b) Vector (or cross) product

These products are elaborated below.

1.4.1 Scalar (or Dot) Products of Two Vectors

Scalar (or dot) product of two vectors $\vec{P}$ and $\vec{Q}$ is the product of their magnitudes and the cosine of the angle between them. It is given as scalar or dot product $\vec{P}\cdot\vec{Q}=|\vec{P}||\vec{Q}|\cos\theta$.

Here θ is the smaller angle between vectors $\vec{P}$ and $\vec{Q}$. Let $\vec{P}=P_x\vec{a}_x+P_y\vec{a}_y+P_z\vec{a}_z$ and $\vec{Q}=Q_x\vec{a}_x+Q_y\vec{a}_y+Q_z\vec{a}_z$.

Their scalar product or dot product is given as

$$\vec{P}\cdot\vec{Q}=P_xQ_x+P_yQ_y+P_zQ_z$$

Let us analyse the above for two cases:

Case 1: For $\theta=90°$; $\cos\theta=\cos 90°=0$

Therefore, $\vec{P}\cdot\vec{Q}=0$

In such a case, the two vectors are said to be "orthogonal" to one another.

Case 2: For $\theta=0°$; $\cos\theta=\cos 0°=1$

Therefore, $\vec{P}\cdot\vec{Q}=|\vec{P}||\vec{Q}|$.

In such a case, the two vectors are said to be "parallel." Some salient features of scalar product and dot products are mentioned below.

Scalar and dot product of unit vectors

1. $\vec{a}_x \cdot \vec{a}_y = |\vec{a}_x||\vec{a}_y|\cos 90° = 1 \cdot 1 \cdot 0$
   Similarly, $\vec{a}_y \cdot \vec{a}_z = 0$ and $\vec{a}_z \cdot \vec{a}_x = 0$
2. Also, $\vec{P} \cdot \vec{Q} = \vec{Q} \cdot \vec{P}$
   Its proof: $\vec{P} \cdot \vec{Q} = |\vec{P}||\vec{Q}|\cos\theta$ (dot or scalar product)
   $= |\vec{Q}||\vec{P}|\cos\theta$ or $= \vec{Q} \cdot \vec{P}$
3. Scalar or dot product follows the "Distributive Law." It is
   $\vec{P} \cdot (\vec{Q} + \vec{R}) = \vec{P} \cdot \vec{Q} + \vec{P} \cdot \vec{R}$

1.4.2 Vector (or Cross) Products of Two Vectors

Vector (or cross) product of two vectors $\vec{P}$ and $\vec{Q}$ is a vector quantity. The resultant's direction is perpendicular to the plane containing vectors $\vec{P}$ and $\vec{Q}$. The resultant's magnitude is given as the product of magnitudes of two vectors to be multiplied along with the sine of the angle between them. It is given as

Vector or cross product $\quad \vec{P} \times \vec{Q} = |\vec{P}||\vec{Q}|\sin\theta \cdot \vec{a}_n$

where $\vec{a}_n$ is a unit vector. This unit vector $\vec{a}_n$ is normal to the plane containing vectors $\vec{P}$ and $\vec{Q}$. The vector or cross product $\vec{P} \times \vec{Q}$ is governed by the "Right Hand Screw Rule." According to this rule, the resultant vector's direction is of movement of the right hand screw as it is turned from vector $\vec{P}$ towards vector $\vec{Q}$ through angle "θ" between them. The direction of $\vec{a}_n$ is the direction of the right thumb when the right-hand fingers rotate from $\vec{P}$ to $\vec{Q}$. It is shown in Figure 1.5.

Let $\quad \vec{P} = P_x\vec{a}_x + P_y\vec{a}_y + P_z\vec{a}_z$ and $\vec{Q} = Q_x\vec{a}_x + Q_y\vec{a}_y + Q_z\vec{a}_z$

$\vec{P} \times \vec{Q}$ is obtained as

$$\vec{P} \times \vec{Q} = \begin{vmatrix} \vec{a}_x & \vec{a}_y & \vec{a}_z \\ P_x & P_y & P_z \\ Q_x & Q_y & Q_z \end{vmatrix}$$

$$= (P_yQ_z - P_zQ_y)\vec{a}_x + (P_zQ_x - P_xQ_z)\vec{a}_y + (P_xQ_y - P_yQ_x)\vec{a}_z$$

Some salient features of the vector (or cross) product are as follows:

1. $\vec{P} \times \vec{Q} \neq \vec{Q} \times \vec{P}$
   However, $\vec{P} \times \vec{Q} = -\vec{Q} \times \vec{P}$ (anti-commutative)
2. It does not obey the Associative Law. Thus,
   $(\vec{P} \times \vec{Q}) \times \vec{R} \neq \vec{P} \times (\vec{Q} \times \vec{R})$
3. It obeys the Distributive Law. Thus,
   $\vec{P}(\vec{Q} + \vec{R}) = \vec{P} \times \vec{Q} + \vec{P} \times \vec{R}$
4. $\vec{P} \times \vec{P} = |\vec{P}||\vec{P}|\sin 0° \cdot \vec{a}_n = 0$ (vector or cross product of the same vector to itself is zero).
5. $\vec{a}_x \times \vec{a}_y = |\vec{a}_x||\vec{a}_y|\sin 90° \cdot \vec{a}_z = 1 \cdot 1 \cdot 1 \cdot \vec{a}_z = \vec{a}_z$ (vector (or cross) product of two unit vectors is the third unit vector).

Similarly,

$$\vec{a}_y \times \vec{a}_z = \vec{a}_x \quad \text{and} \quad \vec{a}_z \times \vec{a}_x = \vec{a}_y$$

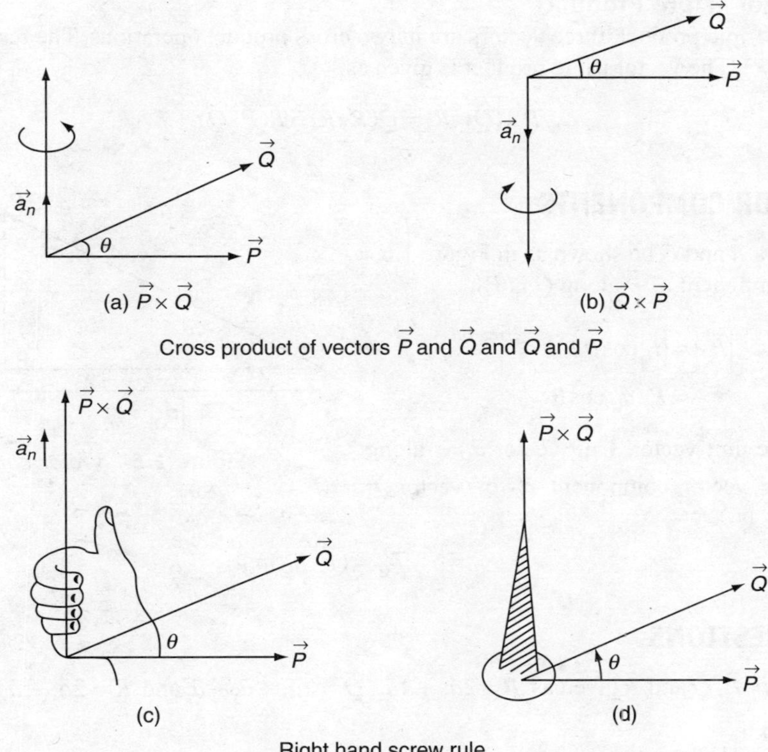

Figure 1.5 *Vector (or cross) product and right-hand screw rule*

1.5 MULTIPLICATION OF THREE VECTORS

Multiplication of three vectors $\vec{P}$, $\vec{Q}$ and $\vec{R}$ can give a scalar or vector resultant. Two ways of multiplication of three vectors are:

(a) Scalar triple product and
(b) Vector triple product

These triple products are elaborated below.

1.5.1 Scalar Triple Product

In this type of triple product, three vectors are there in dot product and cross product. The resultant is scalar. The scalar triple product of vectors $\vec{P}, \vec{Q}$ and $\vec{R}$ is given as

$$\vec{P}\cdot(\vec{Q}\times\vec{R}) = \vec{R}\cdot(\vec{P}\times\vec{Q}) = \vec{Q}\cdot(\vec{R}\times\vec{P})$$

It is solved as

$$\vec{P}\cdot(\vec{Q}\times\vec{R}) = \begin{vmatrix} P_x & P_y & P_z \\ Q_x & Q_y & Q_z \\ R_x & R_y & R_z \end{vmatrix}$$

1.5.2 Vector Triple Product

In this type of triple product, three vectors are in two cross product operations. The resultant is a vector like $\vec{P}\times(\vec{Q}\times\vec{R})$. The vector triple product is given as

$$\vec{P}\times(\vec{Q}\times\vec{R})=\vec{Q}(\vec{P}\cdot\vec{R})-\vec{R}(\vec{P}\cdot\vec{Q})$$

1.6 VECTOR COMPONENTS

Let two vectors $\vec{P}$ and $\vec{Q}$ be shown as in Figure 1.6. The vector component of $\vec{P}$ along $\vec{Q}$ is $\vec{P}_Q$.
$\vec{P}_Q$ is given as

$$\begin{aligned}\left|\vec{P}_Q\right| &= \left|\vec{P}\right|\cos\theta \\ &= \vec{P}\cdot\vec{a}_Q\cos\theta\end{aligned}$$

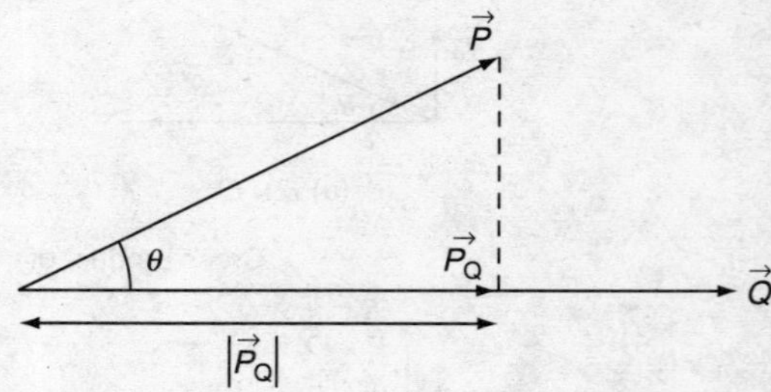

Figure 1.6 *Vector components*

Here, $\vec{a}_Q$ is the unit vector. Unit vector $\vec{a}_Q$ is along vector $\vec{Q}$. The vector component $\vec{P}_Q$ of vector $\vec{P}$ along vector $\vec{Q}$ is

$$\vec{P}_Q=\left|\vec{P}_Q\right|\vec{a}_Q=(\vec{P}\cdot\vec{a}_Q)\vec{a}_Q$$

SOLVED QUESTIONS

1.1 For vectors $\vec{P}$, $\vec{Q}$ and $\vec{R}$ given as $\vec{P}=2\vec{a}_y+4\vec{a}_z$, $\vec{Q}=2\vec{a}_x+\vec{a}_y-\vec{a}_z$ and $\vec{R}=2\vec{a}_x+\vec{a}_y$, find:

(a) $\vec{P}+\vec{Q}$
(b) $\vec{Q}-\vec{R}$
(c) $\left|\vec{Q}\right|$
(d) Unit vector $\vec{a}_R$
(e) $\vec{Q}\cdot\vec{R}$
(f) Angle between $\vec{P}$ and $\vec{Q}$
(g) $\vec{P}\times\vec{R}$
(h) Unit vector normal to $\vec{P}$ and $\vec{Q}$
(i) $\vec{P}\cdot(\vec{Q}\times\vec{R})$
(j) $(\vec{P}\times\vec{Q})\times\vec{R}$

Solution:

(a) $\vec{P}+\vec{Q}=2\vec{a}_x+3\vec{a}_y+3\vec{a}_z$

(b) $\vec{Q}-\vec{R}=-\vec{a}_z$

(c) $\left|\vec{Q}\right|=\sqrt{2^2+1^2+(-1)^2}=\sqrt{6}$

(d) $\vec{a}_R=\dfrac{\vec{R}}{\left|\vec{R}\right|}=\dfrac{2\vec{a}_x+\vec{a}_y}{\sqrt{2^2+1^2}}=\dfrac{1}{\sqrt{5}}(2\vec{a}_x+\vec{a}_y)$

(e) $\vec{Q}\cdot\vec{R}=(2\vec{a}_x+\vec{a}_y-\vec{a}_z)\cdot(2\vec{a}_x+\vec{a}_y)=(2\times2)+(1\times1)=5$

(f) Let the angle between $\vec{P}$ and $\vec{Q} = \theta$

$$\vec{P}\cdot\vec{Q} = |\vec{P}||\vec{Q}|\cos\theta$$

$$\Rightarrow \cos\theta = \frac{(2\vec{a}_y + 4\vec{a}_z)\cdot(2\vec{a}_x + \vec{a}_y - \vec{a}_z)}{|\vec{P}||\vec{Q}|}$$

$$\Rightarrow \cos\theta = \frac{0 + (2\times 1) + (4\times -1)}{\sqrt{2^2+4^2}\sqrt{2^2+1^2+(-1)^2}}$$

$$\Rightarrow \cos\theta = \frac{-2}{\sqrt{20}\sqrt{6}}$$

$$\Rightarrow \cos\theta = -\frac{1}{\sqrt{30}}$$

$$\Rightarrow \theta = \cos^{-1}\left(-\frac{1}{\sqrt{30}}\right)$$

(g) $\vec{P}\cdot\vec{R} = (2\vec{a}_y + 4\vec{a}_z)\times(2\vec{a}_x + \vec{a}_y) = \begin{vmatrix} \vec{a}_x & \vec{a}_y & \vec{a}_z \\ 0 & 2 & 4 \\ 2 & 1 & 0 \end{vmatrix} = -4\vec{a}_x + 8\vec{a}_y - 4\vec{a}_z$

(h) By definition,

$$\vec{P}\times\vec{Q} = |\vec{P}||\vec{Q}|\sin\theta\cdot\vec{a}_n$$

$$\vec{a}_n = \frac{\vec{P}\times\vec{Q}}{|\vec{P}||\vec{Q}|\sin\theta}$$

$$\vec{a}_n = \frac{\vec{P}\times\vec{Q}}{|\vec{P}\times\vec{Q}|}$$

Now finding $(\vec{P}\times\vec{Q})$

$$\begin{vmatrix} \vec{a}_x & \vec{a}_y & \vec{a}_z \\ 0 & 2 & 4 \\ 2 & 1 & -1 \end{vmatrix} = -6\vec{a}_x + 8\vec{a}_y - 4\vec{a}_z$$

$$|\vec{P}\times\vec{Q}| = \sqrt{(-6)^2 + (8)^2 + (-4)^2} = \sqrt{116} = 2\sqrt{29}$$

Unit vector normal to $\vec{P}$ and $\vec{Q} = \dfrac{-6\vec{a}_x + 8\vec{a}_y - 4\vec{a}_z}{2\sqrt{29}}$

$$\vec{a}_n = \frac{1}{\sqrt{29}}(-3\vec{a}_x + 4\vec{a}_y - 2\vec{a}_z)$$

(i) $\vec{P}\cdot(\vec{Q}\times\vec{R}) = \begin{vmatrix} P_x & P_y & P_z \\ Q_x & Q_y & Q_z \\ R_x & R_y & R_z \end{vmatrix} = \begin{vmatrix} 0 & 2 & 4 \\ 2 & 1 & -1 \\ 2 & 1 & 0 \end{vmatrix} = 0 + (-2)\cdot 2 + 4(2-2) = -4$

(j) $\vec{P}\times\vec{Q}\times\vec{R}=(\vec{P}\times\vec{Q})\times\vec{R}=\begin{vmatrix}\vec{a}_x & \vec{a}_y & \vec{a}_z\\ -6 & 8 & -4\\ 2 & 1 & 0\end{vmatrix}=4\vec{a}_x-8\vec{a}_y+(-22)\vec{a}_z=4\vec{a}_x-8\vec{a}_y-22\vec{a}_z$

1.2 If vectors $\vec{P}=\vec{a}_x+3\vec{a}_z$ and $\vec{Q}=5\vec{a}_x+2\vec{a}_y-6\vec{a}_z$, what is the component of vector $\vec{P}$ along $\vec{a}_y$? Also, find the projection of vector $\vec{P}$ on vector $\vec{Q}$.

Solution:

Given vector $\vec{P}=\vec{a}_x+3\vec{a}_z$

Component of $\vec{P}$ along $\vec{a}_y=0$

Finding $\vec{P}\cdot\vec{Q}$ as

$$\vec{P}\cdot\vec{Q}=|\vec{P}||\vec{Q}|\cos\theta$$

$$\Rightarrow|\vec{P}|\cos\theta=\frac{\vec{P}\times\vec{Q}}{|\vec{Q}|}=\frac{(\vec{a}_x+3\vec{a}_z)\cdot(5\vec{a}_x+2\vec{a}_y-6\vec{a}_z)}{|\vec{Q}|}$$

$$=\frac{5-18}{\sqrt{25+4+36}}=\frac{-13}{\pm\sqrt{65}}=\sqrt{\frac{13}{5}}\quad\text{(taking only the positive value)}$$

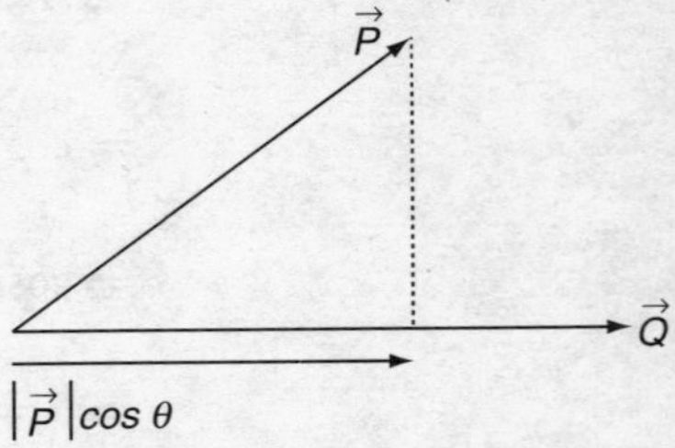

Figure 1.7 *Projection of vector $\vec{P}$ on vector $\vec{Q}$*

1.3 Prove that the cosine of angle "θ" between two vectors $\vec{P}$ and $\vec{Q}$ is the sum of the product of their direction cosines.

Solution:

Let cos α_1, cos α_2 and cos α_3 be the direction cosines of vector $\vec{P}$. Similarly, let cos β_1, cos β_2 and cos β_3 be the direction cosines of vector $\vec{Q}$.

Now
$$\vec{P}=P_x\vec{a}_x+P_y\vec{a}_y+P_z\vec{a}_z$$

$$|\vec{P}|=\sqrt{P_x^2+P_y^2+P_z^2}\quad\text{or}$$

$$\vec{P}=\sqrt{P_x^2+P_y^2+P_z^2}\left[\frac{P_x}{\sqrt{P_x^2+P_y^2+P_z^2}}\vec{a}_x+\frac{P_y}{\sqrt{P_x^2+P_y^2+P_z^2}}\vec{a}_y+\frac{P_z}{\sqrt{P_x^2+P_y^2+P_z^2}}\vec{a}_z\right]\quad\text{or}$$

$$\vec{P}=|\vec{P}|\left[\cos\alpha_1\vec{a}_x+\cos\alpha_2\vec{a}_y+\cos\alpha_3\vec{a}_z\right]$$

Here,

$$P_x=|\vec{P}|\cos\alpha_1$$

$$P_y=|\vec{P}|\cos\alpha_2\quad\text{and}$$

$$P_z=|\vec{P}|\cos\alpha_3$$

Similarly, solving for vector $\vec{Q}$, we get

$$\vec{Q} = |\vec{Q}|\left[\cos\beta_1\vec{a}_x + \cos\beta_2\vec{a}_y + \cos\beta_3\vec{a}_z\right]$$

$\vec{P}\cdot\vec{Q}$ is obtained as

$$\vec{P}\cdot\vec{Q} = |\vec{P}||\vec{Q}|\cos\theta$$
$$= |\vec{P}||\vec{Q}|\left[\cos\alpha_1\cos\beta_1 + \cos\alpha_2\cos\beta_2 + \cos\alpha_3\cos\beta_3\right]$$

Here, θ is the angle between vectors $\vec{P}$ and $\vec{Q}$ and

$$\cos\theta = \cos\alpha_1\cos\beta_1 + \cos\alpha_2\cos\beta_2 + \cos\alpha_3\cos\beta_3$$

1.4 Consider vector $\vec{P} = 2\vec{a}_x - \vec{a}_z$ and vector $\vec{Q} = 2\vec{a}_x - \vec{a}_y + 2\vec{a}_z$. Find the unit vector that will be perpendicular to both vectors $\vec{P}$ and $\vec{Q}$.

Solution:

$$\vec{P}\times\vec{Q} = \begin{vmatrix} \vec{a}_x & \vec{a}_y & \vec{a}_z \\ 2 & 0 & -1 \\ 2 & -1 & 2 \end{vmatrix} = -\vec{a}_x - 6\vec{a}_y - 2\vec{a}_z \quad \text{and}$$

$$|\vec{P}\times\vec{Q}| = \sqrt{1+36+4} = \sqrt{41}$$

Also

$$\vec{P}\times\vec{Q} = |\vec{P}||\vec{Q}|\sin\theta\cdot\vec{a}_n = |\vec{P}\times\vec{Q}|\vec{a}_n$$

$$\vec{a}_n = \frac{\vec{P}\times\vec{Q}}{|\vec{P}\times\vec{Q}|} = \frac{-\vec{a}_x - 6\vec{a}_y - 2\vec{a}_z}{\sqrt{41}}$$
$$= \frac{-1}{\sqrt{41}}\vec{a}_x - \frac{6}{\sqrt{41}}\vec{a}_y - \frac{2}{\sqrt{41}}\vec{a}_z$$

1.5 If vectors $\vec{P} = 2\vec{a}_x + \vec{a}_z$ and $\vec{Q} = 5\vec{a}_x + 4\vec{a}_y - 3\vec{a}_z$, what is the component of vector $\vec{P}$ along $\vec{a}_y$? Also, find the projection of vector $\vec{P}$ on vector $\vec{Q}$.

Solution:
Given,

$$\vec{P} = 2\,\vec{a}_x + \vec{a}_z$$

Component of $\vec{P}$ along $\vec{a}_y = 0$.
We know that

$$\vec{P}\cdot\vec{Q} = |\vec{P}||\vec{Q}|\cos\theta$$

Therefore,

$$|\vec{P}|\cos\theta = \frac{\vec{P}\cdot\vec{Q}}{|\vec{Q}|} = \frac{(2\vec{a}_x + \vec{a}_z)\cdot(5\vec{a}_x + 4\vec{a}_y - \vec{a}_z)}{\sqrt{(5)^2 + (4)^2 + (-3)^2}}$$
$$= \frac{10-3}{\sqrt{50}} = \frac{7}{\sqrt{50}}$$

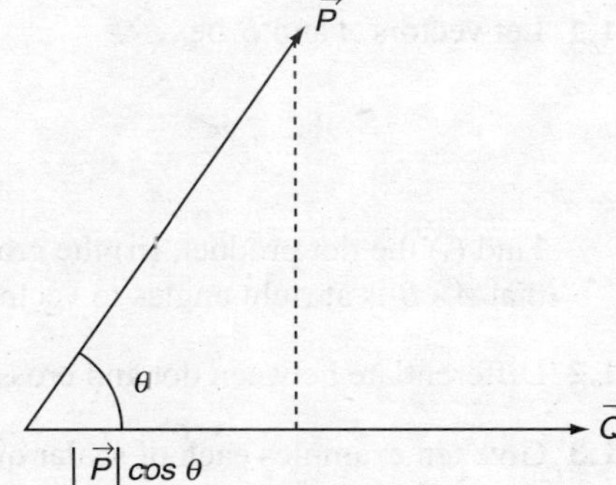

Figure 1.8 *Projection of vector $\vec{P}$ on $\vec{Q}$*

Hence, the projection of $\vec{P}$ on $\vec{Q}$ is $7/\sqrt{50}$.

1.6 Given vector $\vec{A} = 2\vec{a}_x - \vec{a}_y + 2\vec{a}_z$, find the expression of a unit vector $\vec{B}$, which is parallel to $\vec{A}$.

Solution:

Let $\vec{B} = B_x\,\vec{a}_x + B_y\,\vec{a}_y + B_z\,\vec{a}_z$. We know that

$$\left|\vec{B}\right| = (B_x^2 + B_y^2 + B_z^2)^{1/2} = 1 \tag{1}$$

and $\vec{A} = A_x\vec{a}_x + A_y\vec{a}_y + A_z\vec{a}_z$, where $A_x = 2$, $A_y = -1$ and $A_z = 2$. For two vectors to be parallel, we have $\vec{B}\times\vec{A} = 0$. Therefore,

$$\vec{B}\times\vec{A} = \begin{vmatrix} \vec{a}_x & \vec{a}_y & \vec{a}_z \\ B_x & B_y & B_z \\ A_x & A_y & A_z \end{vmatrix}$$

$$= \vec{a}_x(B_yA_z - A_yB_z) - \vec{a}_y(A_zB_x - B_zA_x) + \vec{a}_z(B_xA_y - A_xB_y) = 0 \tag{2}$$

Putting the value of A_x, A_y, A_z and B_x, B_y, B_z in Equation (2), we get

$$2B_y + B_z = 0 \tag{3}$$

$$2B_x - 2B_z = 0 \tag{4}$$

$$-B_x - 2B_y = 0 \tag{5}$$

Equations (3)–(5) are independent; therefore, using Equations (1), (3)–(5), we get

$$B_x = \frac{-2}{3}, \quad B_y = \frac{1}{3} \quad \text{and} \quad B_z = \frac{-2}{3}$$

Therefore,

$$\vec{B} = \frac{1}{0.99}\left(\frac{-2}{3}\vec{a}_x + \frac{1}{3}\vec{a}_y - \frac{2}{3}\vec{a}_z\right)$$

UNSOLVED QUESTIONS

1.1 Let vectors $\vec{A}$ and $\vec{B}$ be

$$\vec{A} = 4\vec{a}_x + 3\vec{a}_y + 0\vec{a}_z$$

$$\vec{B} = 3\vec{a}_x + 5\vec{a}_y + 2\vec{a}_z$$

Find (i) the dot product, (ii) the cross product and (iii) the angle between vectors $\vec{A}$ and $\vec{B}$. Show that $\vec{A}\times\vec{B}$ is at right angles to vector $\vec{A}$.

1.2 Differentiate between dot and cross products with suitable examples.

1.3 Give ten examples each of scalar quantities and vector quantities from your surrounding.

1.4 What is the importance of a unit vector? What is its magnitude? Why?

1.5 Define a vector. Give suitable examples.

1.6 What are vector components? Give a suitable illustration while clearly depicting vectors and vector components.

1.7 Give some numerical examples to show vector addition.

1.8 Give some numerical examples to show vector subtraction.

1.9 Give some numerical examples to show a scalar triple product.

1.10 Give some numerical examples to show a vector triple product.

SUMMARY

1. A scalar quantity has only magnitude.
2. A vector quantity has both magnitude and direction.
3. A unit vector has unity magnitude but its direction is the same as that of the main vector.
4. Coordinates of a vector $\vec{P}$ in Cartesian are represented by (P_x, P_y, P_z).
5. Vectors are added by using the Law of Parallelogram.
6. There are two ways of multiplication of two vectors, i.e. scalar (or dot) product and vector (or cross) product.
7. The scalar or dot product of two vectors $\vec{P}$ and $\vec{Q}$ is given as $\vec{P}\cdot\vec{Q} = |\vec{P}||\vec{Q}|\cos\theta$.
8. The vector or cross product of two vectors $\vec{P}$ and $\vec{Q}$ is given as

$$\vec{P}\times\vec{Q} = |\vec{P}||\vec{Q}|\sin\theta.\vec{a}_n$$

where $\vec{a}_n$ is the unit vector that is normal to the plane containing vectors $\vec{P}$ and $\vec{Q}$.

MULTIPLE-CHOICE QUESTIONS

1. A scalar quantity has __________.
(a) magnitude (b) direction (c) both (a) and (b) (d) none of these

2. Which of the following is/are scalar quantity(s)?
(a) distance (b) density (c) temperature (d) all of these

3. Which of the following is not a scalar quantity?
(a) entropy (b) displacement (c) volume (d) mass

4. A vector quantity has __________.
(a) magnitude (b) direction (c) both (a) and (b) (d) none of these

5. Which of the following is/are vector quantity(s)?
(a) force (b) electric field intensity
(c) acceleration (d) all of these

6. A unit vector has its magnitude as ____________.
(a) 0 (b) 1 (c) ∞ (d) none of these

7. A unit vector has _______ direction to that of the main vector.
(a) same (b) opposite
(c) normal upwards (d) normal downwards

8. A vector $\vec{P}$ in Cartesian coordinates is represented by ____________.
(a) (P_x, P_ϕ, P_z) (b) (P_x, P_y, P_z) (c) (P_x, P_ϕ, P_θ) (d) none of these

9. ________ of two vectors uses the Law of Parallelogram.
(a) multiplication (b) division (c) addition (d) all of these

10. $\vec{P}+(\vec{Q}+\vec{R})=(\vec{P}+\vec{Q})+\vec{R}$ demonstrates ________.
(a) Commutative Law (b) Associative Law
(c) Distributive Law (d) Law of Parallelogram

11. Multiplication of two vectors is __________.
(a) vector (b) scalar (c) either vector or scalar (d) cannot say

12. If two vectors are orthogonal, then their scalar product is __________.
(a) 1 (b) 0 (c) ∞ (d) 100

13. ___________ product is governed by the Right-Hand Screw Rule.
(a) vector (b) scalar (c) simple mathematical (d) none of these

14. The vector product obeys the __________ law.
(a) Commutative (b) Associative (c) Distributive (d) Parallelogram

15. Which of the following is/are in correct?
(a) $\vec{a}_x \times \vec{a}_y = \vec{a}_z$ (b) $\vec{a}_y \times \vec{a}_z = \vec{a}_x$ (c) $\vec{a}_z \times \vec{a}_x = \vec{a}_y$ (d) none of these

16. $\vec{P}\cdot(\vec{Q}\times\vec{R}) =$ ____________.
(a) $\vec{R}\cdot(\vec{P}\times\vec{Q})$ (b) $\vec{Q}\cdot(\vec{R}\times\vec{P})$ (c) both (a) and (b) (d) none of these

17. $\vec{P}\times(\vec{Q}\times\vec{R}) =$ ____________.
(a) $\vec{Q}(\vec{P}\cdot\vec{R})+\vec{R}(\vec{P}\cdot\vec{Q})$ (b) $\vec{Q}\cdot(\vec{P}\cdot\vec{R})-\vec{R}(\vec{P}\cdot\vec{Q})$
(c) $\vec{P}\times\vec{Q}\times\vec{R}$ (d) none of these

18. The cross product of the same vector to itself is __________.
(a) 0 (b) 1 (c) ∞ (d) 100

19. ____________ product of two unit vectors is the third unit vector.
(a) vector (b) scalar (c) both (a) and (b) (d) none of these

20. 'r' in Cylindrical coordinates corresponding to Cartesian coordinate (3, 4, 5) is ____________.
(a) 12 (b) 7 (c) 10 (d) 5

ANSWERS TO MULTIPLE-CHOICE QUESTIONS

(1) a; (2) d; (3) b; (4) c; (5) d; (6) b; (7) a; (8) b; (9) c; (10) b;
(11) c; (12) b; (13) a; (14) c; (15) d; (16) c; (17) b; (18) a; (19) a; (20) d

Coordinate Systems

2

2.1 INTRODUCTION

Coordinate system represents a point $P(x, y, z)$ in a coordinate space. Three orthogonal coordinate systems are discussed below. A point or a vector can be represented in an orthogonal or a non-orthogonal curvilinear coordinate system. In orthogonal coordinate systems, coordinates are mutually perpendicular.

2.2 CARTESIAN COORDINATE SYSTEM

In this, three coordinate axes are mutually perpendicular to each other. The axes are x, y and z. A position vector $\vec{P}$ at the origin of such a Cartesian coordinate system is shown in Figure 2.1. Position vector $\vec{P}$ may be resolved into three component vectors viz. $\vec{P}_x$, $\vec{P}_y$ and $\vec{P}_z$. Vector $\vec{P}$ can be written as

$$\vec{P} = P_x\vec{a}_x + P_y\vec{a}_y + P_z\vec{a}_z$$

where $\vec{a}_x$, $\vec{a}_y$ and $\vec{a}_z$ are unit vectors, $\vec{a}_x$ along the x-axis, $\vec{a}_y$ along the y-axis and $\vec{a}_z$ along the z-axis.

Point A is shown in terms of (x, y, z) with ranges as

$$-\infty < x < \infty$$

$$-\infty < y < \infty \quad \text{and}$$

$$-\infty < z < \infty$$

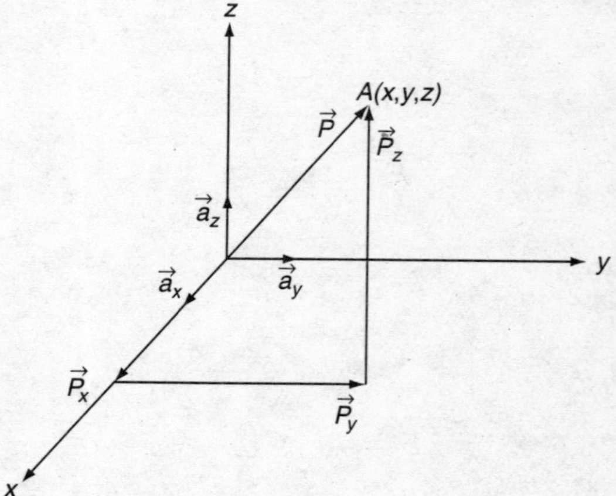

Figure 2.1 *Position vector in a Cartesian coordinate system*

2.3 CYLINDRICAL COORDINATE SYSTEM

Any point A in Cylindrical coordinates is shown as $A(r, \phi, z)$. It is shown in Figure 2.2.

Any vector $\vec{P}$ (in a Cylindrical coordinate system) is written as

$$\vec{P} = P_r\vec{a}_r + P_\phi\vec{a}_\phi + P_z\vec{a}_z$$

where $\vec{a}_r$, $\vec{a}_\phi$ and $\vec{a}_z$ are unit vectors. $\vec{a}_r$ is directed outwards radially, and $\vec{a}_r$ is normal to the cylinder surface at the point of consideration. Thus, $\vec{a}_r$ increases along the r-direction. Unit vector $\vec{a}_\phi$ is perpendicular to ϕ = constant plane. It is directed outwards in the advance ϕ-direction. Unit vector $\vec{a}_z$ is perpendicular to z = constant plane. It increases towards the z-direction. The magnitude of $\vec{P}$ is given as

$$\left|\vec{P}\right| = (P_r^2 + P_\phi^2 + P_z^2)^{1/2}$$

The relationships between Cartesian coordinate system and Cylindrical coordinate system are shown in Figure 2.3.

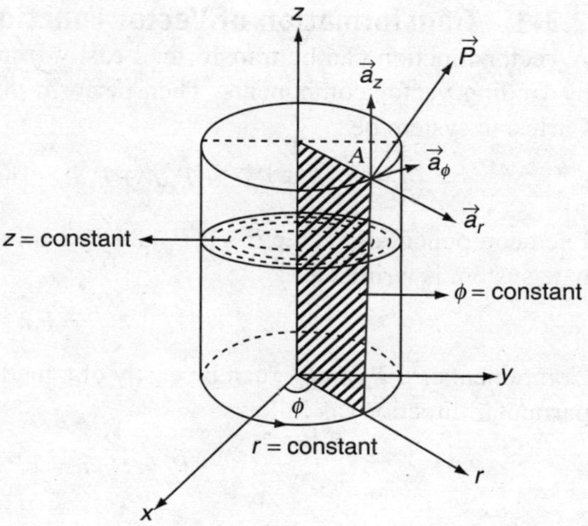

Here, variables r, ϕ and z are given as

Figure 2.2 *Cylindrical coordinate system*

$$r = \sqrt{x^2 + y^2}$$

$$\phi = \tan^{-1}\frac{y}{x} \quad \text{and} \quad z = z \quad \text{or}$$

$$x = r\cos\phi \quad \text{and} \quad y = r\sin\phi$$

$$z = z$$

The ranges are

$$0 \le r < \infty$$
$$0 \le \phi < 2\pi \quad \text{and}$$
$$-\infty < z < \infty$$

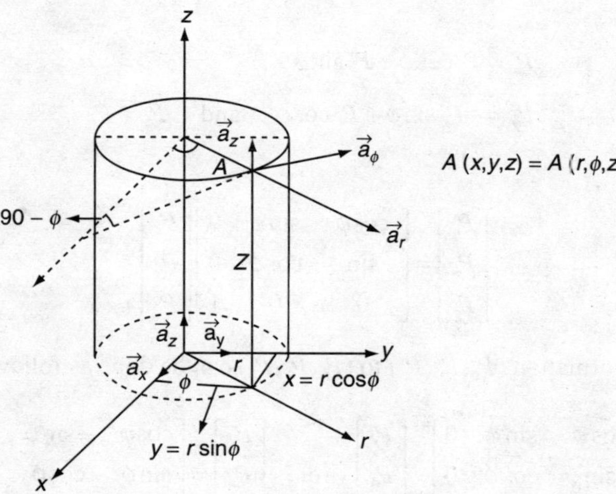

Figure 2.3 *Relationship between Cartesian coordinate system and Cylindrical coordinate system*

2.3.1 Transformation of Vector Function

A vector function can be transformed easily from one coordinate system into another. It can be done by finding vector components. Then perform dot product between unit vectors. Let a vector $\vec{P}$ in a Cartesian system be

$$\vec{P} = P_x\vec{a}_x + P_y\vec{a}_y + P_z\vec{a}_z \quad \text{(in a Cartesian coordinate system)}$$

Each component of vector $\vec{P}$ P_x, P_y and P_z is a function of x, y and z. Vector $\vec{P}$ in a Cylindrical coordinate system is written as

$$\vec{P} = P_r\vec{a}_r + P_\phi\vec{a}_\phi + P_z\vec{a}_z$$

Components P_r, P_ϕ and P_z can be easily obtained by multiplying the vector and the unit vector (in that particular direction) as follows:

$$\begin{aligned} P_r &= \vec{P}\cdot\vec{a}_r = (P_x\vec{a}_x + P_y\vec{a}_y + P_z\vec{a}_z)\cdot\vec{a}_r \\ &= P_x\vec{a}_x\cdot\vec{a}_r + P_y\vec{a}_y\cdot\vec{a}_r \end{aligned}$$

$$\begin{aligned} P_\phi &= \vec{P}\cdot\vec{a}_r = (P_x\vec{a}_x + P_y\vec{a}_y + P_z\vec{a}_z)\cdot\vec{a}_\phi \\ &= P_x\vec{a}_x\cdot\vec{a}_\phi + P_y\vec{a}_y\cdot\vec{a}_\phi \end{aligned} \quad \text{and}$$

$$P_z = P_z$$

Dot multiplication of vectors is done as follows:

$$\vec{a}_x\cdot\vec{a}_r = \cos\phi$$

$$\vec{a}_x\cdot\vec{a}_\phi = \cos(90° + \phi) = -\sin\phi$$

$$\vec{a}_y\cdot\vec{a}_r = \cos(90° - \phi) = \sin\phi \quad \text{and} \quad \vec{a}_y\cdot\vec{a}_\phi = \cos\phi$$

Thus, we get

$$P_r = P_x\cos\phi + P_y\sin\phi$$

$$P_\phi = -P_x\sin\phi + P_y\cos\phi \quad \text{and} \quad P_z = P_z$$

In matrix form

$$\begin{bmatrix} P_r \\ P_\phi \\ P_z \end{bmatrix} = \begin{bmatrix} \cos\phi & \sin\phi & 0 \\ -\sin\phi & \cos\phi & 0 \\ 0 & 0 & 1 \end{bmatrix}\begin{bmatrix} P_x \\ P_y \\ P_z \end{bmatrix}$$

The inverse of the transformation (P_x, P_y, P_z) to (P_r, P_ϕ, P_z) can be done as follows:

$$\begin{bmatrix} P_x \\ P_y \\ P_z \end{bmatrix} = \begin{bmatrix} \cos\phi & \sin\phi & 0 \\ -\sin\phi & \cos\phi & 0 \\ 0 & 0 & 1 \end{bmatrix}^{-1}\begin{bmatrix} P_r \\ P_\phi \\ P_z \end{bmatrix} \quad \text{or} \quad \begin{bmatrix} P_x \\ P_y \\ P_z \end{bmatrix} = \begin{bmatrix} \cos\phi & -\sin\phi & 0 \\ \sin\phi & \cos\phi & 0 \\ 0 & 0 & 1 \end{bmatrix}^{-1}\begin{bmatrix} P_r \\ P_\phi \\ P_z \end{bmatrix}$$

2.4 SPHERICAL COORDINATE SYSTEM

Any point A in a Spherical coordinate system is represented as $A(r,\theta,\phi)$. It is shown in Figure 2.4, where r is the distance from the origin to point A (radius of sphere), θ the angle of elevation, which is between the z-axis and the position vector and ϕ the azimuthal angle, which is measured from the x-axis.

Vector $\vec{P}$ in a Spherical coordinate system is given as

$$\vec{P} = P_r\vec{a}_r + P_\theta\vec{a}_\theta + P_\phi\vec{a}_\phi$$

where $\vec{a}_r$ is a unit vector directed along the radius. Unit vector $\vec{a}_r$ is perpendicular to the spherical surface. It is directed in the increasing r-direction. $\vec{a}_\theta$ is a unit vector perpendicular to the surface of a cone. It is directed in the direction of increasing θ. $\vec{a}_\phi$ is a unit vector perpendicular to the shifted x–z plane. Unit vector $\vec{a}_\phi$ is in the direction of increasing ϕ. The ranges are given as follows:

$$0 \le r \le \infty$$
$$0 \le \theta \le \pi \qquad \text{and}$$
$$0 \le \phi < 2\pi$$

The magnitude of $\vec{P}$ is given as

$$\left|\vec{P}\right| = \sqrt{P_r^2 + P_\theta^2 + P_\phi^2}$$

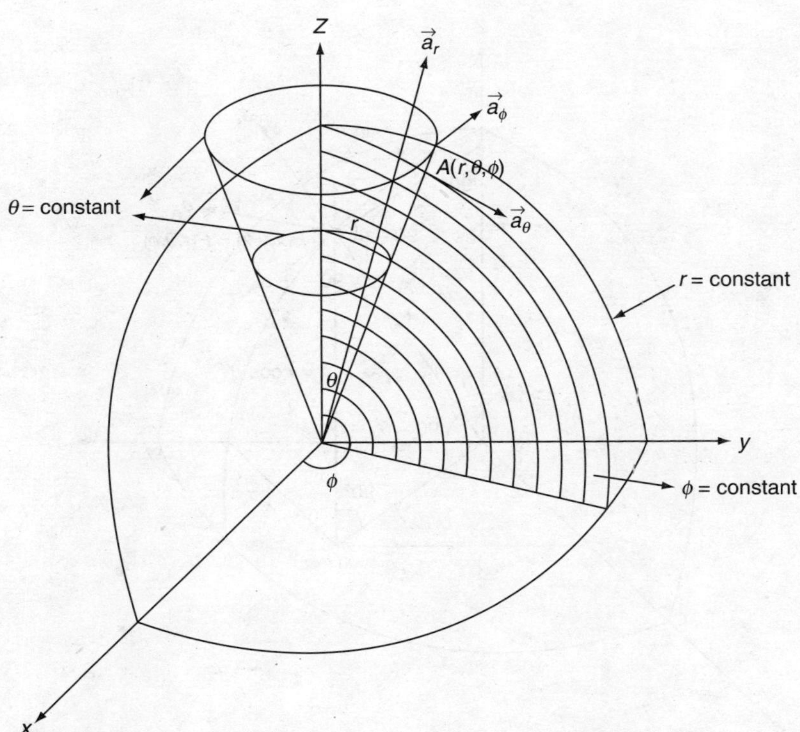

Figure 2.4 *Spherical coordinate system*

Unit vectors $\vec{a}_r, \vec{a}_\theta$ and $\vec{a}_\phi$ are mutually orthogonal. The Spherical coordinate system is a right-handed system, which is shown in Figure 2.5.

Thus,

$$\vec{a}_r \times \vec{a}_\theta = \vec{a}_\phi$$

$$\vec{a}_\theta \times \vec{a}_\phi = \vec{a}_r \quad \text{and}$$

$$\vec{a}_\phi \times \vec{a}_r = \vec{a}_\theta$$

$$\vec{a}_r \cdot \vec{a}_\theta = \vec{a}_\theta \cdot \vec{a}_\phi = \vec{a}_\phi \cdot \vec{a}_r = 0 \quad \text{and}$$

$$\vec{a}_r \cdot \vec{a}_r = \vec{a}_\theta \cdot \vec{a}_\theta = \vec{a}_\phi \cdot \vec{a}_\phi = 1$$

The values of variables are obtained as

$$r = \sqrt{x^2 + y^2 + z^2}$$

$$\theta = \tan^{-1} \frac{\sqrt{x^2 + y^2}}{z} \quad \text{and}$$

$$\phi = \tan^{-1} \frac{y}{x}$$

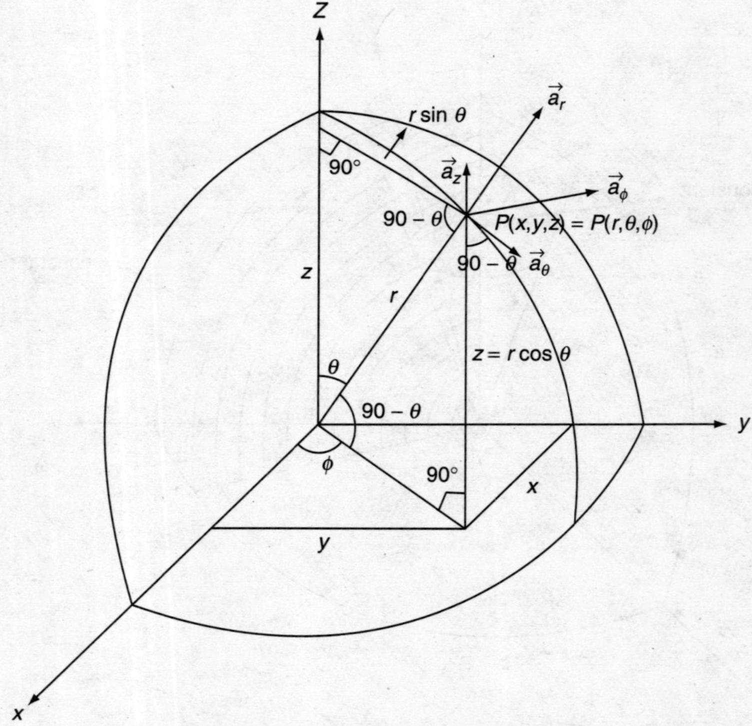

Figure 2.5 *Spherical coordinate system (a right-handed system)*

Values of variables x, y and z are obtained as

$$x = r\sin\theta\cos\phi$$

$$y = r\sin\theta\sin\phi \quad \text{and} \quad z = r\cos\theta$$

2.4.1 Transformation from a Cartesian Coordinate System to a Spherical Coordinate System

For this transformation, obtain the dot product of unit vectors (refer Figure 2.5):

$$\vec{a}_z \cdot \vec{a}_r = \cos\theta$$

$$\vec{a}_z \cdot \vec{a}_\theta = \cos(90° + \theta) = -\sin\theta$$

$$\vec{a}_z \cdot \vec{a}_\phi = \cos 90° = 0$$

$$\vec{a}_x \cdot \vec{a}_r = \sin\theta\cos\phi \quad \text{(projection of } \vec{a}_r \text{ in the } x\text{–}y \text{ plane)}$$

$$\vec{a}_y \cdot \vec{a}_r = \sin\theta\sin\phi$$

$$\vec{a}_x \cdot \vec{a}_\phi = -\sin\phi$$

$$\vec{a}_x \cdot \vec{a}_\theta = \cos\theta\cos\phi$$

$$\vec{a}_y \cdot \vec{a}_\theta = \cos\theta\sin\phi \quad \text{and} \quad \vec{a}_y \cdot \vec{a}_\phi = \cos\phi$$

Relationships of unit vectors between the Cartesian coordinates and the Spherical coordinates are obtained as follows:

$$\vec{a}_x = \sin\theta\cos\phi\,\vec{a}_r + \cos\theta\cos\phi\,\vec{a}_\theta - \sin\phi\,\vec{a}_\phi$$

$$\vec{a}_y = \sin\theta\sin\phi\,\vec{a}_r + \cos\theta\sin\phi\,\vec{a}_\theta + \cos\phi\,\vec{a}_\phi \quad \text{and}$$

$$\vec{a}_z = \cos\theta\,\vec{a}_r - \sin\theta\,\vec{a}_\theta$$

Vector $\vec{P}$ in terms of $\vec{a}_x, \vec{a}_y$ and $\vec{a}_z$ in a Cartesian system is

$$\vec{P} = P_x\vec{a}_x + P_y\vec{a}_y + P_z\vec{a}_z$$

Substituting expressions for $\vec{a}_x, \vec{a}_y$ and $\vec{a}_z$ in the above expression, we get

$$\begin{aligned}\vec{P} = {} & P_x(\sin\theta\cos\phi\,\vec{a}_r + \cos\theta\cos\phi\,\vec{a}_\theta - \sin\phi\,\vec{a}_\phi) \\ & + P_y(\sin\theta\sin\phi\,\vec{a}_r + \cos\theta\sin\phi\,\vec{a}_\theta + \cos\phi\,\vec{a}_\phi) \\ & + P_z(\cos\theta\,\vec{a}_r - \sin\theta\,\vec{a}_\theta)\end{aligned}$$

Rearranging further, we get

$$\begin{aligned}\vec{P} = {} & (P_x\sin\theta\cos\phi + P_y\sin\theta\sin\phi + P_z\cos\theta)\vec{a}_r \\ & + (P_x\cos\theta\cos\phi + P_y\cos\theta\sin\phi - P_z\sin\theta)\vec{a}_\theta \\ & + (-P_x\sin\phi + P_y\cos\phi)\vec{a}_\phi\end{aligned}$$

Writing in a matrix form, we get

$$\begin{bmatrix} P_r \\ P_\theta \\ P_\phi \end{bmatrix} = \begin{bmatrix} \sin\theta\cos\phi & \sin\theta\sin\phi & \cos\theta \\ \cos\theta\cos\phi & \cos\theta\sin\phi & -\sin\theta \\ -\sin\phi & \cos\phi & 0 \end{bmatrix} \begin{bmatrix} P_x \\ P_y \\ P_z \end{bmatrix}$$

Inverse transformation (from a Spherical coordinate system to a Cartesian coordinate system) is obtained as

$$\begin{bmatrix} P_x \\ P_y \\ P_z \end{bmatrix} = \begin{bmatrix} \sin\theta\cos\phi & \cos\theta\cos\phi & -\sin\phi \\ \sin\theta\sin\phi & \cos\theta\sin\phi & \cos\phi \\ \cos\theta & -\sin\theta & 0 \end{bmatrix} \begin{bmatrix} P_r \\ P_\theta \\ P_\phi \end{bmatrix}$$

SOLVED QUESTIONS

2.1 Cartesian coordinates of a point A are (1, 2, 4). Transform it into Cylindrical coordinates.

Solution:

Coordinates of A are $x = 1$, $y = 2$ and $z = 4$ (as given).

Coordinates of point A in a Cylindrical system are obtained as

$$r = \sqrt{x^2 + y^2} = \sqrt{1^2 + 2^2}$$

Thus, $r = 2.23$

$$\phi = \tan^{-1}\frac{y}{x} = \tan^{-1}\frac{2}{1}$$

Thus, $\phi = 63.43°$ and $z = 4$.

Cylindrical coordinates of point A are (2.23, 63.43°, 4).

2.2 Obtain Cylindrical coordinates for a vector $\vec{P} = y\vec{a}_x - x\vec{a}_y - z\vec{a}_z$.

Solution:

The component of vector $\vec{P}$ in r, ϕ and z-directions are

$$P_r = \vec{P}\cdot\vec{a}_r$$

$$P_\phi = \vec{P}\cdot\vec{a}_\phi \quad \text{and}$$

$$P_z = \vec{P}\cdot\vec{a}_z$$

Thus,

$$P_r = (y\vec{a}_x - x\vec{a}_y - z\vec{a}_z)\cdot\vec{a}_r$$

$$= y\vec{a}_x\cdot\vec{a}_r - x\vec{a}_y\vec{a}_r - z\vec{a}_z\vec{a}_r$$

$$P_r = y\cos\phi - x\sin\phi + 0$$

$$= (r\sin\phi)\cos\phi - (r\cos\phi)\sin\phi$$

$$= 0$$

Similarly,

$$P_\phi = (y\vec{a}_x - x\vec{a}_y - z\vec{a}_z)\cdot\vec{a}_\phi$$

$$= y\vec{a}_x\cdot\vec{a}_\phi - x\vec{a}_y\cdot\vec{a}_\phi - z\vec{a}_z\cdot\vec{a}_\phi$$

$$= -y\sin\phi - x\cos\phi - 0$$

$$= -(r\sin\phi)\sin\phi - (r\cos\phi)\cos\phi$$

$$= -(r\sin^2\phi + r\cos^2\phi) \quad \text{and}$$

$$= -r$$

$$P_z = (y\vec{a}_x - x\vec{a}_y - z\vec{a}_z)\cdot\vec{a}_z$$

$$= y\vec{a}_x\cdot\vec{a}_z - x\vec{a}_y\cdot\vec{a}_z - z\vec{a}_z\cdot\vec{a}_z$$

$$= -z$$

Vector $\vec{P}$ in Cylindrical coordinates is given as

$$\vec{P} = P_r\vec{a}_r + P_\phi\vec{a}_\phi + P_z\vec{a}_z$$

or simply $$\vec{P} = -r\vec{a}_\phi - z\vec{a}_z.$$

2.3 A point A in Cartesian coordinates is given as (2, 1, 3). Transform it in Spherical coordinates.

Solution:

Coordinates of point A are $x = 2$, $y = 1$ and $z = 3$ (as given).

Coordinates of A in Spherical coordinates are obtained as

$$r = \sqrt{x^2 + y^2 + z^2} = \sqrt{2^2 + 1^2 + 3^2} = \sqrt{14} = 3.74$$

$$\cos\theta = \frac{z}{\sqrt{x^2 + y^2 + z^2}} = \frac{3}{\sqrt{14}} \quad \text{or}$$

$$\theta = \cos^{-1}\left(\frac{3}{\sqrt{14}}\right)$$

$$\theta = 36.7^\circ \quad \text{and}$$

$$\tan\phi = \frac{y}{x} = \frac{1}{2} = 0.5$$

$$\phi = \tan^{-1}(0.5) \quad \text{or} \quad \phi = 26.56^\circ$$

Spherical coordinates of vector P are (3.74, 36.7°, 26.56°).

2.4 Transform a vector P given as $(4\vec{a}_x + 2\vec{a}_y + 4\vec{a}_z)$ into Spherical coordinates at a point A. Point A is given as $x = -2$, $y = -3$ and $z = 4$.

Solution:

Vector $\vec{P} = 4\vec{a}_x + 2\vec{a}_y + 4\vec{a}_z$ (given).

Point A $(x = -2, y = -3, z = 4)$ (given).

Parameters in Spherical coordinates are obtained as

$$r = \sqrt{x^2 + y^2 + z^2} = \sqrt{(-2)^2 + (-3)^2 + (4)^2} = 5.38$$

$$\cos\theta = \frac{z}{r} = \frac{4}{5.38} \quad \text{or} \quad \theta = 42.03°$$

$$\tan\phi = \frac{y}{x} = \frac{3}{2} \quad \text{or} \quad \phi = 56.31°$$

where y and x are negative. This point lies in the third quadrant. In addition,

$$\phi = -180° + 56.31° = -123.69°$$

Components of vector $\vec{P}$ in the Spherical coordinates are obtained as

$$P_r = \vec{P}.\vec{a}_r = (4\vec{a}_x + 2\vec{a}_y + 4\vec{a}_z)\cdot\vec{a}_r$$

$$= 4\vec{a}_x \cdot \vec{a}_r + 2\vec{a}_y\vec{a}_r + 4\vec{a}_z\vec{a}_r$$

$$= 4\sin\theta\cos\phi + 2\sin\theta\sin\phi + 4\cos\theta$$

$$P_\theta = \vec{P}\cdot\vec{a}_\theta = (4\vec{a}_x + 2\vec{a}_y + 4\vec{a}_z)\cdot\vec{a}_\theta$$

$$= 4\cos\theta\cos\phi + 2\cos\theta\sin\phi - 4\sin\theta \quad \text{and}$$

$$P_\theta = \vec{P}\cdot\vec{a}_\theta = (4\vec{a}_x + 2\vec{a}_y + 4\vec{a}_z)\cdot\vec{a}_\phi$$

$$= -4\sin\phi + 2\cos\phi$$

In addition,

$$\cos\theta = 0.7428 \quad \text{and} \quad \sin\theta = 0.6695$$

$$\cos\phi = 0.5547 \quad \text{and} \quad \sin\phi = -0.8321$$

Thus, component values are $P_r = 0.3715$, $P_\theta = -6.386$ and $P_\phi = 2.219$.

Therefore,

$$\vec{P} = 0.3715\vec{a}_r - 6.386\vec{a}_\theta + 2.219\vec{a}_\phi$$

2.5 In Cartesian coordinates, vector $\vec{P}$ is $z\vec{a}_x + (x+z)\vec{a}_z$, and point A is (–2, 6, 3). Find this vector in Spherical coordinates.

Solution:

Point A is $x = -2$, $y = 6$, $z = 3$ (as given).

Parameters in Spherical coordinates are obtained as

$$r = \sqrt{x^2 + y^2 + z^2} = \sqrt{4+36+9} = 7$$

$$\tan\phi = \frac{y}{x} = \frac{6}{-2} = -3 \quad \text{or} \quad \phi = 108.43° \quad \text{and}$$

$$\tan\theta = \frac{\sqrt{x^2 + y^2}}{z} = \frac{\sqrt{40}}{3} \quad \text{or}$$

$$\theta = 64.62°$$

Components of vectors are obtained as

$$\begin{bmatrix} P_r \\ P_\theta \\ P_\phi \end{bmatrix} = \begin{bmatrix} \sin\theta\cos\phi & \sin\theta\sin\phi & \cos\theta \\ \cos\theta\cos\phi & \cos\theta\sin\phi & -\sin\theta \\ -\sin\phi & \cos\phi & 0 \end{bmatrix} \begin{bmatrix} z \\ 0 \\ x+z \end{bmatrix}$$

(standard relationships, as derived earlier)

In addition,

$$\cos\phi = \frac{-2}{\sqrt{40}}, \quad \sin\phi = \frac{6}{\sqrt{40}} \quad \text{and}$$

$$\cos\theta = \frac{3}{7}, \quad \sin\theta = \frac{\sqrt{40}}{7}$$

Thus,

$$\begin{bmatrix} P_r \\ P_\theta \\ P_\phi \end{bmatrix} = \begin{bmatrix} -\frac{2}{7} & \frac{6}{7} & \frac{3}{7} \\ -\frac{6}{7\sqrt{40}} & \frac{18}{7\sqrt{40}} & -\frac{\sqrt{40}}{7} \\ -\frac{6}{\sqrt{40}} & -\frac{2}{\sqrt{40}} & 0 \end{bmatrix} \begin{bmatrix} 3 \\ 0 \\ 1 \end{bmatrix} \quad \text{or}$$

$$\begin{bmatrix} P_r \\ P_\theta \\ P_\phi \end{bmatrix} = \begin{bmatrix} -\frac{6}{7} + \frac{3}{7} \\ -\frac{18}{7\sqrt{40}} - \frac{\sqrt{40}}{7} \\ -\frac{6}{\sqrt{40}} \end{bmatrix}$$

Thus,

$$P_r = -428$$

$$P_\theta = -1.31 \quad \text{and}$$

$$P_\phi = -9.48$$

Therefore, vector $\vec{P}$ in a Spherical coordinate is

$$\vec{P} = -0.428\vec{a}_r - 1.31\vec{a}_\theta - 0.948\vec{a}_\phi$$

2.6 Transform the vector as given in Question number 2.5 into Cylindrical coordinates.

Solution:

Point A is $x = -2$, $y = 6$ and $z = 3$ (as given).

In Cylindrical coordinates, its parameters are obtained as

$$r = \sqrt{x^2 + y^2} = \sqrt{4 + 36} = \sqrt{40}$$

$$\phi = \tan^{-1}\frac{y}{x} = \tan^{-1}\frac{6}{-2} = 108.43° \quad \text{and}$$

$$z = 3$$

For its transformation, we use the relationship matrix as follows:

$$\begin{bmatrix} P_r \\ P_\phi \\ P_z \end{bmatrix} = \begin{bmatrix} \cos\phi & \sin\phi & 0 \\ -\sin\phi & \cos\phi & 0 \\ 0 & 0 & 1 \end{bmatrix} \begin{bmatrix} P_x \\ P_y \\ P_z \end{bmatrix}$$

In addition, we get

$$\cos\phi = \frac{-2}{\sqrt{40}} \quad \text{and} \quad \sin\phi = \frac{6}{\sqrt{40}}$$

Thus,

$$\begin{bmatrix} P_r \\ P_\phi \\ P_z \end{bmatrix} = \begin{bmatrix} -\frac{2}{\sqrt{40}} & \frac{6}{\sqrt{40}} & 0 \\ -\frac{6}{\sqrt{40}} & -\frac{2}{\sqrt{40}} & 0 \\ 0 & 0 & 1 \end{bmatrix} \begin{bmatrix} 6 \\ 0 \\ 1 \end{bmatrix} \quad \text{or}$$

$$\begin{bmatrix} P_r \\ P_\phi \\ P_z \end{bmatrix} = \begin{bmatrix} -\frac{12}{\sqrt{40}} \\ -\frac{36}{\sqrt{40}} \\ 1 \end{bmatrix}$$

Component values of vector $\vec{P}$ are

$$P_r = -1.89$$

$$P_\phi = -5.69 \quad \text{and}$$

$$P_z = 1$$

Therefore, vector $\vec{P}$ in Cylindrical coordinates is

$$\vec{P} = -1.89\vec{a}_r - 5.69\vec{a}_\phi - \vec{a}_z.$$

2.7 A vector $\vec{P}$ in a Cartesian coordinate is given as

$$\vec{P} = \left(\frac{\sqrt{x^2+y^2}}{x^2+y^2+z^2}\vec{a}_x - \frac{yz}{\sqrt{x^2+y^2+z^2}}\vec{a}_z \right)$$

Transform it into Cylindrical coordinates.

Solution:

Parameters x, y and z (in Cartesian coordinates) into Cylindrical coordinates are given by the following relationships:

$$r = \sqrt{x^2+y^2}$$

$$y = r\sin\phi \quad \text{and} \quad z = z$$

$$\vec{P} = \frac{r}{\sqrt{r^2+z^2}}\vec{a}_x - \frac{r\sin\phi}{\sqrt{r^2+z^2}}\vec{a}_z \quad \text{(given)}$$

Further,

$$\begin{bmatrix} P_r \\ P_\phi \\ P_z \end{bmatrix} = \begin{bmatrix} \cos\phi & \sin\phi & 0 \\ -\sin\phi & \cos\phi & 0 \\ 0 & 0 & 1 \end{bmatrix} \begin{bmatrix} P_x \\ P_y \\ P_z \end{bmatrix} = \begin{bmatrix} \cos\phi & \sin\phi & 0 \\ -\sin\phi & \cos\phi & 0 \\ 0 & 0 & 1 \end{bmatrix} \begin{bmatrix} \frac{r}{\sqrt{r^2+z^2}} \\ 0 \\ \frac{r\sin\phi z}{\sqrt{r^2+z^2}} \end{bmatrix}$$

Thus,

$$P_r = \frac{r\cos\phi}{\sqrt{r^2+z^2}}$$

$$P_\phi = -\frac{r\sin\phi}{\sqrt{r^2+z^2}} \quad \text{and}$$

$$P_z = \frac{zr\sin\phi}{\sqrt{r^2+z^2}}$$

Therefore, vector $\vec{P}$ in Cylindrical coordinates is

$$\vec{P} = \frac{r}{\sqrt{r^2+z^2}}[\cos\phi\vec{a}_r - \sin\phi\vec{a}_\phi - z\sin\phi\vec{a}_z]$$

2.8 In a Cartesian coordinate system, $\vec{P} = (t^2, \cos t, 4)$ and $\vec{Q} = (e^t, t, \sin t)$. Obtain $\vec{P}\cdot\vec{Q}$ and $\vec{P}\times\vec{Q}$.

Solution:
Vectors $\vec{P}$ and $\vec{Q}$ are

$$\vec{P} = t^2\vec{a}_x + \cos t\,\vec{a}_y + 4\vec{a}_z \quad \text{and}$$

$$\vec{Q} = e^t\vec{a}_x + t\,\vec{a}_y + \sin t\,\vec{a}_z$$

Now

$$\vec{P}\cdot\vec{Q} = (t^2\vec{a}_x + \cos t\,\vec{a}_y + 4\vec{a}_z)\cdot(e^t\vec{a}_x + t\,\vec{a}_y + \sin t\,\vec{a}_z)$$

$$\Rightarrow \vec{P}\cdot\vec{Q} = t^2e^t + t\cos t + 4\sin t \quad \text{and}$$

$$\vec{P}\times\vec{Q} = \begin{vmatrix} \vec{a}_x & \vec{a}_y & \vec{a}_z \\ t^2 & \cos t & 4 \\ e^t & t & \sin t \end{vmatrix}$$

$$\Rightarrow \vec{P}\times\vec{Q} = (\sin t\cos t - 4t)\vec{a}_x - (t^2\sin t - 4e^t)\vec{a}_y + (t^3 - e^t\cos t)\vec{a}_z$$

2.9 (a) Write an expression of the vector going from point $P_1(1, 2, 4)$ to $P_2(3, -2, 4)$, in Cartesian coordinates. (b) What is the length of this line?

Solution:

(a) $P_1P_2 = \overline{OP_2} - \overline{OP_1}$

$$= (\vec{a}_x3 - \vec{a}_y2 + \vec{a}_z4) - (\vec{a}_x + 2\vec{a}_y + \vec{a}_z4)$$
$$= \vec{a}_x2 - \vec{a}_y4 + 0 = \vec{a}_x2 - \vec{a}_y4$$

(b) The length of the line is

$$P_1P_2 = \left|\overline{P_1P_2}\right| = \sqrt{(2)^2 + (-4)^2} = \sqrt{20}$$

2.10 In the Cartesian co-ordinate system, $\vec{P} = (t, \cos t, 5)$ and $\vec{Q} = (e^t, t^2, \sin t)$. Obtain $\vec{P}\cdot\vec{Q}$ and $\vec{P}\times\vec{Q}$.

Solution:
Vectors $\vec{P}\times\vec{Q}$ are

$$\vec{P} = t\,\vec{a}_x + \cos t\,\vec{a}_y + 5\vec{a}_z \quad \text{and} \quad \vec{Q} = e^t\vec{a}_x + t^2\vec{a}_y + \sin t\,\vec{a}_z$$

Now

$$\vec{P}\cdot\vec{Q} = \left(t\,\vec{a}_x + \cos t\,\vec{a}_y + 5\vec{a}_z\right)\cdot\left(e^t\vec{a}_x + t^2\vec{a}_y + \sin t\,\vec{a}_z\right)$$

$$\Rightarrow \vec{P}\cdot\vec{Q} = t\cdot e^t + t^2\cos t + 5\sin t \quad \text{and}$$

$$\vec{P}\times\vec{Q} = \begin{vmatrix} \vec{a}_x & \vec{a}_y & \vec{a}_z \\ t & \cos t & 5 \\ e^t & t^2 & \sin t \end{vmatrix}$$

$$\Rightarrow \vec{P}\times\vec{Q} = \vec{a}_x(\cos t\,\sin t - 5t^2) - \vec{a}_y(t\sin t - 5e^t) + \vec{a}_z(t^3 - e^t\cos t)$$

2.11 The Cartesian co-ordinates of a point "*B*" are (4, 6, 2). Transform it in Cylindrical co-ordinates.

Solution:

Coordinates of "*B*" are given as

$$x = 4, \quad y = 6 \quad \text{and} \quad z = 2$$

In a Cylindrical system, co-ordinates of "*B*" are obtained as

$$r = \sqrt{x^2 + y^2} = \sqrt{(4)^2 + (6)^2} = \sqrt{52} = 7.21 \quad \text{and}$$

$$\phi = \tan^{-1}\frac{y}{x} = \tan^{-1}\frac{6}{4} = \tan\frac{-3}{2} \quad \text{and}$$

$$\phi = 56.3°$$

$$z = 2$$

Cylindrical co-ordinates of point "*B*" are (7.21, 56.30°, 2).

2.12 Cartesian co-ordinates of point "*B*" are (1, 4, 3). Transform it in Spherical co-ordinates.

Solution:

Co-ordinates of "*B*" are $x = 1$, $y = 4$ and $z = 3$.

In a Spherical co-ordinate system, we know that

$$r = \sqrt{x^2 + y^2 + z^2} = \sqrt{(1)^2 + (4)^2 + (3)^2} = \sqrt{26} = 5.09$$

$$\cos\theta = \frac{z}{\sqrt{x^2 + y^2 + z^2}} = \frac{3}{5.09} = 0.588 \quad \text{and}$$

$$\Rightarrow \theta = \cos^{-1}(0.588) = 53.96°$$

$$\tan\phi = \frac{y}{x} = \frac{4}{1} = 4$$

$$\Rightarrow \quad \phi = \tan^{-1} 4 = 75.96°$$

Spherical co-ordinates are (5.09, 53.96°, 75.96°).

2.13 Transform vector $\vec{Q}$ given as $(5\vec{a}_x + 3\vec{a}_y + 4\vec{a}_z)$ into Spherical co-ordinates at a point "*B*". Point "*B*" is given as $x = -3$, $y = -2$ and $z = 5$.

Solution:

Vector $\vec{Q} = (5\vec{a}_x + 3\vec{a}_y + 4\vec{a}_z)$

Point *B* is given as $x = -3$, $y = -2$ and $z = 5$.

In a Spherical co-ordinate system

$$r = \sqrt{x^2 + y^2 + z^2} = \sqrt{(-3)^2 + (-2)^2 + (5)^2} = \sqrt{38} = 6.16$$

$$\cos\theta = \frac{z}{r} = \frac{4}{6.16}$$

$$\Rightarrow \theta = 49.54°$$

$$\tan\phi = \frac{y}{x} = \frac{2}{3} \Rightarrow \phi = 33.69°$$

where *y* and *x* are negative; so this point lies in the third quadrant. In addition,

$$\phi = -180° + 33.69° = -146.30°$$

Components of vector $\vec{Q}$ in a Spherical co-ordinate system are obtained as

$$Q_r = \vec{Q} \cdot \vec{a}_r = (5\vec{a}_x + 3\vec{a}_y + 4\vec{a}_z) \cdot \vec{a}_r$$

$$= 5\vec{a}_x\vec{a}_r + 3\vec{a}_y\vec{a}_r + 4\vec{a}_z\vec{a}_r$$

2.14 Find a unit vector, which is perpendicular to the surface given by $\psi = (x^2 z + 3xy - 2)$ at a point (2, −1, 4).

Solution:

We know that the gradient of a function gives a vector, which is normal to a surface.

Thus,

$$\nabla\psi = \frac{\partial\psi}{\partial x}\vec{a}_x + \frac{\partial\psi}{\partial y}\vec{a}_y + \frac{\partial\psi}{\partial z}\vec{a}_z$$

$$= \frac{\partial}{\partial x}(x^2 z + 3xy - 2)\vec{a}_x + \frac{\partial}{\partial y}(x^2 z + 3xy - 2)\vec{a}_y + \frac{\partial}{\partial z}(x^2 z + 3xy - 2)$$

$$\nabla\psi = (2xz + 3y)\vec{a}_x + 3x\vec{a}_y + x^2\vec{a}_z$$

At point (2, −1, 4),

$$\nabla\psi = [16 + (-3)]\vec{a}_x + 6\vec{a}_y + 4\vec{a}_z$$

$$\nabla\psi = 13\vec{a}_x + 6\vec{a}_y + 4\vec{a}_z$$

Its unit vector will be

$$\frac{\nabla\psi}{|\nabla\psi|} = \frac{13\vec{a}_x + 6\vec{a}_y + 4\vec{a}_z}{\sqrt{(13)^2 + (6)^2 + (4)^2}}$$

$$\frac{\nabla\psi}{|\nabla\psi|} = \frac{13\vec{a}_x + 6\vec{a}_y + 4\vec{a}_z}{14.86}$$

UNSOLVED QUESTIONS

2.1 Let point A be (3, −1, −2) and point B be (2, 2, 1). Find the vector extending from A to B for the following coordinate systems:
(a) Cartesian coordinates (b) Cylindrical coordinates and (c) Spherical coordinates

2.2 What are Cylindrical coordinates of field $\vec{E} = 2xyz\,\vec{a}_x - 3xyz\,\vec{a}_z$? In addition, calculate $|\vec{E}|$ at a point A given as ($r = 1$, $\Phi = 45$ and $z = 2$).

2.3 Find the Cartesian and Cylindrical coordinates of vector $\vec{A} = (5/r)\vec{a}_r + r\cos\theta\,\vec{a}_\theta + \vec{a}_\phi$. In addition, obtain $\vec{A}(-4,3,0)$ and $\vec{A}(3,\pi/2,-1)$.

2.4 Transform the Cartesian coordinates vector $\vec{A} = y\vec{a}_r + \vec{a}_y + \left(x^2/\sqrt{x^2+y^2}\right)\vec{a}_z$ into Cylindrical co-ordinates.

SUMMARY

1. A point P in Cartesian coordinates is represented as $P(x, y, z)$.
2. A point P in Cylindrical coordinates is represented as $P(r, \phi, z)$.
3. Cartesian and Cylindrical coordinate systems are related to each other by the following expressions:

$$r = \sqrt{x^2 + y^2},\ \phi = \tan^{-1}\left(\frac{y}{x}\right) \text{ and } z = z$$

4. The Cartesian of Cylindrical transformation can be done as

$$\begin{bmatrix} P_r \\ P_\phi \\ P_z \end{bmatrix} = \begin{bmatrix} \cos\phi & \sin\phi & 0 \\ -\sin\phi & \cos\phi & 0 \\ 0 & 0 & 1 \end{bmatrix} \begin{bmatrix} P_x \\ P_y \\ P_z \end{bmatrix}$$

5. Cylindrical to Cartesian transformation can be done as

$$\begin{bmatrix} P_x \\ P_y \\ P_z \end{bmatrix} = \begin{bmatrix} \cos\phi & -\sin\phi & 0 \\ \sin\phi & \cos\phi & 0 \\ 0 & 0 & 1 \end{bmatrix}^{-1} \begin{bmatrix} P_r \\ P_\phi \\ P_z \end{bmatrix}$$

6. A point P in Spherical coordinates is represented as $P(r, \theta, \phi)$.
7. Cartesian and Spherical coordinate systems are related to each other by the following expressions:

$$r = \sqrt{x^2 + y^2 + z^2},\ \theta = \tan^{-1}\frac{\sqrt{x^2 + y^2}}{z} \text{ and } \phi = \tan^{-1}\frac{y}{x} \quad \text{or}$$

$$x = r\sin\theta\cos\phi,\ y = r\sin\theta\sin\phi \text{ and } z = r\cos\theta$$

8. Cartesian to Spherical transformation can be done as

$$\begin{bmatrix} P_r \\ P_\theta \\ P_\phi \end{bmatrix} = \begin{bmatrix} \sin\theta\cos\phi & \sin\theta\sin\phi & \cos\theta \\ \cos\theta\cos\phi & \cos\theta\sin\phi & -\sin\theta \\ -\sin\phi & \cos\phi & 0 \end{bmatrix} \begin{bmatrix} P_x \\ P_y \\ P_z \end{bmatrix}$$

9. Spherical to Cartesian transformation can be done as

$$\begin{bmatrix} P_x \\ P_y \\ P_z \end{bmatrix} = \begin{bmatrix} \sin\theta\cos\phi & \cos\theta\cos\phi & -\sin\phi \\ \sin\theta\sin\phi & \cos\theta\sin\phi & \cos\phi \\ \cos\theta & -\sin\theta & 0 \end{bmatrix} \begin{bmatrix} P_r \\ P_\theta \\ P_\phi \end{bmatrix}$$

MULTIPLE-CHOICE QUESTIONS

1. In a Cartesian coordinate system, axes x, y and z are at __________ to each other.
(a) 45° (b) 90° (c) 120° (d) 180°

2. Cartesian coordinates are represented in terms of __________.
(a) (r, ϕ, z) (b) (r, θ, ϕ) (c) (x, y, z) (d) all of these

3. Cylindrical coordinates are represented in terms of __________.
(a) (r, ϕ, z) (b) (r, θ, ϕ) (c) (x, y, z) (d) all of these

4. Spherical coordinates are represented in terms of __________.
(a) (r, ϕ, z) (b) (r, θ, ϕ) (c) (x, y, z) (d) all of these

5. In Cylindrical coordinate systems, unit vector $\vec{a}_r$ is __________.
(a) directed outwards radially
(b) normal to cylindrical surface at a point
(c) both (a) and (b)
(d) none of these

6. Unit vector $\vec{a}_\phi$ is at __________ to $\phi =$ constant plane.
(a) 90° (b) 180° (c) 270° (d) 360°

7. Cylindrical coordinate 'r' is related to the Cartesian coordinate as __________.
(a) (x, y) (b) $(x + y)$ (c) (x^2/y^2) (d) $(x^2 + y^2)^{1/2}$

8. Cylindrical coordinate 'z' is related to the Cartesian coordinate as __________.
(a) $\tan^{-1}(y/x)$ (b) z (c) xy/z (d) $\cot z$

9. Dot multiplication $\vec{a}_y \cdot \vec{a}_\phi$ is equal to __________.
(a) $\sin\phi$ (b) $\cos\phi$ (c) $-\sin\phi$ (d) $-\cos\phi$

10. In a Spherical coordinate system, ϕ is __________.
(a) angle of elevation
(b) azimuthal angle
(c) distant from the origin to the point
(d) all of these

11. Spherical coordinate system is a ________.
(a) right-handed system
(b) left-handed system
(c) both (a) and (b)
(d) none of these

12. Cross product $\vec{a}_\phi \times \vec{a}_r$ is equal to ____________.
(a) $\vec{a}_\theta$ (b) $-\vec{a}_\theta$ (c) 1 (d) 0

13. In a Spherical coordinate system, ϕ is given as __________.
(a) y/x (b) x/y (c) $\tan^{-1}(y/x)$ (d) $\tan^{-1}(x/y)$

14. In a Spherical coordinate system, r is given as ___________.
(a) $(x+y+z)$
(b) $(x^2+y^2+z^2)$
(c) $(x^2+y^2+z^2)^{1/2}$
(d) $\sqrt{x^2+y^2}/z$

15. Dot product $\vec{a}_r \cdot \vec{a}_\theta$ is equal to ____________.
(a) $\vec{a}_\phi$ (b) $-\vec{a}_\phi$ (c) 1 (d) 0

16. For transformation from the Cartesian coordinate system to Spherical coordinate system, $\vec{a}_z.\vec{a}_r$ should be equal to ___________.
(a) $\cos\theta$ (b) $-\cos\theta$ (c) $\sin\theta$ (d) $-\sin\theta$

17. In terms of Spherical coordinate system variables, y of Cartesian coordinate system is given as __________.
(a) $r\sin\theta\cos\phi$ (b) $r\sin\theta\sin\phi$ (c) $r\cos\theta\sin\phi$ (d) $r\cos\theta\cos\phi$

18. In a Spherical coordinate system, unit vector $\vec{a}_\phi$ is perpendicular to the shifted ________ plane.
(a) x–y (b) y–z (c) x–z (d) none of these

19. In the Cylindrical coordinate system, z ranges between ___________.
(a) 0 and 1 (b) $-\infty$ and 0 (c) 0 and $-\infty$ (d) $-\infty$ and ∞

20. In the Cylindrical coordinate system, ϕ ranges from _____________.
(a) 0 to less than π
(b) 0 to less than 2π
(c) 0 to less than 3π
(d) 0 to less than 4π

ANSWERS TO MULTIPLE-CHOICE QUESTIONS

(1) b; (2) c; (3) a; (4) b; (5) c; (6) a; (7) d; (8) b; (9) b; (10) b;
(11) a; (12) a; (13) c; (14) c; (15) d; (16) a; (17) b; (18) c; (19) d; (20) b

3 Vector Calculus

3.1 INTRODUCTION

Vector calculus deals with differentiation and integration of vectors. Integration of vectors is of three types. These three types are discussed in the following sections.

3.2 LINE INTEGRALS

The line integral involves a scalar (or dot) product. Integral of the dot product of a vector $\vec{P}$ is evaluated along a specified path. The differential displacement vector $d\vec{l}$ is tangential to the path.

Line integral for a given vector $\vec{P}$ and path L is worked out as

$$\int_L \vec{P} \cdot d\vec{l} = \int_B^A |\vec{P}||d\vec{l}|\cos\theta = \int_B^A |\vec{P}|\cos\theta \cdot dl$$

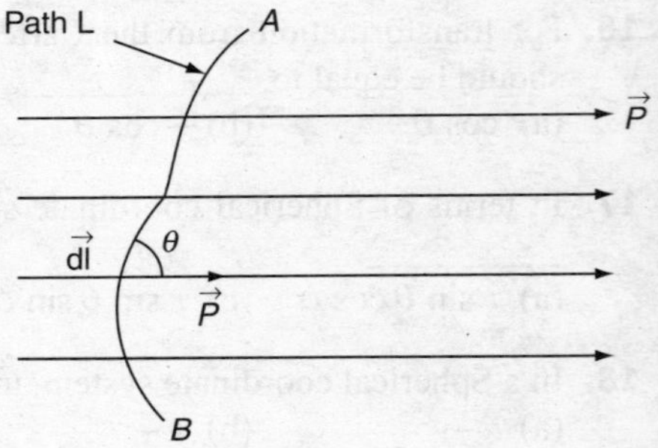

Figure 3.1 *Line integral*

3.3 SURFACE INTEGRALS

A vector $\vec{P}$ on a surface S is shown in Figure 3.2.

Surface integral is given as

$$\varphi = \int_s \vec{P} \cdot d\vec{S} = \int_s \vec{P} \cdot \vec{a}_n ds$$

$$= \int_s |\vec{P}| \cos\theta \, ds$$

where $\vec{a}_n$ is a unit vector on S. It is normal to surface S. Integration is performed for the entire surface S. Surface integral for a closed surface is given as

$$\varphi = \oint_s \vec{P} \cdot ds$$

Surface integral is shown as $\iint or \int_S$.

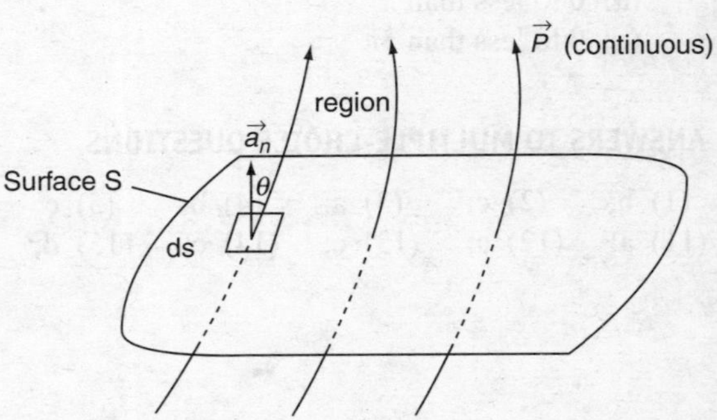

Figure 3.2 *Surface integral*

3.4 VOLUME INTEGRALS

The total mass of a quantity is given as

$$M = \int_v P \cdot dv$$

where P is the mass density of quantity in kg/m^3 and dv is a product of three differential lengths. Volume integral is shown as $\int_v or \iiint$.

3.5 GRADIENTS OF A SCALAR FIELD

Gradient uses a "del operator", which is shown as "∇." Del operator is a vector differential operator. It is given as

$$\nabla = \frac{\partial}{\partial x}\vec{a}_x + \frac{\partial}{\partial y}\vec{a}_y + \frac{\partial}{\partial z}\vec{a}_z \qquad \text{(in Cartesian coordinates)}$$

Del operator "∇" is also called a gradient operator. The gradient of a scalar quantity 'V' (in Cartesian coordinates) is given as

$$\text{grad } V = \nabla V = \frac{\partial V}{\partial x}\vec{a}_x + \frac{\partial V}{\partial y}\vec{a}_y + \frac{\partial V}{\partial z}\vec{a}_z$$

Gradient V is expressed as a vector. Its magnitude is obtained as

$$|\nabla V| = \sqrt{\left(\frac{\partial V}{\partial x}\right)^2 + \left(\frac{\partial V}{\partial y}\right)^2 + \left(\frac{\partial V}{\partial z}\right)^2}$$

Del operator "∇" in Cylindrical coordinates is given as

$$\nabla = \vec{a}_\rho \frac{\partial}{\partial \rho} + \vec{a}_\phi \frac{1}{\rho}\frac{\partial}{\partial \phi} + \vec{a}_z \frac{\partial}{\partial z}$$

Del operator "∇" in Spherical coordinates is given as

$$\nabla = \vec{a}_r \frac{\partial}{\partial r} + \vec{a}_\theta \frac{1}{r}\frac{\partial}{\partial \theta} + \vec{a}_\phi \frac{1}{r\sin\theta}\frac{\partial}{\partial \phi}$$

The gradient of scalar quantity "V" in Cylindrical coordinates is given as

$$\nabla V = \frac{\partial V}{\partial \rho}\vec{a}_\rho + \frac{1}{\rho}\frac{\partial V}{\partial \phi}\vec{a}_\phi + \frac{\partial V}{\partial z}\vec{a}_z$$

The gradient of scalar quantity "V" in Spherical coordinates is given as

$$\nabla V = \frac{\partial V}{\partial r}\vec{a}_r + \frac{1}{r}\frac{\partial V}{\partial \theta}\vec{a}_\theta + \frac{1}{r\sin\theta}\frac{\partial V}{\partial \phi}\vec{a}_\phi$$

"Gradient V represents the magnitude and direction of the maximum space rate of increase in scalar field V."

Let A and B be two scalar quantities. The following relations hold true using "∇" (del operator).

1. $\nabla(A+B) = \nabla A + \nabla B$
2. $\nabla(AB) = B\nabla A + A\nabla B$
3. $\nabla A^n = nA^{n-1}\nabla A$ (where "n" is an integer)
4. $\nabla(A/B) = (B\nabla A - A\nabla B)/B^2$

3.6 DIVERGENCE OF VECTOR FIELDS

Divergence of a vector $\vec{P}$ at a point in a given vector field is a scalar quantity. It is defined as "the net outward flux of $\vec{P}$ per unit volume." The volume about the point tends to zero.

Consider a closed surface S having a volume v. The net flux outflow of a vector field $\vec{P}$ from a closed surface S is $\oint \vec{P}\cdot d\vec{S}$ (integral). For an infinitesimal ∇v, an infinitesimal vector $\vec{P}$ flows out of the surface Δs having volume Δv. In this case, $\oint_S \vec{P}\cdot d\vec{S}$ will tend to zero (limit $\Delta v \to 0$). The quantity given by ratio

$$\frac{\oint_S \vec{P}\cdot d\vec{S}}{\Delta v}$$

will however be a non-zero value as $\Delta v \to 0$. The divergence of $\vec{P}$ is the dot product of "∇," and $\vec{P}$ is given as

$$\text{Divergence } \vec{P} = \nabla\cdot\vec{P} = \lim_{\Delta v\to 0}\frac{\oint_S \vec{P}\cdot d\vec{S}}{\Delta v}$$

Let vector $\vec{P} = P_x\vec{a}_x + P_y\vec{a}_y + P_z\vec{a}_z$ (in Cartesian coordinates)

Now divergence $\vec{P}$ is

$$\text{div } P = \nabla\vec{P} = \left(\frac{\partial}{\partial x}\vec{a}_x + \frac{\partial}{\partial y}\vec{a}_y + \frac{\partial}{\partial z}\vec{a}_z\right).(P_x\vec{a}_x + P_y\vec{a}_y + P_z\vec{a}_z)$$

$$= \frac{\partial P_x}{\partial x} + \frac{\partial P_y}{\partial y} + \frac{\partial P_z}{\partial z}$$

Divergence $\vec{P}$ in Cylindrical coordinates is

$$\text{div } \vec{P} = \nabla.\vec{P} = \frac{1}{r}\frac{\partial}{\partial r}(rP_r) + \frac{1}{r}\frac{\partial P_\phi}{\partial \phi} + \frac{\partial P_z}{\partial z}$$

Divergence $\vec{P}$ in Spherical coordinates is

$$\nabla\cdot\vec{P} = \frac{1}{r^2}\frac{\partial}{\partial r}(r^2 P_r) + \frac{1}{r\sin\theta}\frac{\partial}{\partial\theta}(P_\theta \sin\theta) + \frac{1}{r\sin\theta}\frac{\partial P_\phi}{\partial\phi}$$

Divergence $\vec{P}$ at a point is positive if field lines are coming out from a small volume (diverging and surrounding the point).

Divergence $\vec{P}$ is negative if field lines are coming into a small volume (converging). If the rate at which field lines entering a small volume is equal to that leaving the small volume surrounding the point, then divergence $\vec{P}$ is zero.

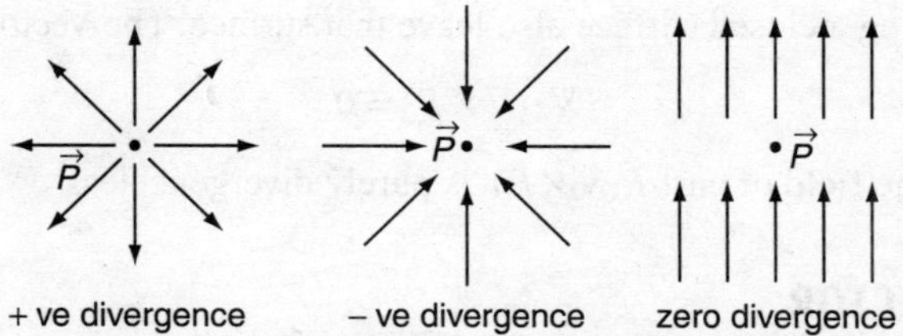

Figure 3.3 *Positive, negative and zero divergence*

3.7 DIVERGENCE THEOREM

The divergence of a vector $\vec{P}$ is given as

$$\nabla\cdot\vec{P} = \lim_{\Delta v\to 0}\frac{\oint_s \vec{P}\cdot d\vec{s}}{\Delta v}$$

Further,

$$\oint_s \vec{\nabla}\cdot d\vec{s} = \oint_v (\vec{\nabla}\cdot\vec{P})\,dv$$

where $\vec{P}$ is a vector field.

As per the divergence theorem, volume integral of divergence $\vec{P}$, taken over a volume 'v' is equal to the surface integral of $\vec{P}$ taken over the closed surface 'S'. The closed surface 'S' bounds the volume 'v'. The divergence theorem is applicable wherever volume v is bounded by a surface S. The direction of $\overrightarrow{dS}$ is always perpendicular to the surface $S\cdot d\vec{S}$, which is directed away from the volume.

3.7.1 Proof of the Divergence Theorem

Let volume v be divided into a large number of small cells. Let the nth cell have volume Δv_n. Also, the nth cell is bounded by surface S_n. The outward flux from surface S_n is

$$\int_{S_n} \vec{P}\cdot d\vec{s}$$

The total outward flux will be the addition of flux through all cells. It is obtained as

$$\int_s \vec{P}\cdot d\vec{s} = \sum_n \int_{S_n} \vec{P}\cdot d\vec{s}$$

$$\Rightarrow \int_s \vec{P}\cdot d\vec{s} = \sum_n \frac{\oint_{S_n} \vec{P}\cdot d\vec{s}}{\Delta v_n}\Delta v_n$$

To limit the above case, i.e. Δv tending to zero, perform integration instead of summation.

We get

$$\int_s \vec{P}\cdot d\vec{s} = \int_v (\nabla\cdot\vec{P})\cdot dv$$

It is the same as the Divergence Theorem. Thus, the Divergence Theorem also relates the surface integral to the volume integral. If $(\nabla\cdot\vec{P}) = 0$, then vector $\vec{P}$ is "divergenceless" also called as "solenoidal". Thus, $\oint_s \vec{P}\cdot d\vec{s} = \int_v (\nabla\cdot\vec{P})\,dv = 0$ (Divergence Theorem).

The flux lines of $\vec{P}$ entering a closed surface also leave that surface. The Vector Theorem gives

$$\nabla \cdot (\nabla \times \vec{P}) = 0$$

Thus, for any vector $\vec{P}$, the field of curl $\vec{P}(\nabla \times \vec{P})$ is purely divergenceless.

3.8 CURL OF A VECTOR

The curl of vector at a point is an 'axial vector'. The curl of a vector measures the rate of change of a vector. Curl measures circular rotation.

The magnitude of curl of a vector $\vec{P}$ is maximum circulation of $\vec{P}$. Per unit area (area tending to zero) and vector direction are in the normal direction of the area. Orientation of area makes maximum circulation in the normal direction.

Let an area Δs be bounded by the curve L. Curl $\vec{P}$ is given as

$$\text{Curl}\,\vec{P} = \nabla \times \vec{P} = \left(\lim_{\nabla s \to 0} \frac{\oint_L \vec{P} \cdot d\vec{l}}{\Delta s} \right)_{\max} \vec{a}_n$$

where $\vec{a}_n$ is a unit vector. Unit vector $\vec{a}_n$ is normal to the surface Δs.

Curl $\vec{P}$ in Cartesian coordinates is given as

$$\nabla \times \vec{P} = \begin{vmatrix} a_x & a_y & a_z \\ \dfrac{\partial}{\partial x} & \dfrac{\partial}{\partial y} & \dfrac{\partial}{\partial z} \\ P_x & P_y & P_z \end{vmatrix}$$

Curl $\vec{P}$ in Cylindrical coordinates is given as

$$\nabla \times \vec{P} = \begin{vmatrix} a_r & ra_\phi & a_z \\ \dfrac{\partial}{\partial r} & \dfrac{\partial}{\partial \phi} & \dfrac{\partial}{\partial z} \\ P_x & rP_\phi & P_z \end{vmatrix}$$

Curl $\vec{P}$ in Spherical coordinates is given as

$$\nabla \times \vec{P} = \frac{1}{r^2 \sin\theta} \begin{vmatrix} a_r & ra_\theta & r\sin\theta a_\phi \\ \dfrac{\partial}{\partial r} & \dfrac{\partial}{\partial \theta} & \dfrac{\partial}{\partial \phi} \\ P_r & rP_\theta & r\sin\theta P_\phi \end{vmatrix}$$

3.9 STOKE'S THEOREM

Figure 3.4 shows a closed path 'L' with surface area 'S'.

Divide the surface S into a large number of cells. Let the nth cell have surface area Δs_n. The nth cell is bounded by path L_n. The line integral as per Stoke's Theorem for the nth cell is given as

$$\oint_{L_n} \vec{P} \cdot d\vec{l}$$

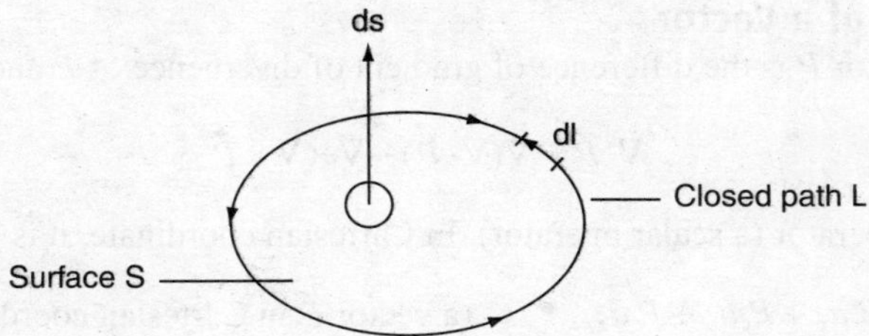

Figure 3.4 *A closed path*

The total line integral of vector $\vec{P}$ for a closed path L is given as

$$\oint_L \vec{P}\cdot d\vec{l} = \sum_n \oint_{L_n} \vec{P}\cdot d\vec{l}$$

$$\Rightarrow \oint_L \vec{P}\cdot d\vec{l} = \sum_n \frac{\oint_{L_n} \vec{P}\cdot d\vec{l}}{\Delta s_n}\Delta s$$

Figure 3.5 shows some cells. Every interior path gets cancelled. Therefore, the sum of line integrals around L_n is the same as the line integral around the bounding curve L.

Consider limiting case Δs_n tending to zero, we get

$$\oint_L \vec{P}\cdot d\vec{l} = \int_s \left(\lim_{\Delta s_n \to 0} \frac{\oint_{L_n} \vec{P}\cdot d\vec{l}}{\Delta s_n} \right). d\vec{s}$$

$$\oint_L \vec{P}\cdot d\vec{l} = \int_S (\nabla \times \vec{P})\cdot d\vec{s}$$

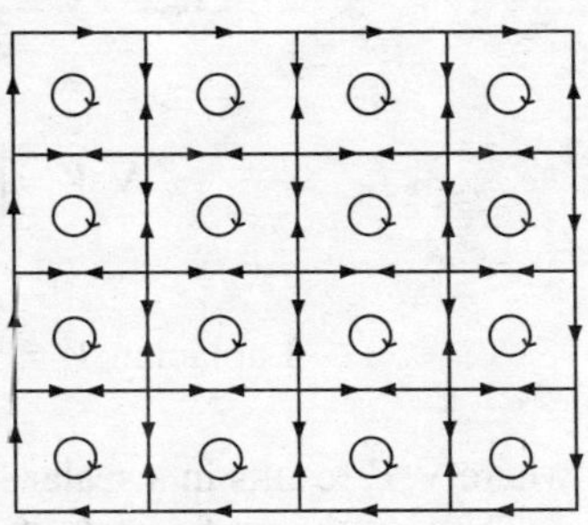

Figure 3.5 *Cells*

This is Stoke's theorem. Stoke's theorem relates a line integral to a surface integral.

If $\nabla \times \vec{P} = 0$, then vector field $\vec{P}$ is said to be "potential" or "irrotational".

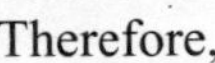

Therefore,

$$\int_S (\nabla \times \vec{P})\cdot d\vec{s} = \int_L \vec{P}\cdot d\vec{l} = 0 \qquad \text{(by Stoke's Theorem)}$$

The line integral of irrotational $\vec{P}$ is independent of the chosen path. It is known as "conservative field". It is similar to electrostatic field and gravitational field. For a scalar field V, by Vector's Theorem, we get

$$\nabla \times (\nabla V) = 0$$

Thus, field of gradient 'V' is purely irrotational.

3.10 THE LAPLACIAN OPERATOR

The Laplacian operator is a composite of gradient and divergence operators. The Laplacian operator can be a scalar or vector function.

3.10.1 Laplacian of a Vector

The Laplacian of a vector $\vec{P}$ is the difference of gradient of divergence of $\vec{P}$ and curl of curl of $\vec{P}$. Thus,

$$\nabla^2 \vec{P} = \nabla(\nabla \cdot \vec{P}) - \nabla \times \nabla \times \vec{P}$$

where ∇^2: Laplacian operator (a scalar operator). In Cartesian coordinate, it is given as

$$\vec{P} = P_x \vec{a}_x + P_y \vec{a}_y + P_z \vec{a}_z \qquad \text{(a vector } \vec{P} \text{ in Cartesian coordinates)}$$

Its Laplacian operator

$$\nabla^2 \vec{P} = \nabla^2 P_x \vec{a}_x + \nabla^2 P_y \vec{a}_y + \nabla^2 P_z \vec{a}_z$$

3.10.2 Laplacian of a Scalar

$\nabla^2 V$ Laplacian of scalar function V is the divergence of gradient of V. Laplacian of a scalar is a scalar. It is because the gradient of V is a vector, whereas the divergence of a vector is a scalar. The Laplacian of scalar function V is given as

$$\nabla^2 \nabla = \nabla \cdot (\nabla V)$$

In Cartesian coordinates, it is

$$\nabla = \frac{\partial}{\partial x}\vec{a}_x + \frac{\partial}{\partial y}\vec{a}_y + \frac{\partial}{\partial z}\vec{a}_z$$

$$\nabla^2 V = \left(\frac{\partial}{\partial x}\vec{a}_x + \frac{\partial}{\partial y}\vec{a}_y + \frac{\partial}{\partial z}\vec{a}_z\right) . \left(\frac{\partial V}{\partial x}\vec{a}_x + \frac{\partial V}{\partial y}\vec{a}_y + \frac{\partial V}{\partial z}\vec{a}_z\right)$$

$$\text{Laplacian } V = \nabla^2 V = \frac{\partial^2 V}{\partial x^2} + \frac{\partial^2 V}{\partial y^2} + \frac{\partial^2 V}{\partial z^2}$$

where $\nabla^2 V$ results in a scalar field.

The Laplacian of a scalar field in Cylindrical coordinates is

$$\nabla^2 V = \frac{1}{r}\frac{\partial}{\partial r}\left(\frac{r\partial V}{\partial r}\right) + \frac{1}{r^2}\frac{\partial^2 V}{\partial \phi^2} + \frac{\partial^2 V}{\partial z^2}$$

The Laplacian of a scalar field in Spherical coordinates is

$$\nabla^2 V = \frac{1}{r^2}\frac{\partial}{\partial r}\left(\frac{r^2\partial V}{\partial r}\right) + \frac{1}{r^2 \sin\theta}\frac{\partial}{\partial \theta}\left(\sin\theta \frac{\partial V}{\partial \theta}\right) + \frac{1}{r^2 \sin^2\theta}\frac{\partial^2 V}{\partial \phi^2}$$

3.11 HARMONIC FIELD

A scalar field V is harmonic in a given region, if the Laplacian of scalar field V is zero. The solution for V in Laplacian equation $\nabla^2 V = 0$ is harmonic.

SOLVED QUESTIONS

3.1 "V" is a scalar function. It is given as $V = xyz + y^2z$. Find the gradient of V.

Solution:

We know grad $V = \nabla V = \frac{\partial V}{\partial x}\vec{a}_x + \frac{\partial V}{\partial y}\vec{a}_y + \frac{\partial V}{\partial z}\vec{a}_z$

$$= \frac{\partial}{\partial x}(xyz + y^2z)\vec{a}_x + \frac{\partial}{\partial y}(xyz + y^2z)\vec{a}_y + \frac{\partial}{\partial z}(xyz + y^2z)\vec{a}_z$$

$$= yz\vec{a}_x + (xy + 2yz)\vec{a}_y + (xy + y^2)\vec{a}_z \quad \text{and}$$

$$\nabla V = yz\vec{a}_x + (x + 2y)z\vec{a}_y + (x + y)y\vec{a}_z$$

3.2 For the scalar fields given as

(a) $V = r^2 z \sin 2\phi$

(b) $V = 10r \sin\phi \cos^2\theta$

Find their gradients.

Solution:

(a) $\nabla V = \frac{\partial V}{\partial r}\vec{a}_r + \frac{1}{r}\frac{\partial V}{\partial \phi}\vec{a}_\phi + \frac{\partial V}{\partial z}\vec{a}_z$

Thus,

$$\nabla V = \frac{\partial}{\partial x}(r^2 z \sin 2\phi)\vec{a}_r + \frac{1}{r}\frac{\partial}{\partial \phi}(r^2 z \sin 2\phi)\vec{a}_\phi + \frac{\partial}{\partial z}(r^2 z \sin 2\phi)\vec{a}_z$$

Therefore,

$$\nabla \cdot V = 2rz \sin 2\phi\, \vec{a}_r + 2rz \cos 2\phi\, \vec{a}_\phi + r^2 \sin 2\phi\, \vec{a}_z$$

(b) We know

$$\nabla V = \frac{\partial V}{\partial r}\vec{a}_r + \frac{1}{r}\frac{\partial V}{\partial \theta}\vec{a}_\theta + \frac{1}{r\sin\theta}\frac{\partial V}{\partial \phi}\vec{a}_\phi$$

Thus,

$$\nabla V = 10\sin\phi\cos^2\theta\vec{a}_r + \frac{10r\sin\phi \cdot 2\cos\theta(-\sin\theta)}{}\vec{a}_\theta + \frac{1}{r\sin\theta}10r\cos^2\theta\cos\phi\,\vec{a}_\phi$$

Therefore,

$$\nabla V = 10\sin\phi\cos^2\theta\vec{a}_r + 10\sin\phi\sin 2\theta\,\vec{a}_\theta + \frac{10\cos^2\theta\cos\phi}{\sin\theta}\vec{a}_\phi$$

3.3 Let $V = y^2 z^2 + xyz$. Find its ∇V. Also, find the direction derivative $\partial V/\partial l$ in the direction $6\vec{a}_x + 8\vec{a}_y + 24\vec{a}_z$ at the point (0, 2, −1).

Solution:
We know

$$\nabla V = \frac{\partial V}{\partial x}\vec{a}_x + \frac{\partial V}{\partial y}\vec{a}_y + \frac{\partial V}{\partial z}\vec{a}_z$$

In this case,

$$\nabla V = yz\,\vec{a}_x + (2yz + xz)\vec{a}_y + (2y^2 z + xy)\vec{a}_z$$

∇V at (0, 2, −1) is obtained as

$$\nabla V = -2\vec{a}_x - 4\vec{a}_y - 8\vec{a}_z$$

Direction derivative $\partial V/\partial l$ is

$$\frac{\partial V}{\partial l} = \nabla V \cdot \vec{a}_l$$

Thus,

$$\frac{\partial V}{\partial l} = (-2\vec{a}_x - 4\vec{a}_y - 8\vec{a}_z)\cdot\frac{(6\vec{a}_x + 8\vec{a}_y + 24\vec{a}_z)}{\sqrt{6^2 + 8^2 + 24^2}} \quad \text{or}$$

$$\frac{\partial V}{\partial l} = \frac{-12 - 32 - 192}{\sqrt{676}}$$

$$\frac{\partial V}{\partial l} = \frac{-118}{13}$$

3.4 For vector $\vec{P} = (x + 2y + z)\vec{a}_x + (y + 3x)\vec{a}_y + (2x^2 + az)\,\vec{a}_z$ to be solenoidal, find constant "a".

Solution:
Vector $\vec{P}$ is solenoidal (given). Its divergence should be zero.
Thus,

$$\nabla \cdot \vec{P} = 0 \quad \text{or}$$

$$\Rightarrow \left[\frac{\partial V}{\partial x}\vec{a}_x + \frac{\partial V}{\partial y}\vec{a}_y + \frac{\partial V}{\partial z}\vec{a}_z\right].[(x + 2y + z)\vec{a}_x + (y + 3x)\vec{a}_y + (2x^2 + az)\vec{a}_z] = 0 \quad \text{or}$$

$$\Rightarrow \frac{\partial}{\partial x}(x + 2y + z) + \frac{\partial}{\partial y}(y + 3x) + \frac{\partial}{\partial z}(2x^2 + az) = 0 \quad \text{or}$$

$$1 + 1 + a = 0$$

Thus,

$$a = -2$$

3.5 Find the divergence of vector $\vec{P}$ if $\vec{P} = 2xz^2\vec{a}_x + 2y^3\vec{a}_z + x^2yz\,\vec{a}_z$.

Solution:

Obtaining the divergence of vector $\vec{P}$ as

$$\nabla\cdot\vec{P} = \left(\frac{\partial}{\partial x}\vec{a}_x + \frac{\partial}{\partial y}\vec{a}_y + \frac{\partial}{\partial z}\vec{a}_z\right).(2xz^2\vec{a}_x + 2y^3\vec{a}_y + x^2yz\,\vec{a}_z) \quad \text{or}$$

$$\nabla\cdot\vec{P} = \frac{\partial}{\partial x}(2xz^2) + \frac{\partial}{\partial y}(2y^3) + \frac{\partial}{\partial z}(x^2yz) \quad \text{or}$$

$$\nabla\cdot\vec{P} = 2z^2 + 6y^2 + x^2y$$

3.6 For the given vector fields

(a) $\vec{P} = r\cos^2\phi\,\vec{a}_r + r^2z^2\sin\phi\,\vec{a}_\phi + z\sin^2\phi\,\vec{a}_z$

(b) $\vec{P} = \dfrac{1}{r^2}\sin^2\theta\,\vec{a}_r + r\sin\theta\cos^2\phi\,\vec{a}_\theta + \sin^2\theta\,\vec{a}_\phi$

Obtain their divergence.

Solution:

(a) Obtaining divergence

$$\nabla\cdot\vec{P} = \frac{1}{r}\frac{\partial}{\partial r}(r^2P_r) + \frac{1}{r}\frac{\partial}{\partial\phi}P_\phi + \frac{\partial}{\partial z}F$$

$$\nabla\cdot\vec{P} = \frac{1}{r}\frac{\partial}{\partial r}(r\cdot r\cdot\cos^2\phi) + \frac{1}{r}\frac{\partial}{\partial\phi}(r^2z^2\sin\phi) + \frac{\partial}{\partial z}(z\sin^2\phi) \quad \text{or}$$

$$\nabla\cdot\vec{P} = 2\cos^2\phi + \frac{r^2z^2}{r}\cos\phi + \sin^2\phi \quad \text{or}$$

$$\nabla\cdot\vec{P} = 2\cos^2\phi + r\cdot z^2\cos\phi + \sin^2\phi$$

(b) Obtaining divergence

$$\nabla\cdot\vec{P} = \frac{1}{r}\frac{\partial}{\partial r}(r^2P_r) + \frac{1}{r\sin\theta}\frac{\partial}{\partial\theta}(P_\theta\sin\theta) + \frac{1}{r\sin\theta}\frac{\partial}{\partial\phi}(P_\phi) \quad \text{or}$$

$$\nabla\cdot\vec{P} = \frac{1}{r^2}\frac{\partial}{\partial r}(\sin^2\theta) + \frac{1}{r\sin\theta}\frac{\partial}{\partial\theta}(r\sin^2\theta\cos^2\phi) + \frac{1}{r\sin\theta}\frac{\partial}{\partial\phi}(\sin^2\theta) \quad \text{or}$$

$$\nabla\cdot\vec{P} = 0 + \frac{1}{r\sin\theta}r\cos^2\phi\,2\sin\theta\cos\theta + 0$$

or simply

$$\nabla\cdot\vec{P} = 2\cos^2\phi\cos\theta$$

3.7 Find the curl of a vector $\vec{P}$ provided $\vec{P} = 4x\vec{a}_x + y^2\vec{a}_y + 5(z^2+z)\vec{a}_z$.

Solution:
Obtaining curl $\vec{P}$ as

$$\text{Curl} \quad \vec{P} = \nabla \times \vec{P} = \begin{vmatrix} \vec{a}_x & \vec{a}_y & \vec{a}_z \\ \dfrac{\partial}{\partial x} & \dfrac{\partial}{\partial y} & \dfrac{\partial}{\partial z} \\ 4x & y^2 & 5(z^2+z) \end{vmatrix}$$

Further,

$$\nabla \times \vec{P} = \left[\frac{\partial}{\partial y}5(z^2+z) - \frac{\partial}{\partial z}y^2\right]\vec{a}_x - \left[\frac{\partial}{\partial x}5(z^2+z) - \frac{\partial}{\partial z}4x\right]\vec{a}_y + \left[\frac{\partial}{\partial y}y^2 - \frac{\partial}{\partial y}(4x)\right]\vec{a}_z$$

Thus,

$$\nabla \times \vec{P} = 0$$

3.8 For the vectors $\vec{P}$ given as

(a) $x^2yz\vec{a}_x + xy\vec{a}_y$

(b) $r\cos\phi\,\vec{a}_r + r^2z\,\vec{a}_\phi + z\sin\phi\,\vec{a}_z$

(c) $\dfrac{1}{r^2}\sin\theta\vec{a}_r + r\sin\theta\cos\phi\vec{a}_\theta + \sin\theta\vec{a}_\phi$

Obtain their curl.

Solution:
We know that curl of $\vec{P}$ is given as

$$\nabla \times \vec{P} = \begin{vmatrix} \vec{a}_x & \vec{a}_y & \vec{a}_z \\ \dfrac{\partial}{\partial x} & \dfrac{\partial}{\partial y} & \dfrac{\partial}{\partial z} \\ P_x & P_y & P_z \end{vmatrix}$$

(a) Obtaining curl as

$$\nabla \times \vec{P} = \left(\frac{\partial P_z}{\partial y} - \frac{\partial P_y}{\partial z}\right)\vec{a}_x + \left(\frac{\partial P_x}{\partial z} - \frac{\partial P_z}{\partial x}\right)\vec{a}_y + \left(\frac{\partial P_y}{\partial x} - \frac{\partial P_x}{\partial y}\right)\vec{a}_z \quad \text{or}$$

$$= (0-0)\vec{a}_x + (x^2y-0)\vec{a}_y + (y - x^2z)\vec{a}_z \quad \text{or}$$

$$= x^2y\vec{a}_y + (y - x^2z)\vec{a}_z$$

(b) Obtaining curl as

$$\nabla \times \vec{P} = \left[\frac{1}{r}\frac{\partial P_z}{\partial \phi} - \frac{\partial P_\phi}{\partial z}\right]\vec{a}_r + \left[\frac{\partial P_r}{\partial z} - \frac{\partial P_z}{\partial r}\right]\vec{a}_\phi + \frac{1}{r}\left[\frac{\partial}{\partial r}(rP_\phi) - \frac{\partial (P_r)}{\partial \phi}\right]\vec{a}_z \quad \text{or}$$

$$= \left[\frac{1}{r} z\cos\phi - r^2\right]\vec{a}_r + [0-0]\vec{a}_\phi + \frac{1}{r}(3r^2 z + r\sin\phi)\vec{a}_z \quad \text{or}$$

$$= \frac{1}{r}[z\cos\phi - r^3]\vec{a}_r + (3rz + r\sin\phi)\vec{a}_z$$

(c) Obtaining curl as

$$\nabla\times\vec{P} = \frac{1}{r\sin\theta}\left[\frac{\partial}{\partial\theta}(P_\phi\sin\theta) - \frac{\partial}{\partial\phi}P_\phi\right]\vec{a}_r + \frac{1}{r}\left[\frac{1}{\sin\theta}\frac{\partial}{\partial\phi}P_r - \frac{\partial}{\partial r}(rP_\phi)\right]\vec{a}_\theta + \frac{1}{r}\left[\frac{\partial}{\partial r}(rP_\theta) - \frac{\partial}{\partial\theta}(P_r)\right]\vec{a}_\phi \quad \text{or}$$

$$\nabla\times\vec{P} = \frac{1}{r\sin\theta}[2\sin\theta\cos\theta - r\sin\theta(-\sin\theta)]\vec{a}_r + \frac{1}{r}[0-\sin\theta]\vec{a}_\theta + \frac{1}{r}\left[2r\sin\theta\cos\phi - \frac{1}{2}\cos\theta\right]\vec{a}_\phi \quad \text{or}$$

$$= \left[\frac{2\cos\theta}{r} + \sin\phi\right]\vec{a}_r + \frac{(-\sin\theta)}{r}\vec{a}_\theta + \left[2\sin\theta\cos\phi - \frac{1}{r^3}\cos\theta\right]\vec{a}_\phi$$

3.9 Vector $\vec{P}$ is given in different coordinates as below. Find the nature of field by obtaining divergence and curl in each case as given below:

(a) $\vec{P} = 30\vec{a}_x + 2xy^2\vec{a}_y + 5x^2z^2\vec{a}_z$ (in Cartesian coordinates)

(b) $\vec{P} = \left(\frac{150}{r^2}\right)\vec{a}_r + 5\vec{a}_\phi$ (in Cylindrical coordinates)

Solution:

(a) Divergence $\vec{P}$ is obtained as

$$\nabla\cdot\vec{P} = \frac{\partial}{\partial x}P_x + \frac{\partial}{\partial y}P_y + \frac{\partial}{\partial z}P_z \quad \text{or}$$

$$= 0 + 4x + 10x^2 \quad \text{or}$$

$$= 2x(2+5x)$$

As $\nabla\cdot\vec{P} \neq 0$, the field is "non-solenoidal".

Now obtaining curl $\vec{P}$ as

$$\text{Curl} \quad \vec{P} = \nabla\times\vec{P} = \begin{vmatrix} \vec{a}_x & \vec{a}_y & \vec{a}_z \\ \frac{\partial}{\partial x} & \frac{\partial}{\partial y} & \frac{\partial}{\partial z} \\ 30 & 2xy^2 & 5xz^2 \end{vmatrix} \quad \text{or}$$

$$= (0 - 0)\vec{a}_x - (5z^2 - 0)\vec{a}_y + (2y^2 - 0)\vec{a}_z \quad \text{or}$$

$$= -5z^2\vec{a}_y + 2y^2\vec{a}_z$$

As curl $\vec{P} \neq 0$, the field is "rotational".

(b) Divergence $\vec{P}$ is obtained as

$$\nabla\cdot\vec{P}=\frac{1}{r}\frac{\partial(rP_r)}{\partial r}+\frac{1}{r}\frac{\partial P_\phi}{\partial\phi}+\frac{\partial P_z}{\partial z}\quad\text{or}$$

$$=\frac{1}{r}\left(-\frac{150}{r^2}\right)+0+0\quad\text{or}$$

$$\nabla\cdot\vec{P}=\frac{-150}{r^3}$$

As $\nabla\cdot\vec{P}=0$, the field is "non-solenoidal".
Now, obtaining curl $\vec{P}$ as

$$\text{Curl}\quad\vec{P}=\nabla\times\vec{P}=\begin{vmatrix}\frac{\vec{a}_r}{r} & \vec{a}_\phi & \frac{\vec{a}_z}{r}\\ \frac{\partial}{\partial r} & \frac{\partial}{\partial\phi} & \frac{\partial}{\partial z}\\ \frac{150}{r^2} & 5r & 0\end{vmatrix}\quad\text{or}$$

$$=(0-0)\vec{a}_r-(0-0)\vec{a}_\phi+(5-0)\frac{\vec{a}_z}{r}\quad\text{or}$$

$$\nabla\times\vec{P}=\frac{5}{r}\vec{a}_z$$

As $\nabla\times\vec{P}\neq 0$, field is "rotational".

3.10 Find a unit vector, which is normal to a surface given $xy^2+3xz^2=4$ at a point (1, −3, 2).

Solution:

We know that gradient of function gives a vector, which is normal to a surface.

Thus, $$\nabla\phi=\frac{\partial\phi}{\partial x}\vec{a}_x+\frac{\partial\phi}{\partial y}\vec{a}_y+\frac{\partial\phi}{\partial z}\vec{a}_z\quad\text{or}$$

$$\nabla\phi=\frac{\partial}{\partial x}(xy^2+3xz^2-4)\vec{a}_x+\frac{\partial}{\partial y}(xy^2+3xz^2-4)\vec{a}_y+\frac{\partial}{\partial z}(xy^2+3xz^2-4)\vec{a}_z\quad\text{or}$$

$$\nabla\phi=(y^2+3z^2)\vec{a}_x+2xy\vec{a}_y+6x\vec{a}_z$$

At point (1, −3, 2), the vector normal to the surface is obtained as

$$\nabla\phi=21\vec{a}_x+(-6)\vec{a}_y+6\vec{a}_z$$

Its unit vector is now obtained as

$$\frac{\nabla\phi}{|\nabla\phi|}=\frac{21\vec{a}_x-6\vec{a}_y+6\vec{a}_z}{\sqrt{21^2+(-6)^2+6^2}}$$

or simply

$$\frac{\nabla\phi}{|\nabla\phi|}=\frac{21}{\sqrt{513}}\vec{a}_x-\frac{6}{513}\vec{a}_y+\frac{6}{\sqrt{513}}\vec{a}_z$$

3.11 Find constants a, b and c for a vector $\vec{P}=(2x+y+az)\vec{a}_x+(bx+4y-3z)\vec{a}_y+(2x+cy+3z)\vec{a}_z$ to be "irrotational".

Solution:

For a vector $\vec{P}$ to be "irrotational," $\nabla\times\vec{P}\neq 0$. Obtaining curl $\vec{P}$ in this case, we get

$$\nabla\times\vec{P}=\begin{vmatrix}\vec{a}_x & \vec{a}_y & \vec{a}_z\\ \frac{\partial}{\partial x} & \frac{\partial}{\partial y} & \frac{\partial}{\partial z}\\ (zx+y+az) & (bx+4y-3z) & (2x+cy+3z)\end{vmatrix}$$

$$\nabla\times\vec{P}=\left[\frac{\partial}{\partial y}(2x+cy+3z)-\frac{\partial}{\partial z}(bx+4y-3z)\right]\vec{a}_x$$

$$-\left[\frac{\partial}{\partial x}(2x+cy+3z)-\frac{\partial}{\partial z}(2x+y+az)\right]\vec{a}_y$$

$$+\left[\frac{\partial}{\partial x}(bx+4y-3z)-\frac{\partial}{\partial y}(2x+y+az)\right]\vec{a}_z \quad \text{or}$$

$$=(c+3)\vec{a}_x-(2-a)\vec{a}_y+(b-1)\vec{a}_z$$

However, $\nabla\times\vec{P}=0$ (given).
Thus,

$$c+3=0$$

$$\Rightarrow c=-3$$

$$2-a=0$$

$$\Rightarrow a=2$$

$$b-1=0 \quad \text{and}$$

$$\Rightarrow b=1$$

3.12 For $\vec{F}=2x^3\vec{a}_x+2y^3\vec{a}_y+2z^3\vec{a}_z$, evaluate $\oint_s \vec{F}\cdot d\vec{s}$ using Gauss' Divergence Theorem. Here S is the surface of sphere given as $x^2+y^2+z^2=r^2$.

Solution:

Gauss' Divergence Theorem gives

$$\int_s F\cdot ds=\int_v \nabla\cdot F\,dv$$

Thus,

$$\oint_s F \cdot ds = \int_v \left[\frac{\partial}{\partial x}\vec{a}_x + \frac{\partial}{\partial y}\vec{a}_y + \frac{\partial}{\partial z}\vec{a}_z\right].\left[2x^3\vec{a}_x + 2y^3\vec{a}_y + 2z^3\vec{a}_z\right]\cdot dv \quad \text{or}$$

$$= \int_v (6x^2 + 6y^2 + 6z^2)\cdot dv \quad \text{or}$$

$$= 6\int_v (x^2 + y^2 + z^2)\cdot dv$$

$$= 6\int_v a^2 \cdot dv$$

$$= 6a^2 \int_v \cdot dv$$

$$= 6a^2 \times \text{ volume of sphere}$$

$$= 6a^2 \times \frac{4}{3}\pi a^3$$

$$= 8\pi a^5$$

Thus,

$$\oint_s F \cdot ds = 8\pi a^5$$

3.13 An electron moves at 20 km/s along $\vec{a}_x$ in a magnetic field. The magnetic flux density is given as $\vec{B} = 0.4\vec{a}_x - 0.6\vec{a}_y + 1.0\vec{a}_z$ Wb/m^2. Find the electric field intensity if

(a) There is no force applied on the electron.

(b) Force is applied on the electron in the presence of both $\bar{B}$ and $\bar{E}$. Consider $\bar{E} = (4\vec{a}_x + 2\vec{a}_y + 3\vec{a}_z)$ kV/m.

Solution:

(a) We know $\vec{F} = q[\vec{E} + \vec{V} \times \vec{B}]$.

If $\vec{F} = 0$, then

$$\vec{E} = -\vec{V} \times \vec{B}$$

$$= -\begin{vmatrix} \vec{a}_x & \vec{a}_y & \vec{a}_z \\ 2\times 10^4 & 0 & 0 \\ 0.4 & -0.6 & 1.0 \end{vmatrix}$$

or simply

$$= (20\vec{a}_y + 12\vec{a}_z) \text{ kV/m}$$

(b) Force on the electron due to both $\bar{E}$ and $\bar{B}$ is given as

$$\vec{F} = q[\bar{E} + \vec{V} \times \bar{B}]$$

Now $$\vec{F} = -1.6\times10^{-19}[4\vec{a}_x + 2\vec{a}_y + 3\vec{a}_z + 20\vec{a}_y + 12\vec{a}_z]\times10^3$$

$$= -1.6\times10^{-19}[4\vec{a}_x + 22\vec{a}_y + 15\vec{a}_z]\times10^3$$

$$= [-6.4\vec{a}_x - 35.2\vec{a}_y - 24\vec{a}_z]\times10^{-16}\text{ Newton}$$

3.14 Find the field at any point for (a) and (b).
(a) Due to an ideal solenoid having an infinite length and a circular shift $\vec{K} = 120\,\vec{a}_\phi$.
(b) A 800-turn solenoid with a finite length of 10 m and current per unit of 8 Amp.
[Hint: Use the concept of field due to two parallel infinite surface sheet currents.]

Solution:

Solenoids are shown in Figure 3.6. By using the concept of field because of two infinite parallel sheets, fields are obtained.

(a) For an ideal solenoid having infinite length, we get

$$\text{'}\vec{K}\text{'} = 120\,\vec{a}_\varphi$$

Also $\vec{H}_R = 120\vec{a}_z$ for $R < a$ and

$$\vec{H} = 0 \quad \text{for} \quad R > a$$

(b) For a 800-turn solenoid having a length of 10 m, we get

Total current = NI (Amp)

= 800 × 8 Amp

= 6400 Amp

Surface current density $\vec{K} = \dfrac{6400}{10}\times120\vec{a}_\phi$ (A/m)

Case 1: ($R \le a$): No current is enclosed. Hence, $\vec{H} = 0$

Case 2: ($a < R < b$): Ampere's Law gives

$$H_R \cdot 2\pi R = 12(\pi R^2 - \pi a^2) \text{ (where '}R\text{' is the radius)} \quad \text{or}$$

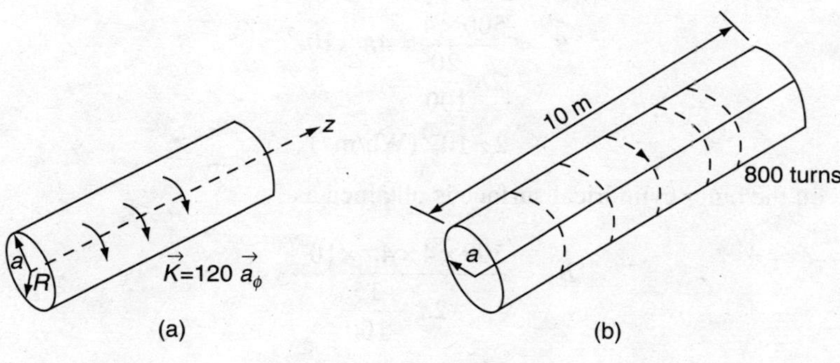

Figure 3.6 *Solenoids*

$$\vec{H}_R = \frac{12\pi(R^2 - a^2)}{2\pi R} \quad \text{or}$$

$$\vec{H}_R = \frac{6(R^2 - a^2)}{R}\vec{a}_\varphi$$

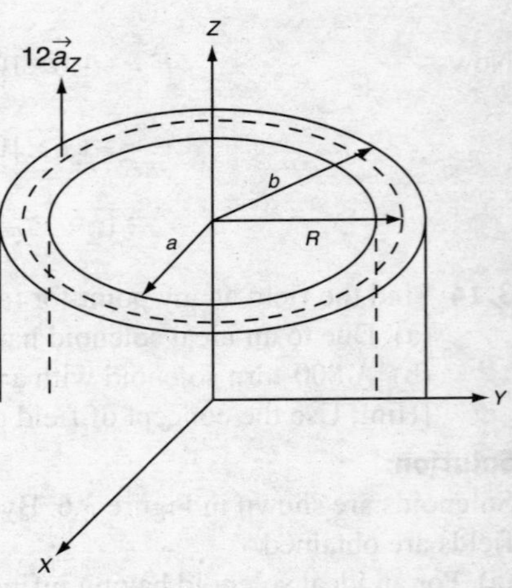

Figure 3.7 *Current distribution*

Current density vector $\vec{J}$ is in the z-direction. Therefore, the field will be in $\vec{a}_\varphi$ direction.

Case 3: ($b \le R < \infty$): As per Ampere's Circuital Law (where full current is enclosed), we get

$$H_R \cdot 2\pi R = 12(\pi b^2 - \pi a^2) \quad \text{or}$$

$$\vec{H}_R = \frac{12\pi(b^2 - a^2)}{2\pi R} \quad \text{or}$$

$$\vec{H}_R = \frac{6(b^2 - a^2)}{R}\vec{a}_\varphi$$

The field inside a solenoid is obtained as

$$\vec{H} = \frac{NI}{d}\vec{a}_z \quad \text{or}$$

$$\vec{H} = \frac{6400}{10}\vec{a}_z \text{(A/m)} \quad \text{or}$$

$$\vec{H} = 640\ \vec{a}_z \text{ A/m}$$

3.15 There is a wooden ring of square section. Its internal diameter is 30 cm. The external diameter is 50 cm. The wooden ring carries a toroidal winding that has 500 turns. It is uniformly distributed. For a coil of 4 *A*, find the magnetic flux density '*B*' on both cylindrical surfaces, viz. inner and outer wooden rings.

Solution:

Flux density 'B_i' on the inner cylindrical surface is obtained as

$$\text{'}B_i\text{'} = \frac{500 \times 4}{2\pi \frac{20}{100}} \times 4\pi \times 10^{-7}$$

$$= 2 \times 10^{-3} \text{(Wb/m}^2\text{)}$$

Flux density 'B_o' on the outer cylindrical surface is obtained as

$$\text{'}B_o\text{'} = \frac{500 \times 4 \times 4\pi \times 10^{-7}}{2\pi \times \frac{15}{100}}$$

$$= 2.66 \times 10^{-3} \text{(Wb/m}^2\text{)}$$

3.16 Find the curl of a vector $\vec{A}$ given $\vec{A} = 4x^2\vec{a}_x + y\vec{a}_y + 6z\,\vec{a}_z$.

Solution:

The curl of a vector $\vec{A}$ is given by

$$\text{Curl } \vec{A} = \nabla \times \vec{A} = \begin{vmatrix} \vec{a}_x & \vec{a}_y & \vec{a}_z \\ \dfrac{\partial}{\partial x} & \dfrac{\partial}{\partial y} & \dfrac{\partial}{\partial z} \\ 4x^2 & y & 6z \end{vmatrix}$$

$$= \vec{a}_x\left[\frac{\partial}{\partial y}(6z) - \frac{\partial}{\partial z}(y)\right] - \vec{a}_y\left[\frac{\partial}{\partial x}(6z) - \frac{\partial}{\partial z}(4x^2)\right]$$

$$+\vec{a}_z\left[\frac{\partial}{\partial x}(y) - \frac{\partial}{\partial y}(4x^2)\right]$$

$$= \vec{a}_x(0) - \vec{a}_y(0) + \vec{a}_z(0)$$

$$\nabla \times \vec{A} = 0$$

3.17 "V" is a scalar function. It is given as $V = x^2yz + yx$; find the gradient of V.

Solution:

We know that

$$\text{grad}\, V = \nabla V = \frac{\partial V}{\partial x}\vec{a}_x + \frac{\partial V}{\partial y}\vec{a}_y + \frac{\partial V}{\partial z}\vec{a}_z$$

$$= \frac{\partial}{\partial x}(x^2yz + xy)\vec{a}_x + \frac{\partial}{\partial y}(x^2yz + xy)\vec{a}_y$$

$$+\frac{\partial}{\partial z}(x^2yz + xy)\vec{a}_z$$

$$= (2xyz + y)\vec{a}_x + (x^2z + x)\vec{a}_y + (x^2y)\vec{a}_z$$

Hence,

$$\nabla V = (2xyz + y)\vec{a}_x + (x^2z + x)\vec{a}_y + (x^2y)\vec{a}_z$$

3.18 The potential scalar field is given as

(a) $V = r\,x^2\cos 2\phi$

(b) $V = 10r\sin^2\phi\cos\theta$

Find their gradients.

Solution:

(a) $\nabla V = \dfrac{\partial V}{\partial r}\vec{a}_r + \dfrac{1}{r}\dfrac{\partial v}{\partial \phi}\vec{a}_\phi + \dfrac{\partial V}{\partial z}\vec{a}_z$

Thus,

$$\nabla V = \frac{\partial}{\partial r}(r^2 x^2 \cos 2\phi)\vec{a}_r + \frac{1}{r}\frac{\partial}{\partial \phi}(r^2 x^2 \cos 2\phi)\vec{a}_\phi + \frac{\partial}{\partial z}(r^2 x^2 \cos 2\phi)\vec{a}_z$$

$$= (x^2 \cos 2\phi)\vec{a}_r + \frac{1}{r}(-2r x^2 \sin 2\phi)\vec{a}_\phi + 0$$

$$\nabla V = (x^2 \cos 2\phi)\vec{a}_r - 2x^2 \sin 2\phi \vec{a}_\phi$$

(b) Given $V = 10r \sin^2 \phi \cos\theta$,

$$\nabla V = \frac{\partial V}{\partial r}\vec{a}_r + \frac{1}{r}\frac{\partial V}{\partial \theta}\vec{a}_\theta + \frac{1}{r\sin\theta}\frac{\partial v}{\partial \varphi}\vec{a}_z$$

$$= \frac{\partial}{\partial r}(10r\sin^2 \phi\cos\theta)\vec{a}_r + \frac{1}{r}\frac{\partial}{\partial \theta}(10r\sin^2 \phi\cos\theta)\vec{a}_\theta + \frac{1}{r\sin\theta}\frac{\partial}{\partial \phi}(10r\sin^2 \phi\cos\theta)\vec{a}_z$$

$$= 10\sin^2 \phi\cos\theta\vec{a}_r + \frac{1}{r}(-10r\sin^2 \phi\sin\theta)\vec{a}_\theta + \frac{1}{r\sin\theta}(10r\cos\theta\sin^2 \phi)\vec{a}_z$$

$$\Rightarrow \nabla V = 10\sin^2 \phi\cos\theta\vec{a}_r - 10\sin^2 \phi\sin\theta\vec{a}_\theta + \frac{1}{r\sin\theta}(10\cos\theta\sin^2 \phi)\vec{a}_z$$

3.19 It is given that $V = x^2 y + zx$. Find the divergence of V. Find the directional derivative $\partial V/\partial l$ in the direction $(2\vec{a}_x + 4\vec{a}_y + 8\vec{a}_z)$ at the point (0, −1, 2).

Solution:
We know that

$$\nabla V = \frac{\partial V}{\partial x}\vec{a}_x + \frac{\partial V}{\partial y}\vec{a}_y + \frac{\partial V}{\partial z}\vec{a}_z$$

In this case,

$$\nabla V = \frac{\partial}{\partial x}(x^2 y + zx)\vec{a}_x + \frac{\partial}{\partial y}(x^2 y + zx)\vec{a}_y + \frac{\partial}{\partial z}(x^2 y + zx)\vec{a}_z$$

$$\Rightarrow \nabla V = (2xy + z)\vec{a}_x + x^2\vec{a}_y + x\vec{a}_z$$

∇V at (0, −1, 2) is given by

$$\nabla V = 2\vec{a}_x$$

Directional derivative $\partial V/\partial l$ is obtained as

$$\Rightarrow \frac{\partial V}{\partial l} = 2\vec{a}_x \cdot \frac{(2\vec{a}_x + 4\vec{a}_y + 8\vec{a}_z)}{\sqrt{(2)^2 + (4)^2 + (8)^2}}$$

$$= 2\vec{a}_x \cdot \frac{(2\vec{a}_x + 4\vec{a}_y + 8\vec{a}_z)}{9.16}$$

$$\frac{\partial V}{\partial l} = \frac{4}{9.16}$$

3.20 Given that $A = (5r^2/4)\vec{a}_r$ (C/m^2) is a Spherical co-ordinate, evaluate both sides of the Divergence Theorem for the volume enclosed by $r = 4m$ and $\theta = \pi/4$.

Solution:

We know that

$$\oint A \cdot ds = \oint (\nabla \cdot A) dv$$

where A has only a radial component. $.A.\ ds$ has a non-zero value on the surface $r = 4$ m.

$$\oint A \cdot ds = \int_0^{2\pi} \int_0^{\pi/4} \frac{5(4)^2}{4} \vec{a}_r .(4)^2 \sin\theta \, d\theta \, d\phi \vec{a}_r$$

$$= 589.1C$$

Divergence theorem gives

$$\nabla \cdot A = \frac{1}{r^2} \frac{\partial}{\partial r}\left(\frac{5r^2}{4}\right) = 5r \quad \text{and}$$

$$\int (\nabla \cdot A) dv = \int_0^{2\pi} \int_0^{\pi/4} \int_0^{4} (5r) r^2 \sin\theta \, dr \, d\varphi = 589.1C$$

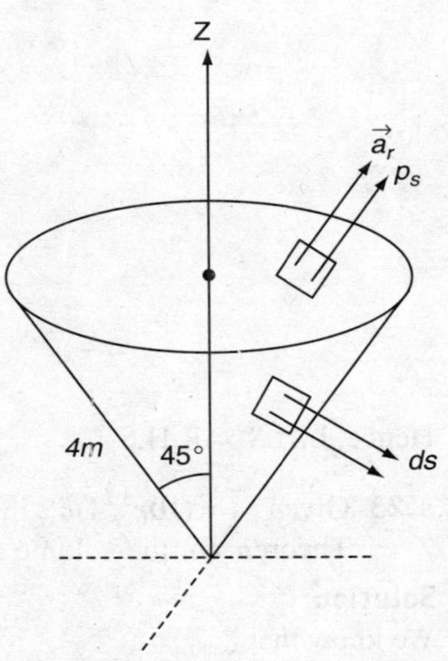

Figure 3.8 *Enclosed volume*

3.21 Given $A = 5\sin\theta \vec{a}_\theta + 5\sin\phi \, \vec{a}_\phi$, find ∇A at $(0.5, \pi/4, \pi/4)$.

Solution:

$$\nabla \cdot A = \frac{1}{r\sin\theta} \frac{\partial}{\partial \theta}(5\sin^2\theta) + \frac{1}{r\sin\theta} \frac{\partial}{\partial \phi}(5\sin^2\phi)$$

$$= 10\frac{\cos\theta}{r} + 5\frac{\cos\theta}{r\sin\theta}$$

$$\Rightarrow \Delta \cdot A\Big|_{0.5,\pi/4,\pi/4} = \frac{10\cos\pi/4}{0.5} + \frac{5\cos(\pi/4)}{0.5\sin(\pi/4)}$$

$$\nabla \cdot A = 24.14$$

3.22 Given $\vec{A} = (10r^3/4)\vec{a}_r$ in a Cylindrical co-ordinate, verify the Divergence Theorem for the volume enclosed by $r = 2$, $z = 0$ to 10.

Solution:

L.H.S.

$$\oint_S \vec{A} \cdot \overline{ds} = \int_0^{10} \int_0^{2\pi} \frac{10r^3}{4} \times r \, d\phi \, dz \Big|_{r=2}$$

$$= \frac{10}{4} \times (2)^4 \times 2\pi \times 10 = \frac{10}{4} \times 16 \times 20\pi$$

$$= 800\pi$$

R.H.S.

$$\int_V (\nabla \cdot \vec{D}) \, dv, \quad \text{where} \quad dv = r \, dr \, d\phi \, dz$$

$$\nabla \cdot \vec{D} = \frac{1}{r} \frac{\partial}{\partial r}(r \cdot Dr) = \frac{1}{r} \times \frac{10}{4} \times 4r^3 = 10r^2$$

$$\int_V (\nabla \cdot \vec{D}) \, dv = \int_0^2 \int_0^{10} \int_0^{2\pi} 10r^2 \times r \, dr \, d\phi \, dz$$

$$= \left[\frac{10r^4}{4}\right]_0^2 \times 2\pi \times 10$$

$$= 10 \times 4 \times 2\pi \times 10 = 800\pi$$

Hence, L.H.S. = R.H.S.

3.23 Given $\vec{A} = (10r^3/4)\vec{a}_x$ in the Cylindrical co-ordinate, evaluate both sides of the Divergence Theorem for the volume enclosed by $r = 1$ and $r = 2$ m, $z = 0$ and $z = 10$ m.

Solution:

We know that

$$\oint A \cdot ds = \int_V (\nabla \cdot \vec{A}) dV$$

Now,

$$\nabla \cdot \vec{A} = \frac{1}{r} \frac{\partial}{\partial r}(r \times \vec{a}_r) + 0 + 0 \qquad (\because \vec{a}_\phi = \vec{a}_z = 0)$$

$$= \frac{1}{r} \frac{\partial}{\partial r}\left(r \times \frac{10}{4} r^3\right) = \frac{1}{r} \times \frac{10}{4} \times 4r^3 = 10r^2$$

$$\int_V (\nabla \cdot \vec{A}) \, dv = 10 \int_1^2 \int_0^{2\pi} \int_0^{10} r^2 \times r (dr \, d\phi \, dz)$$

$$= \frac{10r}{4} \times 2\pi \times 10 = 50\pi(16 - 1) = 750\pi$$

Now at L.H.S.,

$$r = 1, \quad \vec{a}_x = -\vec{a}_r \quad \text{and} \quad r = 2, \quad \vec{a}_x = \vec{a}_r$$

$$\oint A \cdot ds = \int_0^{2\pi}\int_0^{10} \frac{10r^3\vec{a}_r}{4} \times r\, d\phi\, dz\Big|_{r=1}^{(-\vec{a}_r)}$$

$$+\int_0^{2\pi}\int_0^{10} \frac{10(+2)^3\vec{a}_r}{4} \times 2\, d\phi\, dz \cdot \vec{a}_r$$

$$= \frac{-10}{4} \times 2\pi \times 10 + \frac{160}{4} \times 2\pi \times 10 = \frac{-200\pi}{4} + 800\pi$$

$$= \frac{3000\pi}{4} = 750\pi$$

∴ L.H.S. = R.H.S.

3.24 Find the divergence of the vector fields and evaluate them at given points:

(a) $\vec{A} = yz\,\vec{a}_x + 4xy\,\vec{a}_y + y\,\vec{a}_z$ at (1, – 2, 3)

(b) $\vec{A} = \rho z\, \sin\phi\, \vec{a}_\rho + 3\rho z^2 \cos\phi\, \vec{a}_\varphi$ at $(5,\ \pi/2,\ 2)$

Solution:

(a) $\nabla \cdot \vec{A} = \dfrac{\partial A_x}{\partial x} + \dfrac{2A_y}{\partial y} + \dfrac{\partial A_z}{\partial z}$

Given

$$\vec{A} = yz\,\vec{a}_x + 4xy\,\vec{a}_x + y\vec{a}_z$$

$$A_x = yz,\ A_y = 4xy,\ A_z = y$$

$$\frac{\partial A_x}{\partial x} = 0, \quad \frac{\partial A_y}{\partial y} = 4x, \quad \frac{\partial A_z}{\partial z} = 0$$

$$\Rightarrow \Delta \cdot \vec{A} = 0 + 4x + 0 = 4x$$

$$\Delta \cdot \vec{A} \text{ at } (1, -2, 3) = 4$$

(b) $\nabla \cdot \vec{A} = \dfrac{1}{\rho}\dfrac{\partial}{\partial \rho}(\rho A\rho) + \dfrac{1}{\rho}\dfrac{\partial A_\phi}{\partial \phi} + \dfrac{\partial A_z}{\partial z}$

Given

$$\vec{A} = \rho z \sin\phi\, \overrightarrow{a\rho} + 3\rho z^2 \cos\phi\, \overrightarrow{a_\phi}$$

$$A_\rho = \rho z \sin\phi, \quad A_\phi = 3\rho z^2 \cos\phi, \quad A_z = 0$$

$$\nabla \cdot \vec{A} = \frac{1}{\rho}\frac{\partial}{\partial \rho}(\rho^2 z \sin\phi) + \frac{1}{\rho}\frac{\partial}{\partial \phi}(2\rho z^2 \cos\phi) + 0$$

$$= \frac{1}{\rho} \times 2\rho z \sin\phi + \frac{1}{\rho} \times 3\rho z^2(-\sin\phi)$$

$$= 2z\sin\phi - 3z^2 \sin\phi = z\sin\phi(2 - 3z)$$

$$\Rightarrow \nabla \times \vec{A} = z\sin\phi(2 - 3z)$$

Now,

$$\nabla \times \vec{A} \text{ at } (5, \pi/2, 1) = -1$$

3.25 Evaluate $\nabla \cdot (\nabla \times \vec{A})$ if

$$\vec{A} = x^2 y \vec{a}_x + y^2 z \vec{a}_y - 2xz \vec{a}_z$$

Solution:

$$\nabla \times \vec{A} = \begin{vmatrix} \vec{a}_x & \vec{a}_y & \vec{a}_z \\ \partial/\partial x & \partial/\partial y & \partial/\partial z \\ x^2 y & y^2 z & -2xz \end{vmatrix}$$

$$= \vec{a}_x \left[\frac{\partial}{\partial y}(-2x) - \frac{\partial}{\partial z}(y^2 z) \right] - \vec{a}_y \left[\frac{\partial}{\partial x}(-2xz) - \frac{\partial}{\partial z}(x^2 y) \right]$$

$$+ \vec{a}_z \left[\frac{\partial}{\partial x}(y^2 z) - \frac{\partial}{\partial y}(x^2 y) \right]$$

$$= \vec{a}_x (0 - y^2) + \vec{a}_y (-2z - 0) + \vec{a}_z (0 - x^2)$$

$$\Rightarrow \nabla \times \vec{A} = -y^2 \vec{a}_x + 2z \vec{a}_y - x^2 \overrightarrow{az}$$

Now,

$$\nabla \cdot (\nabla \times \vec{A}) = \frac{\partial}{\partial x}(-y^2) + \frac{\partial}{\partial y}(2z) + \frac{\partial}{\partial z}(-x^2) = 0 + 0 + 0 = 0$$

Therefore,

$$\nabla \cdot (\nabla \times \vec{A}) = 0$$

3.26 Given that $A = 30e^{-r}\vec{a}_r - 2z\vec{a}_z$ in Cylindrical co-ordinates, evaluate both sides of the Divergence Theorem for the volume enclosed by $r = 2, z = 0$ and $z = 5$.

Solution:

We know that

$$\oint A \cdot ds = \int (\nabla \cdot A) dv$$

It is noted that $\vec{a}_z = 0$ for $z = 0$ and hence $A.ds$ is zero over that part of the surface.

$$\oint A \cdot ds = \int_0^5 \int_0^{2r} 30 e^{-r} \vec{a}_r 2 d\phi \, dz \, \vec{a}_r + \int_0^{2\pi} \int_0^2 -2(5)\vec{a}_z \cdot r \, dr \, d\phi \, \vec{a}_z$$

$$= 60e^{-2}(2\pi)(5) - 10(2\pi)(2) = 129.4$$

For the right side of the Divergence Theorem,

$$\nabla \cdot A = \frac{1}{r}\frac{\partial}{\partial}(30e^{-r}) + \frac{\partial}{\partial r}(-2r)$$

$$= \frac{30e^{-r}}{r} - 30e^{-r} - 2$$

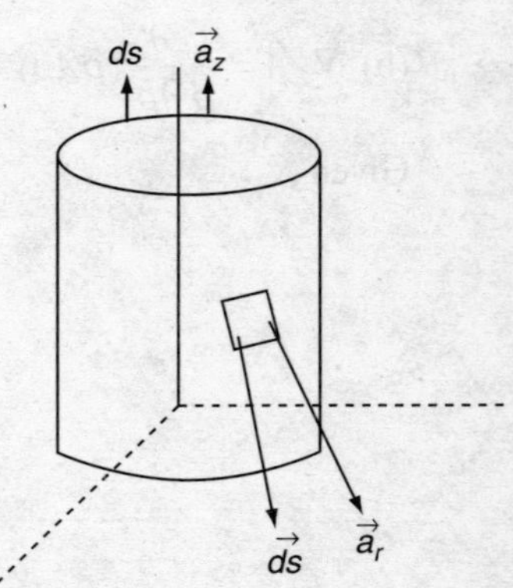

Figure 3.9 *Enclosed volume*

Therefore,

$$\int (\nabla \cdot A)\, dv = \int_0^5 \int_0^{2\pi} \int_0^2 \left(\frac{30e^{-r}}{r} - 30e^{-r} - 2 \right) r\, dr\, d\phi\, dz$$
$$= 129.4$$

3.27 Given $\phi = xy + yz + xz$, find the gradient at (1, 2, 3) and the directional derivative of ϕ at the same point in the direction towards the point (3, 4, 4).

Solution:

$$\nabla \cdot \phi = \vec{a}_x \left(\frac{\partial \phi}{\partial x} \right) + \vec{a}_y \left(\frac{\partial \phi}{\partial y} \right) + \vec{a}_z \left(\frac{\partial \phi}{\partial z} \right)$$
$$= \vec{a}_x (y+z) + \vec{a}_y (x+z) + \vec{a}_z (x+y)$$

$$(\nabla \phi)\big|_{1,2,3} = \vec{a}_x (2+3) + \vec{a}_y (1+3) + \vec{a}_z (1+2)$$

$$= 5\vec{a}_x + 4\vec{a}_y + 3\vec{a}_z$$

Now,

$$d\vec{l} = (3-1)\vec{a}_x + (4-2)\vec{a}_y + (4-3)\vec{a}_z$$

$$d\vec{l} = 2\vec{a}_x + 2\vec{a}_y + \vec{a}_z$$

Directional derivative

$$\frac{d\phi}{dl} = \nabla \phi \cdot \vec{dl} = \nabla \phi \cdot \frac{d\vec{l}}{|dl|} = (5\vec{a}_x + 4\vec{a}_y + 3\vec{a}_z) \cdot \frac{(2\vec{a}_x + 2\vec{a}_y + \vec{a}_z)}{\sqrt{9}} = \frac{10+8+3}{3} = \frac{21}{3} = 7$$

UNSOLVED QUESTIONS

3.1 Let a scalar field 'f' be $f(x, y, z) = xy^2 + e^x$. Find its gradient. At point $P(1, 4, -3)$, what is the magnitude and direction of the gradient?

3.2 In the case of a vector field $\vec{A} = yz\vec{a}_x + zx\vec{a}_y + xy\vec{a}_z$, prove that it is both irrotational and solenoidal.

3.3 A circle with radius 2 cm has its centre at the origin of the xyz axes. Also, the circle rests in the yz plane. Let vector $\vec{A}$ be $\vec{A} = 2y^2\vec{a}_x + 5z\vec{a}_y + 6y\vec{a}_z$. Find the value of $\oint \vec{A}.dl$ if the contour is the circumference of this circle.

3.4 If $V_x = x/(x^2 + y^2), V_y = x/(x^2 + y^2)$ and $V_z = 0$, find the divergence and curl. (Use Cylindrical co-ordinates.)

3.5 Evaluate $\oint F.ds$ using the Gauss–Divergence Theorem. Here, $F = 3xz\,\vec{a}_x - yz\,\vec{a}_y + yz\,\vec{a}_z$. S is the surface of the cube and S is bounded by $x = 0, x = 1, y = 0, y = 1$ and $z = 0, z = 1$.

3.6 Find the divergence of vector fields given as

(a) $\vec{P} = \rho \sin \Phi \vec{a}_\rho + \rho^2 z\, \vec{a}_\phi + z \cos \Phi\, \vec{a}_z$ and

(b) $\vec{P} = \frac{1}{r^2} \cos \theta \vec{a}_r + r \sin \theta \cos \Phi \vec{a}_\theta + 5\vec{a}_\phi$

3.7 Check $\vec{A} = (3x - y)\vec{a}_x - yz^2\vec{a}_y - y^2 z\vec{a}_z$ for Stoke's Theorem. Here, *S* is the upper half surface of the sphere $x^2 + y^2 + z^2 = 1$. Let '*C*' be its boundary.

3.8 Prove that the divergence of the curl of any vector field is zero.

(Help: Prove that for a vector field $\vec{A}, \nabla.\nabla \times \vec{A} = 0$.)

3.9 For the scalar fields as given below, find their Laplacian.

(a) $V = 8r \sin^2\theta \cos^2\phi$

(b) $V = z^2\rho \sin\phi + z \cos^2\phi + \rho^2$

3.10 Let a vector field be $\vec{A} = x\vec{a}_x + y^2\vec{a}_y + 2z\vec{a}_z$. For a cubical volume bounded by planes $x = 0$, $x = 1$, $y = 0$, $y = 1$ and $z = 0$, $z = 1$, prove the Divergence Theorem.

3.11 A volume charge density is $\rho_v = 8z^2 e^{-0.1x} \sin \Pi y$. Find the total charge inside a volume given as $(-1 < x < 2)$, $(0 < y < 1)$ and $(3 < z < 3.6)$.

3.12 Let a vector field be $\vec{A}\,(r, \Phi, z) = 20e^{-r}\,\vec{a}_r - 3z\,\vec{a}_z$. For volume enclosed by $r = 1.5$, $z = 0$ and $z = 4$, verify the Divergence Theorem.

SUMMARY

1. Line integral is integral of the dot product of a vector evaluated along a specified path and is denoted by $\int_L$.
2. Surface integral is performed over an entire surface and is denoted by $\int_s$ or $\iint$.
3. Volume integral is performed over the entire volume and is denoted by $\int_v$ or $\iiint$.
4. Gradient operator is also called Del operator. It is denoted by "∇." The gradient of a scalar quantity '∇' in Cartesian coordinates is given as

$$\text{grad } V = \nabla V = \frac{\partial V}{\partial x}\vec{a}_x + \frac{\partial V}{\partial y}\vec{a}_y + \frac{\partial V}{\partial z}\vec{a}_z$$

5. The gradient of a scalar quantity '*V*' in Cylindrical coordinates is given as

$$\text{grad } V = \nabla V = \frac{\partial V}{\partial \rho}\vec{a}_\rho + \frac{1}{\rho}\frac{\partial V}{\partial \phi}\vec{a}_\phi + \frac{\partial V}{\partial z}\vec{a}_z$$

6. The gradient of a scalar quantity *V* in Spherical coordinates is given as

$$\text{grad } V = \nabla V = \frac{\partial V}{\partial r}\vec{a}_r + \frac{1}{r}\frac{\partial V}{\partial \theta}\vec{a}_\theta + \frac{1}{r\sin\theta}\frac{\partial V}{\partial \phi}\vec{a}_\varphi$$

7. The gradient represents magnitude and direction of maximum space rate of increase in a scalar field.

8. Divergence of a vector $\vec{P}$ is defined as net outward flux of $\vec{P}$ per unit volume.

9. Divergence

$$\vec{P} = \nabla \cdot \vec{P} = \lim_{\Delta v \to \infty} \frac{\int_s \vec{P} \cdot \vec{ds}}{\nabla v}$$

where Δs is the surface and Δv the volume.

10. Divergence in Cartesian coordinates for a vector $\vec{P}$ is given as

$$\text{div}\vec{P} = \frac{\partial P_x}{\partial x} + \frac{\partial P_y}{\partial y} + \frac{\partial P_z}{\partial z}$$

11. Divergence in Cylindrical coordinates for a vector $\vec{P}$ is given as

$$\text{div}\,\vec{P} = \frac{1}{r}\frac{\partial}{\partial r}(rP_r) + \frac{1}{r.}\frac{\partial P_\phi}{\partial \phi} + \frac{\partial P_z}{\partial z}$$

12. Divergence in Spherical coordinates for a vector $\vec{P}$ is given as

$$\text{div}\vec{P} = \frac{1}{r^2}\frac{\partial}{\partial r}(r^2 P_r) + \frac{1}{r\sin\theta}\frac{\partial}{\partial \theta}(P_\theta \sin\theta) + \frac{1}{r\sin\theta}\frac{\partial P_\phi}{\partial \phi}$$

13. Divergence $\vec{P}$ at a point is positive if field lines are diverging and negative if field lines are converging.

14. As per the Divergence Theorem, the volume integral of divergence "$\vec{P}$" taken over a volume "V" is equal to the surface integral of $\vec{P}$ taken over the closed surface "S." This closed surface "S" bounds the volume "V."

15. The curl of a vector measures the rate of change of vector. Curl $\vec{P}$ is given as

$$\text{curl}\ \vec{P} = \nabla \times \vec{P} = \left(\lim_{\Delta s \to 0} \frac{\oint_L \vec{P}.\vec{dl}}{\nabla s} \right)_{\max} \vec{a}_n$$

16. Curl $\vec{P}$ in Cartesian coordinates is given as

$$\nabla \times \vec{P} = \begin{vmatrix} a_x & a_y & a_z \\ \frac{\partial}{\partial x} & \frac{\partial}{\partial y} & \frac{\partial}{\partial y} \\ P_x & P_y & P_y \end{vmatrix}$$

17. Curl $\vec{P}$ in Cylindrical coordinates is given as

$$\nabla \times \vec{P} = \begin{vmatrix} a_r & ra_\phi & a_z \\ \dfrac{\partial}{\partial r} & \dfrac{\partial}{\partial \phi} & \dfrac{\partial}{\partial z} \\ P_x & rP_\phi & P_z \end{vmatrix}$$

18. Curl $\vec{P}$ in Spherical coordinates is given as

$$\nabla \times \vec{P} = \frac{1}{r^2 \sin\theta} \begin{vmatrix} a_r & ra_\theta & r\sin\theta\, a_\phi \\ \dfrac{\partial}{\partial r} & \dfrac{\partial}{\partial \theta} & \dfrac{\partial}{\partial \phi} \\ P_r & rP_\theta & r\sin\theta P_\phi \end{vmatrix}$$

19. Line integral as per Stoke's Theorem for the *n*th cell is given as $\oint_{L_n} \vec{P} \cdot \overrightarrow{dl}$.

20. Stoke's Theorem relates a line integral to a surface integral.

21. If $\nabla \times \vec{P} = 0$, then a vector field $\vec{P}$ is said to be "potential" or "irrotational."

22. For a scalar field V, by vector's theorem, we get $\nabla \times (\nabla V) = 0$. It means that the field of gradient V is purely irrotational.

23. Laplacian of a vector $\vec{P}$ is given as $\nabla^2 \vec{P} = \nabla(\nabla \cdot \vec{P}) - \nabla \times \nabla \times \vec{P}$, where ∇^2 is called the Laplacian operator (a scalar operator).

24. Laplacian of a scalar function V in Cartesian coordinates is given as

$$\nabla^2 V = \frac{\partial^2 V}{\partial x^2} + \frac{\partial^2 V}{\partial y^2} + \frac{\partial^2 V}{\partial z^2}$$

25. Laplacian of a scalar function V in Cylindrical coordinates is given as

$$\nabla^2 V = \frac{1}{r} \frac{\partial}{\partial r}\left(\frac{r \partial V}{\partial r}\right) + \frac{1}{r^2} \frac{\partial^2 V}{\partial \phi^2} + \frac{\partial^2 V}{\partial z^2}$$

26. Laplacian of a scalar function V in Spherical coordinates is given as

$$\nabla^2 V = \frac{1}{r^2} \frac{\partial}{\partial r}\left(\frac{r^2 dV}{\partial r}\right) + \frac{1}{r^2 \sin\theta} \frac{\partial}{\partial \theta}\left(\sin\theta \frac{\partial V}{\partial \theta}\right) + \frac{1}{r^2 \sin^2\theta} \frac{\partial^2 V}{\partial \phi^2}$$

27. A scalar field V is "harmonic" in a given region if its Laplacian is zero.

MULTIPLE-CHOICE QUESTIONS

1. Line integral involves _________ product.
(a) scalar (b) vector (c) both (a) and (b) (d) none of these

2. Del operator is _____________.
(a) same as the gradient operator
(b) vector differential operator
(c) both (a) and (b)
(d) none of these

3. Gradient represents the __________ of maximum space rate of increase of a scalar field.
(a) magnitude (b) direction (c) both (a) and (b) (d) none of these

4. $\nabla(AB)$ is equal to ___________.
(a) $\nabla A.\ \nabla B$ (b) $\nabla A + \nabla B$ (c) $\nabla^2 AB$ (d) $B\nabla A + A\nabla B$

5. ∇A^n is equal to _____________.
(a) $n\nabla A$ (b) $nA^{n-1}\nabla A$ (c) $nA^n\nabla A$ (d) $\nabla^n A$

6. Divergence of a vector at a point in a vector field is __________ quantity.
(a) vector (b) scalar (c) both (a) and (b) (d) none of these

7. Divergence of vector field is net outward flux of a vector per unit ______.
(a) surface area (b) volume (c) length (d) none of these

8. Divergence at a point is positive if field lines are _________
(a) diverging (b) converging (c) both (a) and (b) (d) none of these

9. If the rate at which field lines enter into a small volume and also leave, then divergence is _________.
(a) ∞ (b) 1 (c) 0 (d) -1

10. Divergence Theorem is applicable for a ________ that is bounded by a _________.
(a) volume, surface (b) surface, volume (c) surface, line (d) line, surface

11. Divergence Theorem relates ________ integral to __________ integral.
(a) surface, volume (b) line, surface (c) volume, line (d) all of these

12. A vector $\vec{P}$ is solenoidal, if $(\nabla.\vec{P}) =$ ___________.
(a) 1 (b) -1 (c) 0 (d) ∞

13. Curl measures _________.
(a) rate of change of vector
(b) circular rotation
(c) both (a) and (b)
(d) none of these

14. Stoke's Theorem relates ______ integral to a _______ integral.
(a) volume, surface (b) volume, line (c) line, surface (d) all of these

15. Vector field $\vec{P}$ is irrotational if $\nabla \times \vec{P} =$ ________.
(a) ∞ (b) -1 (c) 1 (d) 0

16. For a scalar field, as per vector's theorem, $\nabla \times (\nabla V)$ __________.
(a) 1 (b) 0 (c) −1 (d) ∞

17. Laplacian operator __________.
(a) is a scalar function (b) is a vector function
(c) can be a scalar or vector function (d) none of these

18. Laplacian of a scalar is __________.
(a) scalar (b) vector (c) both (a) and (b) (d) none of these

19. Laplacian of a vector is __________ of gradient of its divergence and its curl of curl.
(a) addition (b) difference (c) multiplication (d) division

20. A scalar field is harmonic in a given region, if its Laplacian is__________.
(a) ∞ (b) −1 (c) 1 (d) 0

ANSWERS TO MULTIPLE-CHOICE QUESTIONS

(1) a; (2) c; (3) c; (4) d; (5) b; (6) b; (7) b; (8) a; (9) c;
(10) a; (11) a; (12) c; (13) c; (14) c; (15) d; (16) b; (17) c; (18) a;
(19) b; (20) d

Metals, Dielectrics, Resistors, Capacitors, Inductors, Conductors and Magnetic Materials

4

4.1 INTRODUCTION

Materials are described in terms of conductivity (σ). Conductivity depends on temperature and frequency. A material with high conductivity ($\sigma >> 1$) is called a "metal." A material with low conductivity ($\sigma << 1$) is called an "insulator." For semiconductor materials, the conductivity lies between the conductivity of metals and insulators.

4.2 CONDUCTOR OR METALS

A conductor contains moving free charges. An isolated conductor is shown in Figure 4.1. Let an external electric field E be applied on it. A force will get applied on the charges inside the conductor. As a result, charges will get accumulated on the surface. These induced charges generate a counteracting internally induced field E_i. This internally induced field E_i opposes and cancels the externally applied field E.

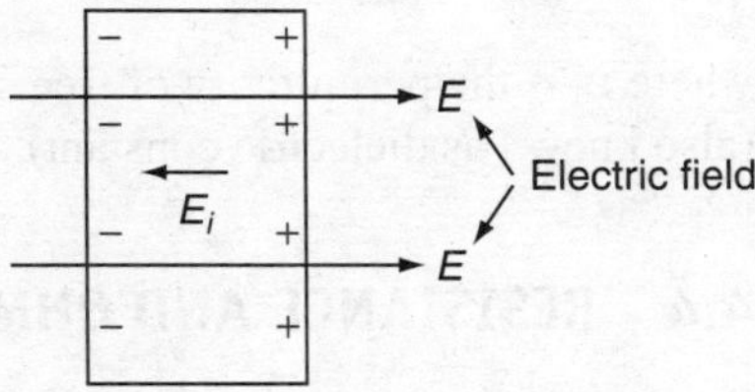

Figure 4.1 *Isolated conductors*

Thus, there does not exist any electrostatic field inside a perfect conductor. As a result $E = -\nabla V = 0$.

It implies that the potential is the same at every point inside the conductor. According to Gauss's Law, the charge density inside a perfect conductor is zero if $E = 0$.

Therefore,

$$\rho_v = 0$$

It is shown in Figure 4.2.

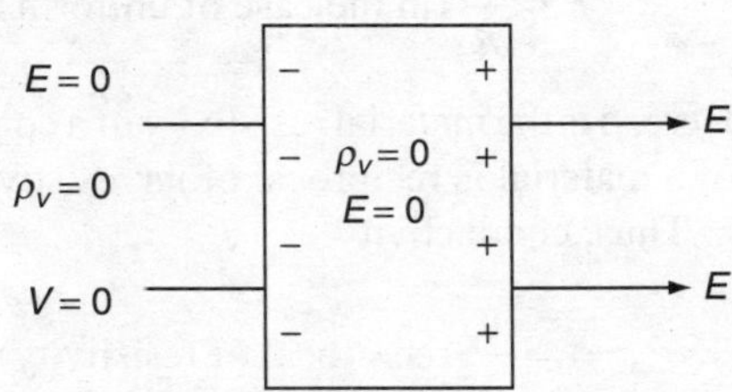

Figure 4.2 *Perfect conductor*

$$E = 0, \quad \rho_v = 0 \quad \text{and} \quad V = 0$$

4.3 DIELECTRICS OR INSULATORS

Conductivity of a dielectric is very low. There are two types of dielectric materials, viz.

(1) Polar dielectrics: Such as oxygen, nitrogen, benzene, etc.
(2) Non-polar dielectrics: Such as water, alcohol, CO_2, etc.

A dielectric material acquires an electric dipole moment if it is placed in an electric field. The direction of electric dipole moment is the same as that of the applied electric field. This is known as "polarization." Induced dipole moment is denoted as p. It is given as

$$p = \alpha E$$

where α is the proportionality constant. The electric polarization of a material is electric dipole moment per unit volume. It is denoted by P and is given as

$$P = \frac{p}{V} \quad \text{or} \quad P = x_e \varepsilon_0 E$$

where x_e is the material's electric susceptibility.

If electric field E is applied on a dielectric material, then in that case electric flux density D is obtained as

$$\begin{aligned} D &= \varepsilon_0 E + P \quad \text{or} \\ &= \varepsilon_0 E + \chi_e \varepsilon_0 E \quad \text{or} \\ &= \varepsilon_0 E(1 + \chi_e) \quad \text{or} \\ &= \varepsilon_0 \varepsilon_r E^2 \end{aligned}$$

or simply

$$D = \varepsilon E$$

where ε_0 is the permittivity of free space, ε the permittivity of dielectric and ε_r the relative permittivity (also known as dielectric constant).

4.4 RESISTANCE AND OHM'S LAW

Resistance R of a uniform conductor of length l having a cross-sectional area A is

$$R = \frac{\rho l}{A} \text{ (in the case of uniform current distribution)}$$

Here, ρ is the material resistivity of a conductor. Conductivity of a material is reciprocal of its resistivity.

Thus, conductivity

$$\sigma = \frac{1}{\rho} \text{ (reciprocal of resistivity)}$$

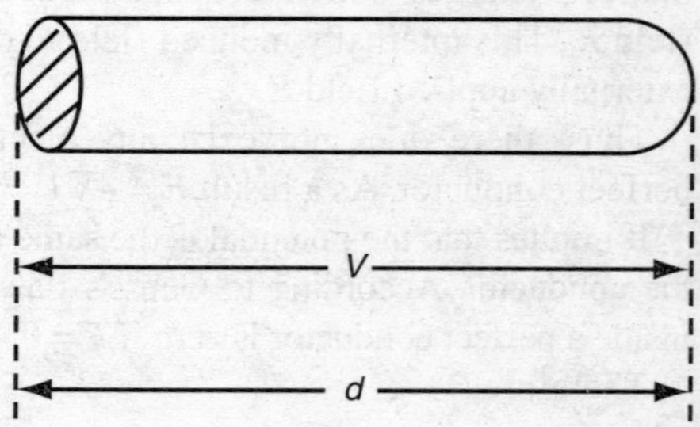

Figure 4.3 *A conductor*

Voltage V across a conductor is $\int E \cdot dl$ and current I through the conductor is $\int_S J \cdot ds$. In the case of a non-uniform current distribution, the resistance of a conductor is

$$R = \frac{V}{I} = \frac{\int E \cdot dl}{\int_S J \cdot ds} \text{ (in the case of non-uniform current distribution)}$$

4.5 CAPACITOR

It is an electrical/electronic component, which stores electrical charge on its plates. Capacitance C expressed in Farads is a measure of its charge storage capacity. Let the voltage difference between two capacitor plates be V. It is given as

$$V = -\int E \cdot dl$$

Let the charge stored on a plate be q. Charge q is given as

Charge $$q = \oint D \cdot ds = \oint \varepsilon E \cdot ds \quad \text{or simply}$$

$$q = \varepsilon \oint E \cdot ds$$

Capacitance C is the ratio of charge magnitude on one of its plates to the difference of potential between them. Capacitance C is given as

Capacitance,

$$C = \frac{q}{V} \quad \text{or}$$

$$= \frac{\varepsilon \oint E \cdot ds}{\int E \cdot dl} \text{ (in Coulomb / Volt or Farads)}$$

4.5.1 Parallel-plate Capacitor

A parallel-plate capacitor is shown in Figure 4.4. The two charges on the two plate capacitors are $+q$ and $-q$.

Let A be the area of each plate. Charge density ρ_s is given as

$$\rho_s = \frac{q}{A}$$

Further, electric field E between the plates of this capacitor and voltage difference V are obtained as

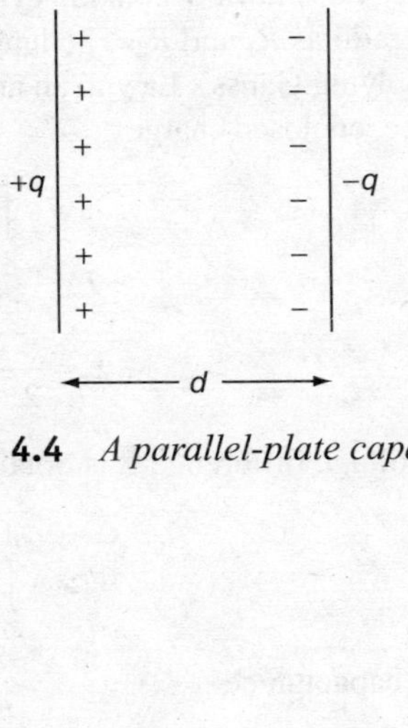

$$E = \frac{\rho_s}{\varepsilon} \quad \text{or} \quad E = \frac{q}{\varepsilon A}$$

$$V = E \cdot d \quad \text{or}$$

$$= \frac{qd}{\varepsilon A} \quad \text{or}$$

$$\frac{q}{V} = \frac{\varepsilon A}{d}$$

Thus, capacitance

$$C = \frac{\varepsilon A}{d} \text{ (in Farads)}$$

+q + + + + + + − − − − − − −q
d

Figure 4.4 *A parallel-plate capacitor*

4.5.2 Spherical Capacitor

Figure 4.5 shows a spherical capacitor.

Applying Gauss' law in an arbitrarily selected Gaussian spherical surface as below:

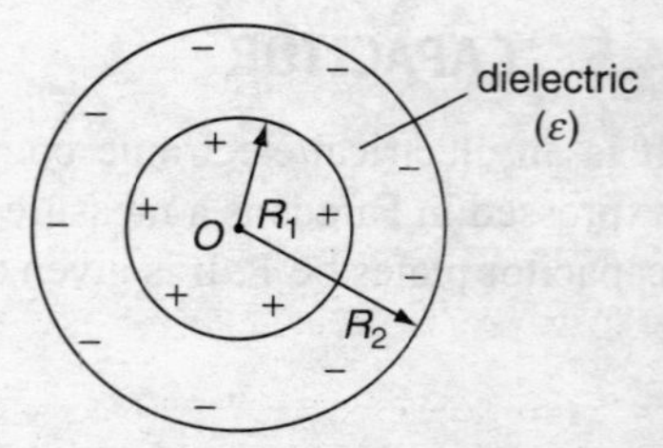

Figure 4.5 *Spherical capacitor*

$$q_{\text{enc}} = \varepsilon \oint E ds \quad \text{or}$$

$$= \varepsilon E \, 4\pi r^2$$

The potential difference (between conductors) is given as

$$V = -\int_{R_2}^{R_1} E \cdot dl \quad \text{or}$$

$$= \frac{q}{4\pi\varepsilon}\left[\frac{1}{R_1} - \frac{1}{R_2}\right]$$

Thus, capacitance

$$C = \frac{q}{V}$$

$$= \frac{4\pi\varepsilon}{\left(\frac{1}{R_1} - \frac{1}{R_2}\right)} \quad \text{(in Farads)}$$

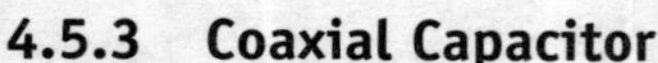

4.5.3 Coaxial Capacitor

Figure 4.6 shows a coaxial cylindrical capacitor. It has inner and outer radii as R_1 and R_2. The length is l.

Applying Gauss's Law in an arbitrarily chosen Gaussian cylindrical surface, enclosed charge

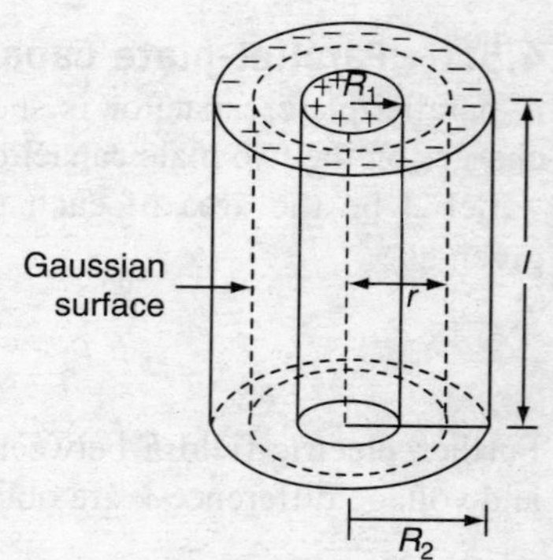

Figure 4.6 *Coaxial capacitor*

$$q = \varepsilon \oint E \cdot ds \quad \text{or}$$

$$q = \varepsilon E 2\pi r l \quad \text{or}$$

$$E = \frac{q}{2\pi\varepsilon r l}$$

The voltage difference V is obtained as

$$V = -\int_{R_2}^{R_1} E \cdot dl \quad \text{or} \quad V = \frac{q}{2\pi\varepsilon l} \ln \frac{R_2}{R_1}$$

Thus, capacitance

$$C = \frac{q}{V} \quad \text{or}$$

$$= \frac{2\pi\varepsilon l}{\ln (R_2/R_1)}$$

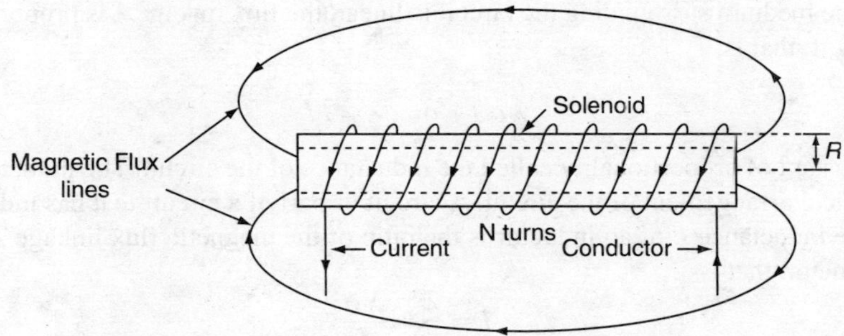

Figure 4.7 *A solenoid coil*

4.6 INDUCTORS, INDUCTANCES AND MUTUAL INDUCTANCES

An inductor stores energy in the magnetic field. An inductor is the magnetic counterpart of a capacitor. A capacitor stores energy in an electric field. Loops, coils, solenoids, etc. are examples of inductors. Inductance is one of the electrical parameters like resistance and capacitance. An inductor's value depends on its physical dimensions. In addition, the value of inductors depends upon the type of magnetic material used in its construction.

Consider a solenoid coil as shown in Figure 4.7. Here, magnetic lines are produced by a current. Each magnetic line passes through the solenoid and links the circuit N times (N: number of turns). If all magnetic lines link all N turns, the total magnetic flux linkage λ of this coil is given as

Flux linkage $\lambda = N\phi_m$ (Wb-turns)

where ϕ_m is the total magnetic flux.

Inductance (L) is the ratio of the total magnetic flux linkage (λ) to the current (I). It is given as

$$L = \frac{\lambda}{I}$$

$$= \frac{N\phi_m}{I}$$

The above expression is valid only for materials with constant permeability. If permeability is not constant, inductance L is given as

$$L = \frac{\partial\lambda}{\partial I}$$

As expressed above, inductance is the ratio of infinitesimal change in flux linkage to the infinitesimal change in current producing this flux. The unit of inductance is Henry (H).

A circuit (or closed conducting path) carrying current I produces a magnetic field B, which causes flux $\phi = \int \vec{B} \cdot d\vec{s}$ to pass through each turn of the circuit as shown in Figure 4.8. If the circuit has N identical turns, flux linkage λ is given as

$$\lambda = N\phi$$

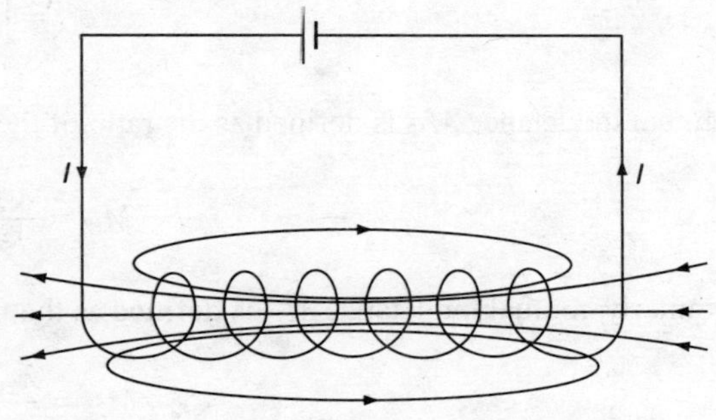

Figure 4.8 *Magnetic field B produced by a circuit*

In addition, if the medium surrounding the circuit is linear, the flux linkage λ is proportional to the current I producing it; that is,

$$\lambda \alpha I \quad \text{or} \quad \lambda = LI$$

where L is a constant of proportionality called the inductance of the circuit. The inductance L is a property of the physical arrangement of the circuit. A circuit or part of a circuit that has inductance is called an inductor. The inductance L of an inductor is the ratio of the magnetic flux linkage λ to the current I through the inductor, that is

$$L = \frac{\lambda}{I} = \frac{N\phi}{I}$$

The unit of inductance is Henry (H), which is the same as Webers/Ampere. Since Henry is a fairly large unit, inductances are usually expressed in milliHenrys (mH).

The inductance defined by the above equation is commonly referred to as self-inductance since the linkages are produced by the inductor itself. Similar to capacitances, inductances are a measure of how much magnetic energy is stored in an inductor.

The magnetic energy (in Joules) stored in an inductor is expressed as

$$W_m = \frac{1}{2}LI^2 \quad \text{or} \quad L = \frac{2W_m}{I^2}$$

Thus, the self-inductance of a circuit may be defined or calculated from energy.

If instead of having a single circuit, there are two circuits carrying current I_1 and I_2 as shown in Figure 4.9, a magnetic interaction exists between the circuits. Four component fluxes ϕ_{11}, ϕ_{12}, ϕ_{21} and ϕ_{22} are produced. The flux ϕ_{12}, for example, is the flux passing through circuit 1 due to current I_2 in circuit 2. If $\vec{B}_2$ is the field due to I_2 and S_1 is the area of circuit 1, then

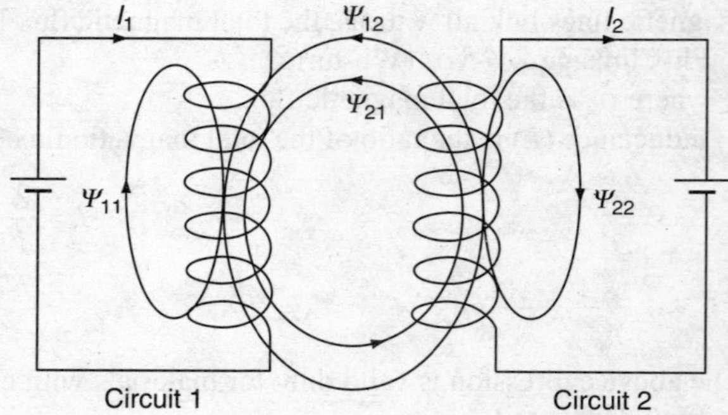

Figure 4.9 *Magnetic interaction between two circuits*

$$\phi_{12} = \int_{S_1} \vec{B}_2 \cdot d\vec{s}$$

Mutual inductance M_{12} is defined as the ratio of the flux linkage $\lambda_{12} (= N_1\phi_{12})$ on circuit 1 to current I_2.

$$M_{12} = \frac{\lambda_{12}}{I_2} = \frac{N_1\phi_{12}}{I_2}$$

Similarly, mutual inductance M_{21} is defined as the flux linkages of circuit 2 per unit current I_1:

$$M_{21} = \frac{\lambda_{21}}{I_1} = \frac{N_2\phi_{21}}{I_1}$$

It can be shown by using energy concepts that if medium surrounding the circuit is linear (in the absence of ferromagnetic material), then

$$M_{12} = M_{21}$$

The mutual inductance M_{12} or M_{21} is expressed in Henrys. It should not be confused with the magnetization vector $\vec{M}$ expressed in Amperes/meter.

The self-inductance of circuits 1 and 2 is defined as

$$L_1 = \frac{\lambda_{11}}{I_1} = \frac{N_1\phi_1}{I_1} \quad \text{and}$$

$$L_2 = \frac{\lambda_{22}}{I_2} = \frac{N_2\phi_2}{I_2}$$

where $\phi_1 = \phi_{11} + \phi_{12}$ and $\phi_2 = \phi_{21} + \phi_{22}$.

The total energy W_m in the magnetic field is the sum of the energies due to L_1, L_2 and M_{12} (or M_{21}).

$$W_m = W_1 + W_2 + W_{12}$$

$$= \frac{1}{2}L_1 I_1^2 + \frac{1}{2}L_2 I_2^2 \pm M_{12} I_1 I_2$$

The positive sign is taken if currents I_1 and I_2 flow such that the magnetic fields of the two circuits strengthen each other. If the currents flow such that their magnetic fields oppose each other, then the negative sign is taken.

As mentioned earlier, an inductor is a conductor arranged in a shape appropriate to store magnetic energy. Typical examples of inductors are toroids, solenoids, co-axial transmission lines and parallel-wire transmission lines. The inductance of each of these inductors can be determined by following a procedure similar to that taken in determining the capacitance of a capacitor. For a given inductor, we find the self-inductance L by taking these steps:

1. Choose a suitable coordinate system.
2. Let, the inductor carry current I.
3. Determine B from Bio-Savart's law and calculate ϕ from $\phi = \int \vec{B}\cdot d\vec{s}$.
4. Finally, find L from

$$L = \frac{\lambda}{I}\left(= \frac{N\phi}{I}\right)$$

The mutual inductance between two circuits may be calculated by adopting a similar procedure.

In an inductor such as a co-axial or a parallel-wire transmission line, the inductance produced by the flux internal to the conductor is called the internal inductance L_{in}, while that produced by the flux external to it is called external inductance L_{ext}. The total inductance L is

$$L = L_{\text{in}} + L_{\text{ext}}$$

As it was shown for capacitors that

$$RC = \frac{\varepsilon}{\sigma}$$

It can also be shown for inductors that

$$L_{ext}C = \mu\,\varepsilon$$

Thus, L_{ext} may be calculated using the above equation, if C is known.

4.6.1 Inductance of a Long Solenoid

A long solenoid has a very large length as compared with its radius. Magnetic field B at any point inside a long solenoid is practically constant. Magnetic field B inside a long solenoid is given as

$$B = \frac{\mu_0 NI}{l}$$

where μ_0 is the absolute magnetic permeability of free space or air, l the length of the solenoid and N the total number of turns in the solenoid.

The field inside a solenoid is almost uniform (expect a little less at the ends). Let B be constant inside the solenoid and be equal to its value at the centre, then the total flux linkage λ (for a long solenoid) is given as

$$\begin{aligned}\lambda &= N\phi_m\\ &= NBA\\ &= N\cdot\frac{\mu NI}{l}\cdot A\end{aligned}$$

The inductance L of a solenoid is given as

$$\begin{aligned}L &= \frac{\lambda}{I}\\ &= \frac{\mu N^2 A}{I}\end{aligned}$$

where A is the area of cross section of the solenoid (m^2).

4.6.2 Inductance of a Toroid

A toroid (toroid coil) is a long solenoid, which is bent into a circle and closed on itself. It is shown in Figure 4.10. It has uniform winding consisting of several turns. Magnetic flux lines in a toroid are confined within its interior. B is zero outside. For very large R/r, the toroid can be assumed to be straightened out into a solenoid.

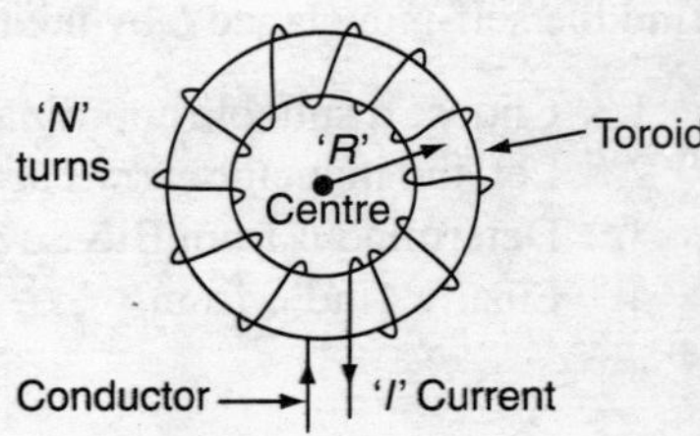

Figure 4.10 *Toroidal coil*

Flux linkage λ

$$\begin{aligned}&= N\phi_m \quad \text{or}\\ &= NBA \quad \text{or}\\ &= N\cdot\frac{\mu NI}{2\pi R}\pi r^2\end{aligned}$$

or simply

$$= \frac{\mu N^2 r^2 I}{2R}$$

Inductance of a toroid is given as

$$L = \frac{\lambda}{I}$$

$$= \frac{\mu N^2 r^2}{2R}$$

where μ is the permeability of material, N the number of turns of the coil, r the radius of cross section of the toroid (m) and R the radius of the toroid (m).

4.6.3 Inductance of a Co-axial Cable or Co-axial Conductor

A co-axial cable (having a common axis) is shown in Figure 4.11. Consider conducting cylinders with radii a and b. Let current I be carried by an inner conductor in the $\vec{a}_z$ direction. Consequently, by the outer conductor, current will be carried in the $-\vec{a}_z$ direction. There is free space between $r = a$ and $r = b$.

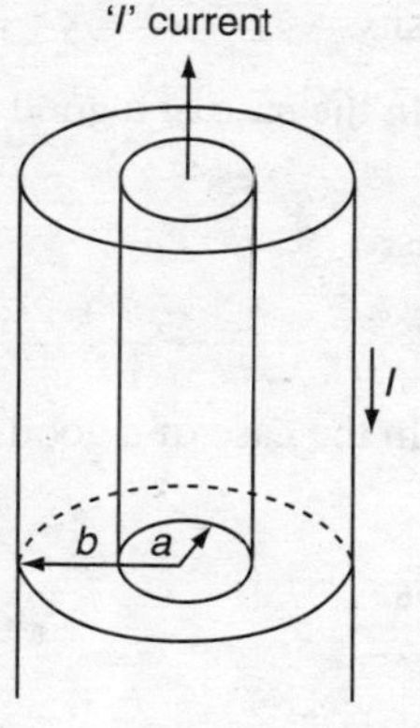

Figure 4.11 *Co-axial cable*

Magnetic flux is given as

$$\phi = \int \vec{B} \cdot d\vec{r}$$

where $\vec{B}$ is the magnetic flux density. Thus,

$$\vec{B} = \frac{\mu_0 I}{2\pi r} \vec{a}_\phi$$

$$\phi = \int_0^l \int_a^b \frac{\mu_0 I}{2\pi r} dr\, dl \text{ (where } l \text{ is the length of cable)}$$

$$= \frac{\mu_0 I l}{2\pi} \int_a^b \frac{dr}{r}$$

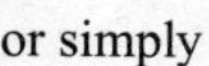
or simply

$$\phi = \frac{\mu_0 I l}{2\pi} \ln \frac{b}{a}$$

Inductance L is given as

$$L = \frac{\phi}{I} \quad \text{or} \quad L = \frac{\mu_0 l}{2\pi} \ln \frac{b}{a}$$

The inductance of a co-axial cable is given as

$$L/l = \frac{\mu_0}{2\pi} \ln \frac{b}{a}$$

In the above expression, L/l is 'inductance per unit length' of the co-axial cable.

4.7 CONDUCTORS AND DIELECTRICS

In electromagnetics, materials or mediums are divided into two categories:

(a) Conductors
(b) Dielectrics or insulators

The dividing line between the above two classes is not very conspicuous. Some materials are considered as conductor in one portion of the radio frequency range and dielectric (with loss) in another portion of the radio frequency range.

4.7.1 Criteria for its Classification

Maxwell's equation is

$$\nabla \times \vec{H} = \sigma \vec{E} + j\omega\varepsilon \vec{E}$$

where $\sigma\vec{E}$ is the conduction of current density ($\vec{J}_C$) and $\omega\varepsilon\vec{E}$ the displacement of current density ($\vec{J}_d$). The ratio $\sigma/\omega\varepsilon$ of the medium is the ratio of conduction current density to the displacement current density.

(1) In the case of a good conductor,

$$\vec{J}_C > \vec{J}_d$$

Thus,

$$\frac{\sigma}{\omega\varepsilon} >> 1$$

(2) In the case of a good dielectric,

$$\vec{J}_C < \vec{J}_d$$

Thus,

$$\frac{\sigma}{\omega\varepsilon} << 1$$

(3) In the case of quasi-conductors,

$$\vec{J}_C = \vec{J}_d \quad \text{and} \quad \omega\varepsilon = \sigma$$

Thus,

$$\frac{\sigma}{\omega\varepsilon} = 1 \text{ (medium is a quasi-conductor)}$$

The term $(\sigma/\omega\varepsilon)$ is called "loss tangent" or "dissipation factor." The key features are summarized below:

(1) In the case of good conductors, σ and ε are almost independent of frequency.
(2) Generally, in dielectrics, σ and ε are functions of frequency. Ratio $(\sigma/\omega\varepsilon)$ is often constant for a given frequency range.
(3) $(\sigma/\omega\varepsilon)$ approximates various relations like that for attenuation constant, phase constant and intrinsic impedance in the case of both, viz.. conductors and dielectric materials.

4.7.2 Solution for Perfect Dielectric Conditions

Consider an electromagnetic phenomenon in a homogeneous perfect dielectric having zero charges and zero conduction current, i.e. $\rho_v = 0$, $\sigma = 0$ and $J = 0$. Field equations in such a case are given below:

$$\nabla \times \vec{H} = \frac{\varepsilon \partial \vec{E}}{\partial t}$$

$$\nabla \times \vec{E} = -\mu \frac{\partial \vec{H}}{\partial t}$$

$$\nabla\cdot\vec{D}=0,\quad \nabla\cdot\vec{E}=0 \quad \text{and}$$
$$\nabla\cdot\vec{B}=0,\quad \nabla\cdot\vec{H}=0$$

According to the above equations, changing electric field produces a magnetic field. In addition, a changing magnetic field produces an electric field.

Thus, a changing electric field produces a changing magnetic field, which in turn produces an electric field, which again produces a magnetic field and so on. It is a series of energy transfer. Thus, energy transfer takes place, whenever electric or magnetic distribution occurs. Energy is transferred from an electric to a magnetic field and so on for indefinitely. Magnetic energy is not confined to one location in space as the electric energy. This magnetic energy is derived from electrical energy. Magnetic energy extends a little beyond in space. If electric energy is derived from magnetic energy, then electrical energy is advanced in space. Thus, energies are also being propagated.

From two equations of $\vec{E}$ and $\vec{H}$, let us get only one equation for $\vec{E}$ and $\vec{H}$.

Taking the curl on both sides, we get

$$\nabla\times\nabla\times\vec{E}=-\mu\nabla\frac{\partial\vec{H}}{\partial t}$$

$$\nabla(\nabla\cdot\vec{E})-\nabla^2\vec{E}=-\mu\nabla\times\frac{\partial\vec{H}}{\partial t} \quad \text{or}$$

$$\nabla^2\vec{E}=-\mu\frac{\partial}{\partial t}(\nabla\times\vec{H}) \qquad (\because \nabla\cdot\vec{E}=0)$$

Substituting for $\nabla\times\vec{H}$, we get

$$\nabla^2\vec{E}=-\mu\frac{\partial}{\partial t}\left(\frac{\varepsilon\partial\vec{E}}{\partial t}\right) \quad \text{or}$$

$$\nabla^2\vec{E}=\mu\varepsilon\frac{\partial^2\vec{E}}{\partial t^2} \quad \text{(general form of wave equation)}$$

This is a general form of wave equation. The above expression can be expanded in Cartesian coordinates as a set of three expressions, as given below:

$$\frac{\partial^2 E_x}{\partial x^2}+\frac{\partial^2 E_x}{\partial y^2}+\frac{\partial^2 E_x}{\partial z^2}=\mu\varepsilon\frac{\partial^2 E_x}{\partial t^2}$$

$$\frac{\partial^2 E_y}{\partial x^2}+\frac{\partial^2 E_y}{\partial y^2}+\frac{\partial^2 E_y}{\partial z^2}=\mu\varepsilon\frac{\partial^2 E_y}{\partial t^2} \quad \text{and}$$

$$\frac{\partial^2 E_z}{\partial x^2}+\frac{\partial^2 E_z}{\partial y^2}+\frac{\partial^2 E_z}{\partial z^2}=\mu\varepsilon\frac{\partial^2 E_z}{\partial t^2}$$

The general form of a wave equation of $\vec{H}$ in the similar manner is obtained as

$$\nabla^2\vec{H}=\mu\varepsilon\frac{\partial^2\vec{H}}{\partial t^2}$$

Electromagnetic wave propagation involves electric and magnetic fields with more than one component. They depend on all three coordinates along with time. A simple and useful electromagnetic wave consists of electric and magnetic fields. These fields are perpendicular to each other. In addition, these fields are uniform in planes, which are perpendicular to the direction of propagation. Such waves are known as "Uniform plane waves". For simplicity, let

$$\vec{E} = \vec{E}_y(x, t)\vec{a}_y$$

Assume that E_x and E_z do not exist. Thus,

$$\frac{\partial^2 E_y}{\partial x^2} = \mu\varepsilon \frac{\partial^2 E_y}{\partial t^2}$$

Solutions of the above expressions are

$$E^-{}_y = f(x - ut) \quad \text{and} \quad E^+{}_y = g(x + ut)$$

where $u = \sqrt{1/\varepsilon\mu}$.

Further,

$$E_y = f(x - ut) + g(x + ut)$$

where f and g are functions of $(x - ut)$ and $(x + ut)$, respectively. Assuming harmonic (or sinusoidal) time dependence $e^{j\omega t}$, we get

$$\frac{\partial^2 E_s}{dx^2} + \beta^2 E_s = 0$$

where $\beta = \omega/u$ and E_s the phasor of E_y.

Its possible solutions are given below:

$$E^+{}_y = A \cdot e^{j(\omega t - \beta x)} \quad \text{and} \quad E^-{}_y = B\, e^{j(\omega t + \beta x)}$$

Thus,

$$E_y = A\, e^{j(\omega t - \beta x)} + B\, e^{j(\omega t + \beta x)}$$

where A and B are real constants. Considering only the imaginary part of the above expression, we get

$$E = A \sin(\omega t - \beta x)$$

4.8 NATURE OF MAGNETIC MATERIALS

Magnetic properties of magnetic materials depend upon their response at atomic, electronic, molecular and microscopic levels to the magnetic fields. Some magnetic material characteristics are analogous to some characteristics of selected dielectric materials. Atoms and molecules give magnetic dipole moments. It is similar to the electric dipole moment. Electric dipole is because of fundamental charges. Magnetic field is due to electric current flowing in a closed loop. The combined effect of electric dipoles can be seen in 'polarization'. The combined effect of magnetic dipoles can be seen in 'magnetization' of magnetic material. Basic magnetic parameters, which help understand the response of magnetic field to magnetic materials, are described below.

4.8.1 Permeability and Magnetic Susceptibility

For a magnetic material

$$\vec{B} = \mu\, \vec{H}$$

where $\vec{B}$ is the magnetic flux density, $\vec{H}$ the magnetic field intensity and μ the proportionality constant known as 'permeability' of medium (or material). μ is measured in Henry/meter. For vacuum $\mu = \mu_0$, μ_0 is the permeability of vacuum equal to 4×10^{-7} H/m. Relative permeability μ_r is given as

$$\mu_r = \frac{\mu}{\mu_0}$$

where μ_r is a dimensionless quantity. μ_r depends on the atomic structure of magnetic material. Factor μ remains constant for many magnetic materials. For some magnetic materials, the direction of vector $\vec{B}$ and vector $\vec{H}$ in space are not same. In such a case, μ is a tensor (where $\vec{B}$ and $\vec{H}$ are not in the same direction).

Quantity $(\mu_r - 1)$ is magnetization per unit magnetic field intensity. $(\mu_r - 1)$ is called magnetic susceptibility (χ_m), and $(\mu_r - 1)$ is a dimensionless quantity. It is dimensionless and has the same units of magnetization and field intensity. Thus,

$$\chi_m = (\mu_r - 1)$$

Normally, all materials are affected by magnetic field. In magnetic fields, materials acquire magnetic moments. For a material/medium, magnetization is the magnitude of this magnetic moment per unit volume. Magnetization is denoted by a vector $\vec{M}$ (A/m). Magnetic induction $\vec{B}$ (flux density) for a material subjected to a magnetic field is given as

$$\vec{B} = \underbrace{\mu_0 \vec{H}}_{\text{Effect in vacuum}} + \underbrace{\mu_r \vec{H}}_{\text{Effect in material}} = \mu_0 \mu_r \vec{H}$$

Further,

$$\vec{M} = (\mu_r - 1)\vec{H} \quad \text{or}$$

$$= \chi_m \vec{H}$$

where $\chi_m = (\mu_r - 1)$ is called 'magnetic susceptibility'. Magnetic susceptibility is a dimensionless quantity. Units of $\vec{M}$ and $\vec{H}$ are (A/m) (same for both).

Magnetization $\vec{M}$ of a material is related to its elementary magnetic dipole moment $\vec{P}_m$ as

$$\vec{M} = N\vec{P}_m \qquad \text{(dipole moment per unit volume)}$$

where N is the number of dipoles per unit volume.

4.9 MAGNETIC DIPOLE

Magnetic field is produced by a small current loop. It is quite similar to the electric field caused by a small electric dipole. A small current loop is also called as 'magnetic dipole'. Dipole moment $\vec{P}_m$ is the product of a plane loop area and circulating current magnitude. It is shown in Figure 4.12.

Vector direction of moment is always perpendicular to the plane of the loop. Dipole moment $\vec{P}_m$ is in the direction of a 'right hand screw', when it is moved in the current's direction in the loop.

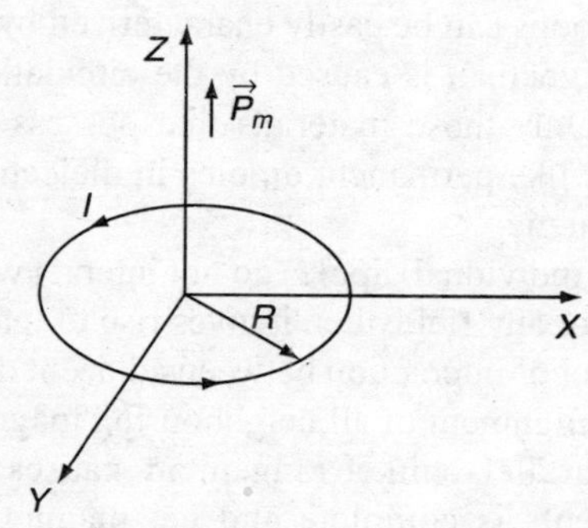

Figure 4.12 *Magnetic dipole and dipole moment*

Dipole moment $|\vec{P}_m|$ is given as

$$|\vec{P}_m| = \text{Current in loop} \times \text{Area of the loop}$$

In the case of a circular loop, the dipole moment $\vec{P}_m$ is given as

$$\vec{P}_m = \pi R^2 I \vec{a}_z (\text{A}\cdot\text{m}^2)$$

A magnetic dipole moment placed in a magnetic field orients itself in the direction of the magnetic field.

4.10 CLASSIFICATION OF MAGNETIC MATERIALS

In materials, a central positive nucleus is surrounded by electrons in orbits. An electron placed in an orbit is quite analogous to a small current loop. This electron experiences a torque in an external magnetic field so as to orient its dipole in the field's direction. This orbital magnetic dipole's direction is opposite to the direction of the applied magnetic field. It results in 'negative susceptibility'. Such materials are called "diamagnetic materials". A second 'moment' is caused by 'electron spin'. This phenomenon is quite difficult to comprehend and cannot be explained easily by a simple model. An electron may possesses 'spin magnetic moment' of $\pm 9\times10^{-24}(\text{A-m}^2)$. Plus and minus signs indicate alignment either aiding or opposing the external magnetic field. In an atom, spins of electrons in shells (which are not completely filled) contribute towards the 'magnetic moment' of an atom. A third 'moment' in atom is caused by 'nuclear spin'. However, it has negligible effect on the overall magnetic properties of material. On the basis of magnetic behaviour, materials are classified as

(1) Diamagnetic
(2) Paramagnetic
(3) Ferromagnetic
(4) Antiferromagnetic
(5) Ferrimagnetic

Magnetic behaviour is chiefly due to 'electron orbital motion', 'electron-spin' and 'nuclear-spin'. These three causes can be represented by equivalent atomic currents, which flow in circular loops. These atomic currents are bound to every individual molecule. Atomic currents produce magnetic fields. Atomic currents in molecules of a material may produce significant 'macroscopic fields'. These equivalent atomic current loops can be easily characterized by their magnetic dipole moments.

Diamagnetism is caused by the circulating charges in the orbits. All the materials exhibit diamagnetism. Only those materials that possess permanent magnetic dipoles (mostly spin contribution by electrons like permanent dipoles in dielectrics) are paramagnetic, ferromagnetic, antiferromagnetic or ferrimagnetic.

When individual dipoles do not interact with each other, and also these are randomly distributed in the absence of any field, then it gives rise to 'paramagnetism'. In the remaining three magnetic behaviours, there is lot of interaction between adjacent dipoles. In ferromagnetic materials, dipole interaction causes parallel alignment of all neighbouring magnetic moments.

Antiparallel spin arrangement causes 'antiferromagnetism' (here compensation of magnetic components is complete and net magnetization is zero) or 'ferrimagnetism' (here compensation is incomplete). Figure 4.13 shows various dipole arrangements for various magnetic behaviours. Ferromagnetic materials are most common (iron, nickel, cobalt, etc.). Ferromagnetic materials have very large and positive susceptibility. Below the Curie temperature, ferromagnetic materials evince hysteresis loop in the B-H curve.

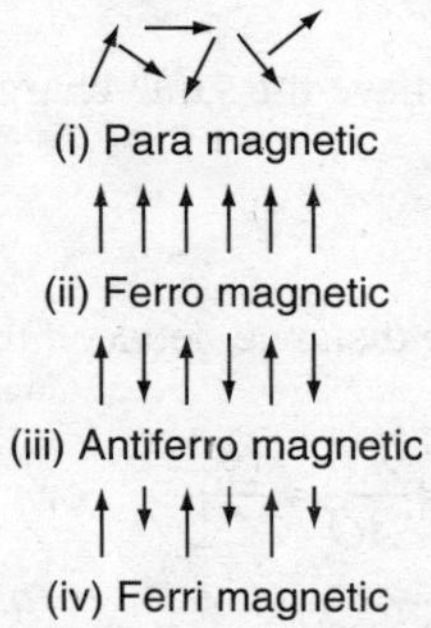

(i) Para magnetic

(ii) Ferro magnetic

(iii) Antiferro magnetic

(iv) Ferri magnetic

Figure 4.13 *Various dipole arrangements and characterization of magnetic materials*

SOLVED QUESTIONS

4.1 In a rectangular coordinate system, a charge $Q_1 = -2\mu C$ is placed at the origin. Another charge $Q_2 = -20\mu C$ is placed 50 cm away from the origin on the x-axis. Find the force on Q_1 due to Q_2 (they are in free space).

Solution:

Given that

$$Q_1 = -2\mu C \quad \text{at } (0, 0, 0)$$

$$Q_1 = -20\mu \text{C} \quad \text{at } (0.5, 0, 0)$$

As per Coulomb's Law, we find

$$\vec{F}_{12} = \frac{Q_1 Q_2}{4\pi\varepsilon_0 r^2}\vec{a}_{12} \tag{1}$$

and

$$r = |\vec{r}| = |(0.a_x + 0.a_y + 0.a_z) - (0.5a_x + 0a_y + 0a_z)|$$
$$= |-0.5a_x|$$
$$= 0.5$$

also

$$\vec{a}_{12} = -\vec{a}_x$$

Thus,

$$\vec{F}_{12} = \frac{(-2\times10^{-6})\times(-2\times10^{-3})\times9\times10^{9}(-\vec{a}_x)}{(0.5)^2} \quad \text{or}$$

$$\vec{F}_{12} = -1440\vec{a}_x \text{ Newton}$$

4.2 Two charged particles of mass M and charge Q each are suspended from the same point by strings of length L each. Prove

$$Q^2 = 16\pi\,\varepsilon_0\, MgL^2 \sin^2\theta \tan\theta$$

where θ is the inclination angle of each thread from the vertical axis at equilibrium.

Solution:
Two particles have the same charge. The force of repulsion will be given by

$$F_r = \frac{Q^2}{4\pi\varepsilon_0 R^2} \tag{1}$$

where R is the distance between the two particles, as shown in Figure 4.14. In triangle OAT,

$$\sin\theta = \frac{AT}{AO} = \frac{(R/2)}{L} \quad \text{or} \quad R = 2L\sin\theta$$

Substituting R in F_r above, we get,

$$F_r = \frac{Q^2}{4\pi\varepsilon_0 (2L\sin\theta)^2}$$

where F_r is the electric force, T the tension in each thread and M_g the weight of each charge.

At points A and B, we get

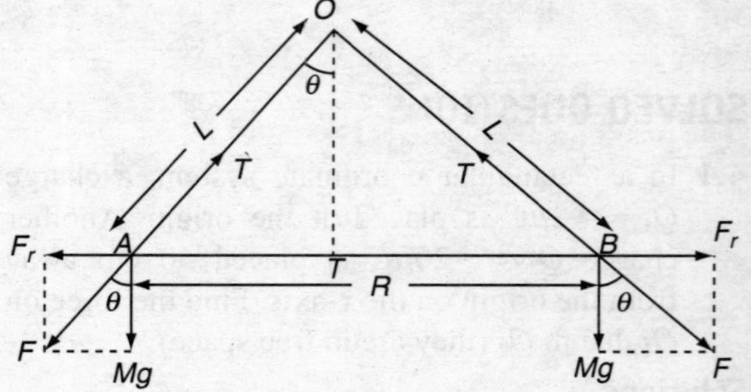

Figure 4.14 *Two charged particles*

$$T\sin\theta = F_r \quad \text{and} \quad T\cos\theta = Mg$$

From above, we get

$$\tan\theta = \frac{F_r}{Mg} \quad \text{or} \quad \tan\theta = \frac{Q^2}{4\pi\varepsilon_0 (2L\sin\theta)^2} \times \frac{1}{Mg}$$

or simply

$$Q^2 = 16\pi\varepsilon_0 \; MgL^2 \sin^2\theta \tan\theta$$

4.3 Four particles of the same charge are placed at the corners of a square. This square has an area of 1 m^2. The force on each charge is 20 N. Find the value of each charge.

Solution: Given that

Area of square = 1 m^2 = 1 × 1 m^2

Force on each charge = 20 N

Side of square = 1 m

Let the charge be $+q$. It is shown in Figure 4.15.

Forces are given as

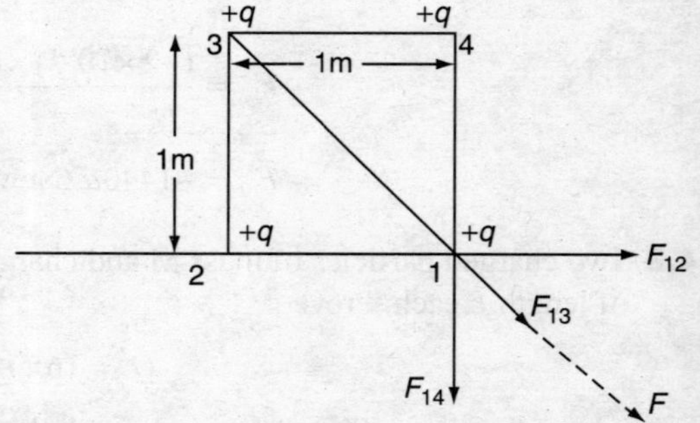

Figure 4.15 *Charges q placed at four corners of a square*

$$F_{12} = \frac{1}{4\pi\varepsilon_0}\frac{q\cdot q}{r^2}$$

$$= 9\times10^9 \times q^2 \text{ Newton}$$

$F_{14} = 9\times10^9 \times q^2$ Newton and

$$F_{13} = 9\times10^9 \times \frac{q^2}{(\sqrt{1^2+1^2})^2} \text{ Newton} \quad \text{or}$$

$$F_{13} = 4.5 \times 10^9 \times q^2 \text{ Newton}$$

The directions of these forces are shown in Figure 4.15. Let F_R be the resultant of forces F_{12} and F_{14}. Thus,

$$F_R = 2\times 9\times 10^9 \times q^2 \times \cos 45° \text{ Newton} \quad \text{or}$$
$$= 9\sqrt{2}\times10^9 \times q^2 \text{ Newton}$$

The resultant of all forces F is given as

$$F = F_R + F_{13} \quad \text{or}$$

$$F = (4.5\times10^9 \times q^2 + 9\sqrt{2}\times10^9 \times q^2)\text{ Newton} \quad \text{or}$$

$$F = (4.5\times10^9 \times q^2[1+2\sqrt{2}])\text{ Newton} \quad \text{or}$$

$$F = (17.2278 \times 10^9 \times q^2)\text{ Newton}$$

As given in the problem statement, the resultant of all forces on charge is 20 N. Thus,

$$20 = 17.2278 \times 10^9 \times q^2 \quad \text{or}$$

$$q^2 = \frac{20}{17.2278\times10^9} \quad \text{or}$$

$$q^2 = \frac{200}{17.2278}\times10^{-10} \quad \text{or}$$

$$q = 3.40 \times10^{-5}\text{ C} \quad \text{or}$$

$$q = 34\ \mu\text{C}$$

4.4 A charge $Q_1 = 242 \times 10^{-9}$ C in vacuum is located at a point P_1 (0.03, 0.01, –0.04). Find the force on Q_1 due to another charge $Q_2 = 50\ \mu$C. Charge Q_2 is located at a point P_2 (0.03, 0.08, 0.02).

Solution: Finding $\vec{R}_{21}$ as below:

$$\vec{R}_{21} = (-0.03-0.03)\vec{a}_x + (-0.01-0.08)\vec{a}_y + (-0.04-0.02)\vec{a}_z \quad \text{or}$$

$$\vec{R}_{21} = (-0.06\vec{a}_x - 0.07\vec{a}_y - 0.06\vec{a}_z)$$

Further,

$$|\vec{R}_{21}| = \sqrt{(0.06)^2 + (.07)^2 + (.06)^2} = 0.11$$

Unit vector $\vec{a}_{R_{21}}$ is obtained as

$$\vec{a}_{R_{21}} = \frac{(-0.06\vec{a}_x - 0.07\vec{a}_y - 0.06\vec{a}_z)}{0.11}$$

Force F_1 on Q_1 is given as

$$F_1 = \frac{Q_1 Q_2}{4\pi\varepsilon_0 \left|\vec{R}_{21}\right|^2} \vec{a}_{R_{21}}$$

Substituting values of Q_1. Q_2, $\left|\vec{R}_{21}\right|^2$ and $\vec{a}_{R_{21}}$ as obtained above, we get

$$F_1 = (-54\vec{a}_x - 63\vec{a}_y - 54\vec{a}_z) \text{ Newton}$$

4.5 Electric dipole consists of positive and negative charges. The charge in the case of a electric dipole is 8 μC. The distance between them is 10 mm. Obtain the dipole moment.

Solution: Given that

Charges $q = +8\ \mu\text{C}$ and $-8\ \mu\text{C}$ or

$$= +8 \times 10^{-6}\text{C and } -8 \times 10^{-6}\text{C}$$

Distance $d = 10$ mm or

$$= 10 \times 10^{-3}\text{ m}$$

Dipole moment $|\vec{p}|$ is obtained as

$$|\vec{p}| = qd \quad \text{or}$$

$$= 8 \times 10^{-6} \times 10 \times 10^{-3}\text{ C–m}$$

or simply

$$= 8 \times 10^{-8}\text{ C–m}$$

The direction of $\vec{p}$ will be from negative charge towards positive charge.

4.6 Two similar charges of 2 μC but of opposite signs are kept at a 2-cm distance. Calculate the electric field at a point, which is 10 cm away from the mid-point of the axial line of this dipole.

Solution: Given that

Charge $q = 2\ \mu\text{C} = 2\times10^{-6}\text{ C}$

Distance $2d = 2\text{ cm} = 2 \times 10^{-2}\text{ m}$

where $R = 10\text{ cm} = 10^{-1}\text{ m}$, electric field E on the axial line of the dipole is given as

$$E = \frac{2|\vec{p}|R}{4\pi\varepsilon_0 (R^2 - d^2)^2} \quad \text{or}$$

$$E = \frac{2 \cdot q \times d \times R}{4\pi\varepsilon_0 (R^2 - d^2)^2}$$

Putting the value, we get

$$E = \frac{9\times10^9 \times 2\times2\times10^{-6} \times 2\times10^{-2} \times10^{-1}}{(10^{-2} - 10^{-4})^2} \quad \text{or}$$

$$= \frac{72}{9.801\times10^{-5}}$$

or simply

$$= 7.34\times10^5\text{ N/C}$$

4.7 In a case $\vec{D} = zr \cdot \cos^2 \phi (\text{C/m}^2)$, find charge density at $(2, \pi/4, 5)$. Also, find the total charge enclosed by a cylinder of 1-m radius with $-3 \le z \le 3$ (meters).

Solution: Charge density

$$\rho_v = \nabla \cdot D \text{ (standard formula)} \quad \text{or}$$

$$= \frac{\partial D}{\partial z} \quad \text{or}$$

$$= \frac{\partial}{\partial_z}(zr \cdot \cos^2 \phi) \quad \text{or}$$

$$= (r \cdot \cos^2 \phi)$$

where in this case $r = 2$, $\phi = \pi/4$ and $z = 5$.
Thus,

$$\rho_v = 2\cos^2 \pi/4 \quad \text{or}$$

$$= 1(\text{C/m}^2)$$

Total charge

$$q = \int_v \rho_v . dv \quad \text{or}$$

$$q = \int_v r.\cos^2 \phi r \, d\phi \, dr \, dz \quad \text{or}$$

$$= \int_{z=-3}^{3} dz \int_{\phi=0}^{2\pi} \cos^2 \phi d\phi \int_{r=0}^{1} r^2 dr \quad \text{or}$$

$$= 6(\pi)(1/3)$$

or simply

$$q = 2\pi \, C$$

4.8 Find the capacitance of an isolated sphere whose radius is 10 cm.

Solution: Capacitance is given as

$$C = 4\pi\varepsilon_0 R \quad \text{or}$$

$$= 4\pi\varepsilon_0 \times 10 \times 10^{-12} \quad \text{or}$$

$$= \frac{10 \times 10^{-2}}{9 \times 10^9} \quad \text{or}$$

$$= 11.1 \times 10^{-12} \text{ F}$$

or simply

$$= 11.1 \text{ pF}$$

4.9 Find the capacitance of a capacitor as shown in Figure 4.16. Here, $\varepsilon_{r1} = 3$, $\varepsilon_{r2} = 7$, $d = 10$ mm and area $A = 60$ cm^2.

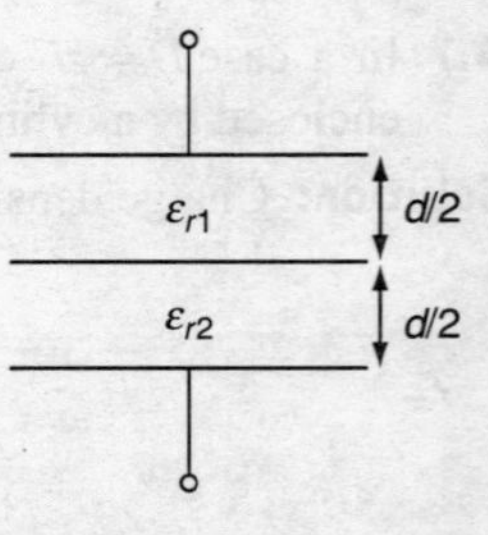

Figure 4.16 *Capacitor*

Solution: $C_1 = \dfrac{\varepsilon_1 A_1}{d_1}$ (standard formula for capacitance) or

$$= \frac{\varepsilon_0 \varepsilon_{r1} A}{(d/2)} (A_1 = A) \quad \text{or}$$

$$= \frac{2\varepsilon_0 \varepsilon_{r1} A}{d}$$

Similarly,

$$C_2 = \frac{\varepsilon_2 A_2}{d_2} \quad \text{or}$$

$$= \frac{\varepsilon_0 \varepsilon_{r2} A}{(d/2)} (A_2 = A) \quad \text{or}$$

$$= \frac{2\varepsilon_0 \varepsilon_{r2} A}{d}$$

Total capacitance C is a parallel combination of C_1 and C_2:

$$C = \frac{C_1 \cdot C_2}{(C_1 + C_2)} \quad \text{or}$$

$$= \frac{2\varepsilon_0 A}{d} \frac{(\varepsilon_{r1}\varepsilon_{r2})}{(\varepsilon_{r1} + \varepsilon_{r2})} \quad \text{or}$$

$$= \frac{2 \times 10^{-9} \times 60 \times 10^{-4} \times 3 \times 7}{36\pi \times 10 \times 10^{-13} \times 10} \quad \text{or}$$

$$= 22.27 \times 10^{-12}\,\text{F} \quad \text{or}$$

$$= 22.27\,\text{pF}$$

4.10 Find the capacitance per unit length of a coaxial transmission line. Consider $\varepsilon_r = 6$, $a = 10$ m and $b = 50$ m.

Solution: Capacitance of a coaxial transmission line is given as

$$C = \frac{2\pi\varepsilon l}{2.303 \log_{10}(b/a)} (\text{F})$$

Capacitance per unit length is obtained as

$$C/l = \frac{2\pi\varepsilon}{2.303 \log_{10}(b/a)} (\text{F/length}) \quad \text{or}$$

$$= \frac{2\pi \times 6}{2.303 \times 36\pi \times 10^9 \log_{10}(50/10)} (\text{F/length}) \quad \text{or}$$

$$= \frac{6\times10^{-9}}{2.303\times10\times0.6990}(\text{F/length}) \quad \text{or}$$

$$= 0.20706\times10^{-9}\ (\text{F/length})$$

or simply

$$= 207.06\ (\text{pF/length})$$

4.11 Concentric metal spheres of radii 4 m and 10 m are shown in Figure 4.17. A charge q is placed on the outer surface of the inner shell. Find the capacitance between these concentric spheres.

Solution: Positive charge $+q$ is there on the outer surface of the inner shell. It produces negative charge $-q$ inside of the outer shell.

Electric field $(\vec{E})$ between spheres is obtained as

$$\vec{E} = \frac{q}{4\pi\varepsilon_0 r^2}\vec{a}_r \quad \text{for } 4<r<10$$

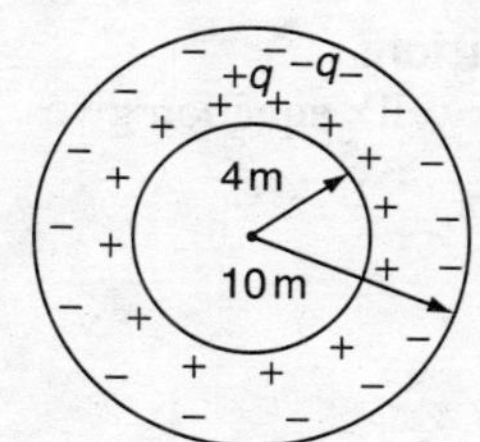

Figure 4.17 *Concentric metal spheres*

Potential difference V between spheres is obtained as

$$V = \frac{q}{4\pi\varepsilon_0}\int_{10}^{4}\frac{dr}{r} \quad \text{or}$$

$$V = \frac{q}{4\pi\varepsilon_0}\left[\frac{1}{4}-\frac{1}{10}\right] \quad \text{or}$$

$$V = \frac{q}{4\pi\varepsilon_0}\times\frac{6}{40}$$

Capacitance between these concentric spheres is obtained as

$$C = \frac{q}{V}$$

$$= 4\pi\varepsilon_0\times\frac{40}{6}$$

or simply

$$C = \frac{80}{3}\pi\varepsilon_0 \text{ Farads}$$

4.12 A parallel plate capacitor has a plate area of 45 cm^2. Plate separation is 12 mm. A voltage of 15 sin $10^2\,t$ Volts is applied to its plate. Find the displacement current (assume $\varepsilon = 3\,\varepsilon_0$).

Solution:

It is given that

$$A = 45\times10^{-4}\,\text{m}^2$$

$$d = 12\times10^{-3}\,\text{m}$$

$$V = 15\sin10^2\,t \text{ Volts} \quad \text{and} \quad \varepsilon = 3\,\varepsilon_0$$

We know that

$$D = \varepsilon E = \varepsilon\frac{V}{d} \quad \text{and}$$

$$J_d = \frac{\partial D}{\partial t} = \frac{\varepsilon}{d}\frac{\partial V}{\partial t}$$

Thus,

$$I_d = J_d A = \frac{A\varepsilon}{d}\frac{\partial V}{\partial t} \quad \text{or}$$

$$= \frac{45\times10^{-4}\times3\times10^{-9}}{12\times10^{-3}\times36\times\pi}\times10^{2}\times15\cos10^{2}t \quad \text{or}$$

$$= 14.9\cos10^{2}t\,(\text{nA})$$

4.13 In free space, $\vec{H} = 60\cos(\omega t - 6x)\vec{a}_z$ (A/M). What is the displacement current density $(\vec{J}_d)$?

Solution:

Maxwell's equation gives

$$\nabla\times\vec{H} = \frac{\partial\vec{D}}{\partial t} \quad \text{or}$$

$$\begin{vmatrix} \vec{a}_x & \vec{a}_y & \vec{a}_z \\ \dfrac{\partial}{\partial x} & \dfrac{\partial}{\partial y} & \dfrac{\partial}{\partial z} \\ H_x & H_y & H_z \end{vmatrix} = \vec{J}_d$$

Here in this case, $\vec{H}$ is in the z direction. Thus,

$$H_x = H_y = 0 \quad \text{or}$$

$$\vec{a}_x\left|\frac{\partial H_z}{\partial y}\right| - \vec{a}_z\left|\frac{\partial H_z}{\partial x}\right| + 0\,\vec{a}_z = \vec{J}_d \quad \text{or}$$

$$\vec{H} = 60\cos(\omega t - 6x)\vec{a}_z$$

Thus,

$$\frac{\partial H_z}{\partial y} = 0$$

$$\vec{J}_d = \left|\frac{\partial \overrightarrow{H_z}}{\partial x}\right|\vec{a}_y$$

$$\vec{J}_d = -360\sin(\omega t - 6x)\vec{a}_y\,(\text{A/m}^2)$$

4.14 Parallel plates in a capacitor have an area of 12 cm^2. These plates are separated by 2 cm. 20 sin 20^2t Volts is applied to the capacitor. Find the displacement current in the case when dielectric material (between plates) has a relative permittivity of 6.

Solution:

We know that displacement current density (J_d) is given as

$$J_d = \frac{\partial}{\partial t}\left(\frac{\varepsilon V}{d}\right) \quad \text{or}$$

$$= \frac{\varepsilon \partial V}{d \partial t} \quad \text{or}$$

$$= \frac{\varepsilon_0 \varepsilon_r}{d} \frac{\partial V}{\partial t}$$

$$\text{Displacement current } = \frac{\varepsilon_o \varepsilon_r . A}{d} \frac{\partial V}{\partial t}$$

Thus,

$$J_d = \frac{8.86 \times 10^{-12} \times 6 \times 12 \times 10^{-4}}{2 \times 10^{-2}} \times 20 \times 20^2 \cos 20^2 t$$

$$J_d = 25.51 \cos 20^2 t \, (\text{nA})$$

4.15 A parallel plate capacitor with a plate area of 15 cm^2 and plate separation of 2 mm has a voltage of 100 sin $10^2 t$ V is placed to its plate. Calculate the displacement current assuming $\varepsilon = 2\varepsilon_0$.

Solution: Given the area of plate = 15 cm^2

Thickness (d) = 2mm

$$V = 100 \sin 10^2 t$$

$$\varepsilon = 2\varepsilon_0$$

$$\vec{D} = \varepsilon \vec{E} \quad \text{but} \quad \vec{E} = \frac{\vec{V}}{d}$$

$$\vec{D} = \frac{\varepsilon \vec{V}}{d}$$

$$I_d = \frac{\partial \vec{D}}{\partial t} = \frac{\varepsilon}{d} \frac{dV}{dt} \quad \text{or}$$

$$I_d = J_d \times A = \frac{\varepsilon A}{d} \frac{dV}{dt} \quad \text{or}$$

$$= C \frac{dv}{dt} \quad \text{where } \varepsilon = 2\varepsilon_0 \quad \text{or}$$

$$I_d = \frac{2\varepsilon_0 A}{d} \frac{d}{dt} (100 \sin 10^2 t)$$

$$= \frac{2\varepsilon_0 A}{d} \times 100 \cos 10^2 t \times 100$$

$$= \frac{2 \times 8.83 \times 10^{-12} \times 15 \times 10^{-4} \times 10^4 \cos 10^2 t}{2 \times 10^{-3}}$$

$$I_d = 1.3 \times 10^{-7} \cos 10^2 \text{ tA}$$

4.16 A magnetic core of uniform cross section of area 16 cm^2 is connected to 100 V, 60 Hz generator as shown in Figure 4.18. Find the induced emf in the secondary coil of the generator.

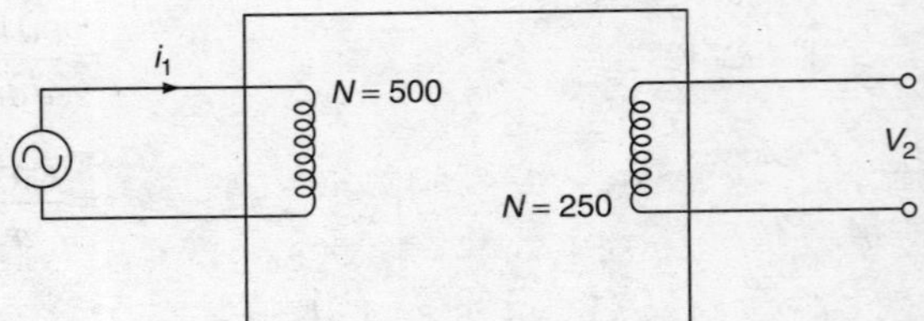

Figure 4.18 *Generator*

Solution:

$$\phi_1 = \vec{B}_1 \cdot \overrightarrow{ds}$$

$$d_1 = \frac{\mu N_1 I}{l} \times A$$

$$V_2 = -N_1 N_2 A \times V_1 \quad \text{or}$$

$$= -500 \times 250 \times 16 \times 10^{-4} \times 100 \quad \text{or}$$

$$V_2 = 20{,}000 \text{ V}$$

4.17 Inside a right circular cylinder, $\mu_{r1} = 900$. The exterior is free space. If $\vec{B}_1 = +5\vec{a}_\phi$ inside the cylinder, find B_2 just outside.

Solution: Given $\mu_{r1} = 900$

$$\mu_2 = \mu_0$$

$$B_1 = +5\vec{a}_\phi$$

By applying Boundary condition, it is clear that

$$\vec{B}_{n1} = \vec{B}_{n2}$$

So,

$$\mu_{r1}\mu_0 H_{n1} = \mu_{r2} H_{n2} = \mu_0 H_{n2}$$

$$900 \times H_{n1} = H_{n2}$$

So,

$$B_{n2} = 5\vec{a}_\phi \text{ Tesla}$$

4.18 In a Spherical coordinate system, Region 1 is $R < a$, Region 2 is $a < R < b$, and Region 3 is $R > b$. Regions 1 and 3 are free space while $\mu_{r2} = 400$. Given $B_1 = 0.5\,\vec{a}_R$, find $\vec{H}$ in each region.

Solution:

$$\vec{B} = \mu \vec{H}$$

$$\vec{H}_1 = \frac{\vec{B}_1}{\mu_0} = \frac{0.5}{\mu_0}\vec{a}_R \text{ (A/m)}$$

$$\vec{H}_2 = \frac{B_{n2}}{\mu_2} = \frac{0.5}{\mu_0 \times \mu_{r2}} = \frac{0.5}{400\mu_0}$$

$$= \frac{1.25 \times 10^{-3}}{\mu_0}\vec{a}_R \text{ (A/m)}$$

$$H_3 = \frac{0.5}{\mu_0}\vec{a}_R \text{(A/m)}$$

4.19 A 150 V voltage generator operating at 40 MHz is connected to the plates of air-dielectric parallel plate capacitor having a plate area of 2.4 cm^2 and a distance of separation 0.4 mm. Find the maximum value of displacement current density and displacement current.

Solution: $\vec{D} = \varepsilon\vec{E}$ and $\vec{E} = \dfrac{V}{d}$ where $V = 150\sin(2\pi\times 40\times 10^6 t)$

Displacement current density

$$\vec{J}_d = \frac{\partial D}{\partial t} = \frac{\varepsilon}{d}\frac{dV}{dt} \quad \text{or}$$

$$J_d = \frac{150\varepsilon}{d}\cos(8\pi\times 10^7 t)\times 8\pi\times 10^7 \quad \text{or}$$

$$= \frac{150\times 8.86\times 10^{-12}\times 8\pi\times 10^7}{0.4\times 10^{-3}}\cos(8\pi\times 10^7 t)\,(\text{A/m}^2) \quad \text{or}$$

$$= 835.03\cos(8\pi\times 10^7 t)\,(\text{A/m}^2)$$

So, displacement current

$$I_d = J_d\times A \quad \text{or}$$

$$I_d = 0.2004\cos(8\pi\times 10^7 t)\,(\text{A})$$

4.20 An area having water has $\mu = \mu_0$, $\varepsilon = 64\,\varepsilon_0$ and $\sigma = 10$ S/m. Find the frequency at which the conduction current density is 10 times the displacement current density (in magnitude).

Solution: Conduction current density $\quad J_c = \sigma\vec{E}$

Displacement current density $\quad J_d = \dfrac{\varepsilon\partial\vec{E}}{\partial t}$

$$\frac{\partial\vec{E}}{\partial t} = \omega E_0\cos(\omega t)$$

Given,

$$J_d = 10J_d$$

$$\sigma\vec{E} = 10\times\varepsilon\frac{\partial\vec{E}}{\partial t}.$$

$$\sigma E_0\sin\omega t = 10\,\omega t\,E_0\cos\omega t$$

$$\frac{\sigma}{64\times\varepsilon_0\times 10} = \frac{10}{640\times 8.86\times 10^{-12}}$$

$$f = 1.76\text{ GHz.}$$

4.21 Copper wire carries the conduction current of 1.1 (one point one) Amp at 50 Hz. Find i_d, assuming $\varepsilon = 2\varepsilon_0$, $\mu = 2\mu_0$ and $\sigma = 4.3\times 10^7$ mho S/m.

Solution: $\vec{J} = \sigma\vec{E}$

$$\vec{E} = \frac{\vec{J}}{\sigma} = \frac{I}{\sigma A} = \frac{1.1}{4.3\times 10^{7} \times A} \text{ (V/m)}$$

$$\vec{J}_d = \omega\varepsilon\vec{E}$$

$$i_d = J_d A \quad \text{or}$$

$$= \omega\varepsilon\vec{E}A \quad \text{or}$$

$$= \omega\varepsilon_0 \times \frac{1}{4.3\times 10^{-7} A} A \quad \text{or}$$

$$= \frac{2\pi \times 50 \times 2 \times 8.86 \times 10^{-12}}{4.3\times 10^{-7}} \quad \text{or}$$

$$= 0.012 \text{ Amp}$$

4.22 Using $\nabla \cdot \vec{D} = \rho$, Ohm's Law and equation of continuity, prove that charge density that exists in a conductor at any instant would decrease to half in time ε/σ seconds.

Solution: As we know $\nabla \cdot \vec{D} = \rho$ where $\vec{D} = \varepsilon\vec{E}$

Also

$$\nabla \cdot \vec{J} = \frac{-\partial\rho}{\partial l}$$

By using Ohm's Law,

$$\vec{J} = \sigma\vec{E}$$

So

$$\nabla(\sigma\vec{E}) = \frac{-\partial}{\partial t}(\nabla \cdot \vec{D}) \quad \text{or}$$

$$\sigma(\nabla.\vec{E}) = \frac{-\partial}{\partial t}(\nabla.\vec{D}) \quad \text{or}$$

$$\sigma\frac{\rho_v}{\varepsilon} = \frac{-\partial\rho_v}{\partial t} \quad \text{or}$$

$$\frac{\partial\rho_v}{\partial t} + \frac{\sigma\rho_v}{\varepsilon} = 0$$

The above equation is of form

$$\frac{dy}{dx} + \rho y = 0$$

Here,

$$\rho = \frac{\sigma}{\varepsilon} \quad \text{and} \quad x = t$$

So, $y = \rho_v$

Now

$$e^{\int \rho dx} = e^{\int (\sigma/\varepsilon)dt} = e^{(\sigma/\varepsilon)t}$$

$$y \cdot e^{\int p dx} = \text{const.}$$

$$\rho_v e^{(\sigma/\varepsilon)t} = \text{const} = \rho_0$$

$$\rho_v = \frac{\rho_v}{e^{\sigma/\varepsilon t}}$$

when
$$\varepsilon = \frac{\varepsilon}{\sigma}$$

then
$$\rho_v = \rho_v \frac{1}{e}$$

Hence, proved.

4.23 In a homogeneous vector region, where $\mu_r = 1$ and $\varepsilon_r = 50$ $\vec{E}_x = 40\pi e^{j(\omega t+\beta z)}$(V/m) and $\vec{B} = 4\mu_0 H_m e^{j(\omega t+\beta z)}\vec{a}_y$. Find ω and H_m if $\lambda = 2.78$ m.

Solution:
$$\beta = \frac{2\pi}{\lambda} = 2.26 \text{ (rad/m)}$$

$$f = \frac{c}{\lambda} = \frac{3\times10^8}{2.78} = 1.079\times10^8 \text{ Hz}$$

The reader is expected to obtain ω and H_m on his or her own.

4.24 A current element $I\,dl = 2\pi(0.8\vec{a}_i - 0.6\vec{a}_j)\ \mu$A is placed at point (4, 3, 2). Find $d\vec{H}$ at the point (2, 2, 3).

Solution:
$$d\vec{H} = \frac{Id\vec{l}\times\hat{a}_R}{4\pi R}$$

$$\hat{a}_R = \frac{(-2\vec{a}_i - \vec{a}_j + \vec{a}_k)}{\sqrt{4+1+1}} = \frac{(-2\vec{a}_i - \vec{a}_j + \vec{a}_k)}{\sqrt{6}}$$

$$d\vec{H} = 2\pi\frac{(0.8\vec{a}_i - 0.6\vec{a}_j)\times(-2\vec{a}_i - \vec{a}_j + \vec{a}_k)\times10^{-6}}{\sqrt{6}\times4\pi\times6} \quad \text{or}$$

$$= \frac{10^{-6}}{12\sqrt{6}}(0.8\vec{a}_k - 0.8\vec{a}_j - 1.2\vec{a}_k - 0.6\vec{a}_i) \quad \text{or}$$

$$= 34.02\times10^{-9}(2.0\vec{a}_k - 0.8\vec{a}_j - 0.6\vec{a}_i)$$

$$d\vec{H} = (20.4\vec{a}_i - 27.2\vec{a}_j + 68.04\vec{a}_k)\text{(n A/m)}$$

4.25 The flux density at a point R distance from a long filament conductor is given by $\vec{B}_\phi = \mu_0 I/2\pi R$. Find the flux crossing the portion of a conductor in a plane $\phi = \pi/6$ defined by $0.02 < R < 0.06$ m and $0 < z < 4$ m for a current of 2 A.

Solution: Flux
$$\phi = \int_s \vec{B}\cdot d\vec{s} = \int_0^4\int_{0.02}^{0.06} \frac{\mu_0 I}{2\pi R} dR\cdot dz$$

$$\phi = \frac{\mu_0 I}{2\pi}\int_0^4 \ln\left(\frac{0.06}{0.02}\right)dz$$

$$= \frac{\mu_0 2}{2\pi}\times \ln\left(\frac{0.06}{0.02}\right)\times 4$$

$$= \frac{4\mu_0}{\pi}\ln(3)$$

$$= 1.3\mu_0 \text{ (Webers)}$$

4.26 Using vector potential, find the magnetic flux density at a point due to a long straight conductor-carrying current in the $\vec{a}_z$ direction.

Solution: From Figure 4.19,

$$r^2 = x^2 + y^2, \quad dl = dz\,\vec{a}_z \quad \text{and} \quad R^2 = r^2 + z^2$$

Magnetic vector potential

$$\vec{A} = \frac{\mu_0}{4\pi}\int_M \frac{Idl}{R}$$

$$\vec{A}_z = \frac{\mu_0}{4\pi}\int_{-M} \frac{Idz}{R}$$

As,

$$r^2 + z^2 = t^2 \quad \Rightarrow 2z\,dz = 2t\,dt \quad \text{or}$$

$$dz = \frac{t\,dt}{z}$$

So

$$\vec{A}_z = \frac{\mu_0}{2\pi}\int_0^M \frac{Id_z}{\sqrt{r^2+z^2}} \quad \text{or}$$

$$= \frac{\mu_0 I}{2\pi}\left(\ln\left(z+\sqrt{r^2+z^2}\right)_0^M\right) \quad \text{or}$$

$$= \frac{\mu_0 I}{2\pi}(\ln\left(M+\sqrt{r^2+l^2}\right) - \text{lnr}) \quad \text{or}$$

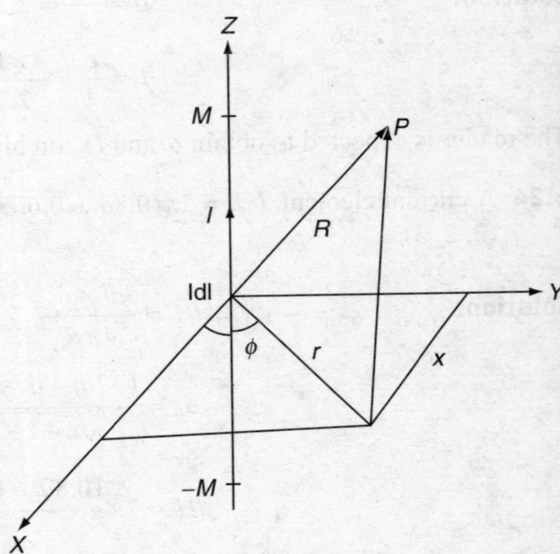

Figure 4.19 *Current-carrying conductor*

$$= \frac{\mu_0}{2\pi}(\ln(2M) - \ln(r)) \quad \text{for} \quad M >> r \quad \text{or}$$

$$\vec{A}_z = \frac{\mu_0}{2\pi}\ln\left(\frac{2M}{r}\right)$$

$$\vec{B} = \nabla\times\vec{A} = \frac{-\partial A_z}{\partial dr} \quad \text{or}$$

$$= \frac{-\mu_0 I}{2\pi}\times\frac{\partial}{\partial r}\left(\ln\left(\frac{2M}{r}\right)\right) \quad \text{or}$$

$$= \frac{\mu_0 I}{2\pi} \times \frac{\partial}{\partial r}(\ln(2M) - \ln r) \quad \text{or}$$

$$= \frac{-\mu_0 I}{2\pi} \times \frac{-1}{r} \vec{a}_\phi \quad \text{or}$$

$$\vec{B} = \frac{\mu_0 I}{2\pi r} \vec{a}_\phi$$

4.27 Calculate the magnetic flux density at the centre of a current-carrying loop with a radius of 12 cms. The loop current is 2 mA and the loop is placed in air.

Solution:

$$\vec{B} = \frac{\mu I}{2\pi R} = \frac{4\pi \times 10^{-7} \times 2 \times 10^{-3}}{2\pi \times 12 \times 10^{-2}}$$

$$= 3.3 \times 10^{-9} \text{ Wb/m}^2$$

$$\vec{B}_{\text{centre}} = \frac{\mu I}{2a} = \frac{4\pi \times 10^{-7} \times 2 \times 10^{-3}}{2 \times 12 \times 10^{-2}}$$

$$= 1.04 \times 10^{-8} \text{ (Tesla)}$$

4.28 A solenoid is 30 cm long and has a 2-cm diameter. It has 500 turns of its winding. It is placed in a uniform magnetic field of strength of 2 (Wb/m^2) and a current of 8 Amp. Calculate the maximum torque on the solenoid.

Solution:

$$l = 300 \text{ cm} = 0.3 \text{ m}$$

$$r = 1 \text{ cm} = 0.01 \text{ m}$$

$$N = 500, \quad I = 8 \text{ Amp}$$

$$B = 2 \text{ Wb/m}^2$$

Torque = $NIAB \sin\theta$

$$T_{\max} = NIAB$$

$$= 500 \times 8 \times 2 \times (0.3 \times 0.01)$$

$$= 24 \text{ (N-m)}$$

4.29 Calculate the inductance of 20-cm long coaxial cable filled with a material for which $\varepsilon_r = 10, \sigma = 0$ and $\mu_r = 88$. It has inner and outer diameters as 2 mm and 6 mm, respectively.

Solution:

$$L = \frac{\mu_0 \mu_r}{2\pi} \ln\left(\frac{b}{a}\right) l$$

$$= \frac{4\pi \times 10^{-7} \times 88}{2\pi} \ln\left(\frac{6}{2}\right) \times 20 \times 10^{-2}$$

$$= 3.86 \times 10^{-6} \text{ (Henry)}$$

4.30 Find the current in a circular wire of radius 4 mm, when the current density in the conductor is $\vec{J} = 40(1-e^{-1000r})\vec{a}_z(\text{A/m}^2)$.

Solution: As given,

$$r = 0.004 \text{ m}$$

$$\vec{J} = 40(1-e^{-1000r})\vec{a}_z$$

As

$$I = \int_s \vec{J}\cdot d\vec{s} \quad \text{or}$$

$$= \int_{\phi=0}^{2\pi}\int_{r=0}^{0.004} 40(1-e^{-1000r})r\,dr\,d\phi$$

$$= 80\pi\int_0^{0.004}(1-e^{-1000r})\,r\,dr$$

$$= 80\pi\left[\int_0^{0.004} r\,dr - \int_0^{0.004} re^{-1000r}dr\right]$$

$$= 80\pi[8\times10^{-6} - 9.08\times10^{-7}]$$

$$= 1.78\times10^{-3}(\text{A})$$

4.31 Two perfectly conducting spherical surfaces are located at $r = 4$ and $r = 12$ cms. The total current passing radially outward through the medium between the sphere is 4.5 A. Find the voltage and resistance between the spheres, and also the electric field in the region between two spheres. Given that $\sigma = 0.02$ S/m.

Solution: It is given that $r = 4$ to 12 cm

$$\sigma = 0.02 \text{ S/m}$$

$$I = 4.5 \text{ A}$$

$$\vec{J} = I/A \quad \text{or}$$

$$= \frac{4.5}{\pi(12^2-4^2)\times10^{-4}} \quad \text{or}$$

$$= 111.90 \text{ A/m}^2$$

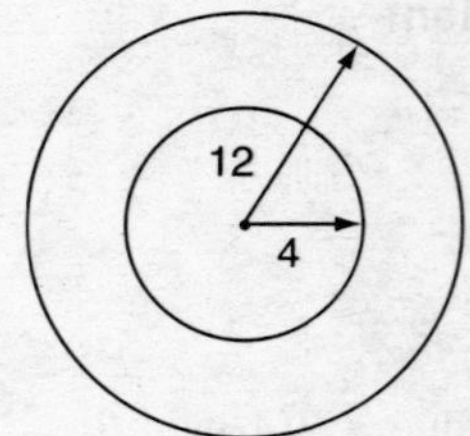

Figure 4.20 *Spherical surface*

4.32 At a plane boundary between two conductors, tangential electric field is 4 V/m in medium '1', and tangential current density in medium '2' is 10 A/m^2. Find σ_2.

Solution:

$$\vec{E}_{t1} = \vec{E}_{t2}$$

$$\vec{J} = \sigma\vec{E}$$

$$\vec{E} = \frac{\vec{J}}{\sigma}$$

Thus,

$$\frac{\vec{J}_{t1}}{\sigma_1} = \frac{\vec{J}_{t2}}{\sigma_2} \quad \text{or}$$

$$4 = \frac{\vec{J}_{t2}}{\sigma_2} \quad \text{or}$$

$$\sigma_2 = \frac{\vec{J}_{t2}}{4} \quad \text{or}$$

$$= \frac{10}{4} \quad \text{or}$$

$$= 2.5 \text{ S/m}$$

4.33 A current $I(A)$ enters a thin right circular cylinder at the top as shown in Figure 4.21. Find $\vec{K}$ if the radius of the cylinder is 6 cm.

Solution: We know that

$$K = \frac{I}{2\pi R}$$

Along the radial direction at the top of the cylinder,

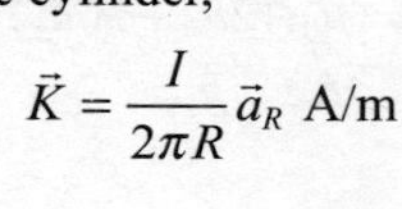

$$\vec{K} = \frac{I}{2\pi R}\vec{a}_R \text{ A/m}$$

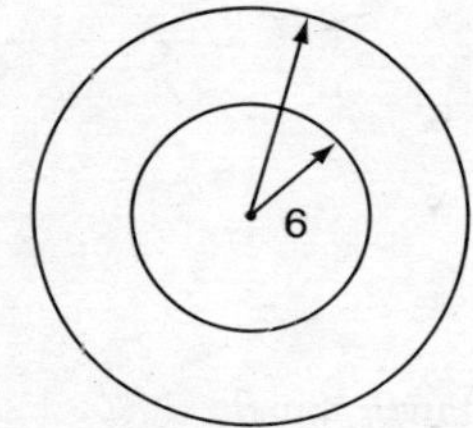

Figure 4.21 *Circular cylinder*

At the surface of cylinder,

$$\vec{K} = \frac{-I}{2\pi \times 0.06}\vec{a}_z \quad \text{or}$$

$$\vec{K} = \frac{-I}{0.12\pi}\vec{a}_z \text{ A/m}$$

4.34 A conductor of radius $r_0 = 2$ cm has an internal field

$$H = \frac{10^4}{R}\left(\frac{1}{a^2}\sin(aR) - \frac{R}{a}\cos(aR)\right)\vec{a}_\phi \text{ A/m},$$

where $a = \pi/2R_0$. Find the total current in the conductor.

Solution: From Ampere's circuital law

$$\oint_L H \cdot dl = I_{\text{enclosed}}$$

$$d\vec{L} = \vec{a}_\phi \cdot d\phi$$

$$I = \frac{10^4}{R_0}\int_0^{2\pi}\left[\frac{1}{\left(\frac{\pi}{2R_0}\right)^2}\sin\left(\frac{\pi}{2R_0}\times R_0\right) - \frac{R_0}{\left(\frac{\pi}{2R_0}\right)}\cos\left(\frac{\pi}{2R_0}\times R_0\right)\right]d\phi \quad \text{or}$$

$$= \frac{10^4}{R_0}\int_0^{2\pi}\left[\frac{4R_0^2}{\pi^2}\times 1 - \frac{2R_0^2}{\pi}\cos\left(\frac{\pi}{2}\right)\right]d\phi \quad \text{or}$$

$$= \frac{10^4}{R_0} \times \frac{4R_0^2}{\pi^2} [\phi]_0^{2\pi} \quad \text{or}$$

$$= \frac{10^4 \times 4R_0}{\pi^2} = \frac{8R_0}{\pi}(10^4) \text{ Amp}$$

As $R_0 = 2$ cm,

$$I = \frac{8}{\pi} \times 2 \times 10^{-4} \times 10^4 \quad \text{or}$$

$$I = \frac{16}{\pi} \text{ Amp}$$

4.35 The interface between two different regions is normal to one of the three Cartesian axes. If

$$\vec{B}_1 = \mu_0(40.5\vec{a}_x + 12\vec{a}_y)$$
$$= \vec{B}_{t1} + \vec{B}_{n1} \qquad \text{and}$$
$$\vec{B}_2 = \mu_0(12\vec{a}_x + 36\vec{a}_z)$$
$$= \vec{B}_{t2} + \vec{B}_{n2}$$

Find $\tan\theta_1 / \tan\theta_2$.

Solution:

$$\vec{H}_{t1} = \frac{\vec{B}_{t1}}{\mu_1}, \quad \vec{H}_{t2} = \frac{\vec{B}_{t2}}{\mu_2} \quad \text{and} \quad \vec{B}_{n1} = \vec{B}_{n2}$$

Here

$$\mu_1 = \mu_2 = \mu_0, \quad H_{t2} = 12 \quad \text{and} \quad H_{t1} = 40.5 \quad \text{or}$$

$$\frac{\tan\theta_1}{\tan\theta_2} = \frac{12}{40.5}$$
$$= 0.296$$

4.36 The field of short vertical element *Idl* is located at the origin of a Spherical coordinate system given as

$$E(R,\theta) = E_\theta(R,\theta)\vec{a}_\theta = \left(j\frac{50\pi I\,dl}{\lambda R}\cos\theta\right)e^{-j\beta R}\,\vec{a}_\theta \text{ (V/m)} \quad \text{and}$$

$$H(R,\theta) = \frac{E_\theta(R,\theta)}{\eta_o}\vec{a}_\phi = \left(j\frac{I\,dl}{2\pi R}\cos\theta\right)e^{-j\beta R}\,\vec{a}_\phi \text{ (A/m)}$$

where λ is the wavelength.

(a) Find out the expression for Poynting vector.
(b) Also, find the average power radiated by current element.

Solution:

(a) As

$$E_\theta / H_\theta = \eta = 120\pi \ (\Omega)$$

So, instantaneous Poynting vector is at time t

$$\vec{P}(R,\theta;t) = Re[E(R,\theta)e^{j\omega t}] \times Re[H(R,\theta)e^{j\omega t}] \quad \text{or}$$

$$= 25\pi\left(\frac{I\,dl}{\lambda R}\right)^2 \cos^2\theta \sin^2(\omega t - \beta R) \quad \text{or}$$

$$= 12.5\pi\left(\frac{I\,dl}{\lambda R}\right)\cos^2\theta[1 - \cos 2(\omega t - \beta r)]\vec{a}_R$$

(b) Average power density vector is given by

$$\vec{P}_{\text{avg}}(R,0) = 12.5\pi\left(\frac{Idl}{\lambda R}\right)^2 \cos^2\theta$$

Therefore, total average power over the surface of radius R is obtained as
Total

$$P_{avg} = \oint_s p_{avg}\,(R,\theta)\cdot ds \quad \text{or}$$

$$= \int_0^{2\pi}\int_0^{\pi} 12.5\pi\left(\frac{Idl}{\lambda R}\right)^2 \cos^2\theta\, R^2 \sin\theta\, d\theta d\phi \quad \text{or}$$

$$= 20.9\left(\frac{I\,dl}{\lambda R}\right)^2 \text{ Watts}$$

4.37 Find the ratio of free space wavelength to wavelength in a conductor with $\sigma = 10^4$ mho/m and $\mu_r = 15$ at 10 kHz.

Solution:

$$\lambda_0 = \frac{c}{f_0} = \frac{3\times10^8}{10\times10^3} = 3\times10^4\,\text{m}$$

Free space wavelength $= 3\times10^4\,\text{m}$
Now,

$$\beta = \sqrt{\frac{\omega\mu\sigma}{2}} = \sqrt{\frac{2\pi\times10\times10^3\times15\times4\pi\times10^{-7}\times10^4}{2}} \quad \text{or}$$

$$= 39.73$$

As

$$\frac{2\pi}{\lambda} = \beta$$

$$\frac{2\pi}{\lambda} = 39.73$$

$$\lambda = \frac{2\pi}{39.73}$$

$$\lambda = 0.158$$

Now

$$\frac{\lambda_0}{\lambda_g} = \frac{3\times10^4}{0.158} = 18.98\times10^4$$

$$E_y = 1.51 \times 240 \quad \text{or}$$

$$E_y = 364.69 \text{ V/m} \quad \text{and}$$

$$E_t = 364.69 \text{ V/m}$$

4.38 Find J_C and J_D for moist soil, which has conductivity $\sigma = 2 \times 10^{-3}$ S/m and $\vec{E} = 5\times10^{-6}$ $\sin(25 \times 10^9\, t)$ V/m.

Solution: Conduction current density $\vec{J}_c = \sigma\vec{E}$

Displacement current density $\vec{J}_D = \varepsilon\dfrac{\partial\vec{E}}{\partial t}$

Now,

$$J_c = 5\times10^{-6}\sin(25\times10^9 t)\,2\times10^{-3}$$

$$J_c = 10\times10^{-9}\sin(25\times10^9 t)(\text{A/m}^2) \quad \text{and}$$

$$\vec{J}_D = \varepsilon\cdot\omega\cdot 5\times10^{-6}\cos(25\times10^9 t)$$

$$= 8\cdot 86\times10^{-12}\times25\times10^9\times5\times10^{-6}\cos(25\times10^9 t)$$

$$\vec{J}_D = 1.1\times10^{-6}\cos(25\times10^9 t)\,(\text{A/m}^2)$$

4.39 For rust at 2 MHz, $\varepsilon_r = 15$, $\sigma = 4\times10^{-12}$ S/m and $\mu_r = 1$. Find the values of α, β, r, v_p and δ.

Solution:

$$\frac{\sigma}{\omega\varepsilon} = \frac{4\times10^{-12}}{2\times\pi\times2\times10^6\times8.85\times10^{-12}}(>>1)$$

Therefore, the material is a good conductor.

$$\alpha = \beta = \sqrt{\frac{\omega\mu\sigma}{2}} \quad \text{or}$$

$$= \sqrt{\frac{2\times\pi\times2\times10^6\times4\times\pi\times10^{-7}\times4\times10^{-12}}{2}}$$

$$\alpha = \beta = 5.61\times10^{-6}\,\text{m}^{-1}$$

$$\delta = \frac{1}{\alpha} = 1.77\times10^5\,\text{m}$$

$$r = 5.61\times10^{-6}(1+j)\,\text{m}^{-1}$$

$$v_p = \frac{\omega}{\beta} = \frac{2\pi f}{5.61\times10^{-6}} = \frac{2\times\pi\times2\times10^6}{5.61\times10^{-6}}$$

$$v_p = 2.23\times10^{12}\,\text{m/s}$$

4.40 Find the force on a wire of length 0.08 m, which is placed inside a long solenoid near its centre. The wire makes an angle of 45° with the axis of the solenoid. It carries a current of 2 A. The magnetic field due to solenoid is 0.25 T.

Solution: It is given that

$$L = 0.08 \text{ M}$$

$$\theta = 45°$$

$$I = 2 \text{ A}$$

$$B = 0.25 \text{ T}$$

Force F is given as $F = ILB \sin \theta$. Substituting the values, we get

$$F = 2 \times 0.08 \times 0.25 \times \sin 45°$$

$$= 0.028284 \text{ Newton}$$

4.41 Current is flowing in a coil. The coil has a self-inductance of 5.0 Henry. The increasing current is given as $i = 5\sin^2 t(\text{Amp})$. What is the amount of energy as the current changes from 0 to 3 Amp?

Solution: It is given that

$$i = 5\sin^2 t \quad \text{and} \quad L = 5.0 \text{ H}$$

Induced emf $(E_{\text{ind}}) = -L\dfrac{\partial i}{\partial t}$

Work done during a small time interval dt is obtained as

$$\frac{\partial W}{\partial t} = E_{\text{ind}} \cdot i$$

Solving it, we get

$$W = -\int_0^t E_{\text{ind}}\, i\, dt$$

$$= -\int_0^t \left(L\frac{di}{dt}\right) i\, dt$$

$$= \int_{i=0}^{t} L\, i\, di$$

$$= L\frac{i^2}{2}\bigg|_0^3$$

Finally,

$$W = \frac{5}{2} \cdot [9] = \frac{45}{2}$$

$$= 22.5 \text{ Joules}$$

4.42 Two parallel conducting plates are placed 2 cm apart. These are situated in air and are connected to a source of constant potential difference of 76 kV. Find the electric field intensity between the plates. If a mica sheet ($\varepsilon_r = 6$) of thickness 1cm is introduced between the plates, determine the field intensities in air and mica, given that the dielectric strengths of air and mica are 40 and 800 kV/cm, respectively.

Solution: The field intensity between the plates in air gap is

$$E = \frac{V}{t} = \frac{76}{2} = 38\,\text{kV/cm}$$

The dielectric strength of air is 40 kV/cm. The above value is safe and dielectric will not break down. Next, with the introduction of mica, there are two dielectrics:

$$\text{Air film: } t_1 = 1 \text{ cm} \quad \text{and} \quad \text{Mica film: } t_2 = 1 \text{ cm}$$

Let the corresponding film intensities be E_1 and E_2 in kV/cm.

Then,

$$E_1 t_1 + E_2 t_2 = V \tag{i}$$

$$E_1 = \frac{D}{\varepsilon_o}$$

$$E_2 = \frac{D}{\varepsilon_o \varepsilon_r} = \frac{D}{6\varepsilon_o} \tag{ii}$$

Hence,

$$E_1 = 6E_2$$

Substituting (ii) in (i),

$$6E_2 \times 1 + E_2 = 76$$

$$7E_2 = 76 \Rightarrow E_2 = 10{\cdot}85\,\text{kV/cm}$$

Field intensity in air gap is obtained as

$$E_1 \times 1 + 10.85 \times 1 = 76$$

$$\Rightarrow \quad E_1 = 76 - 10.85$$

$$E_1 = 65.15 \text{ kV/cm}$$

4.43 A co-axial capacitor of the compressed gas type is designed to have 30×10^{-12} Farad capacitance and to work at 100 kV DC. The maximum voltage gradient is not to exceed 200 kV per cm. If the outside diameter of the inner conductor is 5 cm, determine the inner diameter of the outer conductor and the length of the capacitor. Take relative the permittivity of gas as 1.0.

Solution: Inner conductor radius, $a = 2.5$ cm.
Let b be the inner radius of the outer conductor.

$$C = 30 \times 10^{-12}\,\text{F}$$

$$V = 100 \text{ kV}$$

$$E_{\max} = 200 \text{ kV}$$

We know that

$$E_{max} = \frac{V}{a\ \ln(b/a)}$$

Voltage is maximum for the minimum value of a. Hence,

$$V = E_{max}\, a\left(\ln\frac{b}{a}\right) \qquad \text{(i)}$$

If V is in kV, E_{max} in kV/cm, a should be in cm. Substituting the values in (i), we get

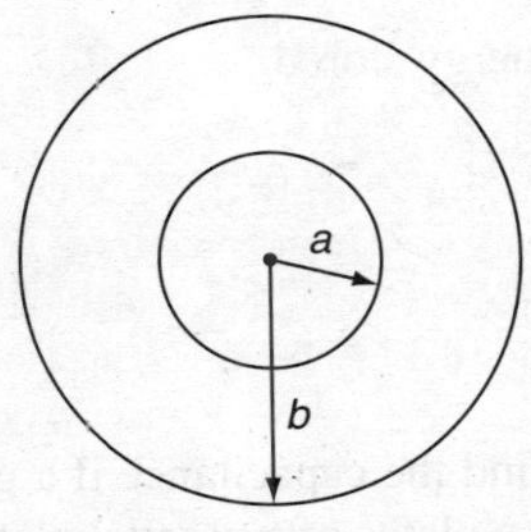

Figure 4.22 *Cross-sectional view of the co-axial cylinder*

$$\ln\left(\frac{b}{a}\right) = \frac{100\times10^3}{200\times10^3\times2.5} = 0.2$$

$$\Rightarrow \quad \log_{10}\frac{b}{a} = 0.2$$

$$\Rightarrow \quad \frac{b}{a} = 1.58$$

4.44 The capacitor $C = 4\ \mu$F is charged from a supply of 200 V. If the supply is removed and another charged capacitor is connected across the capacitor of 4 μF as shown in Figure 4.23, find the potential difference between the plates and the energy charged.

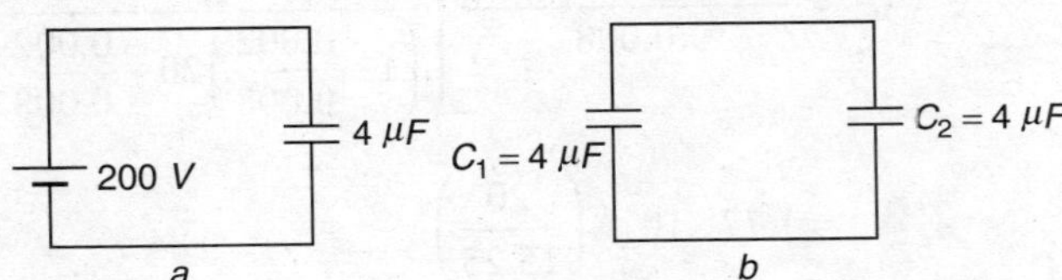

Figure 4.23 *Capacitor charging*

Solution: Charge on C_1 is $Q_1 = C_1 V = 4\times10^{-6}\times200 = 8\times10^{-4}$ C.

$$\text{Energy stored} = \frac{1}{2}CV^2 = \frac{1}{2}\times4\times10^{-6}\times100\times100 = 2\times10^{-2}\ \text{Joule}$$

When C_2 is connected across C_1, i.e. C_1 is parallel to C_2, then

$$\frac{1}{Q} = \frac{1}{Q_1} + \frac{1}{Q_2}$$

$$\Rightarrow Q = \frac{Q_1 Q_2}{Q_1 + Q_2} = \frac{Q_2}{2Q} = \frac{1}{2}Q_1$$

$$\Rightarrow \quad Q_2 = \frac{1}{2}\times8\times10^{-4}\ \text{C} = 4\times10^{-4}\ \text{C}$$

The potential difference across each capacitor is

$$V_1 = V_2 = \frac{Q_1}{C_1} = \frac{Q_2}{C_2} = \frac{4\times10^{-4}}{4\times10^{-6}} = 10^2 = 100\ \text{V}$$

Now, energy stored $= \frac{1}{2}(C_1 + C_2) \times (100)^2$

$$= \frac{1}{2}(8 \times 10^{-6}) \times (100)^2$$

$$= 4 \times 10^{-2} \text{ Joule}$$

4.45 Find the capacitance if a glass sheath of thickness 0.002 m with ε_r of 20 is introduced between the plates of a parallel plate capacitor (0.4 × 0.4) of $d = 0.008$ m and remaining 0.006 m of air.

Solution: Given that

$$d = 0.008 \text{ m}, a = 0.002 \text{ m and } A = 0.16 \text{ m}^2$$

Capacitance when a material is introduced between the plates of a capacitor is obtained as

$$C = \frac{\varepsilon_o A}{d}\left(\frac{\varepsilon_r}{\left(1 - \frac{a}{d}\right)\varepsilon_r + \frac{a}{d}}\right)$$

$$= \frac{8.85 \times 10^{-12} \times 0.16}{0.008}\left(\frac{20}{\left(1 - \frac{0.002}{0.008}\right)20 + \frac{0.002}{0.008}}\right)$$

$$= 1.77 \times 10^{-10}\left(\frac{20}{15.25}\right)$$

$$C = 2.32 \times 10^{-10} \text{ Farad}$$

4.46 Current is flowing in a coil. Coil has a self-inductance of 3.0 Henry. Increasing current is given as $i = 3 \sin t^2$ (Amp.). What is the amount of energy as the current changes from 0 to 4 Amp?

Solution: It is given that

$$i = 3 \sin t^2 \quad \text{and} \quad L = 3.0 \text{ H}$$

$$\text{Induced emf } (E_{\text{ind}}) = -L\frac{\partial i}{\partial t}$$

Work done during a small time interval dt is obtained as

$$\frac{\partial W}{\partial t} = E_{\text{ind}}.i$$

Solving it, we get

$$W = -\int_0^t E_{\text{ind}}\, i\, dt \quad \text{or}$$

$$= -\int_0^t \left(L\frac{di}{dt}\right) i\, dt \quad \text{or}$$

$$= -\int_{i=0}^{t} Li\, di \quad \text{or}$$

$$= L\frac{i^2}{2}\Big|_0^4$$

Finally,

$$W = \frac{3}{2}(4^2 - 0^2) \quad \text{or}$$

$$W = 24 \text{ Joule}$$

UNSOLVED QUESTIONS

4.1 Two small-sized identical conducting spheres have charges of 3.0×10^{-9} Coulomb and -0.6×10^{-9} Coulombs. These are placed 3 cms apart. Find the force between them. If these are brought into contact and then separated by 4 cms, what will be the force between them?

4.2 Obtain the ratio of electrostatic force of repulsion and gravitational force of attraction of two electrons. The electron's charge = 1.602×10^{-19} Coulomb, electron mass = 9.11×10^{-28} gm and gravitational constant $k = 6.66 \times 10^{-8}$ dyne-cm^{2}g^{-2}. (Hint: Assume that Newton's law of gravitation will work.)

4.3 Three concentrated charges of 0.5 μC each are located at the vertices of an equilateral triangle having 8 cms side. Obtain the magnitude and direction of the force on any one charge due to the remaining two charges.

4.4 An infinite sheet is lying in the plane $x - 3y + 4z = 4$. Its surface charge q is $12\varepsilon_0\text{cm}^{-2}$. Derive an expression for the field intensity on the side containing the origin.

4.5 Positive charges are uniformly distributed along a 1-m long line. The charge density is $2(\mu\text{C/m})$. Find the potential at a point P, which is 25 cm away from this line and opposite to the line's centre.

4.6 Three equal point charges of $+q$ each are required to be held in equilibrium at the corners of an equilateral triangle. For this purpose, it is proposed that one places another point charge at the centre of the triangle. Find the charge of this point charge in terms of $+q$.

SUMMARY

1. Materials are categorized into metal and insulator based on conductivity.
2. Inside a perfect conductor, the electrostatic field is zero.
3. In metals, conductivity $\sigma >> 1$.
4. In insulators, conductivity $\sigma << 1$.
5. Dielectric materials are classified into polar and non-polar dielectrics.

6. Polarization refers to a particular direction assumed by the electric dipole material of a dielectric material in an applied electric field.

7. Electric polarization of a material is electric dipole moment per unit volume.

8. Resistance "R" is given as $\rho l/A$, where ρ, l and A are resistivity, length and cross-sectional area of conductor, respectively.

9. Conductivity "σ" is the reciprocal of resistively "ρ."

10. Capacitance defines a capacitor. Its unit is Farads. Capacitance is the ratio of charge magnitude on one of its plates to the difference of potential between them.

11. Capacitors are of various types like parallel plate capacitor, spherical capacitor, coaxial capacitor, etc.

12. Inductance defines an inductor. Inductance is the ratio of total magnetic flux linkage to the current. Its unit is Henry.

13. In a single coil, self-inductance remains. If two current-carrying coils are put in closed vicinity to each other, then mutual inductance should also be considered.

14. Inductors are of various types like long solenoid, toroid, etc.

15. In electromagnetics, materials and medium are divided into two categories, viz. conductors and dielectrics or insulators.

16. Loss tangent or dissipation factor categorizes a good conductor, good dielectric and quasi conductors based on their conduction current density and displacement current density.

17. Changing electric field produces a magnetic field.

18. In addition, a changing magnetic field produces an electric field.

19. The general form of a wave equation is given as $\nabla^2\vec{E} = \mu\varepsilon\left(\partial^2\vec{E}/\partial t^2\right)$.

20. Electromagnetic wave propagation involves electric and magnetic fields with more than one component.

21. Magnetic susceptibility is a dimensionless quantity and is defined as magnetization per unit magnetic field intensity.

22. Magnetic dipole moment is in the direction of a right-hand screw when it is moved in the current's direction in a current loop.

23. Magnetic dipole moment has the same orientation as the magnetic field.

24. On the basis of magnetic behaviour, materials are classified as diamagnetic, paramagnetic, ferromagnetic, anti-ferromagnetic and ferromagnetic.

25. Magnetic behaviour is mainly due to "electron orbital motion," "electron-spin" and "nuclear-spin."

MULTIPLE-CHOICE QUESTIONS

1. Metals have their conductivity "σ" ________.
(a) >>1 (b) <<1 (c) 0 (d) none of these

2. Insulators have their conductivity "σ" ________.
(a) >>1 (b) <<1 (c) 0 (d) none of these

3. Electrostatic field is ________ a perfect conductor.
(a) same inside (b) infinite inside (c) zero inside (d) none of these

4. Potential is ________ a perfect conductor.
(a) same inside (b) infinite inside (c) zero inside (d) none of these

5. As per Gauss' Law, charge density inside a perfect conductor is zero if E is ________.
(a) positive (b) negative (c) unity (d) zero

6. Conductivity of dielectric is ________.
(a) low (b) high (c) both (a) and (b) (d) cannot say

7. Which of the following is not a non-polar dielectric?
(a) water (b) oxygen (c) CO_2 (d) alcohol

8. The direction of electric dipole moment is ________ applied electric field.
(a) orthogonal (b) contrary (c) same (d) none of these

9. Electric polarization of a material is electric dipole moment per unit ________.
(a) length (b) area (c) volume (d) none of these

10. Capacitor stores energy in ________ field.
(a) electric (b) magnetic (c) gravity (d) none of these

11. Inductor stores energy in ________ field.
(a) electric (b) magnetic (c) gravity (d) none of these

12. Which of the following is not an inductor?
(a) toroid (b) transmission line (c) solenoid (d) none of these

13. Magnetic field at any point inside a long solenoid is ________.
(a) zero (b) infinity (c) constant (d) none of these

14. A toroid is________ solenoid.
(a) a long (b) a small (c) not a (d) none of these

15. For a medium to be a quasi-conductor, the dissipation factor should be ________.
(a) infinity (b) zero (c) unity (d) none of these

16. Loss tangent is the same as ________.
(a) mutual inductance (b) resistivity
(c) conductivity (d) dissipation factor

17. Loss tangent is given as ________.

(a) $\sigma/\omega\varepsilon$ (b) $\omega/\sigma\varepsilon$ (c) $\omega\varepsilon/\sigma$ (d) $\sigma\varepsilon/\omega$

18. Loss tangent approximates ________.

(a) attenuation constant (b) phase constant
(c) intrinsic impedance (d) all of these

19. The general form of a wave equation is expressed as ________.

(a) $\nabla^2\vec{E} = \mu\varepsilon\dfrac{\partial^2\vec{E}}{\partial t^2}$ (b) $\nabla^2\vec{E} = \mu\varepsilon\dfrac{\partial\vec{E}}{\partial t}$ (c) $\nabla^2\vec{E} = \dfrac{\mu}{\varepsilon}\dfrac{\partial^2\vec{E}}{\partial t^2}$ (d) $\nabla^2\vec{E} = \dfrac{\mu}{\varepsilon}\dfrac{\partial\vec{E}}{\partial t}$

20. The combined effect of electric dipoles is seen in ________.

(a) polarization (b) magnetization (c) both (a) and (b) (d) none of these

21. The combined effect of magnetic dipoles is seen in ________.

(a) polarization (b) magnetization (c) both (a) and (b) (d) none of these

22. Permeability of vacuum is given as ________.

(a) 4×10^{-7} H/m (b) 14×10^{-7} H/m (c) 4×10^{-9} H/m (d) 14×10^{-9} H/m

23. Magnetic susceptibility is ________.

(a) unitless (b) dimensionless (c) both (a) and (b) (d) none of these

24. Dipole moment is ________ a right-hand screw, when it is moved in the current's direction in a current loop.

(a) orthogonal to the direction of (b) opposite to the direction of
(c) in the direction of (d) none of these

25. On the basis of magnetic behaviour, materials are classified into ________ categories.

(a) 2 (b) 3 (c) 4 (d) 5

ANSWERS TO MULTIPLE-CHOICE QUESTIONS

(1) a; (2) b; (3) c; (4) a; (5) d; (6) a; (7) b; (8) c; (9) b;
(10) a; (11) b; (12) d; (13) c; (14) a; (15) c; (16) d; (17) a; (18) d;
(19) a; (20) a; (21) b; (22) a; (23) b; (24) c; (25) d;

Electrostatics

5

5.1 INTRODUCTION

Discoveries by Coulomb, Ohm, Faraday and Kirchhoff have greatly contributed towards electric circuit theory. Maxwell's equation comprises laws of electricity and magnetism. Maxwell's equation involves magnetic intensity density, electric field intensity, magnetic intensity, current density, charge density, etc.

There are two fundamental laws:

(i) Coulomb's Law
(ii) Gauss's Law

Coulomb's Law deals with electric field due to any charge configuration, whereas it is easier to use Gauss's Law, when charge distribution is symmetrical.

5.2 COULOMB'S LAW AND FIELD INTENSITY

Force F exists between two point charges q_1 and q_2 as per Coulomb's Law along the line joining them. This force is inversely proportional to the square of the distance R between them. Further, this force is directly proportional to the product $q_1 q_2$ of the charges.

In Figure 5.1, force at point P as given by Coulomb's Law is

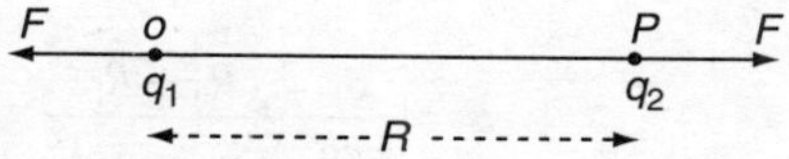

F O P F
q_1 q_2
R

Figure 5.1 *Coulomb's Law*

$$\vec{F} = \frac{1}{4\pi\varepsilon_0} \frac{q_1 q_2}{R^2} \vec{a}_R$$

where $\varepsilon_0 = 8.854 \times 10^{-12}$ F/m, q_1 and q_2 are in Coulombs, distance R in meters and force F in Newtons. Unit vector $\vec{a}_R$ is directed from charge q_1 towards charge q_2. The electric field intensity is also known as electric field strength. It is denoted by $\vec{E}$. Electric field intensity $\vec{E}$ is given as force per unit charge. Charge is placed in an electric field. Thus, electric field intensity $\vec{E}$

$$\vec{E} = \lim_{q \to 0} \frac{\vec{F}}{q} \quad \text{or}$$

$$\vec{E} = \frac{\vec{F}}{q}$$

Because of charge q_1, electric field intensity $\vec{E}$ on charge q_2 is obtained as

$$\vec{E} = \lim_{q \to 0} \frac{\vec{F}}{q_2}$$

or simply

$$\vec{E} = \frac{1}{4\pi\varepsilon_0} \frac{q_1}{R^2} \vec{a}_R$$

Electric field intensity is expressed in Newton per Coulomb (N/C) or alternatively in Volt per Meter (V/M).

In Figure 5.2, the electric field around a point charge q_1, is shown.

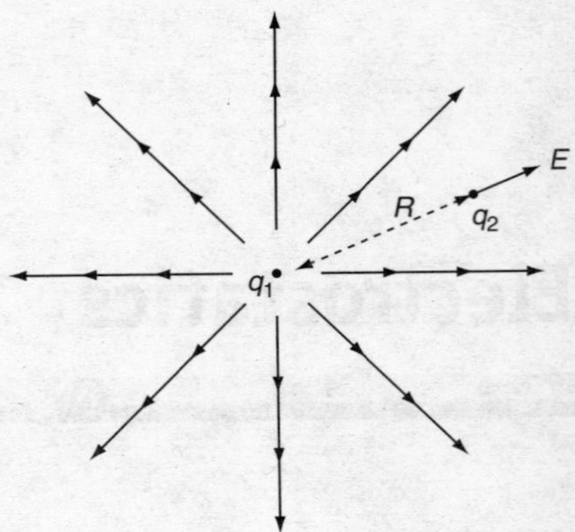

Figure 5.2 *Field around a point charge q_1*

Electric field intensity at a point P at a distance R_o from the origin due to charge q_1 located at a distance R_1 from the origin is depicted in Figure 5.3.

This electric field at P due to q_1 is obtained as

$$\vec{E} = \frac{q_1}{4\pi\varepsilon_0 \left|\vec{R}\right|^2} \vec{a}_R \quad \text{or}$$

$$\vec{E} = \frac{q_1}{4\pi\varepsilon_o \left|\vec{R}\right|^2} \frac{\vec{R}}{\left|\vec{R}\right|} \quad \text{or}$$

$$\vec{E} = \frac{q_1}{4\pi\varepsilon_o} \frac{\vec{R}}{\left|\vec{R}\right|^3} \quad \text{or}$$

$$\vec{E} = \frac{q_1}{4\pi\varepsilon_o} \frac{(\vec{R}_0 - \vec{R}_1)}{\left|(\vec{R}_0 - \vec{R}_1)\right|^3}$$

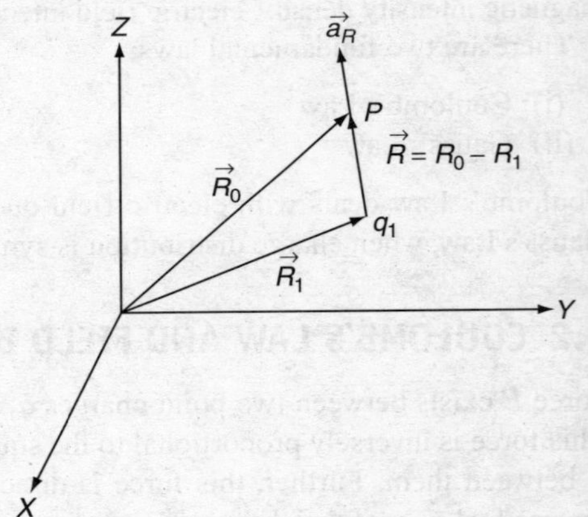

Figure 5.3 *Obtaining electric field intensity at point P due to charge q_1*

5.3 ELECTRIC DIPOLE

Electric dipole is pair of equal and opposite point charges. This charge pair is separated by a very small distance. Dipole moment $\vec{p}$ measures the strength of the electric dipole (Figure 5.4).

The magnitude of the dipole moment is given as

$$\left|\vec{p}\right| = qd \qquad \text{(in Coulomb Meter)}$$

Direction of $\vec{p}$ is from negative charge to the positive charge.

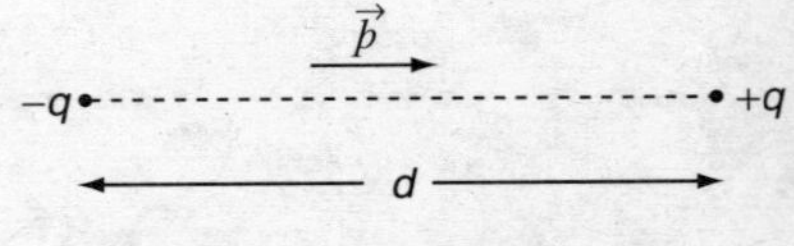

Figure 5.4 *Electric dipole*

5.4 ELECTRIC FIELD INTENSITY ON AXIAL LINE OF ELECTRIC DIPOLE

Figure 5.5 depicts an electric dipole. Electric intensity E at a point P on the axial line of the dipole is obtained from two electric field intensities E_1 and E_2.

E_1 electric intensity at P due to point charge $+q$ is given as

$$E_1 = \frac{q_1}{4\pi\varepsilon_0} \frac{q}{(R+d)^2}$$

E_2 electric intensity at P due to point charge $-q$ is given as

$$E_2 = \frac{q_1}{4\pi\varepsilon_0} \frac{q}{(R-d)^2}$$

Electric field intensities E_1 and E_2 are acting in opposite directions. Let $E_2 < E_1$. E_1 and E_2 are in a line. Resultant E at P is the difference of two electric field intensities E_1 and E_2. It is obtained as below:

$$E = E_2 - E_1$$

where E is the resultant electric field intensity due to E_1 and E_2 or

$$= \frac{1}{4\pi\varepsilon_0} \frac{q}{(R-d)^2} - \frac{1}{4\pi\varepsilon_0} \frac{q}{(R+d)^2} \quad \text{or}$$

$$= \frac{q}{4\pi\varepsilon_0} \left[\frac{1}{(R-d)^2} - \frac{1}{(R+d)^2} \right] \quad \text{or}$$

$$= \frac{q}{4\pi\varepsilon_0} \left[\frac{4dR}{(R^2-d^2)^2} \right]$$

Further,

$$|\vec{p}| = q.2d$$

Thus,

$$E = \frac{|\vec{p}|}{4\pi\varepsilon_0} \frac{2R}{(R^2-q^2)^2}$$

Suppose $2d << R$, then

$$E = \frac{2|\vec{p}|}{4\pi\varepsilon_0 R^3}$$

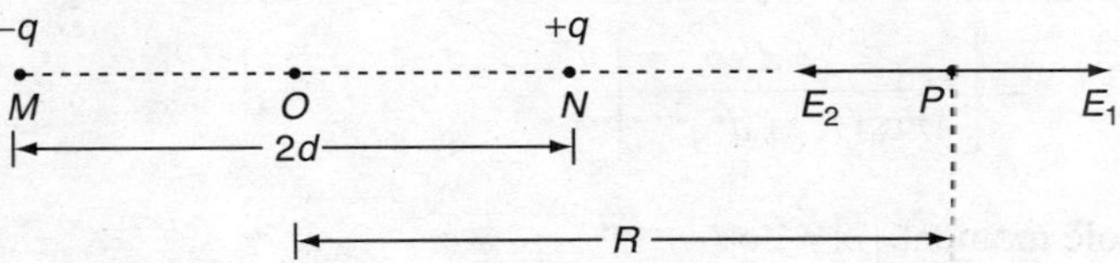

Figure 5.5 *Electric field intensity*

5.5 FIELD INTENSITY ON EQUATORIAL LINE OF ELECTRIC DIPOLE

Electric field intensity E at a point P on the equatorial line of the dipole is depicted in Figure 5.6.

Let E_1 be the electric intensity at P due to charge $+q$. It is given as

$$E_1 = \frac{1}{4\pi\varepsilon_0}\frac{q}{(PN)^2}$$

In $\Delta\, PON$

$$PN^2 = PO^2 + ON^2$$
$$= R^2 + d^2$$

Thus,

$$E_1 = \frac{1}{4\pi\varepsilon_0}\frac{q}{(R^2+d^2)}$$

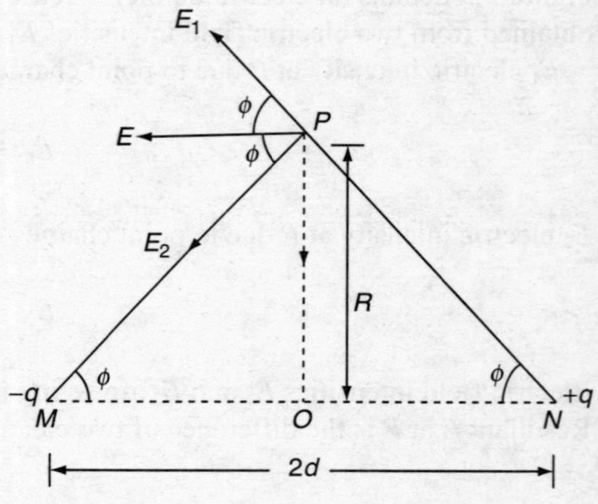

Figure 5.6 *Electric field intensity on equatorial line of dipole*

Let E_2 be the electric intensity at P due to charge $-q$. It is given as

$$E_2 = \frac{1}{4\pi\varepsilon_0}\frac{q}{(PM)^2} \quad \text{or}$$

$$= \frac{1}{4\pi\varepsilon_0}\frac{q}{(R^2+d^2)}$$

In $\Delta\, POM$, $\angle PMO = \Phi$ and

$$\cos\Phi = \frac{d}{(R^2+d^2)^{1/2}}$$

The resultant intensity E at P in such cases is obtained as

$$E = E_1 \cos\Phi + E_2 \cos\Phi \quad \text{or}$$

$$= (E_1 + E_2)\cos\Phi \quad \text{or}$$

$$= \left[\frac{1}{4\pi\varepsilon_o}\frac{1}{(R^2+d^2)} + \frac{1}{4\pi\pi_0}\frac{q}{(R^2+d^2)}\right]\frac{d}{(R^2+d^2)^{1/2}} \quad \text{or}$$

$$= \left[\frac{2q.d}{4\pi\varepsilon_0(R^2+d^2)^{3/2}}\right]$$

Magnitude of the dipole moment $|\vec{p}| = 2qd$

Thus,

$$E = \frac{|\vec{p}|}{4\pi\varepsilon_0 (R^2 + d^2)^{3/2}}$$

Consider $2d << R$ (dipole is short).

Then,

$$E = \frac{|\vec{p}|}{4\pi\varepsilon_0 R^3}$$

5.6 ELECTRIC FIELD DUE TO CONTINUOUS CHARGE DISTRIBUTIONS

Figure 5.7 shows various types of charge distributions.

Charge element dq in various cases are given as

$$dq = \int_v \rho_L dl \quad \text{(for line charge)}$$

$$dq = \int_s \rho_s dl \quad \text{(for surface charge)} \quad \text{and}$$

$$dq = \int_v \rho_v dl \quad \text{(for volume charge)}$$

The electric field intensity E due to charge distribution in each case is given as

$$\vec{E} = \int \frac{1}{4\pi\varepsilon_0} \frac{\rho_L dl}{R^2} \vec{a}_R \quad \text{(for line charge)}$$

$$\vec{E} = \int \frac{1}{4\pi\varepsilon_0} \frac{\rho_S ds}{R^2} \vec{a}_R \quad \text{(for surface charge)} \quad \text{and}$$

$$\vec{E} = \int \frac{1}{4\pi\varepsilon_0} \frac{\rho_V dv}{R^2} \vec{a}_R \quad \text{(for volume charge)}$$

Total electric field intensity $\vec{E}_{\text{total}}$ due to the presence of these three types of charge distributions is simply the vector sum of three types of electric field intensity. It is given as

$$\vec{E}_{\text{total}} = \frac{1}{4\pi\varepsilon_o}\left[\int \frac{\rho_L dl}{4\pi\varepsilon_o R^2}\vec{a}_R + \int \frac{\rho_S ds}{4\pi\varepsilon_o R^2}\vec{a}_R + \int \frac{\rho_V dv}{4\pi\varepsilon_o R^2}\vec{a}_R\right]$$

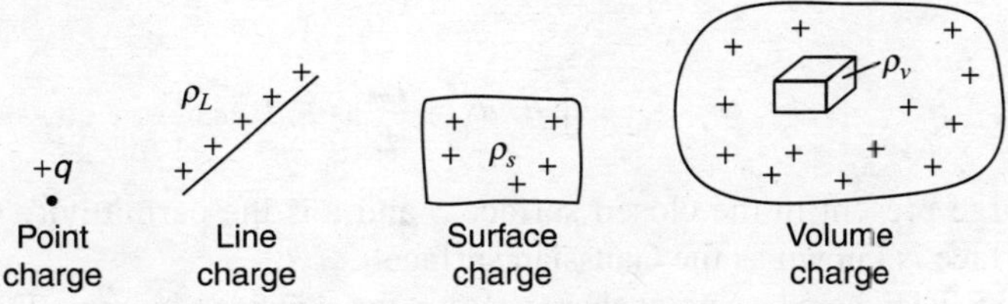

Figure 5.7 *Types of charge distributions*

5.7 ELECTRIC FLUX DENSITY

Electric field intensity depends on medium, where charge is placed. Electric flux density D is a new vector field, which is independent of the medium. It is given as

$$D = \varepsilon_0 E \qquad \text{(valid only in free space)}$$

5.8 ELECTRIC POTENTIAL

Electric potential V_p at a point P in an electric field is given as

$$\text{Electric potential } V_p = -\int_{\infty}^{P} E.dl \text{ (line integral from } \infty \text{ to point } P\text{).}$$

Electric potential difference between two points M and N in an electric field is given as

$$\text{Electric potential difference } (V_N - V_M) = -\int_{M}^{N} E \cdot dl \text{ (line integral from path } N \text{ to M).}$$

An equipotential surface has constant electric potential over its entire surface. Let a charge q be moved from M to N. Then

$$V_N - V_M = \frac{W_{MN}}{q}$$

where W_{MN} is the work done in moving the charge.

If $\qquad V_N = V_M \qquad$ then $\qquad W_{MN} = 0$

Please note that two equipotential surfaces cannot intersect each other. It is because at the point of their intersection, there should be two directions of E, which is not possible.

The electric potential energy V of a system of two charges as shown in Figure 5.8 is given as

$$V = \frac{1}{4\pi\varepsilon_o}\frac{q_1 q_2}{R}$$

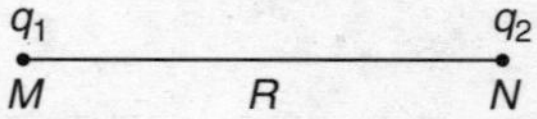

q_1 q_2
M R N

Figure 5.8 *Two-charge system*

5.9 GAUSS'S LAW

According to Gauss's law, the total electric flux Φ through a closed surface is simply the total charge enclosed by that surface.

Thus, the total electric flux

$$\Phi = \oint \vec{D}.\vec{ds} = q_{enc} \qquad \text{or}$$

$$\oint \vec{E} \cdot \vec{ds} = \frac{q_{enc}}{E}$$

where q_{enc} is the charge present in the closed surface S and ε is the permittivity of the medium. This imaginary closed surface is known as the Gaussian surface.

As shown in Figure 5.9, let there be a point charge $+q$ at point A inside a volume. This volume is enclosed by an arbitrary closed surface S, producing an electric field E. This electric flux density is $D = \varepsilon E$.

Consider a small area ds on this surface around a point B. Let the electric field at point B by charge $+q$ be E. Area vector ds is directing outwards normal to area element ds. E is along AB directing outwards. Electric flux $d\Phi$ through the area element ds is given as

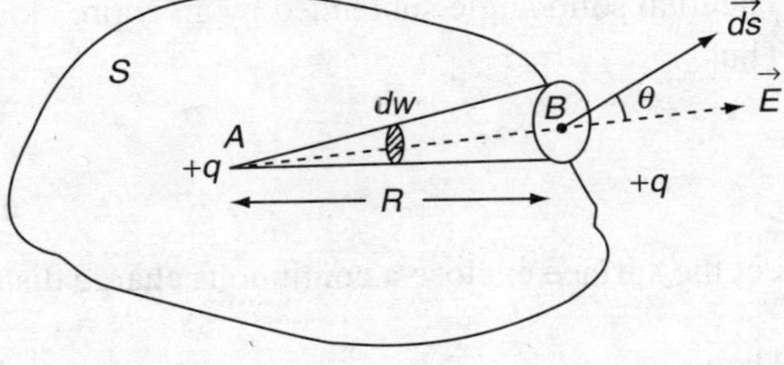

Figure 5.9 *Electric field in a volume enclosed by a surface*

$$d\Phi = \vec{D} \cdot \vec{ds}$$

or simply

$$d\Phi = \left|\vec{D}\right|\left|\vec{ds}\right| \cos\theta$$

Here, θ is the angle between $\vec{E}$ and $\vec{ds}$.

Further,
$$\left|\vec{E}\right| = \frac{1}{4\pi\varepsilon_0}\frac{q_{enc}}{R^2} \quad \text{(for free space)}$$

Also,
$$\left|\vec{D}\right| = \frac{1}{4\pi}\frac{q_{enc}}{R^2}$$

Thus,
$$d\Phi = \frac{1}{4\pi}\frac{q_{enc}}{R^2} ds\cos\theta$$

Total electric flux Φ (through the entire closed surface S) is obtained as

$$\Phi = \int d\Phi$$

On substitution, we obtain

$$\Phi = \int \frac{1}{4\pi}\frac{q_{enc}}{R^2} ds \cos\theta$$

In the case of a small cone, solid angle

$$dw = \frac{ds\cos\theta}{R^2} \quad \text{(by geometry)}$$

On substitution, we obtain

$$\Phi = \int \frac{q_{enc}}{4\pi}\frac{ds\cos\theta}{R^2} \quad \text{or}$$

$$\Phi = \int \frac{q_{enc}}{4\pi} \cdot dw \quad \text{or}$$

$$\Phi = \frac{q_{enc}}{4\pi}\int dw$$

The total solid angle subtended by an entire closed surface S at A is logically 4π (or 360°). Thus,

$$\Phi = \frac{q_{enc}}{4\pi} \cdot 4\pi \quad \text{or}$$

$$\Phi = q_{enc}$$

Let the surface enclose a continuous charge distribution.

Then $$q_{enc} = \int_v \rho_v dv$$

where ρ_v is the volume charge density. The total flux is obtained as

$$\oint_S \vec{D}\, \vec{ds} = \int_v \rho_v dv \quad \text{or}$$

$$\int_v (\nabla \cdot \vec{D})\, dv = \int_v \rho_v dv$$

Thus, $\nabla \cdot \vec{D} = \rho_v$ (Maxwell's first equation). This is the first of four Maxwell's equations.

5.9.1 Application of Gauss's Law

Gauss's Law can be applied to find out the electric field in problems where the charge distribution is symmetrical. Charge distribution symmetry may be of spherical, cylindrical or planar types. Gauss's Law implementation requires a Gaussian surface. Electric flux density D is normal or tangential to such Gaussian surface. Let us consider few cases of application of Gauss's Law.

5.9.1.1 Electric Field Due to a Point Charge

As shown in Figure 5.10, a point charge q is placed at the origin. For an electric field, a spherical surface is shown. The magnitude of electric field E remains the same at every point on the surface of the sphere. The direction of electric field is radially outwards. Such a spherical surface having its centre at the origin is called "Gaussian Surface" for all such cases.

The magnitude of electric flux density $\vec{D}$ is given as

$$\vec{D} = D_R \vec{a}_R$$

where $\vec{a}_R$ is the unit vector of $\vec{D}$ normal to the Gaussian surface. The total flux through the surface is obtained as

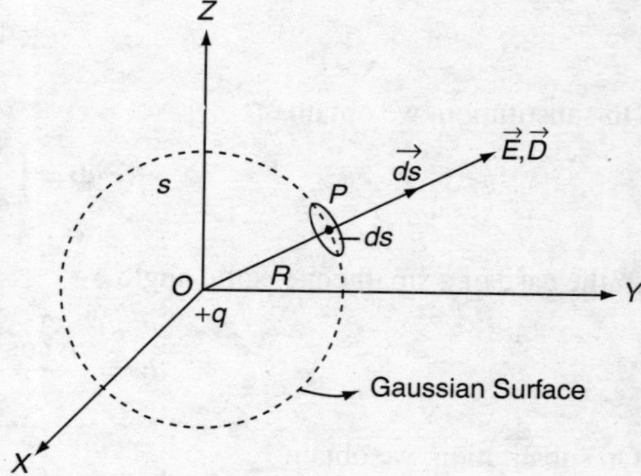

Figure 5.10 *Point charge and spherical Gaussian surface*

$$\Phi = \oint d\Phi \quad \text{or}$$

$$= \oint \vec{D} \cdot \vec{ds} \quad \text{or}$$

$$= \oint D_R\, ds\, \cos 0° \quad \text{or}$$

$$= D_R \oint ds \quad \text{or}$$

$$= D_R \cdot 4\pi R^2$$

As per Gauss's Law,

$$\Phi = q_{enc}$$

Thus,

$$D_R 4\pi R^2 = q \quad \text{or}$$

$$D_R = \frac{q_{enc}}{4\pi R^2} \quad \text{or}$$

$$E = \frac{D_R}{\varepsilon_o} \quad \text{(for free space)}$$

Electric field intensity $\vec{E}$ in this case is given as

$$\vec{E} = \frac{1}{4\pi\varepsilon_o} \frac{q}{R^2} \vec{a}_R \quad \text{(in vector form)}$$

5.9.1.2 Electric Field Due to Infinite Line Charge

Consider an infinite line of charge. This infinite line of charge has uniform linear charge density ρ_L. For electric field $\vec{E}$ at a point *P* a cylindrical surface is delineated. Let this cylindrical surface have radius *R* and length *l*. Charge will be symmetrically distributed. The magnitude of electric field intensity will remain the same on this curved surface everywhere. Its direction will be radially outwards. Such a cylindrical surface is known as "Gaussian surface" for all such cases (Figure 5.11).

The electric field at every point except circular caps (lower and upper) is perpendicular to the area vectors on these caps. Flux through caps is

$$\Phi = D \cdot \vec{ds} = \varepsilon_o \vec{E} \cdot \vec{ds} \quad \text{or}$$

$$= \varepsilon_o \left|\vec{E}\right| \left|\vec{ds}\right| \cos 90°$$

or simply

$$= 0$$

Electric flux through the entire cylindrical surface is obtained as

$$\Phi_2 = \oint \vec{D} \cdot \vec{ds} = \oint \left|\vec{D}\right| \left|\vec{ds}\right| \cos\theta \quad \text{or}$$

$$= \oint D_R ds \quad \text{or}$$

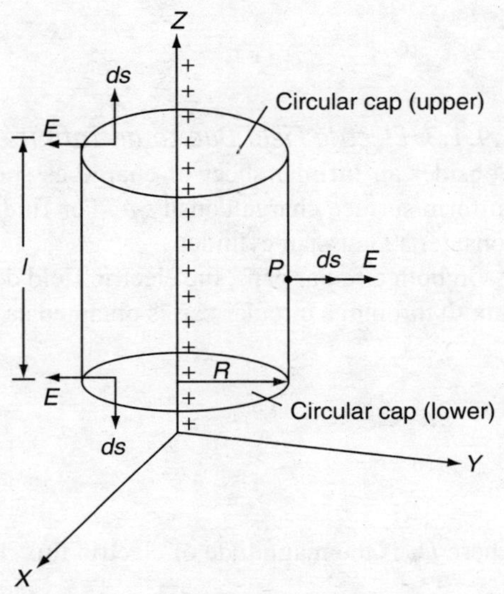

Figure 5.11 *Infinite line charge and cylindrical Gaussian surface*

$$= D_R \oint ds \quad \text{or}$$

$$= D_R\,(2\pi R_1)$$

The total outward flux

$$\Phi = \Phi_1 + \Phi_2 \quad \text{or}$$

$$= 0 + D_R\,(2\pi\,R_1) \quad \text{or}$$

$$= D_R\,2\pi\,R_1$$

The enclosed charge by the cylindrical surface is

$$q_{enc} = \rho_L\,L$$

As per Gauss's Law

$$\Phi = q_{enc}$$

Thus,

$$D_R\,2\pi\,R_1 = \rho_L$$

$$D_R = \frac{\rho_L}{2\pi R_1} \quad \text{or}$$

Further,

$$E = \frac{D_R}{\varepsilon_0} = \frac{\rho_L}{2\pi\,\varepsilon_0 R_1}$$

The electric field at point P in this case is

$$\vec{E} = \frac{\rho_L}{2\pi\varepsilon_0 R_1}\vec{a}_R$$

5.9.1.3 Electric Field Due to an Infinite Sheet of Charge

Consider an infinite sheet of charge as shown in Figure 5.12. Let this infinite sheet of charge have uniform surface charge density ρ_s. For finding electric field at a point P near this infinite sheet, let us consider a Gaussian cylinder.

On both circular caps, the electric field density $\vec{D}$ and area vector $\vec{ds}$ are mutually parallel. Electric flux Φ through a circular cap is obtained as

$$\Phi_1 = \vec{D} \cdot \vec{ds} \quad \text{or}$$

$$= D_R\,ds \cos 0° \quad \text{or}$$

$$= D_R\,ds$$

where D_R is the magnitude of electric flux density $\vec{D}$.

Also,

$$\vec{D} = D_R \vec{a}_Z$$

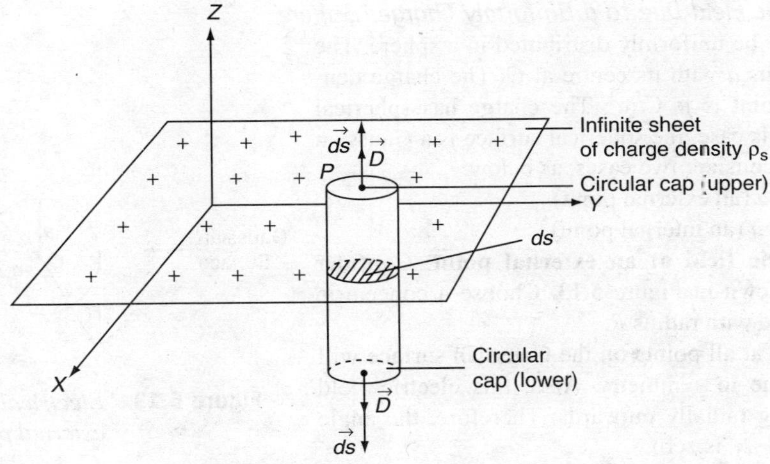

Figure 5.12 *Infinite sheet of charge*

Electric flux density $\vec{D}$ and area vector $\vec{ds}$ make an angle of 90° with each other on every point on the curved surface. Electric flux through the curved surface is obtained as

$$\Phi_2 = \oint \vec{D} \cdot \vec{ds} \quad \text{or}$$

$$= \oint |\vec{D}||\vec{ds}| \cos 90°$$

or simply

$$= 0$$

The total flux through the closed surface is

$$\Phi = \Phi_1 + \Phi_1 + \Phi_2$$

$$= 2\Phi_1 + \Phi_2$$

$$= 2D_R\, ds$$

The total charge enclosed by cylinder is $q_{enc} = \rho_s$ ds
As per Gauss's Law

$$\Phi = q_{enc} \quad \text{or}$$

$$2D_R\, ds = \rho_s \text{ ds} \quad \text{or}$$

$$D_R = \frac{\rho_S}{2}$$

Electric field intensity at a point P near the infinite sheet of charge as in this case is

$$\vec{E} = \frac{\rho_S}{2\varepsilon_o}\vec{a}_z \qquad \text{(for free space)}$$

5.9.1.4 Electric Field Due to a Uniformly Charged Sphere

Let a charge $+q$ be uniformly distributed in a sphere. The sphere has radius a with its centre at O. The charge density at every point is ρ_v C/m^3. The charge has spherical symmetry. In this case, the spherical surface is a Gaussian surface. Let us consider two cases, as below:

Case (1): $R > a$ (an external point)

Case (2): $R < a$ (an internal point)

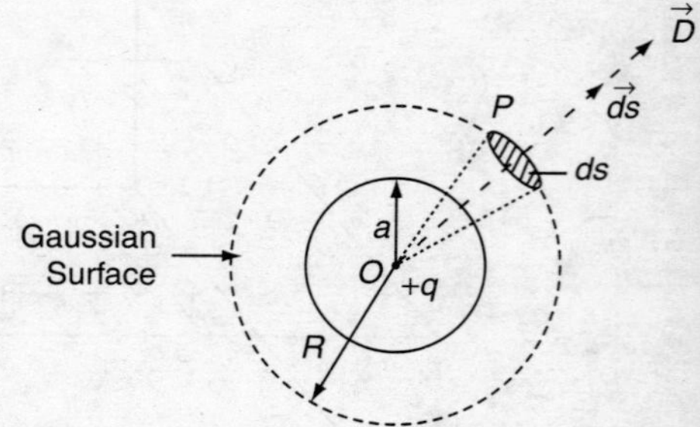

Figure 5.13 *Electrical field at an external point*

Case 1: Electric field at an external point: Consider a point P as shown in Figure 5.13. Choose a concentric spherical surface with radius R.

Electric field at all points on the Gaussian surface will be the same due to symmetry. Also, this electric field will be directing radially outwards. Therefore, the angle between $\vec{E}$ and $\vec{ds}$ is zero.

Total flux (outward) $$\Phi = \oint \vec{D} \cdot \vec{ds} \quad \text{or}$$

$$= \oint D_R ds \cos 0 \quad \text{or}$$

$$= D_R \oint ds \quad \text{or}$$

$$= D_R \, 4\pi R^2$$

The total charge enclosed by such a sphere q_{enc} is given as $\left(\rho_v \cdot \frac{4}{3}\pi a^3\right)$.
As per Gauss's Law,

$$\Phi = q_{enc}$$

Thus,

$$D_R 4\pi R^2 = \rho_v \frac{4}{3}\pi a^3 \quad \text{or}$$

$$D_R = \frac{\rho_v a^3}{3R^2} \quad \text{or}$$

$$E = \frac{D_R}{\varepsilon_o} = \frac{\rho_v a^3}{3R^2}$$

Electrical field intensity is

$$\vec{E} = \frac{\rho_v a^3}{3R^2} \vec{a}_R \qquad \text{(in vector form)}$$

Case 2: Electrical field at an internal point: Consider a point P inside a charged sphere. Choose a spherical surface of radius R through point P as shown in Figure 5.14.

The total electric flux through the Gaussian surface is

$$\Phi = \oint \vec{D} \cdot \vec{ds} \quad \text{or}$$

$$= D_R \oint ds \quad \text{or}$$

$$= D_R \, 4\pi R^2$$

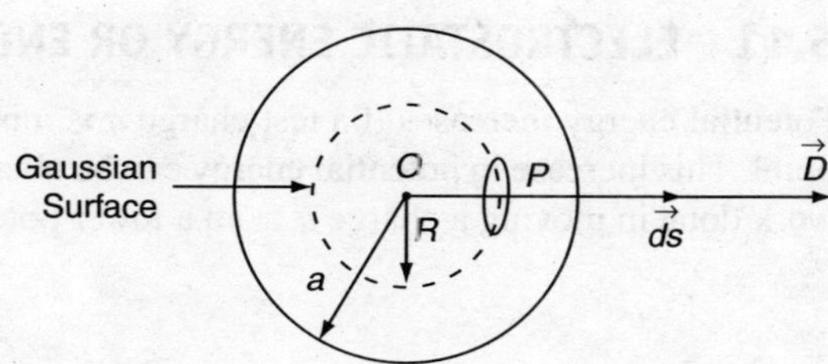

Figure 5.14 *Electric field at an internal point*

The total charge enclosed by the Gaussian surface q_{enc} is given as $\left(\rho_v \frac{4}{3}\pi R^3\right)$. As per Gauss's Law,

$$\Phi = q_{enc}$$

Thus,

$$D_R \, 4\pi R^2 = \rho_v \frac{4}{3}\pi R^3 \quad \text{or}$$

$$D_R = \frac{R}{3}\rho_v \quad \text{or}$$

$$\vec{D} = \frac{R}{3}\rho_v \vec{a}_R \quad \text{(in vector form)}$$

Electric field intensity

$$\vec{E} = \frac{R}{3\varepsilon_o}\rho_v \vec{a}_R$$

5.10 RELATION BETWEEN *E* AND *V*

Consider Figure 5.15. The potential difference between points *M* and *N* is path independent.
where $V_{MN} = -V_{NM}$ or $V_{MN} + V_{NM} = 0$.

Therefore, the net work done in moving a charge along a closed path is zero. It is given as

$$\oint \vec{E} \cdot \vec{dl} = 0$$

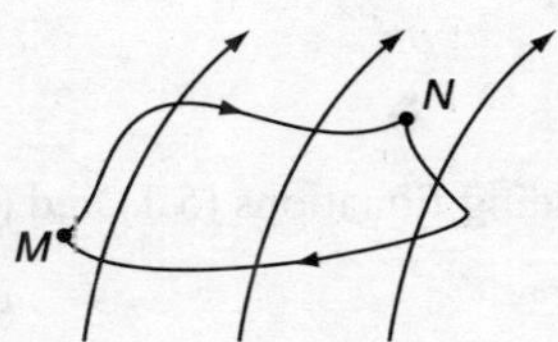

Figure 5.15 *Potential surface*

As per Stokes's Theorem

$$\oint \vec{E} \cdot \vec{dl} = \int (\nabla \times \vec{E}) \cdot \vec{ds} = 0 \quad \text{(in this case)}$$

or simply

$$\nabla \times \vec{E} = 0 \quad \text{(Maxwell's equation)}$$

The curl of the electrostatic field is zero. The electrostatic field is a conservative field as its curl is zero.

5.11 ELECTROSTATIC ENERGY OR ENERGY DENSITY

Potential energy increases if a test charge q is moved from a lower potential point to a higher potential point. This increase in potential energy can be obtained from external work done on the charge. External work done in moving a charge q from a lower potential to a higher potential is given as

$$W = qV \quad \text{or}$$

$$W = q\int E \cdot dl$$

Let us consider a system having three points as shown in Figure 5.16. The three point charges are q_1, q_2 and q_3. Assume that all these charges were initially kept in an empty space. First, q_1 is brought from space, then q_2 and lastly q_3. Initially, space is charge free. Electric field in space is zero. Thus, no work is required for transferring q_1 to position '1' from space due to the absence of electric field.

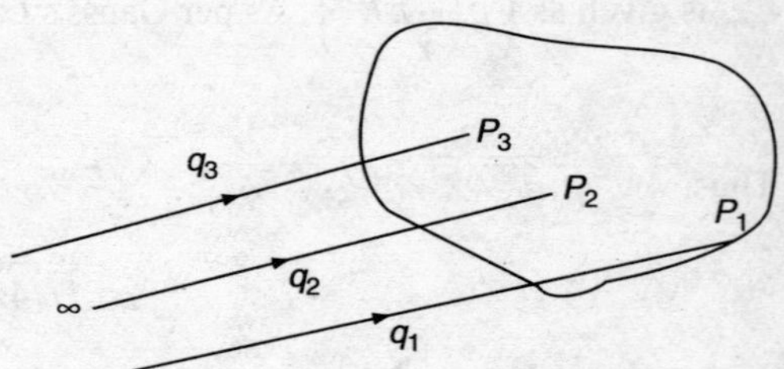

Figure 5.16 *Three point charges*

We get, $W_1 = 0$. Work done in transferring q_2 from space to position '2' is

$$W_2 = q_2V_{21}$$

where V_{21} is the potential difference at the location of q_2 due to q_1. Similarly, work done in bringing q_3 is given as

$$W_3 = (q_3V_{31} + q_3V_{32})$$

The total work done

$$W = (W_1 + W_2 + W_3) \quad \text{or}$$

$$W = (0 + q_2V_{21} + q_3V_{31} + q_3V_{32}) \tag{5.1}$$

If the three charges are brought in an order, which is the reverse of an earlier order, then the total work done is given as

$$W = (W_3 + W_2 + W_1) \quad \text{or}$$

$$W = (0 + q_2V_{23} + q_1V_{22} + q_1V_{13}) \tag{5.2}$$

Adding Equations (5.1) and (5.2), we get

$$2W = q_2V_{21} + q_3V_{31} + q_3V_{32} + q_2V_{23} + q_1V_{12} + q_1V_{13}$$

$$2W = q_1(V_{12} + V_{13}) + q_2(V_{21} + V_{23}) + q_3(V_{31} + V_{32})$$

The above total potential V_1 at q_1 due to q_2 and q_3 is $(V_{12} + V_{13})$. Similarly, V_2 and V_3 are obtained as

$$V_2 = (V_{21} + V_{23}) \quad \text{and} \quad V_3 = (V_{31} + V_{32})$$

Now, the total work is

$$W = \frac{1}{2}(q_1V_1 + q_2V_2 + q_3V_3)$$

For n point charges, the total work done is similarly obtained as

$$W = \frac{1}{2}(q_1V_1 + q_2V_2 + q_3V_3 \cdots q_nV_n) \quad \text{or}$$

$$= \frac{1}{2}\sum_{i=1}^{n} q_iV_i$$

Replace the summation above by integration, if the region has a continuous distribution as shown in Figure 5.17. Let the volume be V, charge density ρ_v (enclosed within a sphere) and sphere radius R. In this case, the total work done is obtained as

$$W = \frac{1}{2}\int_v e_v \cdot V\, dv \qquad \text{(summation replaced by integration) or}$$

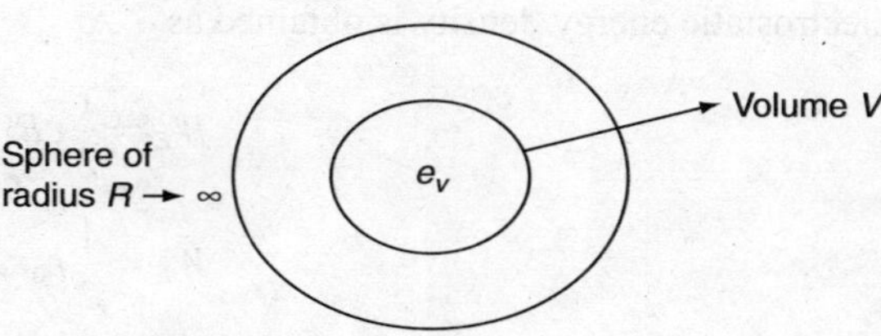

Figure 5.17 *Enclosed region with continuous charge distribution*

$$= \frac{1}{2}\int_v (\nabla \cdot \vec{D}) V dv$$

As per the Vector rule,

$$\nabla \cdot (VD) = (\nabla \cdot D)V + D \cdot (\nabla V)$$

Thus,

$$W = \frac{1}{2}\int_v (\nabla \cdot VD)\, dv - \frac{1}{2}\int (D \cdot \nabla V)\, dv \tag{5.3}$$

As per the Divergence theorem

$$\frac{1}{2}\oint_v (\nabla \cdot VD)\, dv = \frac{1}{2}\oint_S VD \cdot ds$$

For a very large radius R,

$$D \propto \frac{1}{R^2} \quad \text{and} \quad V \propto \frac{1}{R^2} \quad \text{or} \quad \text{Surface} \propto R^2$$

Thus,

$$\frac{1}{2}\oint_{R\to\infty} VD \cdot ds = 0$$

and electric field intensity, $\vec{E} = -\nabla V$.

Substituting all the above in Equation (5.3), we get work done W as

$$W = \frac{1}{2}\int_v (\vec{D} \cdot \nabla\vec{D})\, dv \quad \text{or}$$

$$W = \frac{1}{2}\int_v (\vec{D} \cdot \vec{E})\, dv$$

Also,

$$\vec{D} = \varepsilon\vec{E} \quad \text{or}$$

$$= \varepsilon_0\vec{E} \quad \text{(for free space)}$$

Therefore,

$$W = \frac{1}{2}\int \varepsilon E^2 dv$$

Electrostatic energy density is obtained as

$$W_E = \frac{1}{2}\varepsilon E^2 \quad (\text{in J/m}^3) \quad \text{or}$$

$$W_E = \frac{1}{2}\varepsilon_0\varepsilon_r E^2$$

5.12 CONTINUITY EQUATION

As shown in Figure 5.18, consider a closed volume. This closed volume contains charge with charge density ρ_v. The total charge (enclosed by volume) is given as

$$q_{enc} = \int_v \rho_v \cdot dv$$

Let, I_{out} be the net outward current flowing through the closed surface of the considered volume. I_{out} is the rate of decrease of charge within a considered given volume. I_{out} is given as

$$I_{out} = -\frac{\partial q_{enc}}{\partial t} \quad \text{or}$$

$$\oint_S J \cdot ds = -\frac{\partial}{\partial t}\int_v \rho_v dv$$

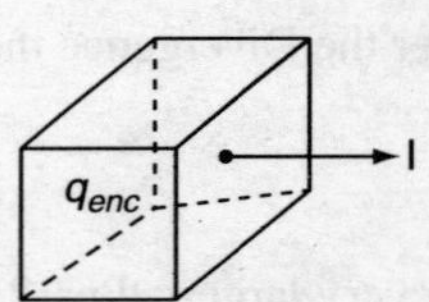

Figure 5.18 *Closed volume*

As per the Divergence theorem,

$$\oint_S J \cdot ds = \int_v (\nabla \cdot J) dv$$

Therefore,

$$\oint_S (\nabla \cdot J) dv = \int_v \frac{\partial \rho_v}{\partial t} dv \quad \text{or}$$

$$\nabla \cdot J = -\frac{\partial \rho_v}{\partial t} \quad \text{(continuity equation)}$$

5.13 POISSON'S AND LAPLACE'S EQUATION

Poisson and Laplace equations help in the study and analysis of classical field. In the case of a linear material medium, these Poisson and Laplace equations are derived easily from Gauss's Law. Applying Gauss's Law in a continuous distribution of charge of volume charge density ρ_v,

$$\nabla \cdot \vec{D} = \rho_v \quad \text{or}$$

$$\nabla \cdot \vec{E} = \frac{\rho_v}{\varepsilon_0} \quad \text{(for free space)}$$

where

$$\nabla = \frac{\partial}{\partial x}\vec{a}_x + \frac{\partial}{\partial y}\vec{a}_y + \frac{\partial}{\partial z}\vec{a}_z$$

Electric field is given as

$$\vec{E} = -\nabla V$$

On substitution, we get

$$\nabla \cdot (-\nabla V) = \frac{\rho_v}{\varepsilon_0} \quad \text{or}$$

$$\nabla \cdot \nabla V = -\frac{\rho_v}{\varepsilon_0} \quad \text{(Poisson's equation)}$$

or simply

$$\nabla^2 V = -\frac{\rho_v}{\varepsilon_0} \quad \text{(Poisson's equation)}$$

Charge density is zero in the case of charge-free regions. That is, $\rho_v = 0$. Therefore,

$$\nabla^2 V = 0 \text{ (Laplace equation)}$$

5.14 BOUNDARY CONDITION AT THE INTERFACE OF TWO MEDIUMS

The electric field must satisfy certain conditions as it passes from medium '1' into another medium '2'. These conditions pertain to the "interface" separating the two mediums. These conditions are termed "Boundary conditions". Mediums can be any, say a 'conductor' or a 'dielectric' or simply a 'free space'. Three different cases are considered and analysed here.

Case 1: Both mediums are dielectric as shown in Figure 5.19.

Consider a small rectangular loop as shown on the boundary in Figure 5.19. The electric field is conservative. For a conservative electric field, the closed path line integral is zero.

Thus, $\oint E \cdot dl = 0$ (Line integral around a closed path is zero.)

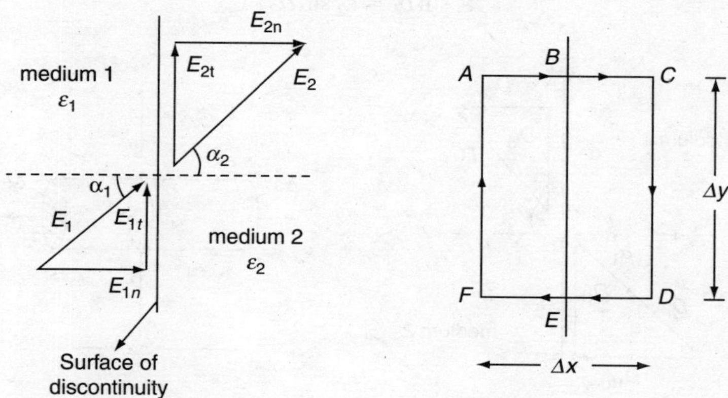

Figure 5.19 *Dielectric medium*

As shown in Figure 5.19, the total electric field $\vec{E}_1$ in medium '1' and $\vec{E}_2$ in medium '2' is given as

$$\vec{E}_1 = \vec{E}_{1n} + \vec{E}_{1t} \quad \text{and}$$

$$\vec{E}_2 = \vec{E}_{2n} + \vec{E}_{2t}$$

$$\therefore \quad \vec{E}_{1n}\frac{\Delta x}{2} + \vec{E}_{2n}\frac{\Delta x}{2} - \vec{E}_{2t}\Delta y - \vec{E}_{2n}\frac{\Delta x}{2} - \vec{E}_{1n}\frac{\Delta x}{2} + \vec{E}_{1t}\,\Delta y = 0$$

For the boundary, let $\Delta x \to 0$; thus,

$$-\vec{E}_{2t}\Delta y + \vec{E}_{1t}\Delta y = 0 \quad \text{or}$$

$$\vec{E}_{1t} = \vec{E}_{2t} \tag{5.4}$$

The tangential component of $\vec{E}$ on the two sides of the boundary are the same. Writing in terms of electric flux density (D), we get

$$\frac{D_{1t}}{\varepsilon_1} = \frac{D_{2t}}{\varepsilon_2}$$

As shown in Figure 5.20, consider a cylinder across medium '1' and medium '2'. The continuity of normal component of D is obtained as below.

We know that $\oint D \cdot ds = q_{enc}$. For $\nabla x \to 0$, we get

$$-D_{1n}\Delta s + D_{2n}\Delta s = \rho_s \Delta s$$

$$D_{2n} - D_{1n} = \rho_s \tag{5.5}$$

where ρ_s is the charge density (at boundary). Consider $\rho_s = 0$ (boundary is charge-free).

Thus,

$$D_{1n} = D_{2n} \quad \text{or}$$

$$D_1 \cos\alpha_1 = D_2 \cos\alpha_2 \tag{5.6}$$

Their tangential components

$$\varepsilon_{1t} = \varepsilon_{2t} \quad \text{or}$$

$$\varepsilon_1 \sin\alpha_1 = \varepsilon_2 \sin\alpha_2 \tag{5.7}$$

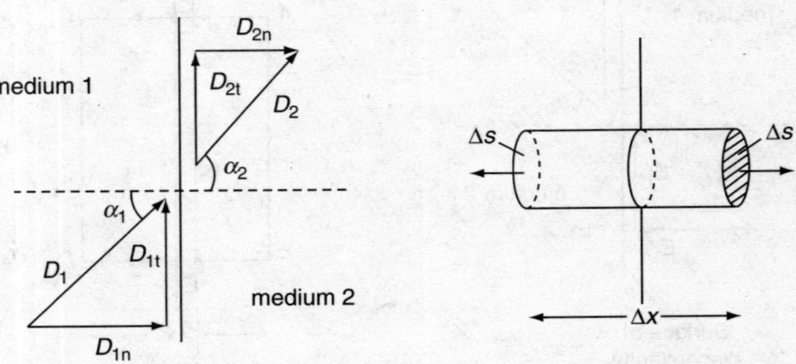

Figure 5.20 *Cylinder across mediums '1' and '2'*

Dividing Equation (5.6) by (5.5), we get

$$\frac{\varepsilon_1}{D_1}\frac{\sin\alpha_1}{\cos\alpha_1} = \frac{\varepsilon_2}{D_2}\frac{\sin\alpha_2}{\cos\alpha_2} \quad \text{or}$$

$$\frac{1}{\varepsilon_1}\tan\alpha_1 = \frac{1}{\varepsilon_2}\tan\alpha_2 \quad \text{or}$$

$$\frac{\tan\alpha_2}{\tan\alpha_1} = \frac{\varepsilon_2}{\varepsilon_1}$$

Case 2: Medium '1' is a conductor and medium '2' is a dielectric.

In a conductor

$$E_1 = 0 \quad \text{(electric field is zero)}$$

and also

$$D_1 = 0$$

but

$$\vec{E}_{2t} = \vec{E}_{1t}$$

Thus,

$$\vec{E}_{2t} = 0 \quad \text{and}$$

$$D_{2n} - D_{1n} = \rho_s$$

$$\Rightarrow D_{2n} = \rho_s$$

$$\Rightarrow E_{2n} = \frac{\rho_s}{\varepsilon_2}$$

Case 3: When medium '1' is a conductor and medium '2' is a free space.

In a conductor,

$$E_1 = 0 \quad \text{(electric field is zero)} \quad \text{and}$$

$$\varepsilon = \varepsilon_r\,\varepsilon_0$$

For medium '2'

$$\varepsilon = \varepsilon_0 \quad \text{(free space)}$$

Thus,

$$E_{2t} = \frac{D_{2t}}{\varepsilon_0} = 0 \quad \text{or}$$

$$D_{2n} = \varepsilon_0 E_{2n}$$

$$= \rho_s \quad \text{or}$$

$$E_{2n} = \frac{\rho_s}{\varepsilon_0}$$

5.15 UNIQUENESS THEOREM

It states that for a large class of boundary conditions, Poisson's equation gives a unique gradient of solution. Thus, if an electric field in electrostatics satisfies boundary conditions then the electric field is complete.

5.15.1 Proof of the Uniqueness Theorem

As per Poisson's equation

$$\nabla\cdot(\vec{E}) = \frac{\rho_v}{\epsilon_o}$$

where $\vec{E}$ is the electric field given as $-\nabla V$, and ρ_v is the volume charge density.

Thus,
$$\nabla\cdot(\nabla V) = -\frac{\rho_v}{\epsilon_o}$$

Let there be two solutions of electric potential viz. V_1 and V_2. Let the difference of two solutions $V_2 - V_1$ be given as ϕ. It is assumed that both V_1 and V_2 satisfy Poisson's equation. In this case,

$$\nabla\cdot(\nabla\phi) = 0$$

Using an identity, it can be written as

$$\nabla\cdot(\phi\nabla\phi) = (\nabla\phi)^2 + \nabla\cdot(\nabla\phi) \quad \text{or}$$

$$\nabla\cdot(\phi\nabla\phi) = (\nabla\phi)^2 [\text{as } \nabla\cdot(\nabla\phi) = 0]$$

Volume integral taken over all space that is satisfied by boundary conditions yields

$$\int_v \nabla\cdot(\phi\nabla\phi)\, d^3r$$

$$= \int_v (\nabla\phi)^2\, d^3r = \sum_i \int_{si} (\phi\nabla\phi)\cdot ds \quad \text{(by the Divergence Theorem)}$$

where S_i refers to boundary surfaces that are specified by the boundary conditions.

As $(\nabla\phi)^2 \geq 0$, $\nabla\phi$ must be zero everywhere. Also the surface integral will become zero. Thus, the gradient of a solution is unique as

$$\sum_i \int_{si} (\phi\nabla\phi)\cdot ds = 0$$

This is true for a Dirichlet boundary condition, a Neumann boundary condition, a modified Neumann boundary condition and for a combination of Dirichlet, Neumann and modified Neumann boundary conditions.

The Dirichlet boundary condition states that $\phi = 0$ at the boundary as $V_1 = V_2$. V is well defined at all boundary surfaces. As $V = 0$, the surface integral is zero. The Neumann boundary condition states that $\Delta\phi = 0$ as $\Delta V_1 = \Delta V_2$. ΔV is well defined at all boundary surfaces. As $\Delta\phi = 0$, the surface integral is zero.

Its modified Neumann boundary condition, boundaries are considered as conductors having charges that are well known.

If an appropriate boundary exists at infinity, then the boundary condition at infinity most often is the value of electric field. Thus, even if the domain of a solution is infinite, the Uniqueness Theorem still holds true.

5.16 METHOD OF IMAGES (MIRROR IMAGES)

It is a mathematical tool for solving differential equations. The method of images is used for solving problems related to electric field distribution of a charge in the vicinity of a conducting surface in electrostatics. The method of images can be used for solving problems related to magnetic field produced by a magnet in the vicinity of a superconducting surface in magnetostatics. The method of images is also used in fluid dynamics where the vortex moving near a wall is calculated with the help of an image vortex.

Figure 5.21 depicts field distribution between the surface and the charge of an infinite flat conducting surface.

Here, charges are opposite. It virtually becomes an electrical dipole. The vector of mirrored dipole has an opposite sign. The force between charge(s) and conducting surface is attractive in nature.

The method of images relies on the fact that the tangential component of electric field to conductor surface is zero. Some electric fields having $\nabla \times E = 0$ and $\nabla \cdot E = 0$ can be uniquely defined by its normal component over the surface confining a particular region. Thus, in another way, it is an extension of the Uniqueness Theorem. In magnetostatics, as per the method of images, the normal component of the magnetic field to superconductor is zero.

Similarly, consider a pair of a magnet and a superconducting surface. The magnetic field does not penetrate into or through a superconductor. If the magnetic dipole is held over the superconducting surface, then the field between the magnet and the superconducting surface will be the same as between the real magnet and the imaginary magnet, and also, the magnetic field will be symmetric (mirror image). However, in this case, the force between the magnet and the superconducting surface will be repulsive.

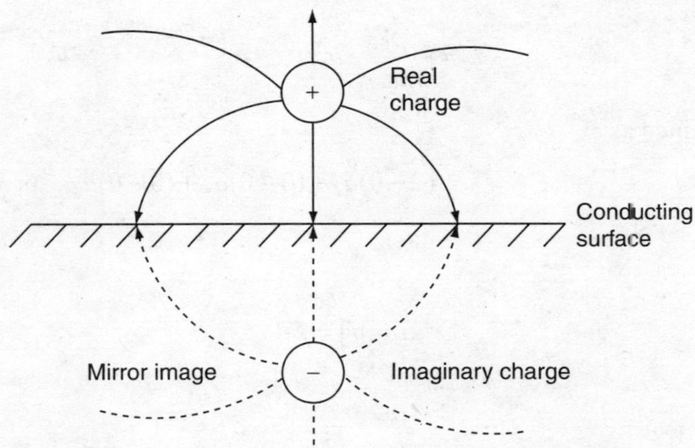

Figure 5.21 *Demonstrating method of images*

SOLVED QUESTIONS

5.1 Obtain the electric field intensity that will balance the gravitational force of an electron.

Solution:

Let the electric field intensity be E and the charge be e. The electric force F_e in this case will be

$$F_e = eE$$

Gravitational force

$$F_R = mg$$

Let m be the mass of an electron. Thus, for balance

$$F_e = F_R \quad \text{or}$$

$$eE = mg \quad \text{or}$$

$$E = \frac{mg}{e} \quad \text{or}$$

$$E = \frac{(9.1\times10^{-31}\,(\text{kg}))(9.8\,(\text{m/s}^2))}{(1.6\times10^{-19}\,(\text{C}))}$$

Thus,

$$E = 5.57\times10^{-12}\,(\text{N/C})$$

5.2 A sphere has charge density of 4.0 pC/m^3. Its volume is 0.2 m^3 and its centre is at (0, 0, 0). Find the electric field intensity at a point (2, 0, 0).

Solution:

Electric field intensity E due to a volume charge is given as

$$\vec{E} = \int \frac{\rho}{4\pi\varepsilon_0 r^2}\, dv \vec{a}_r \quad \text{or}$$

$$\vec{E} = \frac{\rho V}{4\pi\varepsilon_0 r^2}\, \vec{a}_r$$

In this case, $\vec{r}$ is obtained as

$$\vec{r} = (2-0)\vec{a}_x + (0-0)\vec{a}_y + (0-0)\vec{a}_z \quad \text{or}$$

$$\vec{r} = 2\vec{a}_x$$

Further,

$$r = |\vec{r}| = 2$$

Unit vector

$$\vec{a}_r = \frac{\vec{r}}{|\vec{r}|}$$

or simply

$$= \vec{a}_x$$

Thus,

$$\vec{E} = \frac{4\times10^{-12}\times0\cdot2\times9\times10^{9}}{(2)^2}\vec{a}_x \quad \text{or}$$

$$\vec{E} = 1.8\vec{a}_x(\text{mV/m}^3)$$

5.3 A charge with density $\rho_s = 1.5\ \text{nC/m}^2$ uniformly covers an infinite plane given as $-x+3y-6z=6$ on the side containing the origin. Find the electric field intensity.

Solution:

The given plane cuts x, y, z axes at (−6, 0, 0), (0, 2, 0) and (0, 0, −1). It is shown in Figure 5.22.

The electric field intensity in this case is given as

$$\vec{E} = \pm\frac{\rho_s}{2\varepsilon_0}\vec{a}_n \qquad \text{(infinite plane, uniform sheet)}$$

Let us consider the upper side unit normal vector to be positive. Thus, $\vec{a}_x$ will be negative at the origin side. The plane cuts the negative x-axis. Thus,

$$\vec{E} = \frac{1.5\times10^{9}}{2\varepsilon_0}\vec{a}_n(\text{V/m})$$

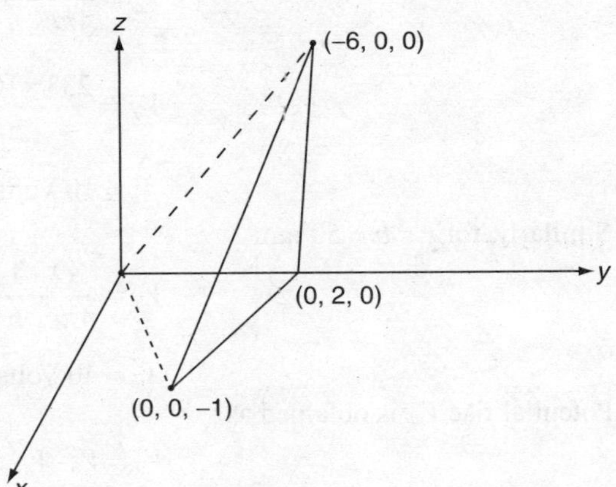

Figure 5.22 *Charge and infinite plane*

The unit normal vector $\vec{a}_n$ for a plane $Ax + By + Cz = D$ is obtained as

$$\vec{a}_n = \pm\frac{A\vec{a}_x + B\vec{a}_y + C\vec{a}_z}{\sqrt{A^2+B^2+C^2}}$$

Here, in this case

$$-\vec{a}_n = -\left[\frac{-\vec{a}_x + 3\vec{a}_y + 6\vec{a}_z}{\sqrt{46}}\right]$$

Thus,

$$\vec{E} = 84.72\left[\frac{\vec{a}_x - 3\vec{a}_y + 6\vec{a}_z}{\sqrt{46}}\right]$$

or simply

$$\vec{E} = (12.48\vec{a}_x - 37.5\vec{a}_y + 75\vec{a}_z)(\text{V/m})$$

5.4 An electric dipole $4.0\,\vec{a}_y$ nC-m is at point (0, 0, 0). Obtain the potential due to this electric dipole at point (1, $\pi/4$, $\pi/4$).

Solution:

The potential due to an electric dipole is given as

$$V = \frac{|\vec{p}|\cos\theta}{4\pi\varepsilon_0 r^2}$$

where $|\vec{p}| = 4.0\times10^{-9}$ C-m, $r = 1.0$, $\theta = \pi/4$ and

$$\frac{1}{4\pi\varepsilon_0} = 9\times10^9$$

Thus,

$$V = \frac{9\times10^9 \times 4\times10^{-9} \times 0.707}{(1)^2} \quad \text{or}$$

$$V = 25.44 \text{ Volt}$$

5.5 A position charge Q of 233 pC is kept in air. Draw equipotential surfaces at radii 200 mm and 50 mm. Find the potential rise between these two radii.

Solution:

Absolute potential V_r due to a point charge at a distance r is given as

$$V_r = \frac{Q}{4\pi\varepsilon_0 r} \qquad \text{(where } r \text{ is the radius)}$$

All the points on this sphere (of radius r) will have the same potential.

For $r = a = 200$ mm, the equipotential curve is shown in Figure 5.23. The absolute potential will be

$$V_a = \frac{Q}{4\pi\varepsilon_0}\frac{1}{a}$$

$$V_a = \frac{233\times10^{-12}\times9\times10^9}{200\times10^{-3}}$$

$$V_a = 10 \text{ Volts}$$

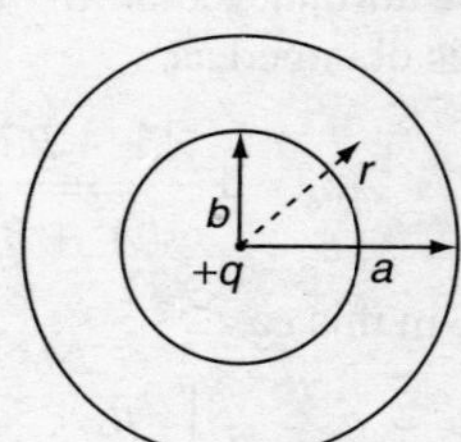

Figure 5.23 *Equipotential surface at radii a and b*

Similarly, for $r = b = 50$ mm

$$V_b = \frac{Q}{4\pi\varepsilon_0}\frac{1}{b}$$

$$V_b = 40 \text{ Volts}$$

Potential rise V_{ab} is obtained as

$$V_{ab} = V_b - V_a \quad \text{or}$$

$$= (40 - 10)$$

or simply

$$= 30 \text{ Volts}$$

5.6 Let an electric field E be

$$E = \frac{-12y}{x^2}\vec{a}_x + \frac{12y}{x}\vec{a}_y + 5\vec{a}_z \text{(V/m)}$$

Find the potential difference V_{AB} between two points A (−7, 2, 1) and B (4, 1, 2).

Solution:

We know that potential difference is

$$V_{AB} = \int_B^A \vec{E}\cdot\vec{dl} \quad \text{or}$$

$$= -\int_B^A \left(\frac{-12y}{x^2}dx + \frac{12y}{x}dy + 5dz\right) \quad \text{or}$$

$$= \int_{x=4,y=1}^{x=-7} \frac{12y}{x^2}dx + \int_{y=1,x=-7}^{y=2} \frac{-12}{x}dy + \int_{z=2}^{1} -10dz \quad \text{or}$$

$$= \int_4^{-7} \frac{12y}{x^2}dx + \frac{6}{7}\int_1^2 dy + \int_2^1 -5dz$$

or simply $= 16.42$ Volts

5.7 Find the flux density at a point A (6, 4, –5) due to
(a) A 40 mC point charge at (0, 0, 0) origin,
(b) A uniform line charge (ρ_l) of 40 mC/m^2 on the z-axis and
(c) A uniform charge density (ρ_s) of 30 mC/m^2 at x =7 (a plane).

Solution:
Flux density can be obtained from field intensity. Field intensity can be obtained at a point due to different charge distributions.
(a) At a point, the electric field due to a point charge is given as

$$\vec{E} = \frac{q}{4\pi\varepsilon_0 r^2}\vec{a}_r$$

Flux density at this point is given as

$$\vec{D} = \varepsilon_0\vec{E} = \frac{q}{4\pi r^2}\vec{a}_r$$

Unit vector

$$\vec{a}_r = \frac{(6\vec{a}_x + 4\vec{a}_y - 5\vec{a}_z)}{\sqrt{77}}$$

As

$$\vec{r} = (6\vec{a}_x + 4\vec{a}_y - 5\vec{a}_z) \qquad \text{(in this case)} \quad \text{and}$$

$$r = |\vec{r}| = \sqrt{6^2 + 4^2 + 5^2} = \sqrt{77}$$

Now, flux density is obtained as

$$\vec{D} = \frac{4\times10^{-3}}{4\pi\times77}\cdot\frac{(6\vec{a}_x + 4\vec{a}_y - 5\vec{a}_z)}{\sqrt{77}}$$

or simply

$$\vec{D} = (28.26\vec{a}_x + 18.84\vec{a}_y - 23.56\vec{a}_z)(\mu\text{C/m}^2)$$

(b) Electric field E at a point due to a line charge on the z-axis will be given as

$$\vec{E} = \frac{\rho_L}{2\pi\varepsilon_0 r}\vec{a}_r$$

Further, electric flux density

$$\vec{D} = \varepsilon_0 \vec{E} \quad \text{or}$$

$$= \frac{\rho_L}{2\pi\varepsilon_0 r^2}\vec{a}_r$$

where $\vec{r} = (6\vec{a}_x + 4\vec{a}_y)$.

Further,
$$r = |\vec{r}| = \sqrt{6^2 + 4^2} = \sqrt{52}$$

Thus,

$$\vec{D} = \frac{40 \times 10^{-6}}{2\pi \times \sqrt{52}} \cdot \frac{(6\vec{a}_x + 4\vec{a}_y)}{\sqrt{52}}$$

or simply
$$\vec{D} = (0.74\vec{a}_x + 0.50\vec{a}_y)(\mu\text{C/m}^2)$$

(c) Electric field E due to an infinite plane of charge density ρ_s is given as

$$\vec{E} = \pm\frac{\rho_s}{2\varepsilon_0}\vec{a}_n$$

Flux density

$$\vec{D} = \varepsilon_0 \vec{E} \quad \text{or}$$

$$= \pm\frac{\rho_s}{2\varepsilon_0}\vec{a}_n$$

The plane is at $x = 8$ (in this case). The point is at $x = 6$. This point is located below the plane. Flux density

$$\vec{D} = -\frac{\rho_s}{2}\vec{a}_n \quad \text{or}$$

$$= -\frac{\rho_s}{2}\vec{a}_n \quad \text{or}$$

$$= -\frac{30}{2}\vec{a}_n \quad \text{or}$$

$$= -15\vec{a}_n(\mu\text{C/m}^2)$$

where $\vec{a}_n = \vec{a}_x$.

5.8 Find the flux density $\vec{D}$ at a point $(2, \pi/2, 0)$, if potential V is $(20/r^2)\sin\theta\cos\phi$.

Solution:

We know that flux density $\vec{D} = \varepsilon_0\vec{E}$, where $\vec{E} = -\nabla V$.

Here
$$\vec{E} = -\left[\frac{\partial V}{\partial r}\vec{a}_r + \frac{1}{r}\frac{\partial V}{\partial \theta}\vec{a}_\theta + \frac{1}{r\sin\theta}\frac{\partial V}{\partial \phi}\vec{a}_\phi\right] \quad \text{or}$$

$$= \frac{40}{r^3}\sin\theta\cos\phi\,\vec{a}_r - \frac{20}{r^3}\cos\theta\cos\phi\,\vec{a}_\theta + \frac{20}{r^3}\sin\phi\,\vec{a}_\phi$$

Flux density D is obtained as

$$D = \frac{20\varepsilon_0}{r^3}\sin\theta\cos\phi\,\vec{a}_r - \frac{20}{r^3}\varepsilon_0\cos\theta\cos\phi\,\vec{a}_\theta + \frac{20}{r^3}\sin\phi\,\vec{a}_\phi$$

Now, in this case

$$r = 2; \qquad \theta = \pi/2 \qquad \text{and} \qquad \Phi = 0$$

Thus,

$$\vec{D} = 5.0\varepsilon_0\vec{a}_r\,(\text{C/m}^2)$$

or simply

$$\vec{D} = 44.2\vec{a}_r\,(\text{pC/m}^2)$$

5.9 Find the flux density $\vec{D}$ at a point (4, 0, 3). There is a point charge -10π mC at (4, 0, 0). Also, there is a line charge of $6\pi\,(\text{mC/m})$ along the y-axis.

Solution:

Let the flux density $\vec{D}$ be

$$\vec{D} = \vec{D}_Q + \vec{D}_L$$

where $\vec{D}_Q$ is the flux density due to point charge and $\vec{D}_L$ the flux density due to a line charge. It is shown in Figure 5.24.

$$\vec{D}_Q = \varepsilon_0\vec{E} = \frac{q}{4\pi r^2}\vec{a}_r$$

where $\vec{r} = (4-4)\vec{a}_x + (0-0)\vec{a}_y + (3-0)\vec{a}_z$ or $\vec{r} = 3\vec{a}_z$.

Unit vector $\qquad \vec{a}_r = \frac{\vec{r}}{|\vec{r}|} = \vec{a}_z$

Thus,

$$\vec{D}_Q = \frac{-10\pi\cdot 10^{-3}}{4\pi(3)^2}\vec{a}_z \quad \text{or}$$

$$= 0.276\vec{a}_z\,(\text{mC/m}^2)$$

and similarly

$$\vec{D}_L = \frac{\rho_L}{2\pi r}\vec{a}_r$$

where $\qquad \vec{r} = (4-0)\vec{a}_x + (0-0)\vec{a}_y + (3-0)\vec{a}_z$

$$= (4\vec{a}_x + 3\vec{a}_z)$$

Unit vector

$$\vec{a}_r = \frac{\vec{r}}{|\vec{r}|} \quad \text{or}$$

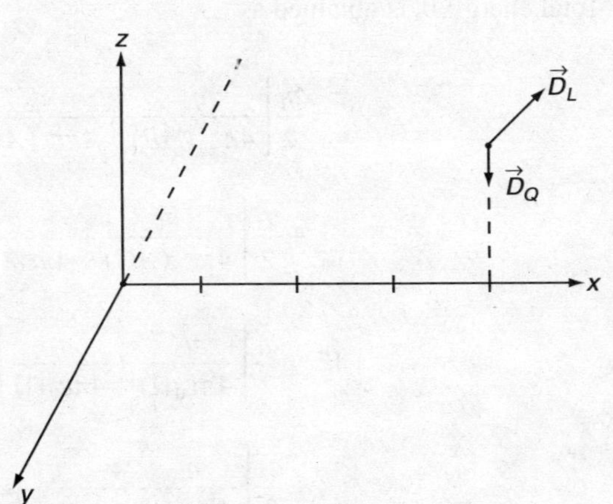

Figure 5.24 *Flux density* $\vec{D}$

$$= \frac{(4\vec{a}_x + 3\vec{a}_z)}{5}$$

Thus,

$$\vec{D}_L = \frac{6\pi}{2\pi \times 5} \times \frac{(4\vec{a}_x + 3\vec{a}_z)}{5}$$

$$= 0.48\vec{a}_x + 0.36\vec{a}_z \ (\text{mC/m}^2)$$

The resultant flux density $\vec{D}$ is obtained as

$$\vec{D} = \vec{D}_Q + \vec{D}_L \quad \text{or}$$

$$= 0.48\vec{a}_x + 0.64\vec{a}_z (\text{mC/m}^2)$$

5.10 Three point charges are −2 nC, 8 nC and 6 nC. These are located at points (0, 0, 0), (0, 0, 1) and (1, 0, 0), respectively. Find the total energy in this system.

Solution:
Total energy is given as

$$W = \frac{1}{2}\sum_{k=1}^{3} q_k V_k \qquad \text{(in this case)} \qquad \text{or}$$

$$= \frac{1}{2}[q_1V_1 + q_2V_2 + q_3V_3]$$

Here, $AB = 1$, $AC = 1$ and $BC = \sqrt{2}$ (Figure 5.25).
Total energy W is obtained as

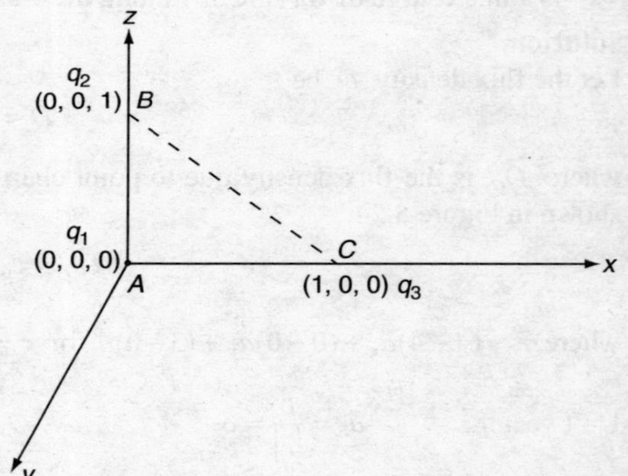

Figure 5.25 *Three point charges*

$$W = \frac{q_1}{2}\left[\frac{q_2}{4\pi\varepsilon_0(AB)} + \frac{q_3}{4\pi\varepsilon_0(AC)}\right] + \frac{q_2}{2}\left[\frac{q_1}{4\pi\varepsilon_0(AB)} + \frac{q_3}{4\pi\varepsilon_0(BC)}\right]$$

$$+\frac{q_3}{2}\left[\frac{q_1}{4\pi\varepsilon_0(AC)} + \frac{q_2}{4\pi\varepsilon_0(BC)}\right]$$

$$W = \frac{q_1}{2}\left[\frac{q_2}{4\pi\varepsilon_0(1)} + \frac{q_3}{4\pi\varepsilon_0(1)}\right] + \frac{q_2}{2}\left[\frac{q_1}{4\pi\varepsilon_0(1)} + \frac{q_3}{4\pi\varepsilon_0(\sqrt{2})}\right]$$

$$+\frac{q_3}{2}\left[\frac{q_1}{4\pi\varepsilon_0(1)} + \frac{q_2}{4\pi\varepsilon_0(\sqrt{2})}\right] \quad \text{or}$$

$$W = \frac{1}{4\pi\varepsilon_0}\left[q_1q_2 + q_1q_3 + \frac{q_2q_3}{\sqrt{2}}\right] \quad \text{or}$$

$$W = 9\times10^{9}\left[(-2)(8)+(-2)(6)+\frac{(8)(6)}{\sqrt{2}}\right]\times10^{-18} \quad \text{or}$$

$$= 36\left[\frac{12}{\sqrt{2}} - 7\right] \text{nJ}$$

or simply

$$= 53.48 \text{ nJ}$$

5.11 Find the stored energy W in a system, where four identical charges with charge $q = 4nC$ are placed at four corners with 1 m sides.

Solution:

Stored energy

$$W = \frac{1}{2}\sum_{k=1}^{4} q_k V_k$$

Due to symmetry

$$q_1V_1 = q_2V_2 = q_3V_3 = q_4V_4$$

Thus,

$$W = \frac{1}{2}[4\times q_1V_1] \quad \text{or}$$

$$W = 2q_1V_1$$

V_1 in this case is obtained as

$$V_1 = \frac{q_2}{4\pi\varepsilon_0 r_{12}} + \frac{q_3}{4\pi\varepsilon_0 r_{13}} + \frac{q_4}{4\pi\varepsilon_0 r_{14}} \quad \text{or}$$

$$= \frac{4\times10^{-9}}{4\pi\varepsilon_0}\left[\frac{1}{1}+\frac{1}{1}+\frac{1}{\sqrt{2}}\right] \quad \text{or}$$

$$= 9\times10^{9}\times4\times10^{-9}\left[\frac{2\sqrt{2}+1}{\sqrt{2}}\right] \quad \text{or}$$

$$= 97.32\ V$$

Thus, stored energy

$$W = 2\times(4\times10^{-9})\times(97.32)$$

$$= 778.56\times10^{-9}\text{ J}$$

$$W = 778.56 \text{ nJ}$$

5.12 Two concentric cylindrical conductors have radii $r_1 = 0.01$ m and $r_2 = 0.05$ m. Electric field between them is

$$\vec{E} = \left(\frac{10^5}{r}\right)\vec{a}_r \text{ V/m}$$

Assuming free space, find energy stored in 1.0 m length.

Solution:

Energy stored W is given as

$$W = \frac{1}{2}\int \varepsilon_0 E^2 dv \quad \text{or}$$

$$= \frac{\varepsilon_0}{2}\int_{0.01}^{0.05}\int_{\phi=0}^{2\pi}\int_{z=0}^{1}\left(\frac{10^5}{r}\right)^2 r\,dr\,d\phi\,dz \quad \text{or}$$

$$= 44.8 \times 10^{-2}$$

or simply

$$= 448 \text{ mJ}$$

5.13 Two point charges are $-8\ \mu$C and $10\ \mu$C. These are located at points (2, −1, 3) and (0, 4, −2), respectively. Find the potential at a point (1, 0, 1) while assuming zero potential at infinity.

Solution:

It is given that $q_1 = -8\ \mu$C and $q_2 = 10\ \mu$C

Potential at point (1, 0, 1) is obtained as

$$V = \frac{q_1}{4\pi\varepsilon_0|\vec{r}_1|} + \frac{q_2}{4\pi\varepsilon_0|\vec{r}_2|} + V_0$$

Given that $V = 0$ at $r = \infty$, it gives $V_0 = 0$. In this case,

$$\vec{r}_1 = (1-2)\vec{a}_x + (0-(-1))\vec{a}_y + (1-3)\vec{a}_z \quad \text{or}$$

$$\vec{r}_1 = -\vec{a}_x + \vec{a}_y - 2\vec{a}_z \quad \text{and}$$

$$|\vec{r}_2| = \sqrt{26}$$

Substituting these values, we get potential V' as

$$V = 9\times10^{+9}\times10^{-6}\left[\frac{-8}{\sqrt{6}} + \frac{10}{\sqrt{26}}\right] \quad \text{or}$$

$$= 9\times10^3[-3.266 + 1.9612]$$

or simply

$$= -11.744\times10^3 \text{ Volt}$$

5.14 In a system, three thousand lines of forces enter a volume of space. Two thousand lines of force leave this volume. Find the charge contained in it.

Solution:

Let q be the charge contained in this volume of space. As per Gauss's Law, net electric flux

$$\phi = \frac{q_{in}}{\varepsilon_0} \quad \text{or}$$

$$= \frac{q}{\varepsilon_0} \quad \text{or}$$

$$q = \varepsilon_0\,\phi$$

The number of lines of force passing normally through the surface is given as

$$= 3000 - 2000$$

$$= 1000$$

Thus,

$$\phi = 1000$$

Now charge

$$q = 8.85 \times 10^{-12} \times 1000 \quad \text{or}$$

$$q = 8.85 \times 10^{-9}\ \text{C}$$

5.15 In the case of a charge distribution, the volume charge density in Spherical coordinates is

$$\rho = \rho_0\left(\frac{a}{r}\right)$$

where $\rho_0 = 5$ and $a = 2$. Find:
(a) The electric flux density,
(b) V, if $V = 0$ at $r = 0$.
(Hint: Use expression of E_r as obtained in part (a))

Solution:

(a) As per Gauss's Law, we find

$$\int D \cdot ds = \int \rho_v \cdot dv = q_{enc}$$

Also,

$$D\int ds = \int \rho_0\left(\frac{a}{r}\right) \cdot 4\pi r^2 dr \quad \text{or}$$

$$D4\pi r^2 = \rho_0 a \int 4\pi r\, dr \quad \text{or}$$

$$D4\pi r^2 = \rho_0 a \frac{4\pi r^2}{2} \quad \text{or}$$

$$D = \frac{\rho_0 a}{2} = 5(\text{C/m}^2)$$

Further,

$$E_r = \frac{D}{\varepsilon_0}$$

$$= \frac{\rho_0 a}{2\varepsilon_0}$$

$$= \frac{5}{\varepsilon_0}(\text{N/C})$$

(b) Now,

$$V = -\int E_r \cdot dr \quad \text{or}$$

$$= -\int \frac{5}{\varepsilon_0} \cdot dr \quad \text{or}$$

$$= -\frac{5}{\varepsilon_0} r + C$$

Given that for $r = 0$, $V = 0$; so $C = 0$.

Thus,

$$V = -\frac{5r}{\varepsilon_0}$$

5.16 There is a large plane sheet of charge. It has surface charge density $\sigma = 4.0 \times 10^{-6}$ C/m^3. It lies in the x–y plane. Find the flux of the field through a circular area with a radius of 2 cm. This circular area lies completely in a region where x, y, z are all positive. Also, its normal makes an angle of 60° with the z-axis.

Solution:

Electric flux Φ through a circular area is given as

$$\Phi = \vec{E} \cdot \Delta\vec{S} \quad \text{or}$$

$$= \left|\vec{E}\right|\left|\Delta\vec{S}\right| \cos\theta$$

The magnitude of intensity of electric field $\left|\vec{E}\right|$ on either side, near a plane sheet is given as

$$\left|\vec{E}\right| = \frac{\sigma}{2\varepsilon_0}$$

The direction of the electric field will be along the z-axis. Flux is obtained as

$$\phi = \frac{\sigma}{2\varepsilon_0} \pi r^2 \cos\theta \quad \text{or}$$

$$= \frac{4 \times 10^{-6} \times 3.14 \times (2 \times 10^{-2})^2 \cos 60°}{2 \times 8.85 \times 10^{-12}}$$

or simply

$$= 141.92 \ (\text{Nm}^2/\text{C})$$

5.17 A spherical volume has a uniform charge density ρ of 10 Coulomb/m^3. It has a radius of 1 meter. Find V at r =50 cm.

Solution:

Here, radius a = 100 cm.

The potential inside a solid sphere with radius a is given as (formula, discussed in text earlier)

$$V = \frac{q}{4\pi\varepsilon_0 a} + \frac{q}{8\pi\varepsilon_0 a^3}(a^2 - r^2)$$

Further,

$$r = \frac{a}{2}$$

$$= 50 \text{ cm}$$

Thus,

$$V = \frac{q}{4\pi\varepsilon_0 a}\left[1 + \frac{3}{8}\right] \quad \text{or}$$

$$= \frac{11}{32}\frac{q}{\pi\,\varepsilon_0 a}$$

Charge density

$$\rho = \frac{q}{4\pi R^3} \quad \text{or} \quad \rho = \frac{q}{(4/3)\pi R^3} \quad \text{or}$$

$$\frac{q}{\pi a} = \frac{4}{3}a^2\rho$$

or simply

$$= \frac{40}{3}$$

Now potential V is obtained as

$$V = \frac{11}{32}\frac{40}{3}\cdot\frac{1}{\varepsilon_0} \quad \text{or} \quad = \frac{110}{24\varepsilon_0}\text{Volts}$$

5.18 In a case, charge distribution with spherical symmetry has density given as

$$\rho_v = \begin{Bmatrix} 10r, & 0 \le r \le a \\ 0, & r > a \end{Bmatrix}$$

where a is the radius of a sphere. Find $\vec{E}$ everywhere.

Solution:

Case 1: Radius; $0 < r < a$.

For the given charge distribution in this case, the Gaussian surface of a uniformly charged sphere is shown in Figure 5.26.

As per Gauss's Law,

$$\int \vec{E}\cdot\vec{ds} = \frac{q_{enc}}{\varepsilon_0}$$

$$= \frac{1}{\varepsilon_0}\int \rho_v\, dv$$

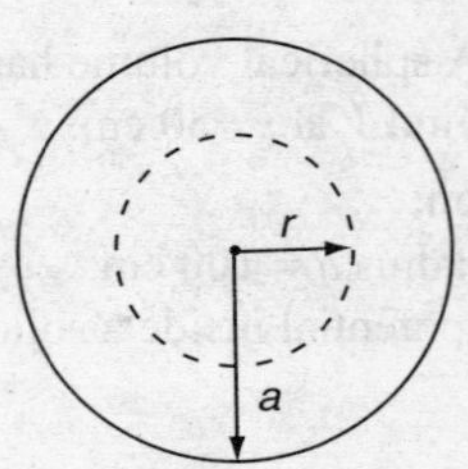

Figure 5.26 *Gaussian surface for a uniformly charged sphere*

Also,

$$\left|\vec{E}\right| 4\pi r^2 = \frac{1}{\varepsilon_0}\int_0^r \int_0^\pi \int_0^{2\pi} \rho_v r^2 \sin\theta\, d\Phi\, d\theta\, dr \quad \text{or}$$

$$\left|\vec{E}\right| 4\pi r^2 = \frac{1}{\varepsilon_0}\int_0^r 4\pi r^2 \cdot dr$$

$$= \frac{10\cdot \pi r^4}{\varepsilon_0 R} \quad \text{or}$$

$$\left|\vec{E}\right| = \frac{10\cdot r^2}{4\varepsilon_0} \quad \text{or}$$

$$\left|\vec{E}\right| = \frac{5r^2}{2\varepsilon_0}\vec{a}_r$$

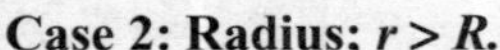

Case 2: Radius; $r > R$.

The Gaussian surface for this case is shown in Figure 5.27.

Applying Gauss's Law, we get

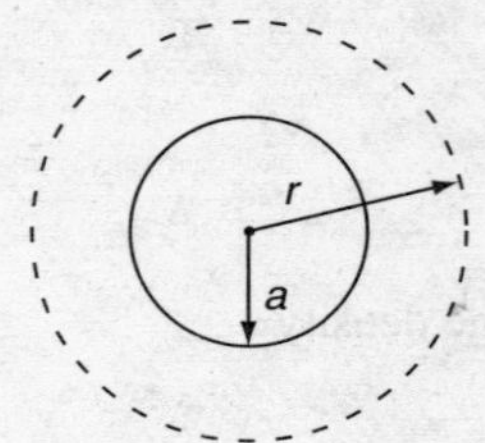

Figure 5.27 *Gaussian surface for $r > R$*

$$\left|\vec{E}\right| 4\pi r^2 = \frac{q_{enc}}{\varepsilon_0} \quad \text{or}$$

$$= \frac{1}{\varepsilon_0}\int_0^r \int_0^\pi \int_0^{2\pi} \rho_v r^2 \sin\theta\, d\Phi\, d\theta\, dr \quad \text{or}$$

$$= \frac{1}{\varepsilon_0}\int_0^a 4\pi r^2 10\cdot r\cdot dr + \frac{1}{\varepsilon_0}\int_a^r 0.4\pi\, r^2 dr \quad \text{or}$$

$$= \frac{10\pi a^4}{\varepsilon_0} \quad \text{or}$$

$$\left|\vec{E}\right| = \frac{5R^4}{2\varepsilon_0 r^2} \quad \text{or}$$

$$\left|\vec{E}\right| = \frac{5R^4}{2\varepsilon_0 r^2}\vec{a}_r$$

5.19 Electric flux density D is

$$\vec{D} = \left[(2y + z^2)\vec{a}_x + 5xy\,\vec{a}_y + (y^2 + x)\vec{a}_z\right](\mu\text{C/m}^2)$$

Find the volume charge density at (a) (0, 0, 0) and (b) (4, 0 , –2).

Solution:
It is given that

$$\vec{D} = (2y + z^2)\vec{a}_x + 5xy\,\vec{a}_y + (y^2 + x)\vec{a}_z$$

Volume charge density ρ_v is given as

$$\rho_v = \nabla \cdot \vec{D} \quad \text{or}$$

$$= \frac{\partial D_x}{\partial x} + \frac{\partial D_y}{\partial y} + \frac{\partial D_z}{\partial z}$$

Thus,

$$\rho_v = (0 + 5x + 0) = 5x$$

(a) at point (0, 0, 0): $\rho_v = 0\ (\mu\text{C/m}^3)$

(b) at point (4, 0, –2): $\rho_v = 20\ (\mu\text{C/m}^3)$

5.20 Find the electrical field in an entire region due to charge distribution in free space given as

$$\rho\,(r, \Phi, z) = \begin{cases} 0, & 0 < r < R_1 \\ 10, & R_1 < r < R_2 \\ 0, & R_2 < r < \infty \end{cases}$$

(Hint: Use Gauss's Law in integral form.)

Solution:
In this case, $\vec{D}$ and $\vec{E}$ will have only radial components in the $\vec{a}_r$ direction. It is shown in Figure 5.28.

Top and bottom surfaces will cancel each other in the calculations of $\vec{D}$ and $\vec{E}$. This is because these vectors will be perpendicular to $\vec{ds}$, as shown in Figure 5.28.

Case 1: $0 < r < R_1$

There is no charge in this region (given in the problem). Applying Gauss's Law, we get

$$\oint \vec{D}.\vec{ds} = q_{enc}$$

$$= 0$$

Thus,

$$\vec{D} = 0 \quad \text{or} \quad \vec{E} = 0$$

Electric field does not exist in this limit of the region.

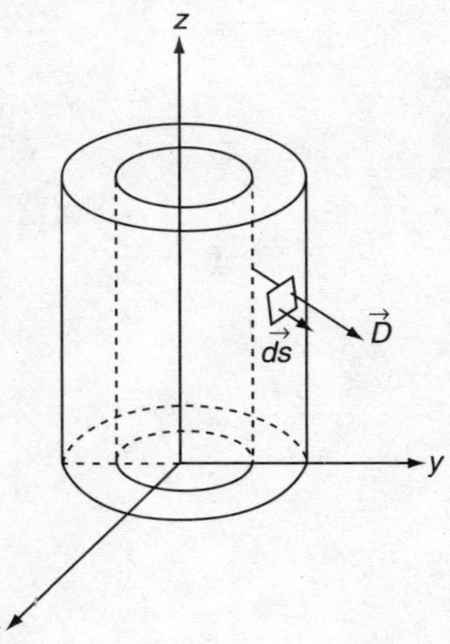

Figure 5.28 *Charge distribution*

Case 2: $R_1 < r < R_2$

Volume charge density is 10 C/m^3 (given in the problem). Charge enclosed q_{enc} between a and r will be given as

$$q_{enc} = 10(\pi r^2 - \pi R_1^2)$$

Applying Gauss's Law, we get

$$\oint \vec{D}.\vec{ds} = q_{enc}$$

$$\int_{z=0}^{1} \int_{\phi=0}^{2\pi} |\vec{D}| r d\phi dz = 10\pi(r^2 - R_1^2)$$

$$|\vec{D}|.r.2\pi = 10\pi(r^2 - R_1^2) \quad \text{or}$$

$$|\vec{D}| = \frac{5(r^2 - R_1^2)}{r}$$

Thus,

$$\vec{D} = \frac{5(r^2 - R_1^2)}{r} \vec{a}_r \quad \text{and}$$

$$\vec{E} = \frac{5(r^2 - R_1^2)}{\varepsilon_0 r} \vec{a}_r$$

Case 3: $R_2 < r < \infty$

Total charge enclosed per unit length will be given as

$$= 10\pi(R_2^2 - R_1^2)$$

Applying Gauss's law, we get

$$\oint \vec{D}.\vec{ds} = q_{enc} \quad \text{or}$$

$$\int_{z=0}^{1} \int_{\phi=0}^{2\pi} \vec{D} r d\phi dz = 10\pi(R_2^2 - R_1^2) \quad \text{or}$$

$$2\pi r|\vec{D}| = 10\pi(R_2^2 - R_1^2) \quad \text{or}$$

$$|\vec{D}| = \frac{5(R_2^2 - R_1^2)}{r} \quad \text{or}$$

$$\vec{D} = \frac{5(R_2^2 - R_1^2)}{r} \vec{a}_r \quad \text{and}$$

$$\vec{E} = \frac{5(R_2^2 - R_1^2)}{\varepsilon_0 r} \vec{a}_r$$

5.21 A positive charge q is uniformly distributed in the air over an imaginary thin spherical shell. The shell has radius R. Find the field intensity at any point (inside and outside) of the shell using Gauss's Law.

Solution:

For $r < R$, there will be no charge inside the shell. The Gaussian surface for $r < R$ is shown in Figure 5.29.

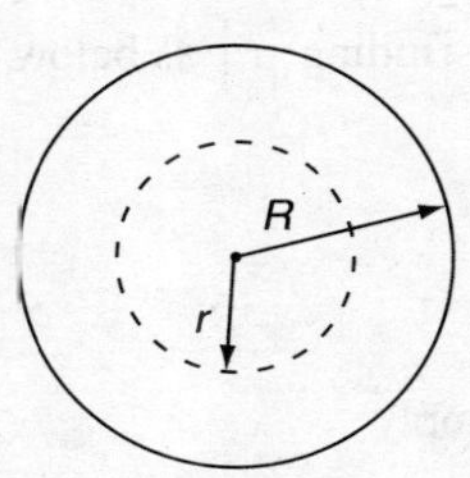

Figure 5.29 *Gaussian surface for $r < R$*

$$q_{enc} = 0 \qquad \text{(enclosed charge is zero)}$$

Applying Gauss's Law, we get

$$\oint \vec{D}.\vec{ds} = q_{enc} = 0 \quad \text{or}$$

$$\vec{D} = \vec{E} = 0 \quad \text{for} \quad r < R$$

For $r \geq R$, the Gaussian surface is shown in Figure 5.30.

Applying Gauss's Law, we get

$$\oint \vec{D}.\vec{ds} = q_{enc} \quad \text{or}$$

$$|\vec{D}|\, 4\pi r^2 = q \quad \text{or}$$

$$|\vec{D}| = \frac{q}{4\pi r^2}$$

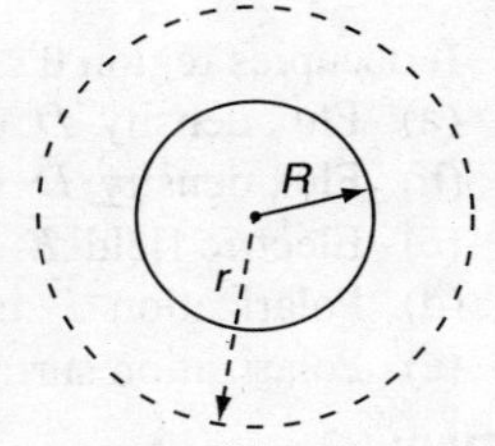

Figure 5.30 *Gaussian surface for $r \geq R$*

Only the radial component of a field will exist due to symmetry. Thus,

$$\vec{D} = \frac{q}{4\pi r^2}\vec{a}_r \qquad \text{at } r \geq R$$

At the surface of the shell, we get

$$\vec{D} = \frac{q}{4\pi r^2}\vec{a}_r \quad \text{and}$$

at $r = R$,

$$\vec{E} = \frac{q}{4\pi\varepsilon_0 r^2}\vec{a}_r$$

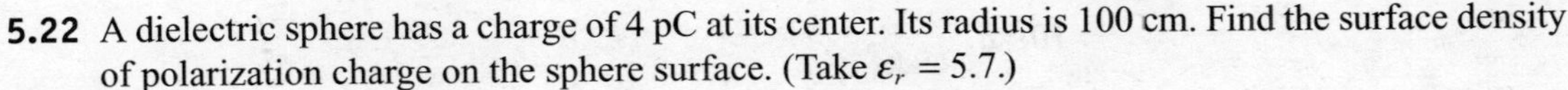

5.22 A dielectric sphere has a charge of 4 pC at its center. Its radius is 100 cm. Find the surface density of polarization charge on the sphere surface. (Take $\varepsilon_r = 5.7$.)

Solution:

As per Coulomb's Law,

$$\vec{E} = \frac{q}{4\pi\varepsilon_0\varepsilon_r r^2}\vec{a}_r$$

also

$$\vec{P} = \chi_e \varepsilon_0 \vec{E}$$

Thus,

$$\vec{P} = \frac{\chi_e q}{4\pi\varepsilon_r r^2}\vec{a}_r$$

Now finding $|\vec{P}|$ as below,

$$|\vec{P}| = \frac{(\varepsilon_r - 1)}{4\pi\varepsilon_r r^2} q\,\vec{a}_r \quad \text{or}$$

$$= \frac{(5.7-1)\times 4\times 10^{-12}}{4\pi\times 5.7\times(100\times 10^{-2})^2}$$

or simply

$$= 26.24\times 10^{-2}\ (\text{pC/m}^2)$$

5.23 There is an infinite plane dielectric slab of thickness h. Its non-uniform permittivity is given as

$$\varepsilon = \frac{5\varepsilon_0}{(1+z/h)^2}$$

It occupies region $0 < z < h$. A uniform field $\vec{E}_a$ of $10\ \vec{a}_z$ is applied. Obtain the following:

(a) Flux density $\vec{D}$ outside the dielectric slab
(b) Flux density $\vec{D}$ inside the dielectric slab
(c) Electric field $\vec{E}$ inside the dielectric slab
(d) Polarization $\vec{P}$ inside the dielectric slab and
(e) Polarization surface charge densities P on surfaces (i) $z = 0$ and (ii) $z = h$.

Solution:

The dielectric slab having infinite length and thickness h is shown in Figure 5.31.

(a) Flux density $\vec{D}$ outside the dielectric slab is obtained as

$$\vec{D} = \varepsilon_0\vec{E} \quad \text{or}$$

$$= \varepsilon_0 \times 10\vec{a}_z$$

or simply

$$= 10\varepsilon_0\vec{a}_z$$

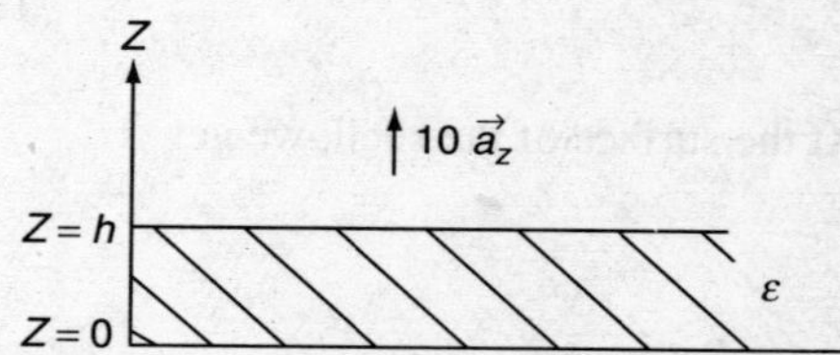

Figure 5.31 *Dielectric slab*

(b) Flux density $\vec{D}$ inside the dielectric slab is obtained as

$$\vec{D} = \varepsilon_0\vec{E} \quad \text{or}$$

$$= 10\varepsilon_0\vec{a}_z$$

$\vec{D}$ is independent of the dielectric medium.

(c) The electric field $\vec{E}$ inside the dielectric slab is obtained as

$$\vec{E} = \frac{\vec{E}_a}{\varepsilon_r} \quad \text{or}$$

$$= \frac{10}{\varepsilon/\varepsilon_0}\vec{a}_z \quad \text{or}$$

$$= \frac{10(1+z/h)^2}{5}\vec{a}_z$$

or simply

$$= (1+z/h)^2\vec{a}_z$$

(d) Polarization $\vec{P}$ inside dielectric is obtained as

$$\vec{P} = \frac{\varepsilon_0 \chi_e \vec{E}_a}{1+\chi_e} \quad \text{or}$$

$$= \frac{\varepsilon_0(\varepsilon_r - 1)}{\varepsilon_r} 10.\vec{a}_z \quad \text{or}$$

$$= \frac{\varepsilon_0(5/(1+z/h)^2 - 1)}{5/(1+z/h)^2} 10.\vec{a}_z \quad \text{or}$$

$$= \frac{\varepsilon_0}{5}(5-(1+z/h)^2)10.\vec{a}_z \quad \text{or}$$

$$= \frac{\varepsilon_0}{5}\left[4 - \frac{z^2}{h^2} - \frac{2z}{d}\right]10\vec{a}_z$$

or simply

$$= 2\varepsilon_0\left[4 - \frac{z^2}{h^2} - \frac{2z}{d}\right]\vec{a}_z$$

(e) Polarization surface charge density ρ is obtained as

$$\rho = 2\varepsilon_0\left[4 - \frac{z^2}{h^2} - \frac{2z}{d}\right]$$

(i) For $z = h$; $\rho = 2\varepsilon_0$ and
(ii) For $z = 0$; $\rho = -p = -8\varepsilon_0$

5.24 Voltages inside a concentric spherical shell are $V = 0$ Volts at $r = 100$ cm and $V = 5$ Volts at $r = 200$ cm. Find $\vec{E}$ and $\vec{D}$.

Solution:
In this case, V is a function of r only and not θ and Φ. As per Laplace's equation,

$$\nabla^2 V = \frac{1}{r^2}\frac{\partial}{\partial r}\left(r^2\frac{\partial V}{\partial r}\right) = 0$$

Integrating the above expression twice so as to obtain V, we get

$$V = -\frac{C_1}{r} + C_2$$

Given that at $r = 100$ cm $= 1$ m; $V = 0$ Volt and at $r = 200$ cm $= 2$ m; $V = 5$ Volts.
Substituting these values and solving for C_1 and C_2, we get

$$-\frac{C_1}{1} + C_2 = 0 \quad \text{and}$$

$$-\frac{C_1}{2}+C_2=5$$

On solving, we get

$$C_1=20 \qquad \text{and} \qquad C_2=20$$

Now

$$V=\left(-\frac{20}{r}+20\right)\text{Volts}$$

But

$$\vec{E}=-\nabla V$$

$$=-\frac{\partial V}{\partial r}\vec{a}_r$$

Thus,

$$\vec{E}=-\frac{20}{r^2}\vec{a}_r(\text{V/m}) \quad \text{and}$$

$$\vec{D}=\varepsilon_0\vec{E} \quad \text{or}$$

$$=\frac{-177.08}{r^2}\vec{a}_r\ (\text{pC/m}^2)$$

5.25 In a case, potential V is $V=2x^2yz+3Ay^3z$. Find the value of A. Also, find electric field $\vec{E}$ at point (0, 1, 1).

Solution:
As per Laplace's equation,

$$\nabla^2 V=0 \quad \text{or}$$

$$\frac{\partial^2 V}{\partial x^2}+\frac{\partial^2 V}{\partial y^2}+\frac{\partial^2 V}{\partial z^2}=0$$

Thus,

$$4yz+18Ayz=0$$

$$\therefore A=-2/9$$

Further,

$$\vec{E}=-\Delta V \quad \text{or}$$

$$=-\left[\frac{\partial V}{\partial x}\vec{a}_x+\frac{\partial V}{\partial y}\vec{a}_y+\frac{\partial V}{\partial z}\vec{a}_z\right] \quad \text{or}$$

$$=-\left[(4xyz)\vec{a}_x+(2x^2z-2y^2z)\vec{a}_y+\left(2x^2y-\frac{2}{3}y^3\right)\vec{a}_z\right] \quad \text{or}$$

$$=\left[-4xyz\vec{a}_x-(x^2-y^2)\vec{a}_y-2y\left(x^2-\frac{y^2}{3}\right)\vec{a}_z\right]$$

Finding $\vec{E}$ at point (0, 1, −1), we get

$$\vec{E} = \left(-2\vec{a}_y + \frac{2}{3}y^3\vec{a}_z\right)$$

5.26 Find the charge density if in a case electric potential V is $(3x^2 + y^2 + 2z^2)$.

Solution:
It is given that $V = (3x^2 + y^2 + 2z^2)$.

Further, $$\nabla^2 V = \frac{\partial^2}{\partial x^2}(3x^2 + y^2 + 2z^2) + \frac{\partial^2}{\partial y^2}(3x^2 + y^2 + 2z^2) + \frac{\partial^2}{\partial z^2}(3x^2 + y^2 + 2z^2) \quad \text{or}$$

$$\nabla^2 V = (3\times 2 + 2 + 2\times 2)$$

or simply $$= 12$$

As per Poisson's equation, we get

$$\nabla^2 V = -\frac{\rho}{\varepsilon_0} \quad \text{or}$$

$$12 = -\frac{\rho}{\varepsilon_0}$$

Finally, $\rho = -12\varepsilon_0 (\text{Coulomb/m}^2)$.

5.27 An infinitesimal insulating gap separates two conducting semi-infinite planes. It is shown in Figure 5.32. Planes are situated at $\phi = 0$ and $\phi = \pi/6$. Voltage (V) is 0 and 200 Volts at these points, respectively. Find V and $\vec{E}$ in the region between the planes.

Solution:
As per Laplace's equation,

$$\nabla^2 V = 0$$

In Cylindrical coordinates, we know that

$$\nabla^2 V = \frac{1}{r^2}\frac{\partial^2 V}{\partial \phi^2} \quad \text{or}$$

$$= 0$$

$$\frac{\partial^2 V}{\partial \phi^2} = 0$$

On solving, we get $V = C_1\Phi + C_2$. Given that at $\Phi = 0$, $V = 0$ and at $\Phi = \pi/6$, $V = 200$ Volts. Putting these values and solving for C_1 and C_2, we get

$$\therefore C_1 = \frac{200}{\pi/6} \quad \text{and}$$

$$= 1200/\pi$$

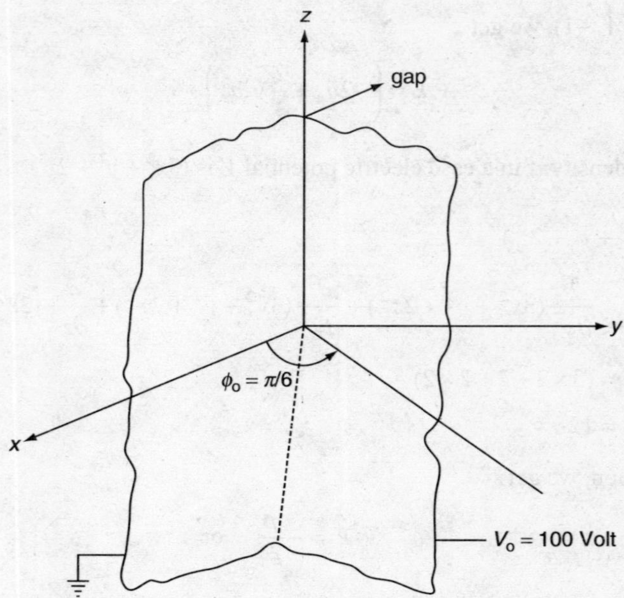

Figure 5.32 *Conducting semi-infinite planes*

$$C_2 = 0$$

Putting values of C_1 and C_2, we get potential V as

$$V = \frac{1200}{\pi}\phi$$

Further,

$$\vec{E} = -\nabla \cdot V \quad \text{or}$$

$$= -\left(\frac{1}{r}\right)\frac{\partial V}{\partial \phi}\vec{a}_\phi \quad \text{or}$$

$$= -\left(\frac{1}{r}\right)\frac{1200}{\pi}\vec{a}_\phi \quad \text{or}$$

$$= \frac{-1200}{\pi r}\vec{a}_\phi$$

5.28 The electric field intensity in a sphere with radius R is given as $E = 5\,r^4$, $r \le R$ and $E = 5\,r^{-2}$, $r > R$. Find the volume charge density ρ_v inside this sphere.

Solution:
The given electric field E only depends on r and not on θ and Φ.

Potential (V) is obtained as

$$V = -\oint \vec{E} \cdot \vec{dl} \quad \text{or}$$

$$= -\oint \vec{E} \cdot [dr\,\vec{a}_r + rd\phi\,\vec{a}_\phi + dz\,\vec{a}_z] \quad \text{or}$$

$$= -\oint E \cdot dr \;\; \vec{a}_r \cdot \vec{a}_r \quad \text{or}$$

$$= -\int_R^r E\,dr \quad \text{or}$$

$$= -\int_R^r 5r^4 dr \quad \text{or}$$

$$= -[r^5]_R^r$$

or simply

$$= [R^5 - r^5]$$

As per Poisson's equation, we get

$$\nabla^2 V = -\frac{\rho_v}{\varepsilon} \quad \text{or}$$

$$\frac{1}{r^2}\frac{\partial^2}{\partial r}\left[r^2 \frac{\partial V}{\partial r}\right] = \frac{-\rho_v}{\varepsilon}$$

Putting the value of V above, we get

$$\frac{1}{r^2}\frac{\partial^2}{\partial r}[r^2(-5r^4)] = -\frac{\rho_v}{\varepsilon} \quad \text{or}$$

$$-\frac{5}{r^2} \times 6r^5 = -\frac{\rho_v}{\varepsilon} \quad \text{or}$$

$$-30r^3 = -\frac{\rho_v}{\varepsilon}$$

or simply

$$\rho_v = 30r^3\varepsilon$$

5.29 An electric field is incident on a metal slab at an angle of 60° with the normal in air (at the boundary between two media). Find the angle of refracted field lines in this metal slab if $\varepsilon_r = 5$.

Solution:

Consider air to be first the medium. It is given that $\alpha_1 = 60°$; $\varepsilon_{r1} = 1$; $\varepsilon_{r2} = 1$ and $\alpha_2 = ?$ (to find out).

Applying the formula, we get

$$\tan\alpha_2 = \frac{\varepsilon_{r2}}{\varepsilon_{r1}} \tan\alpha_1 \quad \text{or}$$

$$\tan\alpha_2 = \frac{5}{1} \tan 60° \quad \text{or}$$

$$\tan \alpha_2 = 5\sqrt{3} \quad \text{or}$$

$$\alpha_2 = \tan^{-1}(5\sqrt{3})$$

or simply
$$\alpha_2 = 83.41°$$

5.30 An electric field makes an angle of 30° to the normal at the inner surface of porcelain. Finally, the electric field emerges into the air. The electric field strength is 500 V/m in a mass of porcelain (in air). Find the angle of emergence of the external field if ε_{r1} of porcelain is 5.

Solution:

It is given that $\varepsilon_{r1} = 5;\ \alpha_1 = 30°$. For $\varepsilon_{r2} = 1$, we have to find $\alpha_2 = ?$

Applying the formula, we get

$$\tan \alpha_2 = \frac{\varepsilon_2}{\varepsilon_1} \tan \alpha_1 \quad \text{or}$$

$$\tan \alpha_2 = \frac{\varepsilon_{r2}}{\varepsilon_{r1}} \tan \alpha_1 \quad \text{or}$$

$$= \frac{1}{5} \tan 30° \quad \text{or}$$

$$= \frac{1}{5} \times \frac{1}{\sqrt{3}}$$

Thus,

$$\alpha_2 = \tan^{-1}(1/(5\sqrt{3})) \quad \text{or}$$

$$\alpha_2 = 6.58°$$

For an electric field, we get

$$E_1 \sin \alpha_1 = E_2 \sin \alpha_2 \quad \text{or}$$

$$E_2 = \frac{E_1 \sin \alpha_1}{\sin \alpha_2} \quad \text{or}$$

$$= \frac{500 \times \sin 30°}{\sin 6.58°}$$

Finally,

$$E_2 = 2179.44 \text{ (Volt/cm)}$$

5.31 A region $Y \geq 0$ consists of a dielectric medium. Region $Y < 0$ is a perfect conductor. For surface charge $4(\text{nC/m}^2)$ on conductor and ε_{r1} (of dielectric medium) as 3, find $\vec{E}$ and $\vec{D}$ at the points
(a) point M (4, –2, 1) and
(b) point N (–3, 1, 4).

Solution:

(a) Point M (4, –2, 1) is located inside the conductor, $Y = -2$ (<0). It is shown in Figure 5.33, where $\vec{E} = 0$ and $\vec{D} = 0$.

(ii) Point N (–3, 1, 4) is located inside the dielectric medium, $Y = 1$ (>0), which is shown in Figure 5.34.

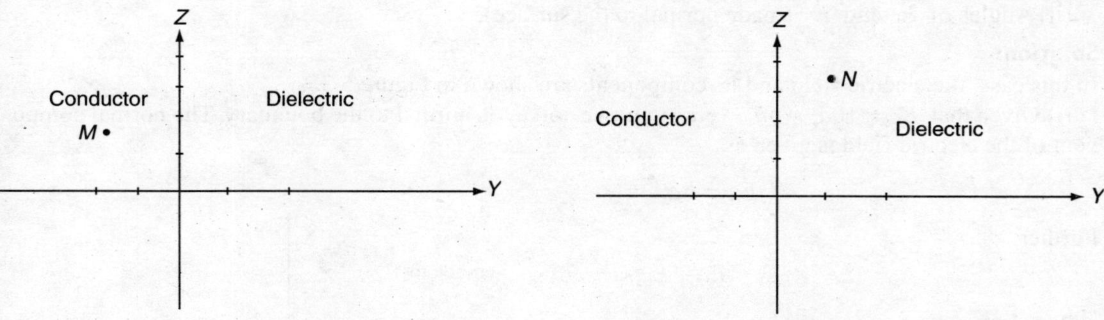

Figure 5.33 *Point M located in the region*

Figure 5.34 *Point N located in the region*

For normal components, we get

$$D_{n2} - D_{n1} = \rho_s \quad \text{or}$$

$$D_n - 0 = 4\ (\text{nC/m}^2) \quad \text{or}$$

$$\vec{D}_n = 4\vec{a}_y\ (\text{nC/m}^2)$$

Further,

$$\vec{E}_n = \frac{\vec{D}_n}{\varepsilon_0 \varepsilon_r} \quad \text{or}$$

$$= 4 \times 10^{-9} \times \frac{36\pi}{3} \times 10^9 \vec{a}_y$$

or simply

$$= 150.8\vec{a}_y\ (\text{Volts/m})$$

For tangential components, we get

$$D_{t2} = D_{t1} = 0$$

$$E_{t2} = E_{t1} = 0$$

Thus,

$$\vec{E} = 150.8\vec{a}_y \text{ (Volt/m)} \quad \text{and}$$

$$\vec{D} = 4\vec{a}_y \text{ (nC/m)}$$

5.32 Two homogeneous isotropic dielectrics separate at $z = 0$. Region $z \geq 0$ has $\varepsilon_{r1} = 3$. Region $z < 0$ has $\varepsilon_{r2} = 2$. Electric field for $z \geq 0$ is

$$\vec{E}_1 = (10\vec{a}_x - 4\vec{a}_y - 6\vec{a}_z)\text{(kV/m)}$$

Find

(i) Electric field in region $z \leq 0$.

(ii) Angles of $\vec{E}_1$ and $\vec{E}_2$ (made normal to the surface).

Solution:

In this case, the electric field and its components are shown in Figure 5.35.

(i) Given that $\vec{E}_1 = (10\vec{a}_x - 4\vec{a}_y - 6\vec{a}_z)$, unit vector $\vec{a}_z$ is normal to the boundary. The normal component of the electric field is given as

$$\vec{E}_{1n} = 6\vec{a}_z$$

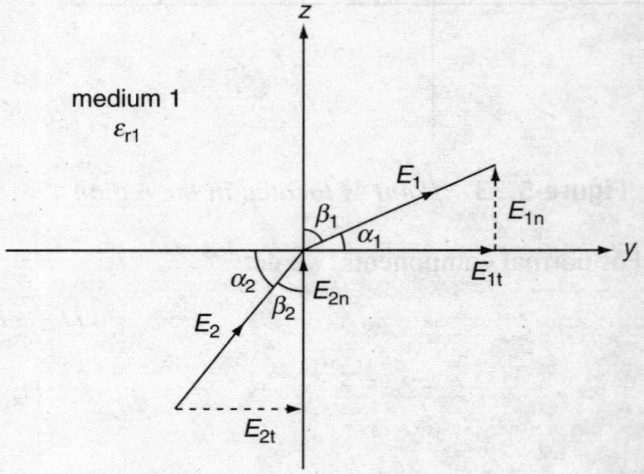

Further,

$$\vec{E}_1 = \vec{E}_{1t} + \vec{E}_{1n}$$

Here,

$$\vec{E}_{1t} = (10\vec{a}_x - 4\vec{a}_y)$$

Thus,

$$\vec{E}_{2t} = \vec{E}_{1t} = (10\vec{a}_x - 4\vec{a}_y)$$

Now, the normal components

$$\vec{D}_{2n} = \vec{D}_{1n} \quad \text{or}$$

$$\varepsilon_{r2}\vec{E}_{2n} = \varepsilon_{r1}\vec{E}_{1n}$$

or simply

$$\vec{E}_{2n} = \frac{3}{2}(6\vec{a}_z)$$

$$= 9\vec{a}_z$$

$\vec{E}$ in medium '2' is obtained as

$$\vec{E}_2 = \vec{E}_{2t} + \vec{E}_{2n}$$

$$= (10\vec{a}_x - 4\vec{a}_y + 9\vec{a}_z)\text{(kV/m)}$$

Figure 5.35 *Electric field and its components*

(i) Suppose β_1 and β_2 are the angles with the normal to the interface for $\vec{E}_1$ and $\vec{E}_2$. Now,

$$\left|\vec{E}_{1t}\right| = \sqrt{100 + 16} = 2\sqrt{29} \quad \text{and}$$

$$\left|\vec{E}_{1n}\right| = 6$$

Thus,

$$\tan\beta_1 = \frac{|\vec{E}_{1t}|}{|\vec{E}_{1n}|} = \frac{2\sqrt{29}}{6} \quad \text{or}$$

$$\beta_1 = 60.9°$$

For medium '2'

$$|\vec{E}_{2t}| = |\vec{E}_{1t}| = 2\sqrt{29} \quad \text{and}$$

$$|\vec{E}_{2n}| = 9$$

Thus,

$$\tan\beta_2 = \frac{|\vec{E}_{2t}|}{|\vec{E}_{2n}|} = \frac{2\sqrt{29}}{9} \quad \text{or}$$

$$\beta_2 = 50.12°$$

5.33 Two homogeneous isotropic dielectric mediums are separated at $z = 0$. ε_{r1} is 2 for region $z \le 0$. ε_{r2} is 3 for region $z > 0$. The electric field (E_1) in medium '1' is

$$\vec{E}_1 = (6\vec{a}_x + 10\vec{a}_y + 14\vec{a}_z).$$

Find

(i) Normal and tangential components of $\vec{E}_1$,
(ii) Angle between $\vec{E}_1$ and the normal (to the surface),
(iii) Normal and tangential components of $\vec{D}_2$ and
(iv) Angle between $\vec{E}_2$ and the normal (to the surface).

Solution:
Electric field and its components are shown in Figure 5.36.
(i) Given that $\vec{E}_1 = (6\vec{a}_x + 10\vec{a}_y + 14\vec{a}_z)$, unit vector $\vec{a}_z$ is normal to the boundary.
Normal component of the electric field is given as

$$\vec{E}_{1n} = 14\vec{a}_z \quad \text{or}$$

$$|\vec{E}_{1n}| = 14(\text{Volt/m})$$

Further,

$$\vec{E}_1 = \vec{E}_{1t} + \vec{E}_{1n} \quad \text{or}$$

$$\vec{E}_{1t} = \vec{E} - \vec{E}_{1n} \quad \text{or}$$

$$= (6\vec{a}_x + 10\vec{a}_y)$$

The tangential component of $\vec{E}_1$ is obtained as

$$|\vec{E}_{1t}| = \sqrt{36+100} = 16\ (\text{Volt/m})$$

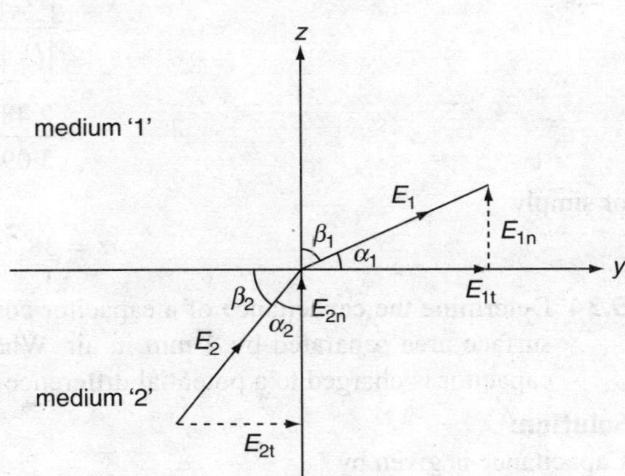

Figure 5.36 *Electric field and its components*

(ii) Let 'α_1' be the angle between $\vec{E}_1$ and normal to surface. We get

$$\tan\alpha_1 = \frac{|\vec{E}_{1n}|}{|\vec{E}_{1t}|}$$

or simply

$$\alpha_1 = 50.2°$$

(iii)

$$\vec{E}_{t_2} = \vec{E}_{t_1} = (6\vec{a}_x + 10\vec{a}_y)$$

Thus,

$$\vec{D}_{t_2} = \varepsilon_{r_2}\varepsilon_0 \vec{E}_{t_2} \quad \text{or}$$

$$= 3\times 8.85\times 10^{-12}(6\vec{a}_x + 10\vec{a}_y) \quad \text{or}$$

$$|\vec{D}_{t_2}| = 3.09\times 10^{-12}\ (\text{C/m}^2) \quad \text{and}$$

$$|\vec{D}_{n_2}| = |\vec{D}_{n_1}| = \varepsilon_0\ \varepsilon_r\ |\vec{E}_{1n}| \quad \text{or}$$

$$|\vec{D}_{n_2}| = 8.854\times 10^{-12}\times 2\times 4 \quad \text{or}$$

$$|\vec{D}_{n_2}| = 2.48\times 10^{-10}\ (\text{C/m}^2)$$

Thus,

$$\vec{D}_2 = (1.59\vec{a}_x + 2.65\vec{a}_y + 2.48\vec{a}_z)10^{-10}\ \text{C/m}^2$$

(iv) Let 'α_2' be the angle between $\vec{E}_2$ and normal to surface, then

$$\tan\alpha_2 = \frac{|\vec{E}_{n_2}|}{|\vec{E}_{t_2}|} \quad \text{or}$$

$$= \frac{|\vec{D}_{n_2}|}{|\vec{D}_{t_2}|} \quad \text{or}$$

$$= \frac{2.48\times 10^{-10}}{3.09\times 10^{-10}}$$

or simply

$$\alpha_2 = 38.75°$$

5.34 Determine the capacitance of a capacitor consisting of two parallel metal plates 30 cm × 30 cm surface area separated by 4 mm in air. What is the total energy stored by the capacitor if the capacitor is charged to a potential difference of 400 V? What is the energy density?

Solution:

Capacitance is given by

$$C = \varepsilon\frac{A}{t} = \varepsilon_o\ \varepsilon_r\ \frac{A}{t} \tag{1}$$

where $\varepsilon_r = 1, \varepsilon_o = 8.05 \times 10^{-12}$, $A = 0.3 \times 0.3 = 0.09\,\text{m}^2$, $t = 4\times10^{-3}$ m. Substituting in Equation (1) and then simplifying, we get

$$C = \frac{8.85\times10^{-12}\times0.09}{4\times10^{-3}} = 1.99\times10^{-10} = 199.125\,\text{pF}$$

Energy stored

$$= \frac{1}{2}CV^2$$

$$= \frac{1}{2}\times199.1\times10^{-12}\times(400)^2 \text{ Joules}$$

$$= 1.59\times10^{-5}\text{ J}$$

Energy density is obtained by dividing the total energy with the volume of dielectric between the plates. Thus,

Energy density $$= \frac{1}{2}\varepsilon_o E^2$$

$$= \frac{1}{2}\times8.85\times10^{-12}\times10^{10} \quad \left[E = \frac{V}{a} = \frac{400}{4}\times10^3 = 10^5\right]$$

$$= 0.044\text{ J/m}^2$$

5.35 A vector field is given by $\vec{F} = 3xy\,\vec{a}_x - y^2\,\vec{a}_y$. Evaluate the integral $\int_c \vec{F}\cdot\vec{d}_r$, where c is the curve in the x–y plane, $y = 2x^2$ from (0, 0) to (2, 2).

Solution:

Curve c is in the x–y plane. $\vec{F}$ is a function of x and y only.

so that $$\vec{r} = x\vec{a}_x + y\vec{a}_y$$

$$\vec{d}_r = dx\,\vec{a}_x + dy\,\vec{a}_y$$

$$y = 2x^2$$

$$dy = 4x\,dx \quad \text{and}$$

$$\vec{F} = 3xy\,\vec{a}_x - y^2\,\vec{a}_z$$

$$\int_c \vec{F}\cdot\vec{d}_r = (3xy\,\vec{a}_x - y^2\,\vec{a}_y)\cdot(c/x\cdot\vec{a}_x + dy\cdot\vec{a}_y)$$

$$= \int_c (3xy\,dx - y^2 dy)$$

$$= \int_{x=0}^{2} [3x(2x^2)\,dx - (2x^2)^2\,4x\,dx]$$

$$= \int_0^2 (6x^3 - 16x^5)\,dx$$

$$= \left[\frac{6x^4}{4} - \frac{16x^6}{6} \right]_0^2$$

$$= \left[\frac{6\times16}{4} - \frac{16\times64}{6} \right]$$

$$= 24 - 170.6$$

$$= -146.6$$

5.36 There is a large plane sheet of charge. It has surface charge density of 8×10^{-6} C/m^3. It lies in the x–y plane. Find the flux of the field through a circular area having radius 3 cm.
This circular area lies completely in a region where x, y, z all are positive. Also, its normal makes an angle of 30° with the z-axis.

Solution:
Electric ϕ through a circular area is given as

$$= \vec{E}\cdot\overrightarrow{\Delta s} \quad \text{or}$$

$$= |E|\left|\overrightarrow{\Delta s}\right|\cos\theta$$

Magnitude of intensity of electric field $|\vec{E}|$ on either side near a plane sheet is given as

$$|\vec{E}| = \frac{\sigma}{2\varepsilon_o}$$

Direction of electric field will be along the z-axis. Flux is obtained as

$$\phi = \frac{\sigma}{2\varepsilon_o}\pi r^2 \cos\theta$$

$$= \frac{8\times10^{-6}\times3.14\times(3\times10^{-2})^2\cos 30°}{2\times8.85\times10^{-12}}$$

or simply $\qquad = 1106.16\ \text{Nm}^2/\text{C}$

5.37 A spherical volume has a uniform charge density ρ of 20 C/m^3. It has a radius of 2 m. Find V at $r = 25$ cm.

Solution:
Here, radius $a = 200$ cm. The potential inside a solid sphere with radius a is given as

$$V = \frac{q}{4\pi\varepsilon_o a} + \frac{q}{8\pi\varepsilon_o a^3}(a^2 - r^2)$$

Further,

$$r = \frac{a}{2} = 100\,\text{cm}$$

Thus,

$$V = \left[\frac{q}{4\pi\varepsilon_o a} + \frac{q}{8\pi\varepsilon_o a^3}\left(a^2 - \frac{a^2}{2}\right)\right]$$

$$= \frac{q}{4\pi\varepsilon_o a} + \frac{q}{8\pi\varepsilon_o a^3}\left(\frac{a^2}{2}\right) = \frac{q}{4\pi\varepsilon_o a}\left[1 + \frac{3}{8}\right] \quad \text{or}$$

$$= \frac{11}{32}\frac{q}{\pi\varepsilon_o a}$$

Charge density

$$\rho = \frac{q}{4\pi R^3} \quad \text{or}$$

$$\rho = \frac{q}{\frac{4}{3}\pi R^3} \quad \Rightarrow \quad \frac{q}{\pi a} = \frac{4}{3}a^2 \cdot \rho$$

$$= \frac{q}{\pi a} = \frac{4}{3} \times (200)^2 \times 10^{-4} \times 20 = 106.66$$

Now, potential V is obtained as

$$V = \frac{11}{32} \times \frac{106.66}{\varepsilon_o} = \frac{36.66}{\varepsilon_o} \text{Volts}$$

5.38 The flux lines of an electric field pass from air into the glass making an angle of 40° with the normal at the air side of the interface. ε for glass is 10. The electric field intensity in air is 260 V/m. Find the flux density in glass and also the angle that the flux lines make with the normal (glass side).

Solution:
Given, $\alpha_1 = 40°$, $\varepsilon = 5$ and $E_1 = 260$ V/m. We know that

$$\frac{\tan\alpha_1}{\tan\alpha_2} = \frac{\varepsilon_1}{\varepsilon_2} \quad \Rightarrow \quad \tan\alpha_2 = \frac{\varepsilon_2}{\varepsilon_1}\tan\alpha_1$$

$$\Rightarrow \alpha_2 = \tan^{-1}\left[\left(\frac{\varepsilon_2}{\varepsilon_1}\right)\tan\alpha_1\right]$$

$$\Rightarrow \alpha_2 = \tan^{-1}\left[\left(\frac{10}{2}\right)\tan(40°)\right]$$

$$\Rightarrow \alpha_2 = 83.20°$$

Also, $D_{n2} = D_{n1}$

$$\Rightarrow \quad D_1\cos\alpha_1 = D_2\cos\alpha_2 \qquad (1)$$

Putting the values in (1), we get

$260 \cos 40° = D_2 \cos (83.2)°$

$$\Rightarrow D_2 = \frac{191.51}{0.118} = 1617.42\,(\text{C/m}^2)$$

5.39 Find the energy stored in a parallel plate capacitor as shown in Figure 5.37.

Solution:
Consider a parallel plate capacitor of separation d and area A. It is charged to a voltage V. Dielectric permittivity is ε. Electric field in the dielectric is given as

$$E = \frac{V}{d} \quad \text{and}$$

$$W_e = \frac{1}{2}\int_v \varepsilon E^2 dv$$

$$= \frac{1}{2}\int \varepsilon \frac{V^2}{d^2} dv = \frac{1}{2}\frac{\varepsilon V^2}{d^2}\cdot v \qquad [\text{Volume } v = A \times d]$$

$$W_e = \frac{1}{2}\frac{\varepsilon V^2 Ad}{d^2} = \frac{1}{2}\frac{\varepsilon V^2 A}{d}$$

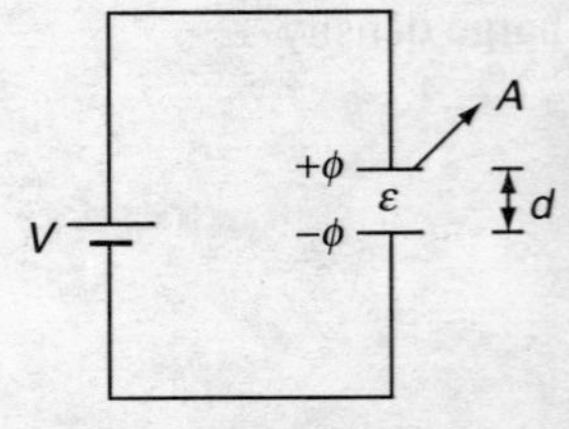

Figure 5.37 *Parallel plate capacitor*

The capacitance of a parallel plate capacitor is given as

$$C = \frac{\varepsilon A}{d}$$

$$W_e = \frac{1}{2} CV^2 \quad \text{but } Q = CV$$

$$W_e = \frac{1}{2} QV \quad \text{or}$$

$$W_e = Q^2/2C$$

$$W_e = \frac{1}{2} QV \text{ Joules} \quad \text{or}$$

$$W_e = \frac{Q^2}{2C} \text{ Joules}$$

5.40 What is the capacitance of the capacitor consisting of two parallel plates 0.4 × 0.4 m separated by 0.008 m air as shown in Figure 5.38? What is the energy stored by the capacitor if it is charged to a potential of 600 V? What is energy density?

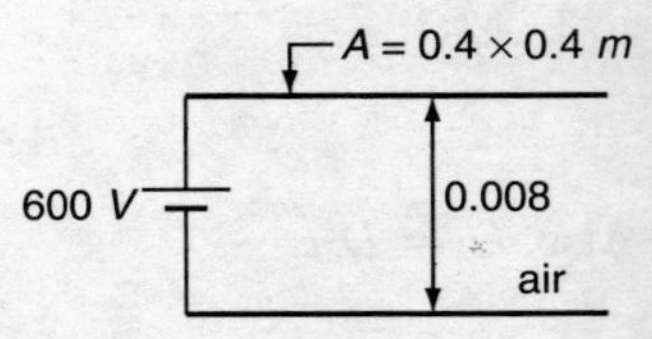

Figure 5.38 *Parallel plate capacitor*

Solution:
Given that $d = 0.008$ m, $\varepsilon_o = 8.85 \times 10^{-12}$ F/m, $A = 0.16$ m^2 and $V = 600$ V.

Capacitance

$$C = \frac{\varepsilon_o A}{d} = \frac{8.85\times10^{-12}\times0.16}{0.008} = 1.77\times10^{-12}\,\text{F}$$

Energy stored

$$= \frac{1}{2}CV^2$$

$$= \frac{1}{2}\times1.77\times10^{-12}\times600\times600$$

$$= 3.18\times10^{-5}\text{ Joules}$$

Energy density

$$= \frac{\text{Energy}}{\text{Volume}}$$

$$= \frac{3.18\times10^{-5}}{0.16\times0.008}$$

$$= 0.024\ \text{J/m}^3$$

5.41 A position charge Q of 140 pC is kept in air. Draw equipotential surfaces at radii 100 mm and 25 mm. Find the potential rise between these two radii.

Solution:

Absolute potential V_r due to a point charge at distance r is given as

$$V_r = \frac{Q}{4\pi\varepsilon_0 r}$$

All the points on this sphere (of radius r) will have the same potential. For $r = a = 100$ mm, the equipotential curve is shown in Figure 5.39.

The absolute potential is obtained as

$$V_a = \frac{Q}{4\pi\varepsilon_0}\frac{1}{a} = \frac{140\times10^{-12}\times9\times10^9}{100\times10^{-3}}$$

$$= \frac{1260\times10^{-3}}{100\times10^{-3}} = 12.6\text{Volts}$$

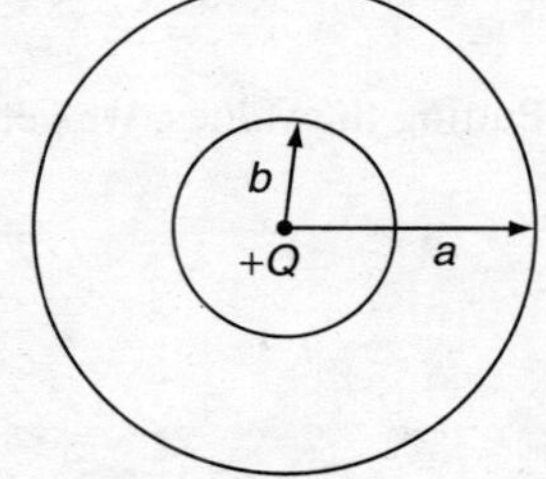

Figure 5.39 *Charge*

Similarly, for $r = b = 25$ mm

$$V_b = \frac{Q}{4\pi\varepsilon_0}\frac{1}{b} = \frac{140\times10^{-12}\times9\times10^9}{25\times10^{-3}} = 50.4\text{Volts}$$

Now,

$$V_{ab} = V_b - V_a$$

$$= 50.4 - 12.6 = 37.8\text{ Volts}$$

5.42 Calculate the force of attraction between $Q_1 = 2\times10^{-6}\,\text{C}$ and $Q_2 = 3\times10^{-5}\,\text{C}$, which are spaced by 5 cm in vacuum. Find the force if these are placed in the medium with $\varepsilon_r = 4$.

Solution:

Force is obtained as

$$F = \frac{Q_1Q_2}{4\pi\varepsilon_0 r^2} = \frac{2\times10^{-6}\times3\times10^{-5}}{4\times\pi\times8.85\times10^{-12}\times(5\times10^{-2})^2}$$

$$= \frac{6\times10^{-11}}{2778.9\times10^{-14}}\,\text{N}$$

$$= 2.15\times10^{-3}\times10^{3}\,\text{N}$$

$$= 2.15\,\text{N}$$

In the medium of $\varepsilon_r = 4$, force F is obtained as

$$F = \frac{2.15}{4} = 0.537\,\text{N}$$

5.43 Two similar charges of 4 μC but of opposite signs are kept at 1 cm distance from each other. Calculate the electric field at a point, which is 10 cm away from the mid-point of the axial line of the dipole.

Solution:

Given that charge $q = 4\,\mu\text{C} = 4\times10^{-6}\,\text{C}$, distance $d = 1\,\text{cm} = 1\times10^{-2}\,\text{m}$. Here, $R = 10\,\text{cm} = 10^{-1}\,\text{m}$. The electric field E on the axial line of dipole is given as

$$\vec{E} = \frac{2|\vec{p}|R}{4\pi\varepsilon_0(R^2 - cl^2)^2}$$

$$= \frac{2\times q\times d\times R}{4\pi\varepsilon_0(R^2 - d^2)^2}$$

Putting the values, we get

$$\vec{E} = \frac{9\times10^{9}\times2\times4\times10^{-6}\times1\times10^{-2}\times10^{-1}}{(10^{-2} - 0.25\times10^{-4})^2} = \frac{72}{9.95\times10^{-5}}$$

$$= 72\times10^{5}\,\text{N/C}$$

5.44 In the region $r \le 1$, $A = (25r^3/4)\vec{a}_r$ and for $r > 1$, $A = (20/r^2)\vec{a}_r$ in Spherical co-ordinates. Find the change density.

Solution:

For $r \le 1$,

$$\rho = \frac{1}{r^2}\frac{\partial}{\partial r}\left(\frac{25r^3}{4}\right)$$

$$= \frac{1}{r^2}\left(\frac{25\times 3 r^2}{4}\right)$$

$$\rho = \frac{75}{4}$$

for $r > 1$,

$$\rho = \frac{1}{r^2}\frac{\partial}{\partial r}(20)$$
$$= 0$$

5.45 A sphere has a charge density of 9 pC/m^3. Its volume is 0.4 m^3 and its center is at (0, 2, 1). Find the electric field intensity at a point (4, 0, 1).

Solution:
Electric field intensity E due to a volume charge is given as

$$\vec{E} = \int \frac{\rho}{4\pi\varepsilon_o r^2} dv.\vec{a}_r \quad \text{or}$$

$$\vec{E} = \frac{\rho v}{4\pi\varepsilon_o r^2}\vec{a}_r$$

In this case, $\vec{r}$ is obtained as

$$\vec{r} = (4-0)\vec{a}_x + (0-2)\vec{a}_y + (1-1)\vec{a}_z$$
$$= (4\vec{a}_x - 2\vec{a}_y)$$

Further,

$$|\vec{r}| = \sqrt{(4)^2 + (2)^2} = 4.47$$

Unit vector,

$$\vec{a}_r = \frac{\vec{r}}{|\vec{r}|} = \frac{(4\vec{a}_x - 2\vec{a}_y)}{4.47} \quad \text{or}$$

$$\vec{E} = \frac{9\times 10^{-12}\times 0.4\times 9\times 10^{9}}{(4.47)^2}\frac{(4\vec{a}_x - 2\vec{a}_y)}{4.47}$$
$$= 0.36\times 10^{-3}(4\vec{a}_x - 2\vec{a}_y)$$

$$\vec{E} = (1.44\vec{a}_x - 0.72\vec{a}_y)\ \text{mV/m}^3$$

5.46 A charge $Q_1 = 42\times 10^{-9}$ C in vacuum is located at a point P_1 (0.02, 0.01, – 0.03). Find the force on Q_1 due to another charge Q_2. Charge Q_2 is located at a point P_2 (0.03, 0.04, 0.02).

Solution:
Finding $\vec{R}_{21}$ as below,

$$\vec{R}_{21} = (-0.02 - 0.03)\vec{a}_x + (-0.01 - 0.04)\vec{a}_y + (-0.03 - 0.02)\vec{a}_z$$
$$= -0.05\vec{a}_x - 0.05\vec{a}_y - 0.05\vec{a}_z$$

Further,

$$\left|\vec{R}_{21}\right| = \sqrt{(0.05)^2 + (0.05)^2 + (0.05)^2} = 0.086$$

$$\vec{a}_{R_{21}} = \frac{-0.05\vec{a}_x - 0.05\vec{a}_y - 0.05\vec{a}_z}{0\cdot 086}$$

Force F_1 on Q_1 is given as

$$F_1 = \frac{Q_1\,Q_2}{4\pi\varepsilon_o\left|\vec{R}_{21}\right|^2}\vec{a}_{R_{21}}$$

Substituting values of Q_1, $Q_2\left|\vec{R}_{21}\right|^2$ and $\vec{a}_{R_{21}}$ as obtained above, we get

$$F_1 = \frac{42\times10^{-9}\times25\times10^{-6}\times9\times10^3}{(0.086)^2}\left(\frac{-0.05\vec{a}_x - 0.05\vec{a}_Y - 0.05\vec{a}_z}{0.086}\right)$$

$$= 0.74(-\vec{a}_x - \vec{a}_y - \vec{a}_z)\,N$$

5.47 A point charge $Q = 10$ nC is located at the origin in Cartesian coordinates. Find the charge density $\vec{D}$ at $(2, -3, 1)$.

Solution:

We know that,

$$\vec{D} = \varepsilon_o\vec{E} = \frac{Q}{4\pi r^2}\vec{a}_r$$

where $\vec{r} = (2\vec{a}_x - 3\vec{a}_y + \vec{a}_z)$.

$$|r| = \sqrt{(2)^2 + (3)^2 + (4)^2} = \sqrt{14} = 3.74$$

$$\vec{D} = \frac{10\times10^{-9}}{4\pi\times14}\;\frac{(2\vec{a}_x - 3\vec{a}_y + \vec{a}_z)}{\sqrt{14}}$$

$$= \frac{10\times10^{-9}}{657.93}(2\vec{a}_x - 3\vec{a}_y + \vec{a}_z)$$

$$= 1.51\times10^{-11}(2\vec{a}_x - 3\vec{a}_y + \vec{a}_z)$$

$$\left|\vec{D}\right| = 5.64\times10^{-11}\ \text{C/m}^2$$

5.48 An electric dipole $6.0\,\vec{a}_y$ n C-m is at point (0, 0, 0). Obtain the potential due to this electric dipole at point $\left(\frac{1}{2}, \frac{\pi}{3}, \frac{\pi}{3}\right)$.

Solution:

Potential due to an electric dipole is given as

$$V = \frac{\left|\vec{P}\right|\cos\theta}{4\pi\varepsilon_o r^2}$$

where $|\vec{P}| = 6\times10^{-9}$ C-m

$$r = \frac{1}{2} = 0.5 \quad \text{and} \quad \theta = \frac{\pi}{3}$$

Putting the values in the above equation,

$$V = \frac{9\times10^{9}\times6\times10^{-9}\times0.5}{(0.5)^2} = \frac{2.7}{0\cdot025} = 108$$

$$V = 108 \text{ V}$$

5.49 The potential distribution is of the form $V = (5y^2 + 10x)$. Find $\vec{E}$ at (2, 4) and (12, 10).

Solution:

$$\vec{E} = -\nabla V = -(10y\vec{a}_y + 10\vec{a}_x)$$

$$\vec{E}(2,4) = -(40\vec{a}_y + 10\vec{a}_x) \quad \text{and}$$

$$\vec{E}(12,10) = -(100\vec{a}_y + 10\vec{a}_x)$$

5.50 Find the flux density at a point $A(2, 4, -3)$ due to a uniform line charge (ρ_l) of 20 μC/m^2 on the z-axis.

Solution:

Electric field E at a point due to a line charge on the z-axis will be given as

$$\vec{E} = \frac{\rho_l}{2\pi r}\vec{a}_r$$

and the electric flux density

$$\vec{D} = \varepsilon_o\vec{E} \quad \text{or} \quad \frac{\rho_l}{2\pi r^2}\vec{a}_r$$

Here, $\vec{r} = 2\vec{a}_x + 4\vec{a}_y$ and $r = \vec{r} = \sqrt{(2)^2 + (4)^2} = \sqrt{20} = 4.47$

Thus,

$$\vec{D} = \frac{20\times10^{-6}}{2\pi\times20}\cdot\frac{(2\vec{a}_x + 4\vec{a}_y)}{4.47}$$

$$= 0.03\times10^{-6}(2\vec{a}_x + 4\vec{a}_y) = (0.06\vec{a}_x + 0.12\vec{a}_y)\,\mu\text{C/m}^3$$

$$\vec{D} = (0.06\vec{a}_x + 0.12\vec{a}_y)\,\mu\text{C/m}^3$$

5.51 It is given that $\nabla\cdot\vec{D} = \partial D_x/\partial x = 25$ nC/m^3. Find D_x at x = 2 cm. Also, find Q for $0 \le x \le 6$, $0 \le y \le 12$ and $z \le 12$ cm.

Solution:

Given,

$$\frac{\partial D_x}{\partial x} = 25 \text{ nC/m}^3 \qquad \text{(given)}$$

Now, D_x at $x = 2$ cm is given by

$$D_x = 25\times\frac{2}{100}\text{nC/m}^2 = 0.5 \text{ nC/m}^2$$

$$\Rightarrow Q = \int_V \frac{25\,nC}{m^3}\,dv = \frac{10^{-9}}{10^{-6}} \times \int_0^6 \int_0^{12} \int_0^{12} 25\,dx\,dy\,dz$$

After solving, we get

$$\Rightarrow Q = 25 \times 12 \times 12 \times 6 \times 10^{-3} = 21.6\text{ C}$$

5.52 For the above question, find $\vec{E}$ at point P, which is 5 cm from the origin along the y-axis.

Solution:

$$\vec{E}_y = \frac{Q}{4\pi\varepsilon_o r^2}\vec{a}_y = \frac{21.6 \times 9 \times 10^9}{(5 \times 10^{-2})^2}$$

$$= \frac{194.4 \times 10^9}{2.5 \times 10^{-3}}$$

$$= 77.76 \times 10^{12}\text{ V/m}$$

5.53 Find the flux density $\vec{D}$ at a point $r(4,\ \pi/4,\ 0)$ if potential V is $5/r^2 \sin\theta\ \cos\phi$.

Solution:

We know that flux density $\vec{D} = \varepsilon_0 \vec{E}$, where $\vec{E} = -\nabla V$.

Here
$$\vec{E} = -\left[\frac{\partial V}{\partial r}\vec{a}_r + \frac{1}{r}\frac{\partial V}{\partial \theta}a_\theta + \frac{1}{r\sin\theta}\frac{\partial V}{\partial \phi}\vec{a}_\phi\right] \quad \text{or}$$

$$= \frac{10}{r^3}\sin\phi\cos\theta\,\vec{a}_r + \frac{5}{r^3}\sin\phi\sin\theta\,\vec{a}_\theta - \frac{1}{r\sin\theta}\left(\frac{5}{r^2}\cos\phi\cos\theta\right)\vec{a}_\phi$$

$$= \frac{10}{r^3}\sin\phi\cos\theta\,\vec{a}_r + \frac{5}{r^3}\sin\phi\sin\theta\,\vec{a}_\theta - \frac{5}{r^3\sin\theta}\cos\phi\cos\theta\,\vec{a}_\phi$$

$$\vec{D} = \frac{10\varepsilon_o}{r^3}\sin\phi\cos\theta\,\vec{a}_r + \frac{5\varepsilon_o}{r^3}\sin\phi\sin\theta\,\vec{a}_\theta - \frac{-5\varepsilon_o}{r^3\sin\theta}\cos\phi\cos\theta\,\vec{a}_\phi$$

Now in this case

$$r = 4,\quad \theta = \frac{\pi}{4},\quad \phi = 0$$

So,

$$\vec{D} = \frac{10}{4^3}\sin(0°)\cos\left(\frac{\pi}{4}\right)\vec{a}_r + \frac{5\varepsilon_o}{4^3}\sin(0°)\sin\left(\frac{\pi}{4}\right)\vec{a}_\theta$$

$$-\frac{5\varepsilon_0}{4^3}\cos(0°)\cos\left(\frac{\pi}{4}\right)\vec{a}_\phi$$

$$\vec{D} = 0.078\vec{a}_\phi\frac{C}{\text{m}^2}$$

5.54 Four particles of the same charge are placed at the corners of a square. This square has an area of 4 m^2. The force on each charge is 10 N. Find the value of each charge.

Solution:

Given that area of the square = 4 m^2 = 2 × 2 m^2, force on each charge = 10 N and side of square = 2 m, let charge $+q$ be placed at four corners. The forces are given as

$$f_{12} = \frac{1}{4\pi\varepsilon_o}\frac{q \cdot q}{r^2} = \frac{9\times10^9}{4}\times q^2\text{N} = 2.25\times10^9 q^2\text{N}$$

$$f_{14} = \frac{1}{4\pi\varepsilon_o}\frac{qq}{r^2} = \frac{9\times10^9\times q^2}{4}\text{N} = 2.25\times10^9 q^2\text{N}$$

$$f_{13} = \frac{9\times10^9\times q^2}{(\sqrt{2^2+2^2})^2}\text{N} = 1.125\times10^9 q^2\text{N}$$

The directions of these forces are such that f_R, the resultant of forces, is given as

$$f_R = 4.5\times9\times10^9\ q^2\ \cos 45°\text{N}$$

$$= 40.5\times10^9\ q^2\frac{1}{\sqrt{2}}\text{N} = 28.63\times10^9 q^2\ \text{N}$$

The resultant of all forces f is given as

$$f = f_R + f_{13}$$

$$= (28.63\times10^9 q^2 + 1.125\times10^9 q^2)\text{N} = 29.76\times10^9 q^2\text{N}$$

$$\Rightarrow f = (29.76\times10^9\times q^2)\ \text{N}$$

As given in the problem statement, the resultant of all forces on the charge is 10 N.
Thus,

$$10 = 29.76\times10^9 q^2$$

$$\Rightarrow q^2 = \frac{10}{29.76\times10^9}$$

$$\Rightarrow q^2 = 3.35\times10^{-10}$$

$$\Rightarrow q = 1.83\times10^{-5}\text{C}$$

5.55 In a Rectangular coordinate system, a charge $Q_1 = -4\ \mu\text{C}$ is placed at the origin. Another charge $Q_2 = -10\ \mu\text{C}$ is placed 30 cm away from the origin on the y-axis. Find the force on Q_1 due to Q_2 (both are in free space).

Solution:

Given that $Q_1 = -4\ \mu\text{C}\,\text{at}\,(0,0,0)$ and $Q_2 = -10\ \mu\text{C}\,\text{at}\,(0,0.3,0)$. As per Coulomb's Law, we find

$$\vec{F}_{12} = \frac{Q_1\,Q_2}{4\pi\varepsilon_0 r^2}\vec{a}_{12}$$

$$r^2 = |\vec{r}| = |(0.\vec{a}_x + 0.\vec{a}_y + 0.\vec{a}_z) - (0.\vec{a}_x + 0.3\vec{a}_y + 0.\vec{a}_z)|$$

$$= \left|-0.3\vec{a}_y\right|$$
$$= 0.3$$

Also, $\vec{a}_{12} = -\vec{a}_y$.

Thus, $$\vec{F}_{12} = \frac{(-4\times10^{-6})\times(-10\times10^{-6})\times9\times10^{9}(-\vec{a}_y)}{(0.3)^2}$$

$$\vec{F}_{12} = -4\vec{a}_y \text{ Newton}$$

5.56 Find the normal and tangential components of $\vec{D}_2$. Given that $\varepsilon_{r1} = 4, \varepsilon_{r2} = 6, \vec{E}_{t1} = \vec{E}_{t2} = (8\vec{a}_x + 9\vec{a}_y)$.

Solution:
Given,

$$\vec{E}_{t2} = \vec{E}_{t1} = 8\vec{a}_x + 9\vec{a}_y$$

Thus,

$$\vec{D}_{t2} = \varepsilon_{r2}\varepsilon_0\vec{E}_{t2} \quad \text{or}$$
$$= 6\times8.85\times10^{-12}(8\vec{a}_x + 9\vec{a}_y) \quad \text{or}$$
$$\left|\vec{D}_{t2}\right| = 6.39\times10^{-12}\left(\text{C/m}^2\right) \quad \text{and}$$
$$\left|\vec{D}_{n2}\right| = \left|\vec{D}_{n1}\right| = \varepsilon_0\varepsilon_{r1}\left|\vec{E}_{1n}\right|$$
$$\left|\vec{D}_{n2}\right| = 8.854\times10^{-12}\times4\times12$$
$$= 4.24\times10^{-10}\left(\text{C/m}^2\right)$$

5.57 Four concentrated charges are located at the vertices of a plane rectangle shown in Figure 5.40. Find the magnitude and direction of the resultant force on q_1.

Solution:
The resultant force on q_1 = ((force of attraction between q_3 and q_1) + (force of repulsion between q_2 and q_1) + (force of repulsion q_4 and q_1)). Coulomb's Law determines the forces $\vec{f}_{31}$, $\vec{f}_{21}$ and $\vec{f}_{41}$. The resultant force on q_1 can be obtained by vector addition of forces (Figure 5.41).
On substitution of values in Coulomb's Law equation, we get

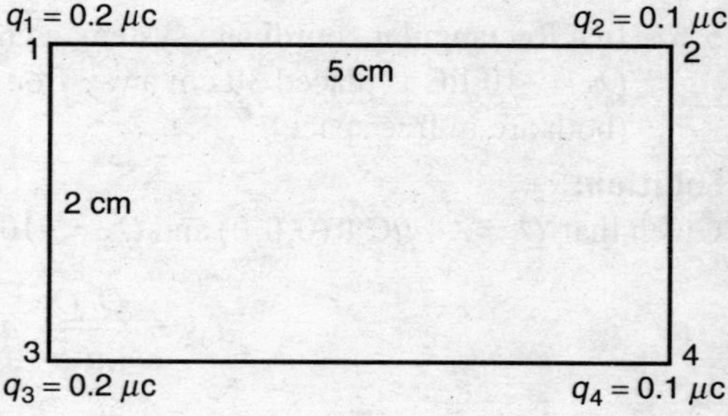

Figure 5.40 *Plane rectangle*

$$\left|\vec{f}_{21}\right| = -0.072 \text{ N (along the } x\text{-axis)} \quad \text{(i)}$$
$$\left|\vec{f}_{31}\right| = -0.9 \text{ N (along the } y\text{-axis)} \quad \text{(ii)}$$
$$\left|\vec{f}_{41}\right| = -0.0577 \text{ N (along the } x\text{-axis)} \quad \text{(iii)}$$
$$\left|\vec{f}_{41}\right| = +0.023 \text{ N (along the } y\text{-axis)} \quad \text{(iv)}$$

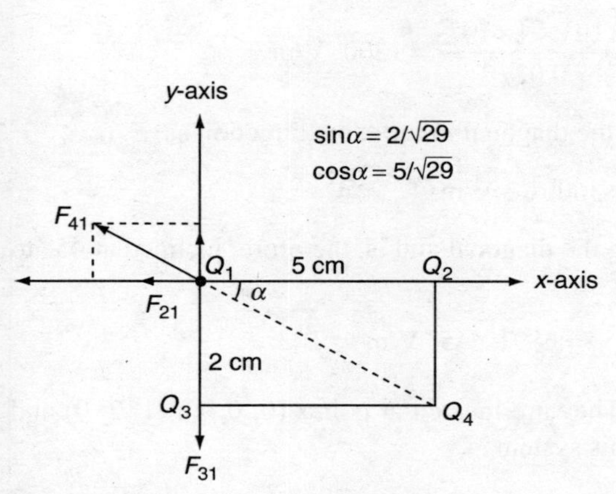

Figure 5.41 *Vector addition of forces*

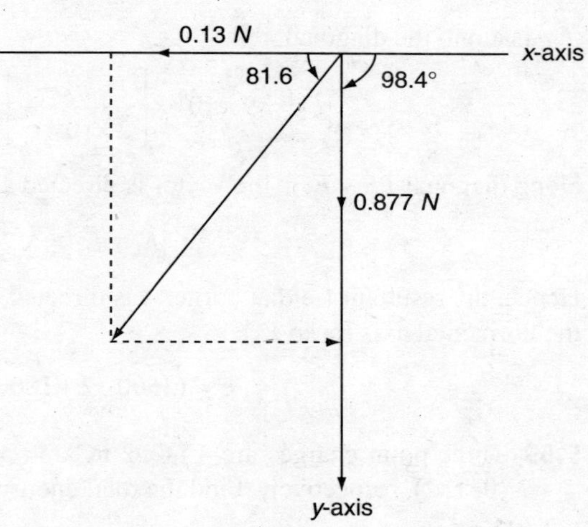

Figure 5.42 *Resultant of forces*

Thus, the resultant is therefore from (i) to (iv) (Figure 5.42).

$$\vec{f_1} = (-0.13 - 0.877j)\text{ N}$$

$$|\vec{f_1}| = 0.887\angle -98.4°\text{ N}$$

5.58 Three equal positive charges of 4×10^{-9} C each are located at three corners of a square of side 10 cm. Determine the magnitude and direction of the electric field at the vacant corner point of square.

Solution:

The three charges q_1, q_2 and q_3 are located at corners 1, 2 and 3, respectively, of the square, with corner 4 vacant. We have to find the resultant field $\vec{E}$ at 4.

The principle of superposition is applied for the evaluation of the resultant electric field intensity $\vec{E}$ at vacant corner 4. Thus,

$$\vec{E} = \vec{E}_1 + \vec{E}_2 + \vec{E}_3$$

The fields E_1 and E_3 are equal in magnitude but perpendicular to each other and directed along 1–4 and 3–4, respectively (Figure 5.43). Here,

$$E_1 = \frac{q_1}{4\pi\varepsilon_0 r^2} = 9\times10^9 \times \frac{q}{(0.1)^2}$$

$$= \frac{9\times10^9 \times 4\times10^{-9}}{(0.1)^2} = 3600\text{ V/m}$$

$$|\vec{E}_1| = |\vec{E}_3| = 3600\text{ V/m}$$

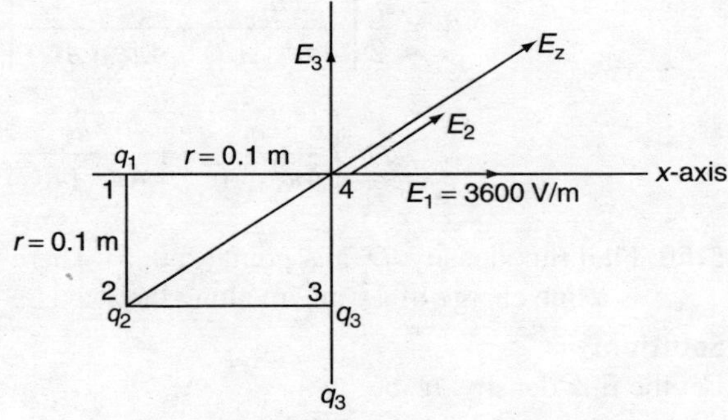

Figure 5.43 *Fields E_1 and E_2*

$\vec{E}_2$ is along the diagonal

$$\left|\vec{E}_2\right| = 9\times10^9 \times\left[\frac{q_2}{2\times(0.1)^2}\right] = \frac{9\times10^9\times4\times10^{-9}}{0.02} = 1800 \text{ V/m}$$

along diagonal $(\vec{E}+\vec{E}_3)$, the vector is directed along the diagonal in the same direction as $\vec{E}_2$.

$$\left|\vec{E}_1+\vec{E}_3\right| = \sqrt{2}\,E_1 = 3600\sqrt{2} \text{ V/m}$$

Hence, the resultant field at corner 4 is directed along the diagonal and is, therefore, inclined at 45° to the horizontal axis (or to E_1).

$$\vec{E} = (3600\sqrt{2}+1800)\angle 45^\circ = 6891\angle 45^\circ \text{ V/m}$$

5.59 Three point charges are 4 nC, 2 nC and 5nC. They are located at points (0, 0, 1), (1, 2, 0) and (0, 1, 2), respectively. Find the total energy in this system.

Solution:

Total energy is given as

$$W = \frac{1}{2}\sum_{K=1}^{3} q_K V_K$$

$$= \frac{1}{2}[q_1V_1+q_2V_2+q_3V_3]$$

Let $A = (0, 0, 1)$, $B = (1, 2, 0)$, $C = (0, 1, 2)$.

Here,

$$AB = \sqrt{(0-1)^2+(0-2)^2+(1-0)^2}$$

$$= \sqrt{1+4+1} = \sqrt{6} = 2.44$$

$$AC = \sqrt{(0-0)^2+(0-1)^2+(1-2)^2} = \sqrt{1+1} = 1.414 \quad \text{and}$$

$$BC = \sqrt{(1-0)^2+(2-1)^2+(0-2)^2} = \sqrt{1+1+4} = 2.44$$

So,

$$W = \frac{q_1}{2}\left[\frac{q_2}{4\pi\varepsilon_0(AB)}+\frac{q_3}{4\pi\varepsilon_0(AC)}\right]+\frac{q_2}{2}\left[\frac{q_1}{4\pi\varepsilon_0(AB)}+\frac{q_3}{4\pi\varepsilon_0(BC)}\right]$$

$$+q_3\left[\frac{q_1}{4\pi\varepsilon_0(AC)}+\frac{q_2}{4\pi\varepsilon_0(BC)}\right]$$

5.60 Find flux density $\vec{D}$ at a point (2, 0, 5). There is a point charge -20π mC at (2, 4, 0). Also, there is a line charge of 4π mC/m along the y-axis.

Solution:

Let the flux density $\vec{D}$ be

$$\vec{D} = \vec{D}_Q + \vec{D}_L$$

where $\vec{D}_Q$ is the flux density due to point charge and $\vec{D}_L$ the flux density due to a line charge.

$$\vec{D}_Q = \varepsilon_0 \vec{E} = \frac{q}{4\pi r^2}\vec{a}_r$$

$$\vec{r} = (2-0)\vec{a}_x + (0-4)\vec{a}_y + (5-0)\vec{a}_z \quad \text{or}$$

$$\vec{r} = (2\vec{a}_x - 4\vec{a}_y + 5\vec{a}_z)$$

Unit vector

$$\vec{a}_r = \frac{\vec{r}}{|\vec{r}|} = \frac{(2\vec{a}_x - 4\vec{a}_y + 5\vec{a}_z)}{6.7}$$

Thus,

$$\vec{D}_Q = \frac{-20\pi \times 10^{-3}}{4\pi \times (6.7)^2} \times \frac{(2\vec{a}_x - 4\vec{a}_y + 5\vec{a}_z)}{6.7}$$

$$= 1.66 \times 10^{-5}(2\vec{a}_x - 4\vec{a}_y + 5\vec{a}_z)\,\text{C/m}^2$$

$$= (332\,\vec{a}_x - 664\,\vec{a}_y + 830)\,\text{mC/m}^2 \quad \text{and}$$

similarly

$$\vec{D}_L = \frac{\rho_L}{2\pi r}\vec{a}_r$$

Here,

$$\vec{r} = (2-0)\vec{a}_x + (0-0)\vec{a}_y + (5-0)\vec{a}_z$$

$$= (2\vec{a}_x + 5\vec{a}_z)$$

Unit vector

$$\vec{a}_r = \frac{\vec{r}}{|\vec{r}|}$$

$$= \frac{(2\vec{a}_x + 5\vec{a}_z)}{5.3}$$

$$\vec{D}_L = \frac{4\pi}{2\pi \times 5.3} \times \frac{(2\vec{a}_x + 5\vec{a}_z) \times 10^{-3}}{5.3}$$

$$= 7.1 \times 10^{-5}(2\vec{a}_x + 5\vec{a}_z)$$

$$\vec{D}_L = (14.2\vec{a}_x + 35.5\vec{a}_z)\,\text{mC/m}^2$$

5.61 In a case, the potential is $V = (4x^2y + 3By^2z)$ V. Find the value of B. Also find electric field $\vec{E}$ at point (1, 1, 1).

Solution:

As per Laplace's Equation,

$$\nabla^2 V = 0 \quad \text{or}$$

$$\frac{\partial^2 V}{\partial x^2}+\frac{\partial^2 V}{\partial y^2}+\frac{\partial^2 V}{\partial z^2}=0$$

Thus,

$$8y+3B\times 2z+0=0$$

$$\Rightarrow 8y+6Bz=0$$

$$\Rightarrow 8+6B=0 \quad \Rightarrow \quad 6B=-8 \quad \Rightarrow \quad B=\frac{-8}{6}=\frac{-4}{3}$$

So,

$$B=\frac{-4}{3}$$

Further,

$$\vec{E}=-\nabla V$$

$$=-\left[\frac{\partial V}{\partial x}\vec{a}_x+\frac{\partial V}{\partial y}\vec{a}_y+\frac{\partial V}{\partial z}\vec{a}_z\right]$$

$$=-\left[8xy\,\vec{a}_x+(4x^2+6B)z\vec{a}_y+3B\vec{a}_z\right]$$

Putting the value, we get

$$=-\left[8\vec{a}_x+\left(4+6\times\left(\frac{-4}{3}\right)\right)\vec{a}_y+3\left(\frac{-4}{3}\right)\vec{a}_z\right]$$

$$=-\left[8\vec{a}_x+(4-8)\,\vec{a}_y-4\vec{a}_z\right]$$

$$=-\left[8\vec{a}_x-4\vec{a}_y-4\vec{a}_z\right]=(-8\vec{a}_x+4\vec{a}_y+4\vec{a}_z)$$

$$\vec{E}=(-8\vec{a}_x+4\vec{a}_y+4\vec{a}_z)$$

5.62 Find the charge density if in a case electric potential V is $(4x^3 + 2y^2)$ at point (1, 0, 2).

Solution:

Given, $V=(4x^3+2y^2)$.

Further, $$\nabla^2 V=\frac{\partial^2}{\partial x^2}(4x^3+2y^2)+\frac{\partial^2}{\partial y^2}(4x^3+2y^2)+\frac{\partial^2}{\partial z_2}(4x^3+2y^2) \quad \text{or}$$

$$\nabla^2 V=24x+4+0$$

$$=24x+4$$

$$=24\times 1+4=28 \quad (\because x=1)$$

According to Poisson's Equation, we get

$$\nabla^2 V=\frac{-\rho}{\varepsilon_o} \quad \text{or}$$

$$28 = \frac{-\rho}{\varepsilon_o}$$

Finally,

$$\rho = -28\varepsilon_o \text{ (Coulomb/m}^2\text{)}$$

5.63 Consider a magnetic flux density $\vec{B} = 5\vec{a}_x + 12\vec{a}_y + 4\vec{a}_z$ Wb/m^2 with a current element $0.30\vec{a}_z$ (mA – m) placed in it. What will be the 'force' per metre in this current element?

Solution:

Force $\vec{F}$ in a current element in the presence of magnetic field is given as

$$\vec{F} = I(\vec{L} \times \vec{B})$$

Here $I\vec{L} = 0.30\vec{a}_z(\text{mA} - \text{m})$ or $= 0.30 \times 10^{-3}\vec{a}_z(\text{A} - \text{m})$

Given that $\vec{B} = (5\vec{a}_x + 12\vec{a}_y + 4\vec{a}_z)\text{Wb/m}^2$

Thus,

$$\vec{F} = 0.30 \times 10^{-3}\vec{a}_z \times (5\,\vec{a}_x + 12\,\vec{a}_y + 4\,\vec{a}_z) \quad \text{or}$$

$$\vec{F} = -3.6\vec{a}_x + 1.5\vec{a}_y \text{ mN}$$

5.64 A magnetic flux through a coil is perpendicular to its plane. It is directed into paper. It is varying as per relationship $\phi = (40t^2 + 15t + 16)$ (milli Weber). Find the induced emf in the loop at time $t = 15$ sec.

Solution:

Magnetic flux ϕ is given as

$$\phi = (40t^2 + 15t + 16) \text{ milli Weber}$$

$$= (40t^2 + 15t + 16) \times 10^{-3} \text{ Weber}$$

Induced emf is obtained as

$$e = -N\frac{d\phi}{dt}$$

where $N = 1$.

$$e = \frac{-d\phi}{dt}$$

$$= (-40t - 15) \times 10^{-3} \text{ V}$$

At $t = 15$ sec

$$e = (-40 \times 15 - 15) \times 10^{-3}$$

$$= (-700 - 15) \times 10^{-3}$$

$$= -715 \times 10^{-3} \text{(V)}$$

5.65 A single phase circuit comprises two parallel conductors. Parallel conductors A and B are of 6-cm diameter each. Conductors A and B are spaced 2 m apart. Conductors A and B carry a current of + 40 and –40 Amps, respectively. Find, the field intensity at the surface of each conductor. Also, find the field intensity in space exactly midway of conductors A and B.

Solution:

Field $\vec{H}$ at a point between parallel conductors A and B is obtained as

$H = H_A + H_B$ (because currents are in opposite directions)

$$= \frac{1}{2\pi R_1} + \frac{1}{2\pi R_2}$$

Each conductor produces a field, which will be directed downwards at that point. Therefore, we get (assume two cases)

Case 1: (H at the conductor surface). We know

$$R_1 = \frac{6}{100 \times 2}\,(\text{m}) \quad \text{and}$$

$$R_2 = 2\text{m}$$

Thus,

$$H = \frac{I}{2\pi}\left[\frac{1}{R_1} + \frac{1}{R_2}\right] = \frac{40}{2}\left[\frac{2\times 100}{6} + \frac{1}{2}\right]$$

$$= 218.57\ (\text{A/m})$$

Case 2: H at midway between parallel conductors A and B is obtained. We know that

$$R_1 = \frac{2}{2} = 1\text{ m}$$

$$R_2 = \frac{2}{2} = 1\text{ m}$$

Thus,

$$H = \frac{40}{2\pi}[1+1]$$

$$= 2.73\ (\text{A/m})$$

5.66 Three infinite plane sheets lie in $z = -2$, $z = 0$ and $z = 2$. These infinite plane sheets of charge have uniform surface charge densities as ρ_{s_1}, ρ_{s_2} and ρ_{s_3} C/m^2, respectively. If the resulting electric field intensities at points (4, 5, −1), (5, −3, 1) and (5, 7, 5) are 0, $-2\vec{a}_z$ and $\vec{a}_z$ (V/m), then find charge densities ρ_{s_1}, ρ_{s_2} and ρ_{s_3}.

Solution:

Electric field $\vec{E}$ is given as

$\vec{E} = \pm\dfrac{\rho_s}{2\varepsilon_0}\vec{a}_n$ (infinite plane sheet with charge density ρ_s). Electric fields in different regions for the points are shown in Figure 5.44.

These electric fields are

$$\vec{E}_1 = \frac{\rho_{s_1}}{2\varepsilon_0}, \quad \vec{E}_2 = \frac{\rho_{s_2}}{2\varepsilon_0} \quad \text{and} \quad \vec{E}_3 = \frac{\rho_{s_3}}{2\varepsilon_0}$$

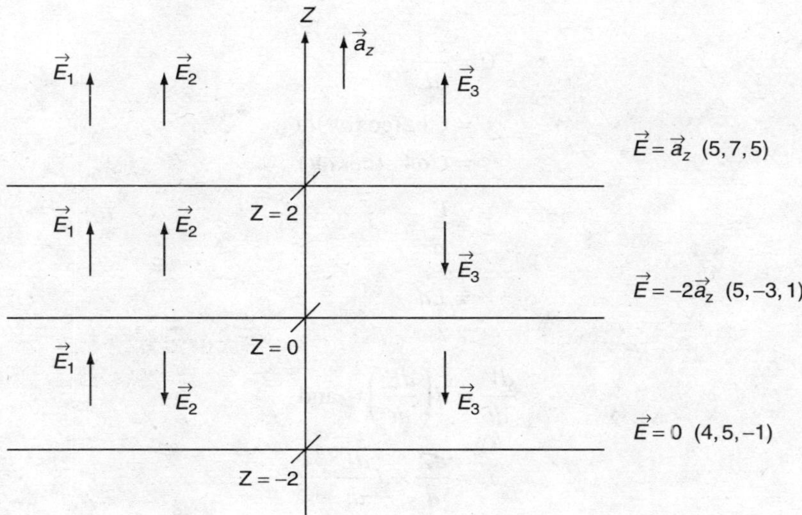

Figure 5.44 *Electric field in different regions of points*

Adding and equating the fields for these points, we get

$$\frac{\rho_{s_1}}{2\varepsilon_0}\vec{a}_z + \frac{\rho_{s_2}}{2\varepsilon_0}\vec{a}_z + \frac{\rho_{s_3}}{2\varepsilon_0}\vec{a}_z = \vec{a}_z$$

$$\frac{\rho_{s_2}}{2\varepsilon_0}\vec{a}_z + \frac{\rho_{s_2}}{2\varepsilon_0}\vec{a}_z - \frac{\rho_{s_3}}{2\varepsilon_0}\vec{a}_z = -2\vec{a}_z \quad \text{and}$$

$$\frac{\rho_{s_2}}{2\varepsilon_0}\vec{a}_z - \frac{\rho_{s_2}}{2\varepsilon_0}\vec{a}_z - \frac{\rho_{s_3}}{2\varepsilon_0}\vec{a}_z = 0$$

Solving these, we find

$$\rho_{s_1} = \varepsilon_0\,(\text{C/m}^2)$$

$$\rho_{s_2} = -2\varepsilon_0(\text{C/m}^2) \quad \text{and}$$

$$\rho_{s_3} = 3\varepsilon_0(\text{C/m}^2)$$

5.67 Prove that in Figure 5.45, displacement current is equal to the conduction current.

Solution:

In Figure 5.45, current through the capacitor is

$$i_c = C\frac{dv}{dt}$$

$$= C\frac{d}{dt}(V_m \sin \omega t)$$

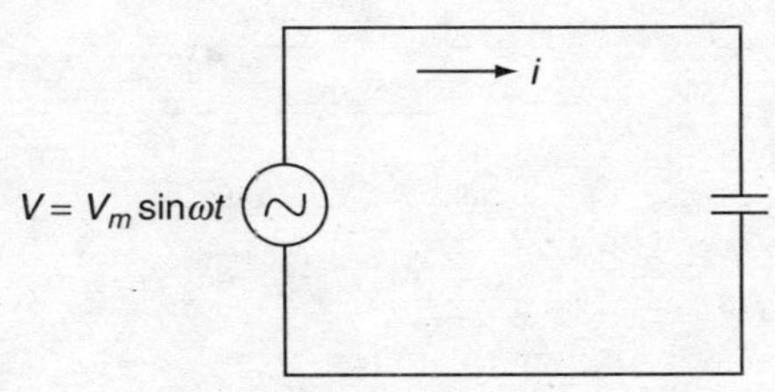

Figure 5.45 *Electrical circuit*

where

$$C = \frac{\varepsilon A}{d}$$

$$i_c = CV_m(\cos\omega t)\omega$$
$$= C\omega V_m(\cos\omega t)$$

$$\vec{E} = \frac{V}{d}$$

$$V = \vec{E}d$$

So,

$$\frac{dV}{dt} = d\left(\frac{dE}{dt}\right) \quad \text{and}$$

$$i_c = \frac{\varepsilon A}{d} \times d\frac{d\vec{E}}{dt} \quad \text{or}$$

$$i_c = A\frac{d}{dt}(\varepsilon\vec{E}) \quad \text{or}$$

$$i_c = A\frac{d\vec{D}}{dt}$$

So,

$$\frac{i_c}{A} = \frac{d\vec{D}}{dt}$$

We know that

$$I_d = \frac{d\vec{D}}{dt}$$

$$I_d = \left(\frac{d\vec{D}}{dt}\right)A \quad \text{or}$$

$$= A\frac{d}{dt}(\varepsilon\vec{E}) \quad \text{or}$$

$$= \frac{\varepsilon A}{d}\frac{d}{dt}(V_m \sin(\omega t)) \quad \text{or}$$

$$= C\frac{dv}{dt} \quad \text{or}$$

$$= i_c$$

As

$$\vec{E} = \frac{V}{d} \quad \text{and}$$

$$D = \varepsilon \frac{V_m \sin \omega t}{d}$$

$$I_d = \frac{\partial \vec{D}}{\partial t} A \quad \text{or}$$

$$= \frac{\varepsilon A}{d} \frac{d}{dt}(V_m \sin \omega t) \quad \text{or}$$

$$= \frac{\varepsilon A}{d} \omega V_m \cos(\omega t) \quad \text{or}$$

$$= C\omega V_m \cos \omega t \quad \text{or}$$

$$= i_c$$

Hence, proved.

UNSOLVED QUESTIONS

5.1 A positive charge $+q$ of 150 pC is placed in air as shown in Figure 5.46. Equipotential surfaces at radii of 20 mm and 50 mm are also shown. Obtain the potential rise between the two radii.

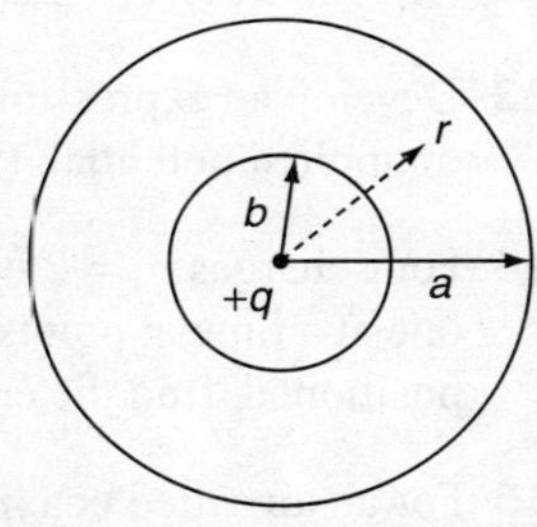

Figure 5.46 *Equipotential surface*

5.2 Find the potential at the centre of a square with each side of 2.0 meter. Four charges q_1, q_2, q_3 and q_4 placed at four corners of this square are $+2.0 \times 10^{-8}$ C, -3.0×10^{-8} C, $+4 \times 10^{-8}$ C and $+ 3.0 \times 10^{-8}$ C, respectively.

5.3 Obtain the electric field due to the following charge distribution (in Spherical coordinates):

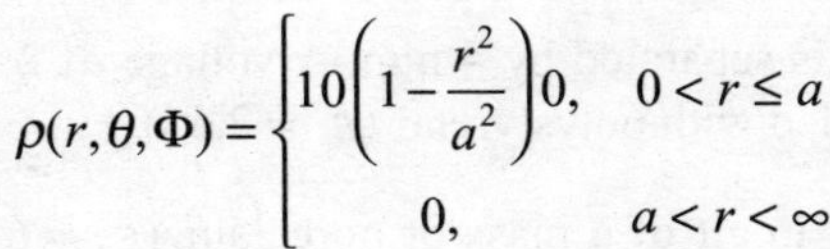

$$\rho(r, \theta, \Phi) = \begin{cases} 10\left(1 - \dfrac{r^2}{a^2}\right)0, & 0 < r \le a \\ 0, & a < r < \infty \end{cases}$$

(Hint: Use Gauss's Law.)

5.4 An infinite long wire is horizontally stretched 3 m above the earth. The charge per length is 1 m coulomb per cm. Find the electric field at a point vertically below the wire on the earth.

5.5 Two charges of $+40\ \mu$C and $-40\ \mu$C are kept at a distance of 3 cm. Find the electric field at a point on the equatorial line. This point is at a distance of 40 cm from the centre of the dipole.

5.6 In a case, two charges $q_A = 1.5 \times 10^{-7}$ C and $q_B = -1.5 \times 10^{-7}$ C are located at points $A(0, 0, -15)$ cm and $B(0, 0, +15)$ cm. Find the total charge and electric dipole movement.

5.7 An electric dipole at the origin is $2.1\, \vec{a}_y$ nC-m. Find the potential at point P (0, 10, 0).

5.8 An infinite sheet of charge has a surface charge density σ of 2.0×10^{-7} Coulomb/m^2. What is the distance between two equipotential surfaces carrying a potential difference of 8.0 Volt? (Consider $\varepsilon_0 = 8.88 \times 10^{-12}\ \text{C}^2/\text{Nm}^2$)

5.9 A potential V of −2.0 V exists on a conductor at 0.04 m. V_0 is 10.0 V at $r = 0.25$ m. A dielectric material with $\varepsilon_r = 3.0$ separates the conductors. Find the surface charge density on the conductors.

5.10 A circular ring carries a uniform charge ρ_L (C/m). Its radius is a. It is placed on the x–y plane with its axis as the z-axis.

(a) Prove $\vec{E}(0,0,h) = \dfrac{\rho_L a\, h}{2\varepsilon_0[h^2 + a^2]^{3/2}}\vec{a}_z$.

(b) Find the value of h giving the maximum value of $\vec{E}$.

(c) Find $\vec{E}$ as $a \to 0$.
The total charge on the ring is q.

5.11 A point charge of 25 nC is placed at the origin. Plane $y = 3$ carries a charge of $8\,(\text{nC/m}^2)$. Find $\vec{D}$ at a point (0, 3, 4).

5.12 Satisfy Laplace's equation for the following potential functions:

(i) $V = x^2 - y^2 + z$ and
(ii) $V = 3x^2 + y^2 - 2z^2$.

5.13 Develop an expression for the potential difference at a point between two spherical shells in terms of applied potential. (Hint: Use Laplace's Equation.)

5.14 Four charges $q_1 = 2$ nC, $q_2 = -3$ nC, $q_3 = 1$ nC and $q_4 = -2$ nC are positioned in the same order, one at a time at points (0, 0, 0), (1, 0, 0), (0, 0, −1) and (0, 0, 1), respectively. After each charge is positioned, find the energy in the system.

5.15 For an air-filled coaxial cable, calculate the capacitance per km of length. The inner conductor is of a 3-mm diameter and the outer conductor is of a 1-cm diameter.

5.16 A parallel plate capacitor is separated by 4 mm. A voltage of 2 kV is applied to its plates. The space between plates is filled with polystyrene ($\varepsilon_r = 2.55$). Find $\vec{E}, \vec{P}$ and ρ_{es}.

5.17 The electric field strength in air of a mass of porcelain ($\varepsilon_{r_1} = 6$) is 100 (V/cm). At the inner surface of this mass, the electric field makes an angle of 60° to the normal and finally emerges in the air. Find the angle of emergence of the external electric field.

5.18 A two-plate capacitor of size 10 cm square is placed parallel. Plates are kept 2 mm apart. A dielectric with relative permittivity of 2.5 fills the space between the plates. A potential difference of 300 V is held between the plates. Find capacitance of this capacitor.

5.19 Polarization $\vec{P}$ is $3\vec{a}_x$ (nC/m^2) in a homogeneous and isotropic dielectric material. Considering η_e to be 5.5, find $\vec{E}$.

5.20 Obtain a general equation of electric flux lines representing the electric field given as

$\vec{E} = \dfrac{p}{4\pi\varepsilon_0 r^3}(2\cos\theta\vec{a}_r + \sin\theta\vec{a}_\theta)$ (in a Spherical coordinate system)

where p is the dipole moment.

5.21 Find the surface charge density at the plane boundary between two media. The current of density 8 (A/m^2) is flowing normal to the surface across the boundary. Consider $\sigma_1 = 80$ (s/m), $\varepsilon_1 = 10\varepsilon_0$, $\mu_1 = \mu_0$ and $\sigma_2 = 300$ (s/m), $\varepsilon_2 = 15\varepsilon_0$ and $\mu_2 = \mu_0$ for two media.

5.22 The tangential electric field in a medium at a plane boundary between two conductors is 3 (V/m). The tangential current density in another medium is 4 (A/m^2). Find the conductivity of the second medium.

5.23 Define and explain the Uniqueness Theorem. Prove it.

5.24 Explain the method of images in detail.

SUMMARY

1. As per Coulomb's law,

$$\vec{F} = \frac{1}{4\pi\varepsilon_0}\frac{q_1 q_2}{R^2}\vec{a}_R \text{ Newton}$$

2. Electric field intensity $\vec{E}$ is given as

$$\vec{E} = \frac{\vec{F}}{q}$$

3. Electric field at a point due to charge q_1 is given as

$$\vec{E} = \frac{q_1}{4\pi\varepsilon_0\left|\vec{R}\right|^2}\vec{a}_R$$

4. Electric dipole is a pair of equal and opposite point charges.

5. Electric field intensity "E" because of line charge is given as

$$\vec{E} = \int \frac{1}{4\pi\varepsilon_0}\frac{\rho_L \, dl}{R^2}\vec{a}_R$$

6. Electric field intensity "E" because of surface charge is given as

$$\vec{E} = \int \frac{1}{4\pi\varepsilon_0}\frac{\rho_s \, ds}{R^2}\vec{a}_R$$

7. Electric field intensity "E" because of volume charge is given as

$$\vec{E} = \int \frac{1}{4\pi\varepsilon_0}\frac{\rho_v \, dv}{R^2}\vec{a}_R$$

8. Electric flux density D in free space is given as

$$D = \varepsilon_0 E$$

9. Electric potential V_ρ is given as

$$V_\rho = -\int_{\infty}^{\rho} E.dl \quad \text{or} \quad V_N - V_M = -\int_M^N E \cdot dl$$

10. Electric potential and work done are related as

$$V_N - V_M = \frac{W_{MN}}{q}$$

11. As per Gauss's law, total electric flux ϕ is given as

$$\phi = \oint \vec{D}.d\vec{s}$$

12. The implementation of Gauss's law requires a Gaussian surface. It is applied to find out electric field in the cases where charge distribution is symmetrical.

13. Electric field at point P due to infinite line charge is given as

$$\vec{E} = \frac{\rho_L}{2\pi\varepsilon_0 R} \; \vec{a}_R$$

14. Electric field at point P in free space due to infinite sheet charge is given as

$$\vec{E} = \frac{\rho_s}{2\varepsilon_0} \; \vec{a}_z$$

15. Curl of electrostatic field is zero. Therefore, this field is a conservative field.

16. For "n" point charges, total work done W is obtained as

$$W = \frac{1}{2}\sum_{i=1}^{n} q_i \; V_i$$

17. Continuity equation is given as

$$\nabla \cdot J = -\frac{\partial \rho_v}{\partial t}$$

18. Poisson's equation is given as

$$\nabla \cdot \nabla V = \nabla^2 V = \frac{-\rho_V}{\varepsilon_0}$$

19. Laplace's equation is given as $\nabla^2 V = 0$.

20. Boundary conditions at the interface of two mediums if both mediums are dielectric are given as

$\vec{E}_{1t} = \vec{E}_{2t}$ (Tangential components of $\vec{E}$ on two sides of the boundary) and

$$\frac{\tan\alpha_2}{\tan\alpha_1} = \frac{\varepsilon_2}{\varepsilon_1}$$

21. Boundary conditions at the interface of two mediums, if medium '1' is the conductor and medium '2' is a dielectric are given as

$$D_{2n} = \rho_s \quad \text{and}$$

$$E_{2n} = \frac{\rho_s}{\varepsilon_2}$$

22. Boundary condition at the interface of two mediums, if medium '1' is the conductor and medium '2' is free space is given as

$$E_{2n} = \frac{\rho_s}{\varepsilon_0}$$

23. According to the Uniqueness theorem, for a large class of boundary conditions, Poisson's equation gives a unique gradient of solution. It is given as

$$\sum_i \int_{S_i} (\phi \nabla \phi) \cdot ds = 0$$

24. Method of images is also called mirror images. It is a mathematical tool for solving problems related to electric field distribution of a charge in the vicinity of a conducting surface and the problems related to magnetic field produced by a magnet in the vicinity of a superconducting surface.

MULTIPLE-CHOICE QUESTIONS

1. Maxwell's equations involve __________.
(a) Charge density (b) Current density
(c) Magnetic intensity (d) All of these

2. As per Coulomb's Law, force between two point charges is __________.
(a) $\propto$ (distance)^2 (b) $\propto$ (distance) (c) $\propto \dfrac{1}{\text{(distance)}}$ (d) $\propto \dfrac{1}{\text{(distance)}^2}$

3. ε_0 is __________ F/m.
(a) 8.854×10^{-12} (b) 6.654×10^{-12} (c) 8.854×10^{-10} (d) 6.654×10^{-10}

4. Electric field intensity $\vec{E}$ is related to force $\vec{F}$ and charge q as __________.
(a) $\vec{F}q$ (b) $\vec{F}/q$ (c) $\vec{F}/\sqrt{q}$ (d) $\sqrt{|F|/q}$

5. Electric dipole is a pair of __________ and __________ point charges.
(a) Equal and same (b) Unequal and same
(c) Equal and opposite (d) Unequal and opposite

6. Magnitude of dipole moment $|\vec{p}|$ is given as __________.
(a) qd (b) q/d (c) d/q (d) None of these

7. Electric field intensity $\vec{E}$ due to volume charge is given as __________.
(a) $\int \frac{1}{4\pi\varepsilon_0} \frac{\rho_v}{R^2} dv$ (b) $\int \frac{1}{4\pi\varepsilon_0} \frac{\rho_v}{R^2} dv\ \vec{a}_R$ (c) $\int \frac{1}{4\pi\varepsilon_0} \frac{\rho_v}{R} dv$ (d) $\int \frac{1}{4\pi\varepsilon_0} \frac{\rho_v}{R} dv\ \vec{a}_R$

8. Electric flux density __________ medium.
(a) Depends on (b) Independent of (c) Both (a) and (b) (d) None of these

9. Electric potential energy V is given as __________.
(a) $\frac{1}{4\pi\varepsilon_0}\frac{q_1q_2}{R^2}$ (b) $\frac{1}{4\pi\varepsilon_0}\frac{R^2}{q_1q_2}$ (c) $\frac{1}{4\pi\varepsilon_0}\frac{q_1q_2}{R}$ (d) $\frac{1}{4\pi\varepsilon_0}\frac{R}{q_1q_2}$

10. As per Gauss's Law, the total electric flux ϕ through a closed surface and the total charge q_{enc} by that surface are related as __________.
(a) $\phi = \sqrt{q_{enc}}$ (b) $\phi = \frac{1}{\sqrt{q_{enc}}}$ (c) $\phi = \frac{1}{q_{enc}}$ (d) $\phi = q_{enc}$

11. If work done W_{MN} is zero, then V_N and V_M are related as __________.
(a) $V_N + V_M = 0$ (b) $V_N - V_M = 0$ (c) $2V_N + V_M = 0$ (d) $V_N + 2V_M = 0$

12. According to Maxwell's first equation, __________.
(a) $\nabla \cdot \vec{D} = \rho_v$ (b) $\nabla \times \vec{D} = \rho_v$ (c) $\nabla \cdot \vec{D} = \rho_s$ (d) $\nabla \times \vec{D} = \rho_s$

13. Charge distribution symmetry can be of __________ type.
(a) Spherical (b) Cylindrical (c) Planar (d) All of these

14. Curl of electrostatic field is __________.
(a) ∞ (b) 1 (c) 0 (d) None of these

15. As per Stoke's Theorem, $\oint \vec{E}.d\vec{l} =$ __________.
(a) ∞ (b) 1 (c) 0 (d) None of these

16. Potential energy __________ if a test charge is moved from a lower potential point to a higher potential point.
(a) Remains the same (b) Increases
(c) Decreases (d) Becomes zero

17. Continuity equation is given as __________.
(a) $\Delta \cdot J = -\frac{\partial \rho_v}{\partial t}$ (b) $\nabla \times J = -\frac{\partial \rho_v}{\partial t}$ (c) $\nabla \cdot J = \frac{\partial \rho_v}{\partial t}$ (d) $\nabla \times J = \frac{\partial \rho_v}{\partial t}$

18. If both the mediums are dielectrics, then boundary condition is given as __________.
(a) $\vec{E}_{1t} + \vec{E}_{2t} = 0$ (b) $\vec{E}_{1t} - \vec{E}_{2t} = 0$ (c) $\vec{E}_{1t} \cdot \vec{E}_{2t} = 0$ (d) None of these

19. Poisson's equation is given as __________.
(a) $\nabla^2 V = \frac{-\rho_v}{\varepsilon_0}$ (b) $\nabla^2 V = \frac{\rho_v}{\varepsilon_0}$ (c) $\nabla^2 V = \rho_v \varepsilon_0$ (d) $\nabla^2 V = -\rho_v \varepsilon_0$

20. In the case of a linear material medium, __________ equation can be derived easily from Gauss' law.
(a) Poisson (b) Laplace (c) Both (a) and (b) (d) None of these

21. According to the method of images, the tangential component of electric field to conductor surface is __________.
(a) Unity (b) Zero (c) Infinity (d) None of these

ANSWERS TO MULTIPLE-CHOICE QUESTIONS

(1) d	(2) d	(3) a	(4) b	(5) c	(6) a	(7) b	(8) b	(9) c
(10) d	(11) b	(12) a	(13) d	(14) c	(15) c	(16) b	(17) a	(18) b
(19) a	(20) c	(21) b						

6 Magnetostatics

6.1 INTRODUCTION

Current produces a magnetic field around it because of the movement of free charges. Steady magnetic fields are similar to steady electrostatic fields. Two laws governing steady magnetic fields are the Biot–Savart's Law and Ampere's Circuital Law, which help in obtaining magnetic fields due to different current distributions. Chapter 5 dealt with static electric field ($\vec{E}$ or $\vec{D}$). Static magnetic fields are denoted by $\vec{H}$ or $\vec{B}$. Static electric field and static magnetic field are quite similar as well as dissimilar. A constant current flow produces a magnetostatic field. This current flow is different in different cases like magnetization current in permanent magnets or current due to an electron beam in vacuum tubes or conduction currents as in current-carrying conductors. Magnetic field in free space is produced due to direct current. In addition, direct current produces magnetostatic fields in material space. The various things involving magnetic phenomenon are transformers, motors, microphones, television focusing controls, telephone bell ringers, advertising displays, magnetically levitated high-speed vehicles (MAGLEV), magnetic separators, memory stores, etc. Basic laws and techniques are used to find the magnetic field due to current-carrying elements. A magnetic field exerts a force on charged particles. It also affects current elements and loops. Such magnetic fields and their analysis help in solving the problems pertaining to electrical devices. Various electrical devices are voltmeters, ammeters, galvanometers, plasmas, cyclotrons, transformers, motors, magneto hydrodynamic generators, etc.

6.2 MAGNETIC FORCES

Any two current-carrying conductors kept in close vicinity exert force on each other. This force between two straight parallel wires (conductors) carrying currents I_M and I_N is always proportional to $I_M I_N / d$ where d is the distance between two conductors. Force of attraction will be produced if the two currents are in the same directions. Force of repulsion will be produced if the two currents are in opposite directions.

The force due to a magnetic field can be experienced in two ways, viz.

(1) *Force between two elements carrying currents*: Consider two current-carrying conductors as shown in Figure 6.1. The force produced between two circuits carrying currents I_M and I_N is given as

$$F_{MN} = \frac{\mu_0}{4\pi} I_M I_N \oint_M \oint_N \frac{dI_N \times (dI_M \times \vec{a}_r)}{r^2} \text{ Newtons}$$

where F_{MN} is the force exerted by current I_M on current I_N. Line integrals are used for two circuits.

Force F_{MN} can also be conveniently expressed as

$$F_{MN} = I_N \oint_N dI_N \times \left(\frac{\mu_0}{4\pi} I_M \oint_M \frac{dI_M \times \vec{a}_r}{r^2} \right) \text{Newtons} \quad \text{or}$$

$$F_{MN} = I_N \oint_N dI_N \times B_M \text{ Newtons}$$

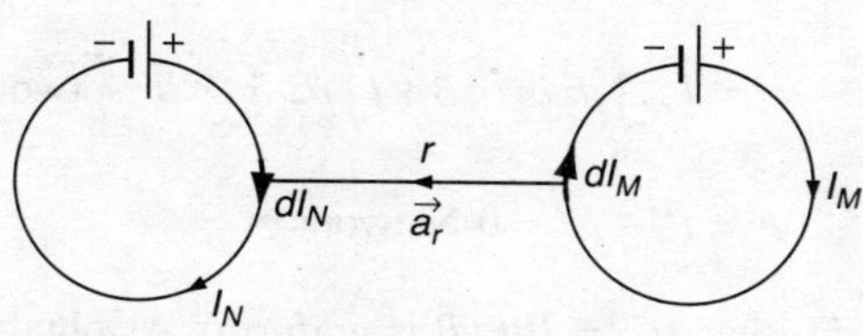

Figure 6.1 *Two current-carrying conductors*

Here,

$$B_M = I_M \frac{\mu_0}{4\pi} I_M \oint_M \frac{dI_M \times \vec{a}_r}{r^2}$$

where B_m is the magnetic induction (magnetic flux density). It is due to current in the first circuit M at a position of an element dI_N in the second circuit N. The unit of B_r is in tesla or weber/m^2.

(2) *Force on a charged particle*: An electric force $\vec{F}_e$ on a stationary or moving electric charge q kept in an electric field is governed by Coulomb's experimental law. Electric force $\vec{F}_e$ is related to electric field intensity $\vec{E}$ as

$$\vec{F}_e = q\vec{E} \text{ Newtons}$$

where q is the charge. If q is positive, then $\vec{F}_e$ and $\vec{E}$ will have the same directions.

A magnetic field exerts force on a moving charge only. Magnetic force $\vec{F}_m$ due to a charge q moving with a velocity $\vec{u}$ in a magnetic field $\vec{B}$ is given as

$$\vec{F}_m = q\,\vec{u} \times \vec{B} \text{ Newtons}$$

$\vec{F}_m$ is always perpendicular to $\vec{u}$ and $\vec{B}$. $\vec{F}_e$ is independent of the velocity of charge. Electric force $\vec{F}_e$ works on the charge and brings changes in its kinetic energy. Unlike $\vec{F}_e$, $\vec{F}_m$ depends on charge velocity. $\vec{F}_m$ is normal to velocity. $\vec{F}_m$ cannot work on charge as it is at right angles orthogonal to the direction of charge motion ($\vec{F}_m \cdot \vec{dl} = 0$); it does not cause any increment in the charge's kinetic energy. In the case of moving charge q in the presence of both fields, viz. electric and magnetic, the total force (on charge) is

$$\vec{F} = \vec{F}_e + \vec{F}_m \quad \text{or}$$

$$\vec{F} = q(\vec{E} + \vec{u} \times \vec{B}) \qquad \text{(Lorentz force equation)}$$

This is known as the Lorentz force equation. It relates a mechanical force to an electrical force.

6.2.1 Magnetic Torque and Moment

Force in a current loop placed in a magnetic field also generates torque. If the current loop is kept in parallel to a magnetic field (along with it), then the force thus generated tries to rotate the current loop. The magnetic torque $\vec{T}$ (or mechanical moment of force) on the current loop is given as the vector product of force $\vec{F}$ and moment arm $\vec{r}$. Magnetic torque $\vec{T}$ is given as

$$\text{Magnetic torque } \vec{T} = \vec{r} \times \vec{F} \text{ (N-m)}$$

Consider a rectangular loop of length l and width w kept in a uniform magnetic field $\vec{B}$ as shown in Figure 6.2. The length $\vec{dl}$ is parallel to magnetic field $\vec{B}$. Along the sides ab and cd of this current loop, no force will be exerted. Here, force on the loop

$$\vec{F} = I\int_d^c \vec{dl} \times \vec{B} + I\int_d^a \vec{dl} \times \vec{B} \text{ Netwons}$$

$$= I \times \int_0^1 dz\, \vec{a}_z \times \vec{B} + I \int_1^0 dz\, \vec{a}_z \times \vec{B} \text{ Netwons}$$

$$F = F' - F' = 0 \text{ Netwons}$$

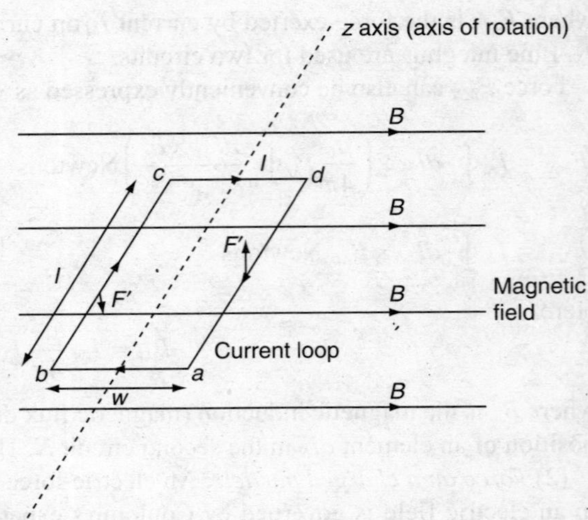

Figure 6.2 *Rectangular loop in magnetic field*

Further, $|F'| = IBl$ ($\vec{B}$ is uniform). As obtained above, no force is exerted on the current loop as a whole. Forces F_0 and $-F_0$ act at different points on the current loop. It creates a couple. Let loop normal to the plane of current makes an angle α with magnetic field $\vec{B}$. Magnetic torque $\vec{T}$ of loop is given as

$$|\vec{T}| = |\vec{F}|\, w \sin\alpha \text{ (N-m)} \quad \text{or}$$

$$|\vec{T}| = |\vec{F}|\, lwI \sin\alpha \text{ (N-m)} \quad \text{or}$$

$$|\vec{T}| = |\vec{B}|\, IS \sin\alpha \text{ (N-m)}$$

where $S = lw$ (area of loop). Further, $\vec{m} = IS\vec{a}_n$, where $\vec{m}$ is the magnetic dipole moment. Magnetic dipole moment is a product of current, area of the loop and its direction. $\vec{m}$ is normal to the current loop. Magnetic torque $\vec{T}$ is given as

$$\vec{T} = \vec{m} \times \vec{B} \text{ (N-m)}$$

This expression helps in determining the magnetic torque on a planar loop. The loop can be of any arbitrary shape. No matter that here $\vec{T}$ was obtained using a rectangular loop, it is applicable on any loop shape. The condition is that the magnetic field must be uniform.

6.3 FARADAY'S LAWS OF ELECTROMAGNETIC INDUCTION

There are two Faraday's Laws of electromagnetic induction:

1. **Faraday's First Law:** As the magnetic flux linked with a circuit changes, it induces an emf in the circuit. This is known as induced emf . As long as the changes in magnetic flux happen, the induced emf persists.
2. **Faraday's Second Law:** A static magnetic field does not cause current flow. A time-varying field induces a voltage in a closed circuit. This induced voltage cause current flow. This is Faraday's Law. Induced voltage V_{emf} is given as

$$V_{\text{emf}} = -N \frac{d\phi}{dt} \text{ Volts}$$

where N is the number of turns in the circuit and ϕ the flux through each turn. The negative sign indicates that the induced voltage acts in opposition to the very cause due to which flux is produced:

$$\frac{d\phi}{dt} \neq 0,$$

if

(1) a time-changing flux links a stationary closed path or
(2) there is relative motion between steady flux and closed path or
(3) combination of the above two possibilities.

The emf existing in a path of integration of electric fields is given as

$$V = \oint \vec{E} \cdot d\vec{l} \text{ Volts}$$

Ampere's Law in a magnetic field holds a similar relation. As per Ampere's law, current is enclosed by a path of integration and is given as

$$I = F \oint \vec{H} \cdot d\vec{l} \text{ Amperes}$$

Quantity F as mentioned above is called "magnetomotance" or "magnetomotive force (mmf)".

Let a path of integration enclose N turns of wire with each turn carrying a current I in the same direction, then the F (mmf) is given as

$$\text{MMF} \quad F = NI = \oint \vec{H} \cdot d\vec{l}$$

6.4 LENZ'S LAW

Lenz's Law provides the direction of induced current flowing in a circuit. Induced current always has a direction that opposes the change in magnetic flux. It is so because changes in mmf are responsible for induced current production. As per Lenz's Law, induced current acts to produce an opposing flux. Consider a wire loop as shown in Figure 6.3(a). As a bar magnet moves up in the loop, a current will be induced in the loop. This induced current will produce a magnetic field that will oppose the field of bar magnet. B flux in the loop increases. Figure 6.3(b) shows the reverse process. The magnet is taken down; as a result, the current in the loop reverses. An opposing field is produced that opposes magnet motion. The B flux in the loop decreases.

The magnet's up and down movements induce an alternating current (AC) in the loop. This arrangement becomes a simple AC generator. As per Lenz's Law, current in the loop holds a direction in such a way that it opposes the changes that produced it.

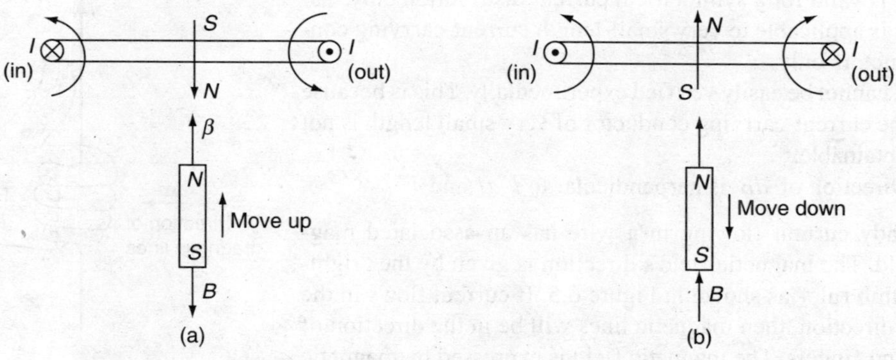

Figure 6.3 *Wire loop and bar magnet*

6.5 BIOT–SAVART LAW

As per the Biot–Savart Law, the magnetic field intensity $d\vec{H}$ due to a current element $Id\vec{l}$ (where current I is passing through an infinitesimal small conductor length $d\vec{l}$) as shown in Figure 6.4 is given as

$$d\vec{H} = \frac{Id\vec{l} \times \vec{a}_r}{4\pi r^2}$$

where $\vec{a}_r$ is the unit vector. Unit vector $\vec{a}_r$ is directed from the current element Idl towards the point of interest. r is the distance between the current element and the point of interest. Total field $\vec{H}$ is given as

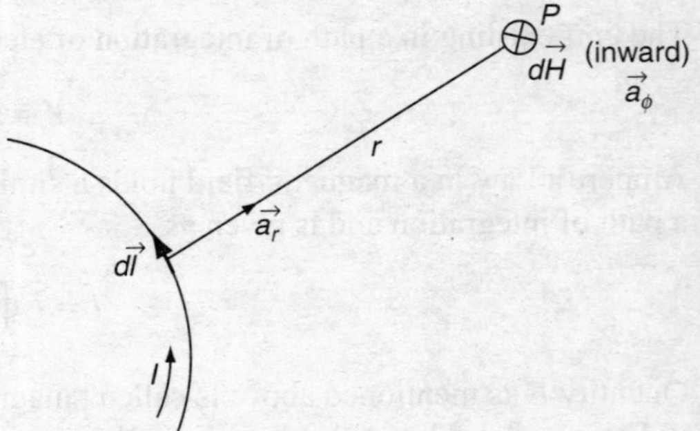

Figure 6.4 *Arrangement evincing Biot–Savart law*

$$\vec{H} = \oint \frac{I\, d\vec{l} \times \vec{a}_r}{4\pi r^2} \qquad \text{(Integral form of Biot–Savart law)}$$

This is the integral form of the Biot–Savart Law. Here, current is considered in a closed circuit.

The Biot–Savart law is alternatively expressed as

$$\vec{H} = \int_S \frac{\vec{K} \times \vec{a}_r\, ds}{4\pi r^2} \qquad \text{(also an expression of the Biot–Savart Law)}$$

where $\vec{K}$ is the surface current density. It is given as

$$\vec{K} = \frac{I}{2\pi r}\vec{a}_z$$

The Biot–Savart Law is also expressed as

$$\vec{H} = \int_v \frac{\vec{J} \times \vec{a}_r}{4\pi r^2} \cdot dv \qquad \text{(one more expression of the Biot–Savart Law)}$$

where the current density J is in Ampere/m^2.

The main features of the Biot–Savart Law are as follows:

1. It is analogous to Coulomb's Law (in electrostatics).
2. It is valid for a symmetrical current distribution only.
3. It is applicable to very small-length current-carrying conductors only.
4. It cannot be easily verified experimentally. This is because the current-carrying conductor of very small length is not obtainable.
5. Direction of $d\vec{B}$ is perpendicular to $I\, d\vec{l}$ and $\vec{r}$.

A steady current flowing in a wire has an associated magnetic field. The magnetic field's direction is given by the "right-hand thumb rule" as shown in Figure 6.5. If current flows in the thumb's direction, then magnetic lines will be in the direction of the closing fingers. The magnetic field is expressed by magnetic field intensity denoted as $\vec{H}$ at a given point.

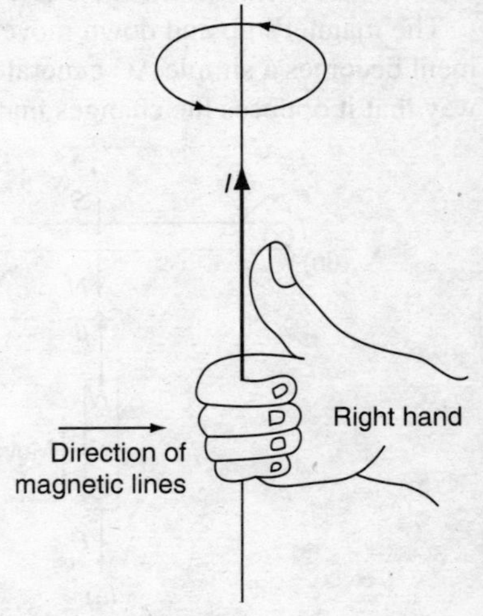

Figure 6.5 *Right-hand thumb rule*

6.5.1 Application of the Biot–Savart Law

Let us consider the following two cases of applications of the Biot–Savart Law:

Case 1: Magnetic field at the centre of a circular current-carrying coil: Let a circular coil have a radius R with its centre at O. Let I be the current flowing in the circular coil. It is shown in Figure 6.6.

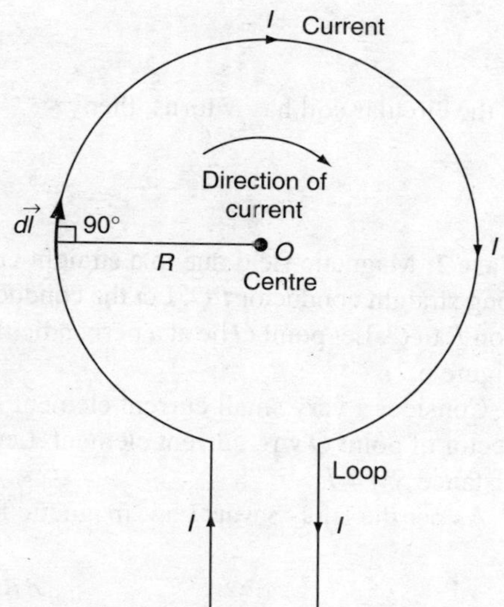

Figure 6.6 *Circular current-carrying coil*

Consider the circular coil to be made up of a large number of current elements. Each element has length dl. As per the Biot–Savart Law, the magnetic field (at centre of circular coil) due to current element (Idl) will be given as

$$d\vec{B} = \frac{\mu_0}{4\pi} I\left(\frac{d\vec{l} \times \vec{R}}{R^3}\right)$$

where R is the position vector of point O from the current element. In Figure 6.6, the angle between $d\vec{l}$ and $\vec{R}$ is 90°, i.e. $\theta = 90°$.

Thus,

$$dB = \frac{\mu_0}{4\pi}\frac{I\ dl\ R\sin\theta}{R^3} \quad \text{or}$$

$$= \frac{\mu_0}{4\pi}\frac{I\ dl\ R\sin\theta}{R^2}(\sin 90° = 1) \quad \text{or}$$

$$= \frac{\mu_0}{4\pi}\frac{I\ dl}{R^2}$$

The direction of $d\vec{B}$ is perpendicular to the plane of the current loop. $d\vec{B}$ is directed inwards. Current through every element of circular coil contributes magnetic field in the same direction. Total magnetic field B at a point O due to current in the circular coil can be very easily obtained by integrating the above expression as

$$B = \int dB \quad \text{or}$$

$$= \int \frac{\mu_0}{4\pi}\frac{I\ dl}{R^2} \quad \text{or}$$

$$= \frac{\mu_0}{4\pi}\frac{I}{R^2}\int dl$$

where $\int dl$ is the total length of the circular coil or

= circumference of current loop

$= 2\pi R$ or

Further,

$$B = \frac{\mu_0}{4\pi}\cdot\frac{I}{R^2}\cdot 2\pi R \quad \text{or}$$

$$= \frac{\mu_0 I}{2R}$$

If the circular coil has N turns, then

$$B = \frac{\mu_0 NI}{2R}$$

Case 2: Magnetic field due to a straight current-carrying conductor: As shown in Figure 6.7, consider a long straight conductor PQ. Let the conductor be lying in a plane of the paper carrying current I in direction P to Q. Let point O be at a perpendicular distance a from the straight conductor. That is $OS = a$ (refer Figure 6.7).

Consider a very small current element Idl of the straight conductor at point M. Let R be the position vector of point O wrt. current element. Let α be the angle between Idl and R as shown in Figure 6.7. Let distance $SM = l$.

As per the Biot–Savart Law, magnetic field $d\vec{B}$ induced at a point P due to a current element Idl is

$$d\vec{B} = \frac{\mu_0}{4\pi} \cdot \frac{Id\vec{l} \times \vec{R}}{R^3} \quad \text{or}$$

$$\left| d\vec{B} \right| = \frac{\mu_0}{4\pi} \cdot \frac{I\, dl \cdot R \sin\alpha}{R^3}$$

In ΔOMS, $\angle\alpha + \angle\theta = 90°$ (refer Figure 6.7). Thus, $\sin\alpha = \sin(90° - \theta) = \cos\theta$. Also from Figure 6.7, we get

$$\cos\theta = \frac{a}{R} \quad \text{or}$$

$$R = a/\cos\theta \quad \text{and}$$

$$\tan\theta = \frac{l}{a}$$

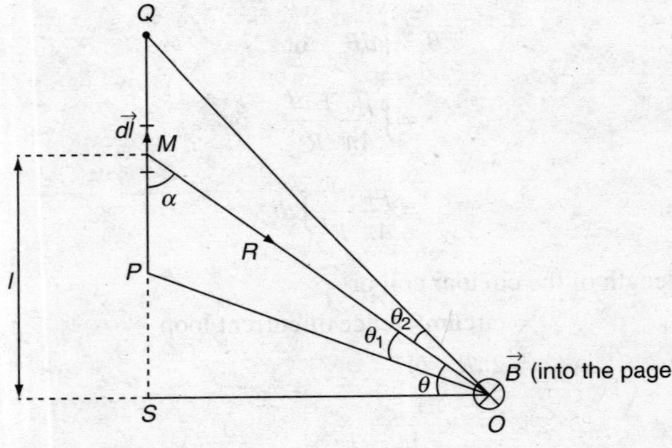

Figure 6.7 *A long straight current-carrying conductor*

Thus, $l = a \tan\theta$. Differentiating above, we get

$$dl = a \sec^2\theta \, d\theta$$

Substituting dl above and simplifying, we get

$$\left|d\vec{B}\right| = \frac{\mu_0}{4\pi} \frac{I(a\sec^2\theta \, d\theta)\cos\theta}{(a/\cos\theta)^2} \quad \text{or}$$

$$\left|d\vec{B}\right| = \frac{\mu_0}{4\pi} \frac{I}{a} \cos\theta \, d\theta$$

The direction of $d\vec{B}$, (as per the right-hand thumb rule), is perpendicular to the plane of paper. $d\vec{B}$ will be directed inwards. The total magnetic field $\vec{B}$ at a point O due to a current through the straight conductor PQ is obtained as

$$\left|\vec{B}\right| = \int_{-\theta_1}^{\theta_2} \left|d\vec{B}\right| \quad \text{(integrating with limits } -\theta_1 \text{ and } \theta_2\text{)} \quad \text{or}$$

$$= \int_{-\theta_1}^{\theta_2} \frac{\mu_0}{4\pi} \cdot \frac{I}{a} \cos\theta \, d\theta \quad \text{or}$$

$$= \frac{\mu_0}{4\pi} \frac{I}{a} \int_{-\theta_1}^{\theta_2} \cos\theta \, d\theta \quad \text{or}$$

$$= \frac{\mu_0}{4\pi} \frac{I}{a} [\sin\theta]_{-\theta_1}^{\theta_2} \quad \text{or}$$

$$= \frac{\mu_0}{4\pi} \frac{I}{a} [\sin\theta_2 - \sin(-\theta_1)]$$

or simply

$$= \frac{\mu_0}{4\pi} \frac{I}{a} [\sin\theta_1 + \sin\theta_2]$$

Consider the following two special cases:

Special Case 1: When the conductor is semifinite (with respect to O), point P is S (0, 0, 0) and Q is (0, 0, α). Thus, $\theta_1 = 0°$ and $\theta_2 = 90°$. On solving, we get

$$\left|\vec{B}\right| = \frac{\mu_0}{4\pi} \frac{I}{a} \quad \text{(Its direction will be into the page)}$$

Special Case 2: When the conductor is infinite in length, point P is (0, 0, −8) and Q is (0, 0, 8).

Thus, $\theta_1 = 90°$ and

$$\theta_2 = 90°$$

On solving, we get

$$\left|\vec{B}\right| = \frac{\mu_0}{4\pi}\frac{I}{R} \qquad \text{(Its direction will also be into the page)}$$

Let the length of the conductor be finite L. Point O lies at the right bisector of the conductor. Thus, $\theta_1 = \theta_2 = \theta°$. Further,

$$\sin\theta = \frac{L/2}{\sqrt{a + (L/2)^2}} \quad \text{or}$$

$$= \frac{L}{\sqrt{4a^2 + L^2}}$$

On solving, we get

$$\left|\vec{B}\right| = \frac{\mu_0}{4\pi}\frac{I}{a}[\sin\theta + \sin\theta] \quad \text{or}$$

$$\left|\vec{B}\right| = \frac{\mu_0}{4\pi}\frac{I}{a}\frac{L}{\left[4a^2 + L^2\right]^{1/2}}$$

6.6 AMPERE'S CIRCUITAL LAW

Field intensity $\vec{H}$ at a point at distance R from a very long straight filament conductor-carrying current I is given as

$$\vec{H} = \frac{I}{2\pi R}\vec{a}_\phi$$

The integral of H around a closed path having radius R enclosing the conductor is $\oint \vec{H} \cdot d\vec{l}$.

It is given as

$$\oint \vec{H} \cdot d\vec{l} = \frac{I}{2\pi R} 2\pi R$$

$$= I$$

This relation holds true for every case of a single closed path. This relation is exactly "Ampere's Circuital Law". As per Ampere's Circuital Law, the line integral of $\vec{H}$ along a closed path is the same as the direct current, which is enclosed by that path. It is given as

$$\oint \vec{H} \cdot d\vec{l} = I$$

The "positive" current is considered in the direction of the advancing right-handed screw, when being turned in the direction in which the considered closed path is traversed. It is shown in Figure 6.8. Ampere's Circuital Law is analogous to Gauss's Law in electrostatics. Ampere's Circuital

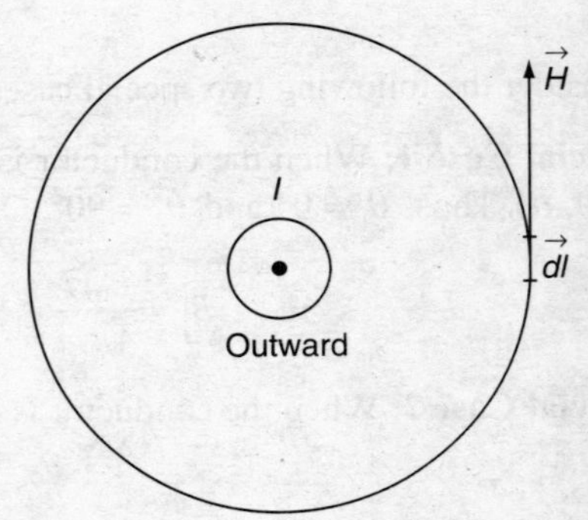

Figure 6.8 *Direction of positive current*

Law helps determining the magnetic field around a current-carrying conductor applicable for symmetrical configurations, only. Figure 6.9 shows the direction of current density and magnetic field.

$\vec{H}$ can be determined using Ampere's Circuital Law. For this, there must be symmetry in the problem. Two essential conditions to be fulfilled are:

(1) In a closed path on each point, $\vec{H}$ is either tangential or normal to the considered path.

(2) Wherever $\vec{H}$ is tangential, it has the same value at all points on the path.

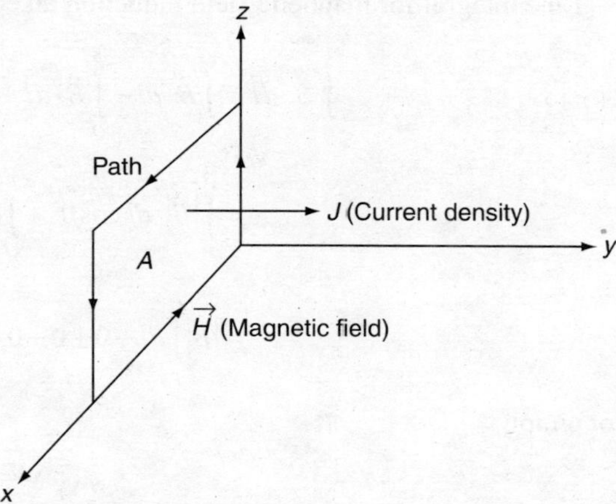

Figure 6.9 *Direction of current density and magnetic field*

Ampere's Circuital Law can be applied either inside the conductor or outside the conductor. The total current that is enclosed by the path must be known. Ampere's Circuital Law is applicable on line currents or sheet currents or volume currents. Figure 6.10 shows a path in a conducting medium. As per Ampere's Circuital Law, a path in general form is given as

$$\oint \vec{H} \cdot d\vec{l} \quad \text{or}$$

$$= \int_s J \cdot ds \ \text{ or}$$

$$= I$$

This is "Ampere's Circuital Law" in integral form.

6.6.1 Applications of Ampere's Circuital Law

Consider the following two applications cases of Ampere's Circuital Law.

Case 1: Magnetic field due to current-carrying solenoid: A solenoid is basically an insulated long wire wound closely in helix form. The length of a solenoid is very large as compared to its diameter.

Consider a long straight solenoid. Let there be N turns per unit length. Let the solenoid be carrying a current I. Consider a rectangle $PQRS$. This is shown in Figure 6.10. The magnetic field is zero at a point situated outside the solenoid. The magnetic field is uniform and parallel to the length at a point situated inside the long solenoid.

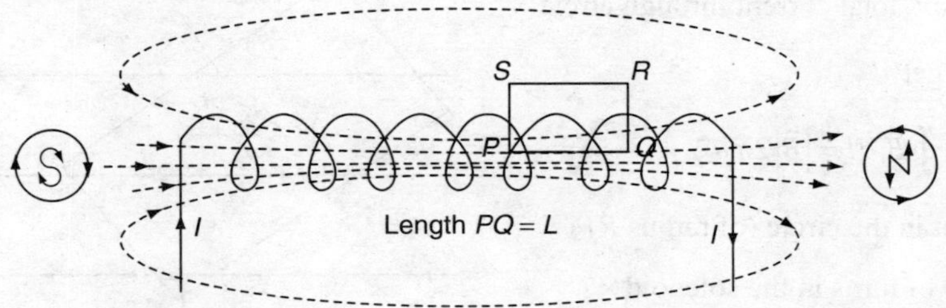

Figure 6.10 *Path in conducting medium*

Line integral for magnetic field induction taken over closed path *PQRS* is given as

$$\oint \vec{B}\cdot d\vec{l} = \int_P^Q \vec{B}\cdot dl + \int_Q^R \vec{B}\cdot d\vec{l} + \int_R^S \vec{B}\cdot d\vec{l} + \int_S^P \vec{B}\cdot d\vec{l} \quad \text{or}$$

$$= \int_P^Q \left|\vec{B}\right|\cdot dl\cos 0° + \int_Q^R \left|\vec{B}\right|\cdot dl\ \cos 90° + \int_R^S 0 + \int_S^P \left|\vec{B}\right|\cdot dl\cos 90° \quad \text{or}$$

$$= \left|\vec{B}\right| \int_P^Q dl + 0 + 0 + 0$$

or simply

$$= \left|\vec{B}\right| L$$

According to Ampere's Circuital Law,

$$\oint \vec{B}\cdot d\vec{l} = \mu_0 \times \text{total current throught rectangle } PQRS \quad \text{or}$$

$$= \mu_0 \times N\,L\,I$$

Further,

$$\left|\vec{B}\right| L = \mu_0\ N\ L\ I \quad \text{or}$$

$$\left|\vec{B}\right| = \mu_0\ NI$$

This is the expression for magnetic field induction at a point, which is rotated inside the solenoid. At the points nearing the end of solenoid, the magnetic field induction is equal to $(\mu_0 NI/2)$.

Case 2: Magnetic field due to current in a toroid: A toroid is an endless solenoid. Toroid is in the form of a ring. A toroid is shown in Figure 6.11. Let *N* be the number of turns per unit length of the toroid. Let current *I* be flowing through the toroid. A magnetic field will get set up inside the turns of the toroid. The magnetic lines of force will be concentric circles inside the toroid. Let us consider a point *M* at a distance *r* from *O*. This point *M* is situated inside the turns of the toroid. At point *M*, magnetic field induction $\vec{B}$ has to be determined. As per Ampere's Circuital Law, the line integral of $\vec{B}$ (magnetic field induction) along the circular path with radius *R* is given as

$$\oint \vec{B}\cdot d\vec{l} = \mu_0 \times \text{ total current through circle}$$

In this case, we get

$$\oint \vec{B}\cdot d\vec{l} = \left|\vec{B}\right| 2\pi R$$

The total current in the circle (of radius *R*)

$= \text{Number of turns in the solenoid} \times I$

$= 2\pi RNI$

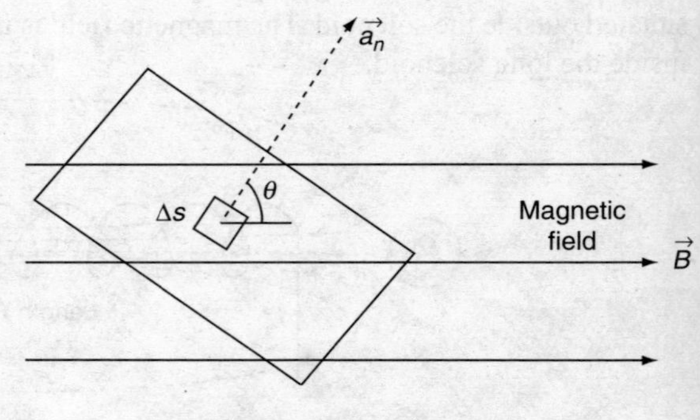

Figure 6.11 *A toroid*

Further, $$\oint \vec{B}\cdot d\vec{l} = \mu_0 \times 2\pi\ RNI$$

It gives $$\left|\vec{B}\right| 2\pi R = \mu_0 \times 2\pi\ RNI \quad \text{or}$$

$$\left|\vec{B}\right| = \mu_0\ NI$$

$\vec{B}$ is independent of radius R.

6.7 MAGNETIC FLUX

Magnetic flux ϕ through a surface held in a magnetic field $\vec{B}$ is determined from the total number of magnetic lines of force, which cross that surface. Thus, magnetic flux

$$\phi = \int_S \vec{B}\ d\vec{s}$$

where $d\vec{s}$ is the small area element. Through $d\vec{s}$, magnetic lines of force are passing.

6.7.1 Magnetic Flux Density

It is magnetic flux per unit area. It is also termed as magnetic induction or alternatively magnetic field strength. Magnetic flux density

$$B = \frac{\phi}{A}\ \text{(Webers)}$$

The S.I. units of magnetic flux are webers (Wb). One weber is defined as the magnetic flux amount over an area of 1 meter^2. The area is held normal to the uniform magnetic field of one tesla.

Thus, we get

$$1\ \text{weber} = 1\ \text{tesla} \times 1\ \text{m}^2$$

The cgs unit of magnetic flux ϕ is Maxwell (Mx). Please note that 1 weber = 10^8 Maxwell. Magnetic flux density $\vec{B}$ is analogous to electric flux density $\vec{D}$.

$$\vec{D} = \varepsilon_0 \vec{E} \qquad \text{(in free space)}$$

Magnetic flux density $\vec{B}$ is given as

$$\vec{B} = \mu_0 \vec{H}$$

where $\vec{H}$ is the magnetic field intensity and μ_0 is the constant (μ_0 is known as the permeability of free space).

$$\mu_0 = 4\pi \times 10^{-7}\ \text{(Henry/meters)}$$

Magnetic flux lines due to a straight very long wire are shown in Figure 6.12. Magnetic flux lines are determined in the same way as the electric flux lines are determined. Each flux line is closed. Flux lines have no beginning or end. The direction of $\vec{B}$ is "North", as indicated by the needle of a magnetic compass. In the case of a straight, current-carrying conductor, magnetic flux lines are closed. These magnetic flux lines do not cross each other regardless of current distribution.

In an electrostatic field, flux passing through a closed surface is the amount of charge enclosed.

$$d\phi = \oint \vec{D} \cdot \vec{ds}$$

$$= Q$$

It is possible to have an isolated electric charge. Thus, electric flux lines necessarily need not be closed. Unlike electric flux lines, always the magnetic flux lines close upon themselves. It is shown in Figure 6.13.

It is not possible to have an isolated magnetic pole (or magnetic charges) "because an isolated magnetic charge does not exist."

Total flux passing through a closed surface held in a magnetic field is zero. It is given as

$$\oint \vec{B} \cdot \vec{ds} = 0 \qquad \text{(Gauss's Law)}$$

It is Gauss's Law for magnetostatic field. It is also known as the law of magnetic flux conservation.

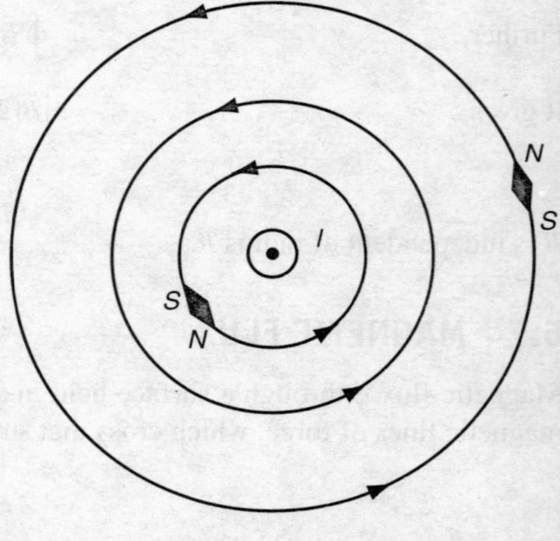

Figure 6.12 *Magnetic flux lines due to a straight long wire*

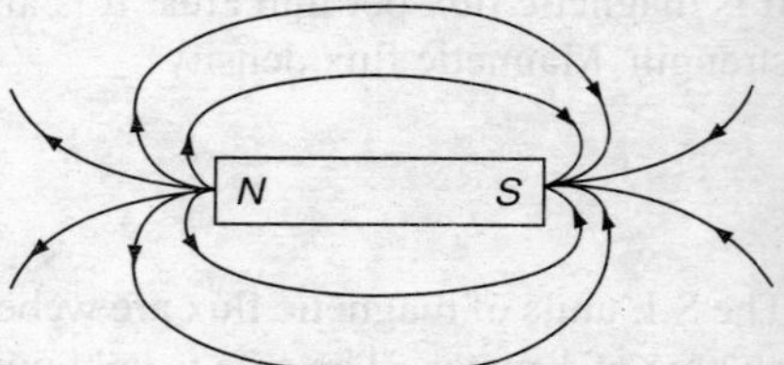

Figure 6.13 *Magnetic flux lines*

6.8 MAGNETIC SCALAR POTENTIAL

In electrostatics, the scalar potential V determines E. The two most important relations are

$$\vec{E} = -\nabla V \quad \text{and}$$

$$\nabla^2 V = 0$$

Firstly, we find the potential for a charge configuration and then from it, we determine the electric field. For static electric field $\vec{E}$ in electrostatics, we know that

$$\oint \vec{E} \cdot d\vec{l} = 0$$

A related scalar potential function V is given as

$$E = -\nabla V$$

Between two points a and b of a path in the field, we get

$$\int_a^b \vec{E} \cdot d\vec{l} = V_a - V_b = -\int_b^a \vec{E} \cdot d\vec{l}$$

The magnetic scalar potential is denoted as V_m. Negative gradient of V_m gives magnetic field intensity $\vec{H}$:

$$\vec{H} = -\nabla V_m \qquad \text{(by analogy from electrostatics)}$$

Substituting it, we get

$$\vec{J} = \nabla \times \vec{H} \quad \text{or}$$

$$\vec{J} = \nabla \times (-\nabla V_m)$$

or simply

$$\vec{J} = 0$$

Therefore, magnetic scalar potential V_m is defined only in a region, where $\vec{J} = 0$. V_m satisfies Laplace's equation just as V does it for the electrostatic field. In the case of homogeneous magnetic materials, we get

$$\nabla \cdot \vec{B} = 0 \quad \text{and}$$

$$\nabla \cdot \vec{H} = 0$$

Further,

$$\nabla \cdot (-\nabla V_m) = 0 \quad \text{or}$$

$$\nabla^2 V_m = 0$$

When no current is enclosed, magnetic field is

$$\oint \vec{H} \cdot d\vec{l} = 0$$

In the above, $\vec{H}$ can be derived from a scalar magnetic potential V_m.

Thus,

$$H = -\nabla V_m$$

Between any two points a and b of a path in the field, we get

$$\int_a^b \vec{H} \cdot d\vec{l} = V_{ma} - V_{mb}$$

Scalar potential V_m has dimensions of current. If there is an emf in an electric field, then we find

$$\oint \vec{E} \cdot d\vec{l} = V_e \qquad \text{(Path of integration of } \vec{E}\text{)}$$

where V_e is the total emf of the path. In a magnetic field also, an analogous relation based on Ampere's Law can be obtained. It is

$$\oint \vec{H} \cdot d\vec{l} = I \qquad \text{(current is enclosed in the path of integration)}$$

$$= F_e$$

where F_e is the magnetomotive force (mmf). F_e is equal to the total current enclosed.

6.9 MAGNETIC VECTOR POTENTIAL

In electrostatics, electric potential depends on the charges, which are responsible for establishing the field. Electric potential is a scalar function. Electric field is the gradient of potential function. It is given as

$$\vec{E} = -\nabla V$$

Thus, E can be obtained from V. V can be obtained from the solution of relevant Laplace's equations using boundary conditions. In magnetics, magnetic field is produced due to a "current element" similar to charge in electrostatic. Potential depends on current element $I\,d\vec{l}$ (a vector quantity).

Thus,

$$d\bar{h} \propto I d\vec{l}$$

Potential is proportional to $I\,d\vec{l}$. Magnetic field intensity is inversely proportional to r^2 ($H \propto 1/r^2$). Also, electric field $\vec{E}$ is inversely proportional to r^2. However, scalar electric potential is inversely proportional to r, only. Therefore, magnetic potential should also be inversely proportional to distance ($\propto 1/r^2$; based on analogy/similarity). Vector magnetic potential is denoted as $\vec{A}$. $\vec{A}$ is a vector. $\vec{E}$ is the negative gradient of potential. $\vec{B}$ is the space derivative of $\vec{A}$. This space derivative can be divergence or curl. Divergence gives scalar quantity. However, "curl" gives a vector quantity. Thus, "curl" is chosen. Thus, it is

$$\vec{B} = \nabla \times \vec{A}$$

where $\vec{A}$ is the "Magnetic Vector Potential". $\vec{A}$ is given as

$$\vec{A} = \int_L \frac{\mu_0 I\, dl}{4\pi r} \quad \text{(for line current)} \quad \text{or}$$

$$\vec{A} = \int_S \frac{\mu_0\, k\, ds}{4\pi r} \quad \text{(for surface current)} \quad \text{or}$$

$$\vec{A} = \int_S \frac{\mu_0\, J dv}{4\pi r} \quad \text{(for volume current)}$$

where r is the distance from the current element to a point of interest at which magnetic vector potential is being calculated. Vector magnetic potential is an intermediate quantity. $\vec{B}$ or $\vec{H}$ can be easily obtained from vector magnetic potential.

Therefore,

$$\nabla \times \vec{B} = \mu \vec{J} \quad \text{and}$$

$$\nabla \times \nabla \times \vec{A} = \nabla(\nabla \cdot \vec{A}) - \nabla^2 \vec{A}$$

For steady current,

$$\nabla \cdot \vec{A} = 0$$

As

$$\nabla \times \nabla \times \vec{A} = -\nabla^2 \vec{A},$$

$$-\nabla^2 \vec{A} = \mu \vec{J}$$

6.10 MAGNETIC FIELD AND MAGNETIZATION

Consider an infinite plane slab of magnetic material having uniform magnetic susceptibility χ_m and thickness d. Assume that it occupies the region $0 < z < d$ as shown in Figure 6.14. A uniform magnetic field $B_0 a_x$ is applied in the x-direction. Magnetic field $B_0 a_x$ begets magnetic dipole moments in the material. These generated magnetic dipole moments are oriented along the field. Magnetization vector $\vec{M}$ will be uniform because χ_m and $\vec{B}$ are constants. It is shown in Figure 6.15(a).

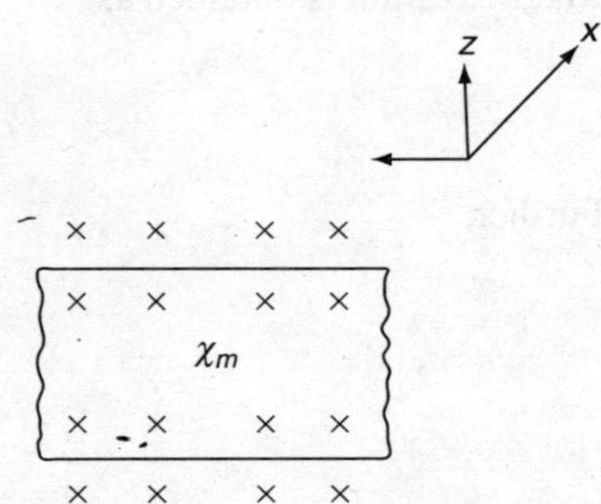

Figure 6.14 *Region*

Dipole moments (loop currents) are arranged is such a manner that they cause cancellation of currents everywhere except at boundaries of material. It is so because for each current segment at the surface, there is no opposite current. It causes surface currents at the material boundaries, viz. $z = 0$ and $z = d$. It is shown in Figure 6.15(b). K is the surface current density at $z = d$. Its value is $-k\vec{a}_y$ at $z = d$ and $k\vec{a}_y$ at $z = 0$. These surface currents are also called "magnetization surface currents."

Figure 6.16 shows a vertical column of infinitesimal rectangular cross-section area $\Delta A(= \Delta x\, \Delta y)$. This column that has been cut from a magnetic material of thickness d surface current is $k\Delta x$. Surface

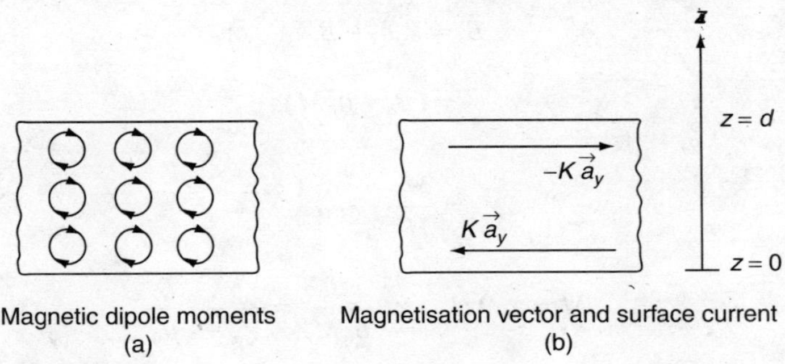

Figure 6.15 *Magnetic field and magnetization*

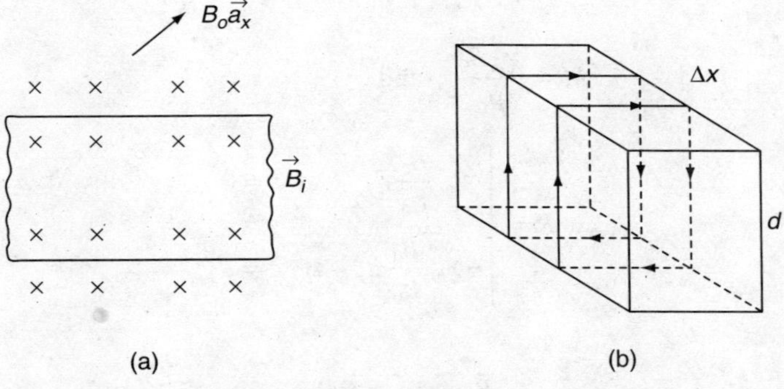

Figure 6.16 *Column of magnetic material*

currents form a dipole moment given as $(k\Delta x)(\Delta yd)$. The dipole moment will be in the $\vec{a}_x$ direction. Magnetization is obtained as

$$M = \frac{\text{dipole moment}}{\text{volume}}$$

Further,

$$\vec{M} = \frac{k\Delta x(\Delta yd)}{(\Delta x)(\Delta yd)}$$

$$= K$$

$$\vec{M} = k\vec{a}_x$$

As shown in the expression above, surface "current density" is linked with magnetization "vector". Magnetization produces a self field inside the magnetic material. Thus,

$$\vec{B}_s = \mu_0 \vec{M} = \begin{cases} \mu_0 k\vec{a}_x & 0 < z < d \\ 0 & \text{otherwise} \end{cases}$$

The field inside the magnetized material is given as

$$\vec{B}_i = B_0\vec{a}_x + B_s\vec{a}_x \quad \text{or}$$

$$= (B_0 + \mu_0 M)\vec{a}_x$$

we know

$$\chi_m = \frac{M}{H} \quad \text{or}$$

$$M = \chi_m \cdot H = \chi_m \frac{B_i}{\mu_0 M_r} = \frac{\chi_m}{\chi_m + 1}\frac{B_i}{\mu_0}$$

Also,

$$M = \frac{\chi_m}{1+\chi_m}\frac{B_0 + \mu_0 M}{\mu_0} \quad \text{or}$$

$$M = \frac{\chi_m B_0}{\mu_0} \quad \text{and}$$

$$\vec{K} = \begin{cases} \chi_m \dfrac{B_0}{\mu_0}\vec{a}_y & \text{at } z = 0 \\ -\chi_m \dfrac{B_0}{\mu_0}\vec{a}_y & \text{at } z = d \end{cases}$$

Thus,

$$\vec{B}_i = (1+\chi_m)B_0\vec{a}_x$$

Flux density increases due to magnetization of material. Secondary field is produced outside the magnetic material by "surface current distribution". This field is zero. The total field outside the magnetic material does not change. The total field is equal to the applied field B_0. Either a non-uniform magnetic field (applied) or a non-uniform magnetic susceptibility of material results in magnetization volume current density in the magnetic material. Basically, magnetization volume current density is due to the imperfect cancellation of dipole currents. Magnetization vector is related to volume current density (due to magnetization $\vec{J}_m$) given as

$$\vec{J}_m = \nabla \times \vec{M}$$

This is quite analogous to

$$\vec{J} = \nabla \times \vec{H}$$

The units of both $\vec{M}$ and $\vec{H}$ are (A/m). $\vec{M}$ and $\vec{H}$ possess the same direction. "Magnetization volume current density" in the magnetic material is always such that the "total volume current" through a given cross-sectional area of the magnetic material is opposite to the "total surface current", which crosses the contour of the area.

6.11 MAGNETIC BOUNDARY CONDITIONS

In the case of a single medium, "magnetic field" is continuous. At the boundary of two magnetic materials, the magnetic field may however change abruptly. This change can be in both viz. magnitude and direction. Boundary relationships are obtained from field components "normal" and "tangential" to the boundary. The boundary relation for B and H can be obtained in a similar manner as is obtaining boundary relations for D and E (in electrostatics). Figure 6.17 shows the boundary of two isotropic

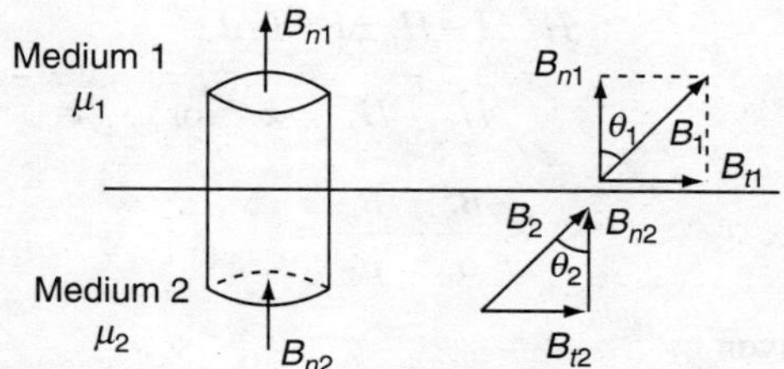

(a) Two isotropic homogenous liner materials

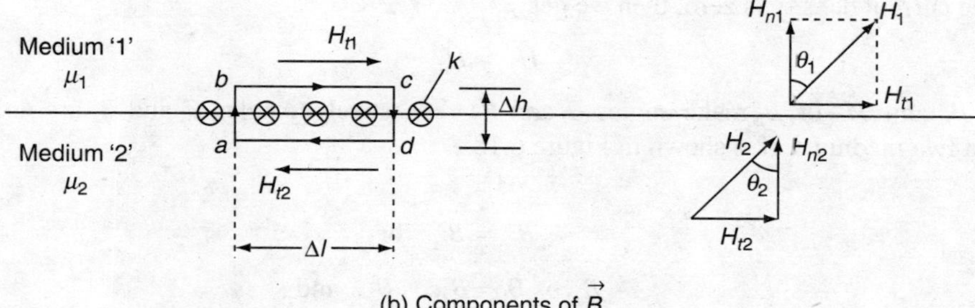

(b) Components of $\vec{B}$

Figure 6.17 *Boundary condition determination*

homogeneous liner materials. They possess permeabilities μ_1 and μ_2, respectively. Boundary conditions for the normal component of $\vec{B}$ can be obtained by letting the surface to cut a small cylindrical Gaussian surface.

By applying Gauss's Law (of magnetic field), we get

$$\nabla \cdot \vec{B} = 0 \quad \text{or}$$

$$\oint \vec{B} \cdot d\vec{s} = 0$$

Let $\vec{B}_{n_1}$ and $\vec{B}_{n_2}$ be the normal components of flux densities. Thus,

$$\vec{B}_{n_1} \cdot \vec{a}_{n_1} \Delta s - \vec{B}_{n_2} \vec{a}_{n_2} \Delta s = 0 \quad \text{(refer Figure 6.17(a))}$$

$$\vec{B}_{n_1} = \vec{B}_{n_2}$$

The normal components of magnetic flux densities are the boundary between two magnetic media. Thus, in terms of field intensity, we get

$$\mu_1 \vec{H}_{n_1} = \mu_2 \vec{H}_{n_2}$$

The normal component of $\vec{B}$ is continuous. The normal component of $\vec{H}$ is discontinuous. Tangential components of $\vec{H}$ are shown is Figure 6.17(b). At the boundary, a sheet current of density k (Amp/m) exists. By applying Ampere's Law at the boundary, we get

$$\oint \vec{H} \cdot d\vec{l} = I$$

For closed path *abcd* as shown in Figure 6.17(b), we get

$$H_{t_1} \Delta l - H_{t_2} \Delta l = k \Delta l$$

$$H_{t_1} - H_{t_2} = k \quad \text{or}$$

$$\frac{B_{t_1}}{\mu_1} - \frac{B_{t_2}}{\mu_2} = k$$

The direction of vectors is given by

$$(\vec{H}_1 - \vec{H}_2) \times \vec{a}_{n_{12}} = \vec{k}$$

where $\vec{a}_{n_{12}}$ is a unit vector. $\vec{a}_{n_{12}}$ is normal to the interface. $\vec{a}_{n_{12}}$ is directed from medium 1 towards medium 2.

It surface current density is zero, then we get

$$\vec{H}_{t_1} = \vec{H}_{t_2}$$

Let $\vec{H}_1$ or $\vec{B}_1$ and $\vec{H}_2$ or $\vec{B}_2$ make angles θ_1 and θ_2, respectively. Angles θ_1 and θ_2 are normal to the interface of two mediums. It is shown in Figure 6.18.

Further,

$$B_{n_1} = B_{n_2} \quad \text{or}$$

$$B_1 \cos \theta_1 = B_2 \cos \theta_2 \quad \text{and}$$

$$H_{t_1} = H_{t_2} \quad \text{or}$$

$$H_1 \sin\theta_1 = H_2 \sin\theta_2$$

Thus,

$$\frac{H_1}{B_1}\tan\theta_1 = \frac{H_2}{B_2}\tan\theta_2 \quad \text{or}$$

$$\frac{\tan\theta_1}{\tan\theta_2} = \frac{H_2/B_2}{H_1/B_1} \quad \text{or}$$

$$\frac{\tan\theta_1}{\tan\theta_2} = \frac{\mu_1}{\mu_2} \quad \text{(Law of refraction)}$$

The above expression gives the "law of refraction". Magnetic flux lines have been considered at the boundary with zero surface current.

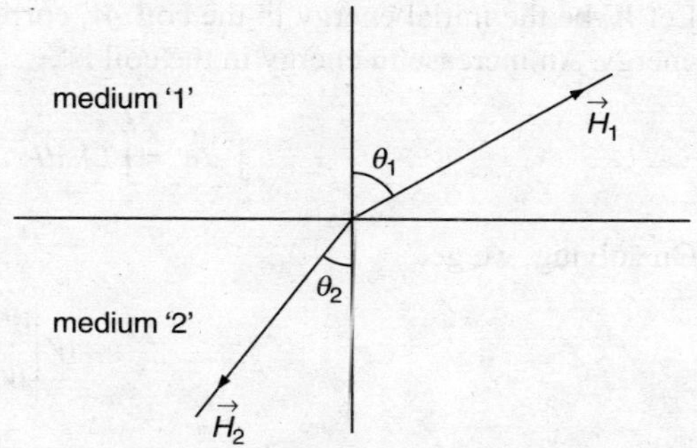

Figure 6.18 *Directions and angles of $\vec{H}$ for mediums 1 and 2*

6.12 ENERGY STORED IN A MAGNETIC FIELD

Induced voltage V as per Faraday's law of induction is given as

$$V = -\frac{Nd\phi}{\partial t}$$

For increased current, the source must supply energy. Thus, we get

$$\frac{\partial w}{\partial t} = -p$$

$$= -VI$$

$$dw = -VI \cdot dt \quad \text{or}$$

$$dw = \frac{Nd\phi}{\partial t} \cdot I\ dt$$

or simply

$$\frac{dw}{I} = Nd\phi$$

In the case of a linear magnetic circuit, we find

$$L = \frac{Nd\phi}{\partial I} \quad \text{or}$$

$$L\ dI = Nd\phi \quad \text{or}$$

$$\frac{dw}{I} = L\ dI \quad \left(\text{as}\ \frac{dw}{I} = Nd\phi\right) \quad \text{or}$$

$$dw = LI\ dI$$

Let W_i be the initial energy in the coil. W_i corresponds with the initial zero current. Let W_f be the final energy. An increase in energy in the coil is

$$\int_{w_i}^{W_f} dw = \int_{o}^{I} LI\, dI \qquad \text{(where } I \text{ is the current)}$$

On solving, we get

$$W\Big|_{W_i}^{W_f} = \frac{LI^2}{2} \quad \text{or}$$

$$W = \frac{1}{2} L\, I^2$$

In the case of linear circuits, we get

$$L = \frac{N\phi}{I} \quad \text{or}$$

$$W = \frac{1}{2} N\,\phi\, I$$

Magnetic flux ϕ is given as

$$= \int B \cdot dA$$

$$= BA$$

where A is the cross-sectional area of the coil. The total current enclosed by a path as per Ampere's Circuital Law is given as

$$NI = \oint \vec{H} \cdot d\vec{l}$$

$$= Hl$$

Further,

$$W = \frac{1}{2} NI\phi$$

Thus,

$$W = \frac{1}{2} Hl \;\; BA \quad \text{or}$$

$$= \frac{1}{2} HBlA \quad \text{or}$$

$$= \frac{1}{2} H\, BV$$

where V is the space volume inside the coil.

Energy density W_H is given as

$$W_H = \frac{W}{V}$$

$$= \frac{1}{2} HB \quad \text{or}$$

$$W_H = \frac{1}{2}\mu_0 H^2 \text{ (Joules/m}^2\text{)}$$

In vector form, it is expressed as

$$W_H = \frac{1}{2}\vec{B}\cdot\vec{H} \text{ (J/m}^3\text{)}$$

SOLVED QUESTIONS

6.1 An electron is moving at 10 km/s along $\vec{a}_x$ in a magnetic field. the magnetic flux density is given as $\vec{B} = 0.4\vec{a}_x - 0.6\vec{a}_y + 1.0\vec{a}_z$ Wb/m^2.
(a) Find the electric field intensity if there is none applied on the electron.
(b) Find the force on the electron in the presence of both $\vec{B}$ and $\vec{E}$.
Consider $\vec{E} = (2\vec{a}_x + \vec{a}_y + 3\vec{a}_z)$ kV/m.

Solution:

(a) We know $\vec{F} = q[\vec{E} + \vec{V}\times\vec{B}]$.
If

$$\vec{F} = 0$$

Further,

$$\vec{E} = -\vec{V}\times\vec{B} \quad \text{or}$$

$$= -\begin{vmatrix} \vec{a}_x & \vec{a}_y & \vec{a}_z \\ 10^4 & 0 & 0 \\ 0.4 & -0.6 & 1.0 \end{vmatrix}$$

or simply

$$= (10\vec{a}_y + 16\vec{a}_z) \text{ kV/m}$$

(b) Force on the electron due to both $\vec{E}$ and $\vec{B}$ is given as

$$\vec{F} = q[\vec{E} + \vec{V}\times\vec{B}]$$

Now

$$\vec{F} = -1.6\times10^{-19}\,[2\vec{a}_x + \vec{a}_y + 3\vec{a}_z + 10\vec{a}_y + 6\vec{a}_z]\times10^3 \quad \text{or}$$

$$= -1.6\times10^{-19}[2\vec{a}_x + 11\vec{a}_y + 9\vec{a}_z] \quad \text{or}$$

$$= [-3.2\vec{a}_x - 17.6\vec{a}_y - 14.4\vec{a}_z]\times10^{-16} \text{ Newton}$$

6.2 Find the force on a wire of length 0.04 m, which is placed inside a long solenoid near its centre. The wire makes an angle of 60° with the axis of solenoid. It carries a current of 3 A. The magnetic field due to the solenoid is 0.25 T.

Solution:
It is given that

$$L = 0.04 \text{ m}$$

$$\theta = 60°$$

$$I = 3 \text{ A} \quad \text{and}$$

$$B = 0.25 \text{ T}$$

Force F is given as

$$F = I\, l\, B \sin\theta$$

Substituting the value, we get

$$= 3 \times 0.04 \times 0.25 \times \sin 60° \quad \text{or}$$

$$= 0.10392 \text{ Newton}$$

6.3 Consider a magnetic flux density $\vec{B} = 2.0\vec{a}_x + 6.0\vec{a}_y \text{Wb/m}^2$ with a current element $0.30\vec{a}_z$ (mA-m) placed in it. What will be the force per metre on this current element?

Solution:
Force $\vec{F}$ on a current element in the presence of a magnetic field is given as

$$\vec{F} = I(\vec{L} \times \vec{B})$$

where $I\ \vec{L} = 30\vec{a}_z (\text{mA-m})$ or

$$= 30 \times 10^{-3} a_z \text{ (A-m)}$$

Given that $\vec{B} = (2.0\vec{a}_x + 6.0\vec{a}_y)\,(\text{Wb/m}^2)$

Thus,

$$\vec{F} = 30 \times 10^{-3} \vec{a}_z \times (2.0\vec{a}_x - 6.0\vec{a}_y)$$

or simply

$$\vec{F} = -180\,\vec{a}_x + 60\vec{a}_y \text{ (mN)}$$

6.4 There is a rectangular loop of area A. It has uniform flux density $\vec{B}$, which is normal to its plane (plane of loop). It is shown in Figure 6.19. The flux density varies harmonically with respect to the time. Flux density is given as $B = B_0 \cos \omega t$. What will be the total emf induced in the loop?

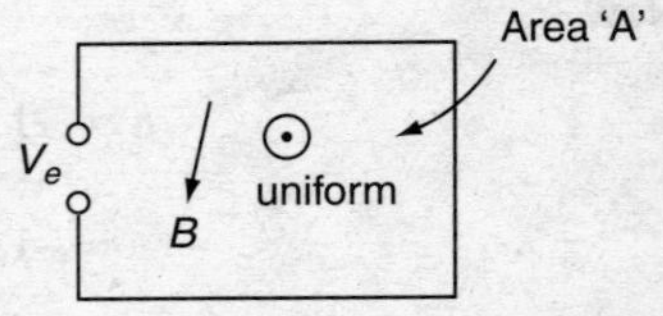

Figure 6.19 *Rectangular loop*

Solution:
It is the case of stationary coil, where $\vec{B}$ is changing with respect to time.

We know that

$$V_{\text{emf}} = -\frac{\partial \phi}{\partial t} \quad \text{or}$$

$$= -\int \frac{\partial B}{\partial t} \cdot ds \quad \text{or}$$

$$= -\int \frac{\partial (B_0 \cos \omega t)}{\partial t} \cdot ds \quad \text{or}$$

$$= \int_s -\omega B_0 \sin \omega t \cdot ds$$

or simply

$$= A\omega B_0 \sin \omega t$$

where A is the area of the loop.

6.5 Figure 6.20 shows a variable loop, which consists of two stationary parallel conductors. The loop is connected at one end to a voltmeter and at the other end to a moving bar. It is moving with a uniform velocity of v (m/s). This loop is kept in a uniform flux density $B\vec{a}_z$ (tesla). Obtain an expression for induced emf in this loop. Consider $L = 5$ cm, $B = 0.8\vec{a}_z$ tesla and velocity $v = 10\sqrt{y}\,\vec{a}_y$ (m/s). Let $y = 4$ cm at $t = 0$. At $t = 0.05$ s, find

(a) Velocity of the bar,
(b) Position of the bar and
(c) emf that is induced in this loop, and also find its polarity.

Solution:
y is the position of a moving bar. Flux (through the surface of the loop) at any time t is given as

$$\phi = B\,y\,L$$

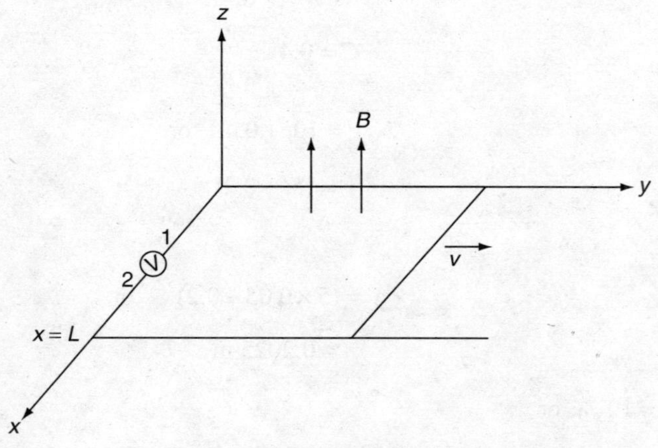

Figure 6.20 *Variable loop*

Thus,

$$V_{\text{emf}} = \oint \vec{E} \cdot d\vec{l}$$

$$= -\frac{\partial \phi}{\partial t}$$

$$= -BL\frac{\partial y}{\partial t} \quad \text{or}$$

$$= -B\,L\,v$$

where v is the velocity of the bar. As per Lenz's Law, end 2 will have a higher potential than end 1. Voltmeter reading

$$V_{\text{emf}_{12}} = -BLv \text{ volts}$$

It is given that $L = 5$ (cm) $B = 0.8\vec{a}_z$(T) and $v = 10\sqrt{y}$ (m/s). Also, at $t = 0$, $y = 4$ cm. Thus,

$$v = \frac{\partial y}{\partial t}$$

$$= 10\sqrt{y}$$

Further,

$$\int \frac{dy}{\sqrt{y}} = \int 10\,dt \quad \text{or}$$

$$2\sqrt{y} = 10t + C$$

where C is the constant. At, $t = 0$, we get

$$2\sqrt{4\times10^{-2}} = C \quad \text{or}$$

$$C = 0.4$$

$$2\sqrt{y} = 10t + 0.4 \quad \text{or}$$

$$y = (5t + 0.2)^2$$

At $t = 0.05$, we get

$$y = (5\times0.05 + 0.2)^2 \quad \text{or}$$

$$= 0.2025 \text{ m}$$

(a) Velocity of bar $v = 10\sqrt{y}$ or

$$= 10\sqrt{0.2025} \quad \text{or}$$

$$= 4.5\,\vec{a}_y \text{ (m/s)}$$

(b) Position of bar $y = 0.2025$ m
(c) Induced emf

$$V_{\text{emf}_{12}} = -BLv \quad \text{or}$$

$$= -0.8 \times 5 \times 10^{-2} \times 4.5 \quad \text{or}$$

$$= -0.18 \text{ Volts}$$

Induced emf $V_{\text{emf}_{12}}$ is negative; therefore, end 2 will be at a higher voltage than end 1.

6.6 There is a straight conductor of length 0.2 m. It is lying on the x-axis. Its one end is at the origin. This conductor is kept in magnetic flux density $\vec{B} = 2.5\,\vec{a}_y$ (T) and velocity $v = 0.04 \sin 10^3\, t\,\vec{a}_z$ (m/s). Find the motional electric field intensity and induced emf in this conductor.

Solution:
It is given that

$$\vec{B} = 2.5\,\vec{a}_y \text{ T}$$

$$\vec{v} = 0.04 \sin 10^3\, t\,\vec{a}_z \text{ (m/s)}$$

Motional emf producing field intensity given as

$$\vec{E} = \vec{v} \times \vec{B} \quad \text{or}$$

$$= 0.10 \sin 10^3\, t\,(-\vec{a}_x) \text{ (V/m)}$$

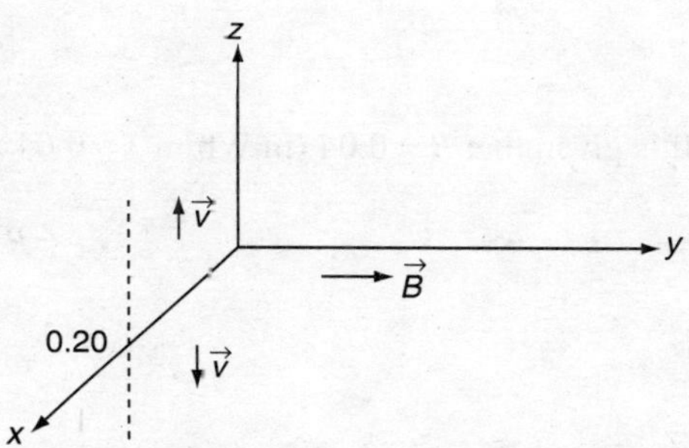

Figure 6.21 *Conductor and directions of $\vec{v}$ and $\vec{B}$*

Induced emf V_{emf} in the conductor is obtained as

$$V_{emf} = \int (\vec{v} \times \vec{B}) \cdot \vec{dl} \quad \text{or}$$

$$= \int_0^{0.20} 0.10 \sin 10^3 t\,(-\vec{a}_z)\, dx\, \vec{a}_x$$

or simply

$$= -0.20 \sin 10^3\, t \text{ (V)}$$

6.7 A magnetic flux through a coil is perpendicular to its plane. It is directed into paper. It is varying as per relationship $\phi = (10t^2 + 5t + 6)$ (mili Weber). Find the induced emf in this loop at time $t = 15$ s.

Solution:
Magnetic flux ϕ is given as

$$\phi = (10t^2 + 5t + 6) \text{ (mili Weber)}$$

$$= (10t^2 + 5t + 6) \times 10^{-3} \text{ (Weber)}$$

Induced emf is obtained as

$$= \frac{\partial \phi}{\partial t} \quad \text{or}$$

$$= -(20t + 5) \times 10^{-3} \text{ Volts}$$

Induced emf at $t = 15$ is obtained as

$$= (20 \times 15 + 5) \times 10^{-3} \text{ Volts or}$$

$$= 0.305 \text{ Volts}$$

6.8 A copper disc has a diameter of 0.50 m. This disc is rotated at 300 rpm constant speed. It is rotated on a horizontal axis, which is perpendicular and passes through the disc's centre. The axis lies in the magnetic medium. Two brushes making contact with the disc one at brush at the edge and other brush is at disc centre. One of the horizontal components of the earth's field is 0.04 (mili Wb/m^2). Obtain induced emf between these brushes.

Solution:

We know

$$V = \int \vec{E} \cdot \vec{dr} \quad \text{or}$$

$$V = \frac{1}{2} \omega B R^2$$

It is given that $B = 0.04$ (m Wb/m^2) $= 0.04 \times 10^{-3}$ (Wb/m^2)

$$R = \frac{0.50}{2} = 0.25(\text{m}) \quad \text{and}$$

$$\omega = 2\pi f = 2\pi \cdot \frac{N}{60}$$

$$= 2\pi \times \frac{3000}{60} \text{ (rad/sec)}$$

Now finding

$$V = \frac{1}{2} \times 2\pi \times \frac{3000}{60} \times 0.04 \times 10^{-3} (0.25)^2 \quad \text{or}$$

$$= 0.3924 \text{ mili Volts}$$

6.9 A current element $I\,\Delta\vec{l}$ is given as $2\pi(1.2\vec{a}_x - 1.6\vec{a}_y)$ µA. This current element is situated at (5, –1, 4). What is the incremental field $\Delta\vec{H}$ at (2, 4, 3)?

Solution:

By using the Biot–Savart Law, we get

$$\Delta\vec{H} = \frac{Id\vec{l} \times \vec{a}_r}{4\pi R^2} \quad \text{and}$$

$$I\Delta\vec{l} = 2\pi\,(1.2\vec{a}_x - 1.6\vec{a}_y)$$

$$\vec{R} = (2-5)\vec{a}_x + (4+1)\vec{a}_y + (3-4)\vec{a}_z \quad \text{or}$$

$$\vec{R} = (-3\vec{a}_x + 5\vec{a}_y - \vec{a}_z)$$

Now

$$\vec{a}_R = \frac{\vec{R}}{|\vec{R}|}$$

$$= \frac{(-3\vec{a}_x + 5\vec{a}_y - \vec{a}_z)}{\sqrt{35}}$$

Thus,

$$\Delta\vec{H} = \frac{2\pi(1.2\vec{a}_x - 1.6\vec{a}_y)\times 10^{-6} \times (-3\vec{a}_x + 5\vec{a}_y - \vec{a}_z)}{4\pi \times 35 \times \sqrt{35}} \quad \text{or}$$

$$\Delta\vec{H} = \frac{10^{-6}}{35\sqrt{35}}(0.8\vec{a}_x + 0.6\vec{a}_y + 0.6\vec{a}_z) \quad \text{or}$$

$$\Delta\vec{H} = 3.86\vec{a}_x + 2.88\vec{a}_y + 2.88\vec{a}_z \text{ (mA/metre)}$$

6.10 A circular loop of wire has a radius a. This loop is lying in the x–y plane. The loop's centre is at the origin. The loop carries a current I. Current I is in the $+\theta$ direction. Find $\vec{H}$ $(0, 0, z)$ and $\vec{H}$ $(0, 0, 0)$ using the Biot–Savart Law.

Solution:

It is shown as Figure 6.22

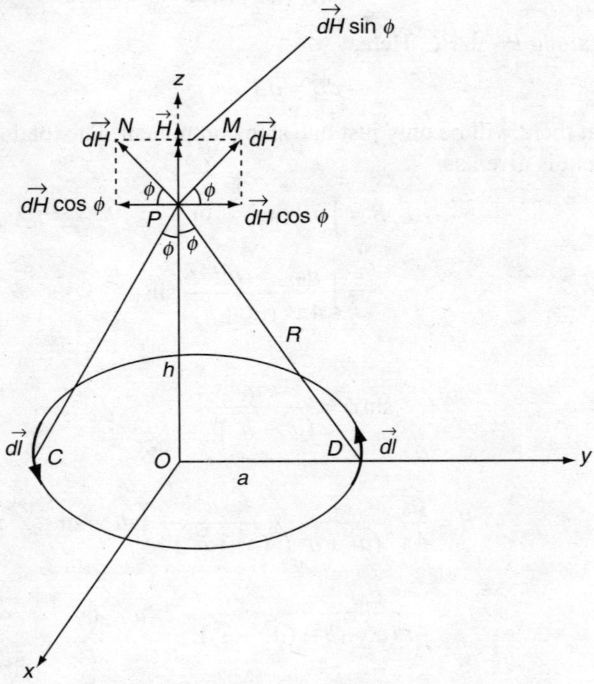

Figure 6.22 *Circular loop*

Consider two small elements of coil. Each element is of length dl at C and D. Two points C and D are situated at diametrically opposite edges. Here, $PC = PO = R = \sqrt{a^2 + h^2}$

Let

$$\angle CPO = \angle DPO = \phi$$

As per the Biot–Savart Law, the magnitude of the magnetic field induction at a point P due to a current element Idl at another point C is given as

$$dB = \frac{\mu_0}{4\pi} \cdot \frac{Idl \sin 90°}{R^2} \quad \text{or}$$

$$= \frac{\mu_0}{4\pi} \cdot \frac{Idl}{(a^2 + h^2)}$$

The direction of $d\vec{B}$ will be in a plane, which is perpendicular to both $d\vec{l}$ and $\vec{R}$. $d\vec{B}$ is directed and is given by the right-hand screw rule. $d\vec{B}$ is acting along $PH \perp PD$.

The magnitude of the magnetic field induction at a point P due to the current element of length Idl at another point D is given as

$$dB = \frac{\mu_0}{4\pi} \cdot \frac{Idl \sin 90°}{R^2} \quad \text{or}$$

$$= \frac{\mu_0}{4\pi} \cdot \frac{I\,dl}{(a^2 + h^2)}$$

The direction of dB is along $PN \perp PC$. Here,

$$dB = dB'$$

On solving, we find that there will be only just one sine component. The total magnetic field induction at a point P due to current is given as

$$B = \int dB \sin \phi \quad \text{or}$$

$$= \int \frac{\mu_0}{4\pi} \cdot \frac{Idl}{(a^2 + h^2)} \sin \phi$$

Also,

$$\sin \phi = \frac{a}{(a^2 + h^2)^{1/2}}$$

Thus,

$$B = \frac{\mu_0}{4\pi} \cdot \frac{I}{(a^2 + h^2)} \cdot \frac{a}{(a^2 + h^2)^{1/2}} \int dl \quad \text{or}$$

$$= \frac{\mu_0 I}{4\pi(a^2 + h^2)} \frac{a}{(a^2 + h^2)^{1/2}} 2\pi a \quad \text{or}$$

$$= \frac{\mu_0 I a^2}{2(a^2 + h^2)^{3/2}} \quad \text{(in the positive } z\text{-axis)}$$

6.11 A circular coil has 60 turns, the coil has a radius of 4.5 cm. The coil carries a current I of 6.0 Amp. Find the magnitude of the magnetic field:

(a) At the centre of the coil

(b) At a point lying on the axis of this coil at a distance from the coil centre, which is equal to the radius of coil.

Solution:

It is given that $I = 6.0$ Amp, $R = 4.5$ cm $= 0.045$ m, $h = R = 0.045$ m and $N = 120$.

(a) We know that

$$B = \frac{\mu_0 NI}{2R} \quad \text{or}$$

$$= 10^{-7} \times 2 \times \frac{22}{7} \times \frac{60 \times 6}{0.045} \quad \text{or}$$

$$= 5.028 \times 10^{-3}\ \text{T}$$

(b) Again

$$B = \frac{\mu_0}{4\pi} \frac{2\pi NIa^2}{(a^2 + h^2)^{3/2}} \quad \text{or}$$

$$= 10^{-7} \times 2 \times \frac{22}{7} \times \frac{60 \times 6 \times (0.045)^2}{[(0.045)^2 + (0.045)^2]^{3/2}} \quad \text{or}$$

$$= 17.6 \times 10^{-4}\ \text{T}$$

6.12 Find the contribution to the magnetic field at a point O in Figure 6.23 caused by semicircular section BC, two horizontal conductors AB and CD and short vertical section DA.

Solution:

Semi-circular section (BC) contribution is obtained as

$$B1 = \frac{\mu_0 I}{4R}$$

$$= 37.68\ (\mu\text{Wb/m}^2)$$

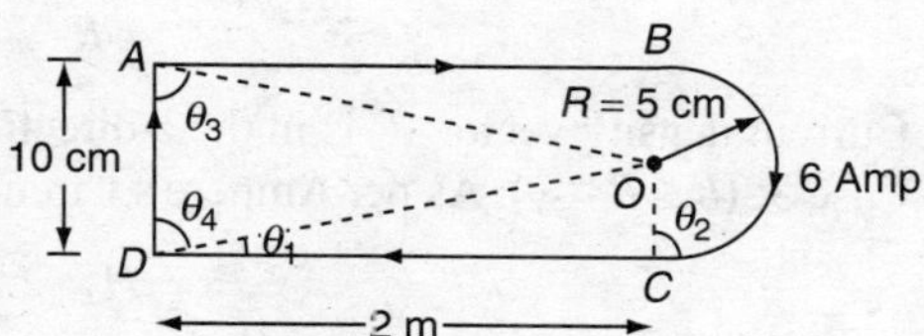

Figure 6.23 *Conductor*

The contribution of two horizontal conductors (AB and CD) is obtained as

$$B2 = 2 \cdot \frac{\mu_0}{4\pi} \frac{I}{R} (\cos\theta_1 - \cos\theta_2) \quad \text{or}$$

$$= 2 \times \frac{\mu_0 \times 6}{4\pi \times .5} (1 - 0) \qquad [\theta_1 \approx 0°, \theta_2 \approx 90°] \quad \text{or}$$

$$= 25\ (\mu\text{Wb/m}^2)$$

The contribution of the short vertical section (DA) is obtained as

$$B = \frac{\mu_0}{4\pi} \frac{I}{2} (\cos\theta_2 - \cos\theta_1) \quad \text{or}$$

$$= \frac{\mu_0 I}{4\pi \times 2}(\cos\theta_4 + \cos\theta_2) \quad \text{or}$$

$$= \frac{\mu_0}{4\pi} I \cos\theta_3 \quad \text{or}$$

$$= 0.015\ (\mu\text{Wb/m}^2)$$

6.13 Find $\vec{H}$ everywhere due to volume current density $\vec{J}$ (in Cylindrical coordinates).

$$\text{Volume charge density } \vec{J} = \begin{cases} 0, & 0 < R \le a \\ 10\vec{a}_z, & a < R < b \\ 0, & b \le R < \infty \end{cases}$$

[Hint: Use Ampere's Circuital Law in integral form.]

Solution:

The current distribution is shown in Figure 6.24. As per Ampere's Circuital Law, we find

$$\int \vec{H} \cdot d\vec{l} = I_{\text{enclosed}}$$

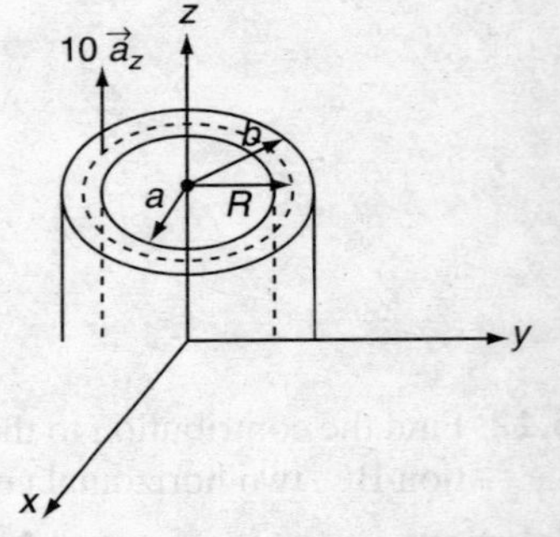

Figure 6.24 *Current distribution*

Consider the following three cases:
Case 1: ($R \le a$): No current is enclosed. Hence, $\vec{H} = 0$.
Case 2: ($a < R < b$): Ampere's Law gives

$$H_R \cdot 2\pi R = 10(\pi R^2 - \pi a^2) \quad \text{(where } R \text{ is the radius)} \quad \text{or}$$

$$H_R = \frac{10\pi(R^2 - a^2)}{2\pi R} \quad \text{or}$$

$$\vec{H}_R = \frac{5(R^2 - a^2)}{R}\vec{a}_\phi$$

Current density vector $\vec{J}$ is in the z-direction. Therefore, the field will be in the $\vec{a}_\phi$ direction.
Case 3: ($b \le R < \infty$): As per Ampere's Circuital Law (where full current is enclosed), we get

$$H_R \cdot 2\pi R = 10(\pi b^2 - \pi a^2) \quad \text{or}$$

$$H_R = \frac{10\pi(b^2 - a^2)}{2\pi R} \quad \text{or}$$

$$H_R = \frac{5(b^2 - a^2)}{R}\vec{a}_\phi$$

6.14 Find the field at any point:
(a) Due to an ideal solenoid having an infinite length and a circular current shift $\vec{K} = 20\vec{a}_\phi$ and
(b) A 500-turn solenoid with a finite length of 5 m and current per turn of 10 Amp.
[Hint: Use the concept of field due to two parallel infinite surface sheet currents.)

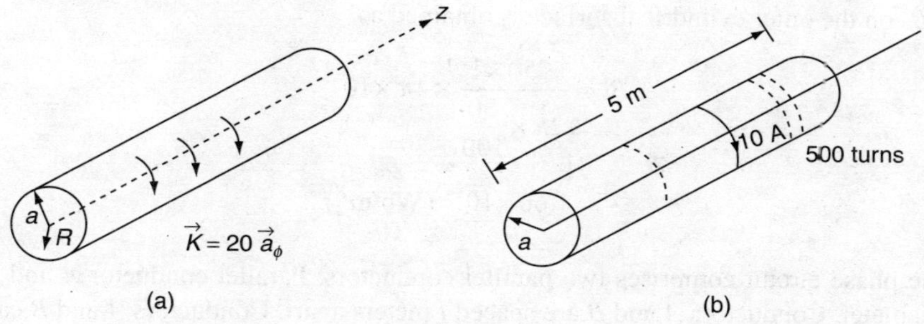

Figure 6.25 *Solenoids*

Solution:

Solenoids are shown in Figure 6.25. By using the concept of field because of two infinite parallel sheets, fields are obtained as below:

(a) For an ideal solenoid having an infinite length, we get

$$\vec{K} = 20\vec{a}_\phi$$

Also,

$$\vec{H}_R = 20\vec{a}_z \quad \text{for } R < a \quad \text{and}$$

$$\vec{H} = 0 \quad \text{for } R > a$$

(b) For a 500-turn solenoid having a length of 5 m, we get

Total current = $N\,I$ (Amp) or

$$= 500 \times 10 \text{ Amp} \quad \text{or}$$

$$= 5000 \text{ Amp}$$

Surface current density $\vec{K} = \dfrac{5000}{5}\, 20\vec{a}_\phi$ (A/m)

The field inside a solenoid (as per the concept used in part (a)) is obtained as

$$\vec{H} = \frac{NI}{d}\vec{a}_z \quad \text{or}$$

$$\vec{H} = 1000\,\vec{a}_z \text{ (A/m)}$$

6.15 There is a wooden ring of square section. Its internal diameter is of 40 cm. The external diameter is of 60 cm. The wooden ring carries a toroidal winding having 250 turns. It is uniformly distributed. For a coil width of 1 A, find the magnetic flux density B on both cylindrical surfaces, viz. inner and outer wooden rings.

Solution:

Flux density B_i on the inner cylindrical surface is obtained as

$$B_i' = \frac{250 \times 1}{2\pi \dfrac{20}{100}} \times 4\pi \times 10^{-7}$$

$$= 25 \times 10^{-4} \text{ (Wb/m}^2\text{)}$$

Flux density B_0 on the outer cylindrical surface is obtained as

$$B_o = \frac{250\times 1}{2\pi\times\frac{30}{100}}\times 4\pi\times 10^{-7}$$

$$= 1.66\times 10^{-4}\ (\text{Wb/m}^2)$$

6.16 A single phase circuit comprises two parallel conductors. Parallel conductor *A* and *B* each has 3 cm diameter. Conductors *A* and *B* are spaced *l* meters apart. Conductors *A* and *B* carry current of +50 and −50 Amps, respectively. Find the field intensity at the surface of each conductor. Also, find the field intensity in space exactly midway of conductors *A* and *B*.

Solution:
Field *H* at a point between parallel conductors *A* and *B* is obtained as

$$H = H_A + H_B \qquad \text{(Because current is in the opposite direction)}$$

$$= \frac{I}{2\pi R_1} + \frac{I}{2\pi R_2}$$

Each conductor produces field, which will be directed downwards at that point. Therefore, we get (assume two cases)
Case 1: (*H* at the conductor surface.) We know, $R_1 = 3/100\times 2$ (m) and $R_2 \approx 1$ (m).
Thus,

$$H\frac{1}{2\pi}\left[\frac{1}{R_1}+\frac{1}{R_2}\right] = \frac{50}{2\pi}\left[\frac{2\times 100}{3}+1\right] \quad \text{or}$$

$$= 538.47\ (\text{A/m})$$

Case 2: (*H* is midway between parallel conductors *A* and *B*.) We know that

$$R_1 = \frac{1}{2} = 0.5(\text{m}) \quad \text{and}$$

$$R_2 = \frac{1}{2} = 0.5(\text{m})$$

Thus,

$$H' = \frac{50}{2\pi}\left[\frac{1}{0.5}+\frac{1}{0.5}\right] \quad \text{or}$$

$$= 31.83\ (\text{A/m})$$

6.17 A magnetic vector potential is

$$\vec{A} = \frac{r^2}{6}\vec{a}_z\ (\text{Wb/m})$$

Find total magnetic flux, which crosses surface $\phi' = \pi/2\,(2\le r\le 4\text{m})$ and $(0\le z\le 5\text{m})$.

Solution:

Let L be the path, which bounds the surface S as shown in Figure 6.26.

Here,

$$\phi = \phi_1 + \phi_2 + \phi_3 + \phi_4$$

$$= \oint_L \vec{A} \cdot d\vec{l}$$

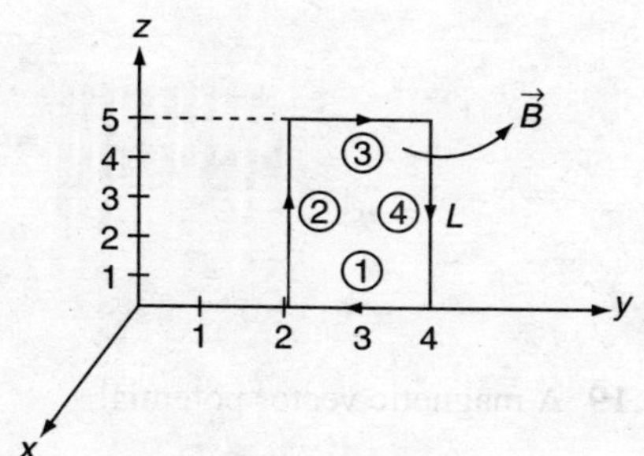

Figure 6.26 *Surface and magnetic flux*

where $(\phi_1 + \phi_2 + \phi_3 + \phi_4)$ is the evaluation of $\int \vec{A} \cdot d\vec{l}$ along segments of path, L labelled as 1, 2, 3 and 4 as shown in Figure 6.26. Here,

$$\phi_1 = \phi_3 = 0$$

Now,

$$\phi = \phi_2 + \phi_4$$

$$= -\frac{1}{6}\left[(2)^2 \int_0^5 dz + (4)^2 \int_5^0 dz\right] \quad \text{or}$$

$$= -\frac{1}{6}\left[\{(2)^2 - (4)^2\}\left\{\int_0^5 dz\right\}\right] \quad \text{or}$$

$$= -\frac{1}{6}[(4-16)(5)] \quad \text{or}$$

$$= -\frac{1}{6}[(-12)\times(5)] \quad \text{or}$$

$$= 10 \text{ (Wb)}$$

6.18 A direct current i of 20 Amp flows through a long and straight round conductor. What will be the magnetic flux through half of the wire's cross-section per meter length?

Solution:

The shaded portion depicts half of the wire's cross-section as shown in Figure 6.27.

Assume B to be the magnetic induction at a distance r from the axis and R the radius of the conductor. By using Ampere's Circuital Law, we get

$$B \cdot 2\pi r = \mu_0 \, i \frac{\pi r^2}{\pi R^2} \quad \text{or}$$

$$B = \frac{\mu_0}{2\pi} \frac{ir^2}{R^2}$$

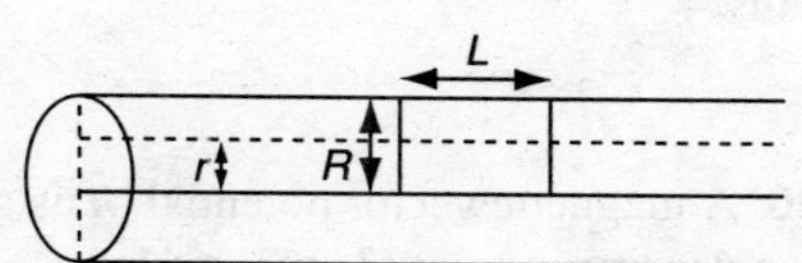

Figure 6.27 *Magnetic flux determination through half the wire's cross-section*

Magnetic flux ϕ through the shaded region is obtained as

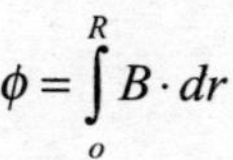

$$\phi = \int_o^R B \cdot dr$$

$$= \int_{o}^{R} \frac{\mu_0}{2\pi} \frac{ir}{R^2} dr \quad \text{or}$$

$$= \frac{\mu_0}{4\pi} i \quad \text{or}$$

$$= 2 \times 10^{-6} \text{ (Wb/m}^2) \quad \text{or}$$

6.19 A magnetic vector potential

$$\vec{A} = -\frac{\rho^2}{8} \vec{a}_z \text{ (Wb/m)}$$

Find the total magnetic flux, which crosses surface $\phi = \pi/2 (3 \leq \rho \leq 4\text{m})$ and $(0 \leq z \leq 10\,\text{m})$.

Solution:

We know that

$$\vec{B} = \nabla \times \vec{A}$$

$$= -\frac{\partial A_z}{\partial \rho} \vec{a}_\phi \quad \text{or}$$

$$\vec{B} = +\frac{\rho}{4} \vec{a}_\phi \quad \text{and}$$

$$\vec{ds} = d\rho\, dz\, a_\phi$$

Hence, the total flux ϕ is given as

$$= \int \vec{B} \cdot \vec{ds} \quad \text{or}$$

$$= \frac{1}{4} \int_{z=0}^{10} \int_{\rho=3}^{4} \rho \cdot dl\, dz \quad \text{or}$$

$$= \frac{1}{4} \frac{\rho^2}{2} \Big|_3^4 (10) \quad \text{or}$$

$$= \frac{1}{8} \times (16 - 9) \times (10) \quad \text{or}$$

$$= \frac{70}{8}$$

Finally,

$$\phi = 8.75 \text{ (Wb)}$$

6.20 A magnetic vector potential $\vec{A}$ is given as 15 $\sin\theta\ \vec{a}_\theta$ (in Spherical coordinates). Obtain the flux density at point (3, $\pi/2$, $\pi/4$).

Solution:

We know that

$$\vec{B} = \nabla \times \vec{A} \quad \text{or}$$

$$\vec{B} = \nabla \times \vec{A} \quad \text{or}$$

$$= \begin{vmatrix} \dfrac{\vec{a}_r}{r^2 \sin\theta} & \dfrac{\vec{a}_\theta}{r\sin\theta} & \dfrac{\vec{a}_\phi}{r} \\ \dfrac{\partial}{\partial r} & \dfrac{\partial}{\partial \theta} & \dfrac{\partial}{\partial \phi} \\ 0 & 15r\sin\theta & 0 \end{vmatrix} \quad \text{or}$$

$$\vec{B} = \frac{\partial}{\partial r}(15r\sin\theta)\cdot\frac{\vec{a}_\phi}{r} \quad \text{or}$$

$$\vec{B} = \frac{15\sin\theta}{r}\vec{a}_\phi$$

so $\vec{B}$ at point (3, $\pi/2$, $\pi/4$)

$$= \frac{15\sin\pi/2}{3}\vec{a}_\phi \quad \text{or}$$

$$\vec{B} = 5\vec{a}_\phi$$

6.21 A magnetizing field H of 400 (Am^{-1}) causes a magnetic flux ϕ of 1.2×10^{-5} (Weber) in an iron bar. The iron bar has a cross-sectional area S of 1 cm^2. Find the permeability and susceptibility of this iron bar.

Solution:

It is given that $H = 400$ (Am^{-1}), $\phi = 1.2 \times 10^{-5}$ (Weber) and cross-sectional area = 1 $\text{cm}^2 = 10^{-4}$ (m^2). We have to find μ = ? and χ_m = ? We know that

$$B = \frac{\phi}{S}$$

$$= \frac{1.2\times10^{-5}}{10^{-4}} \quad \text{or}$$

$$= 0.12 \text{ (Weber/m}^2\text{)}$$

Also,

$$\mu = \frac{B}{H}$$

$$= \frac{1.2}{400} \quad \text{or}$$

$$= 3\times10^{-4} \text{ T/Am}$$

Also,

$$\mu = \mu_0(1+\chi_m) \quad \text{or}$$

$$\chi_m = \frac{\mu}{\mu_0} - 1 \quad \text{or}$$

$$\chi_m = \frac{3\times10^{-4}}{4\pi\times10^{-7}} - 1 \quad \text{or}$$

$$\chi_m = 238.63 - 1 \quad \text{or}$$

$$\chi_m = 237.63$$

6.22 A magnetic material possesses $\mu_r = 4/p$. This magnetic material is in a magnetic field of strength $H = 3\rho^2\vec{a}_\phi$ (A/m). Obtain magnetization.

Solution:
Magnetic flux density is given as

$$\vec{B} = \mu\vec{H} \quad \text{or}$$

$$= \mu_r\mu_0\ \vec{H} \quad \text{or}$$

$$= \frac{4}{\pi}\times 4\pi\times10^{-7}\times 3\rho^2\vec{a}_\phi \text{ (A/m)}$$

Finally,

$$\vec{B} = 8\times10^{-6}\times\rho^2\vec{a}_\phi \text{ (Wb/m}^2\text{)}$$

Also,

$$\vec{B} = \mu_0(\vec{H}+\vec{M}) \quad \text{or}$$

$$= \mu_0\mu_r\vec{H} = \mu_0(\vec{H}+\vec{M}) \quad \text{or}$$

$$\vec{M} = \vec{H}(\mu_r - 1) \quad \text{or}$$

$$= \vec{H}\left(\frac{4}{\pi} - 1\right) \quad \text{or}$$

$$= 3\rho^2\cdot 27\vec{a}_\phi \text{ (A/m)}$$

Finally,

$$\vec{M} = 0\cdot818\rho^2\vec{a}_\phi \text{ (A/m)}$$

6.23 A toroid case has 1500 turns. It has inner and outer radii of 10 cm and 13 cm, respectively. Find the relative permeability of the core. A current of 2.1 Amp produces a magnetic field intensity of 1.5 (T) in this core.

Solution:
It is given that total number of turns = 1500

$$I = 2.1 \text{ (Amp)} \quad \text{and}$$

$$B = 1.5 \text{ (T)}$$

Average radius

$$r = \frac{10+13}{2} \quad \text{or}$$

$$= 11.5 \text{ cm}$$

Number of turns/length $(n) = \dfrac{1500}{2\pi r}$ or

$$= \frac{1500}{2\pi \times 11.5 \times 10^{-2}} \quad \text{or}$$

$$= 2075$$

Further,

$$B = \mu n I \quad \text{or}$$

$$= \mu_0 \mu_r n I \quad \text{or}$$

$$\mu r = \frac{B}{\mu_0 n I} \quad \text{or}$$

$$\mu_r = \frac{1.5}{4\pi \times 10^{-7} \times 2075 \times 2.1}$$

Finally,

$$\mu_r = 273.8$$

6.24 There is a boundary at $z = 0$ between two magnetic materials. Permittivities are $\mu_1 = z\mu_0$ (H/m) for region 1 ($z > 0$) and $\mu_2 = 14\mu_0$ (H/m) for region 2 ($z < 0$). Surface current density $\vec{K}$ is $50\,\vec{a}_x$ (A/m) at the boundary ($z = 0$). There is a field $\vec{B}_1$ $(= 4\vec{a}_x - 6\vec{a}_y + 4\vec{a}_z)$ (mT) in region 1. What is the flux density $(\vec{B}_2)$ in region 2?

Solution:

Given that $\vec{B}_1 = (4\vec{a}_x - 6\vec{a}_y + 4\vec{a}_x)$ (mT). It is shown in Figure 6.28.

Normal component of $\vec{B}_1$ is obtained as

$$\vec{B}_{n_1} = (\vec{B}_1 \cdot \vec{a}_n)\vec{a}_n \quad \text{or}$$

$$= \lfloor (4\vec{a}_x - 6\vec{a}_y + 4\vec{a}_z) \cdot \vec{a}_z \rfloor \vec{a}_z \quad \text{or}$$

$$= 4\vec{a}_z \text{ (mT)}$$

And also,

$$\vec{B}_{n_2} = \vec{B}_{n_1} = 4\vec{a}_z \text{ (mT)}$$

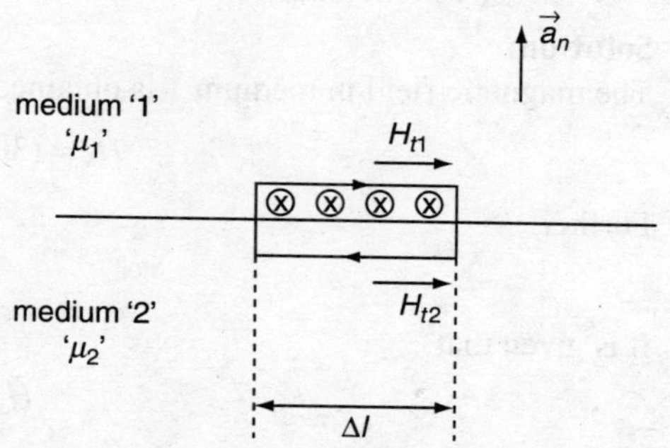

Figure 6.28 *Two magnetic materials*

Tangential component of $\vec{B}_1$ is obtained as

$$\vec{B}_{t_1} = (\vec{B}_1 - \vec{B}_{n_1})$$

$$= (4\vec{a}_x - 6\vec{a}_y)(\text{mT}) \quad \text{or}$$

$$= (1592\,\vec{a}_x - 2388\,\vec{a}_y)\,(\text{A/m})$$

Also,

$$\vec{H}_{t_1} = \frac{\vec{B}_{t_1}}{\mu_1}$$

$$= \frac{(4\vec{a}_x - 6\vec{a}_y)\times 10^{-3}}{2\times 4\pi\times 10^{-7}} \quad \text{or}$$

$$= (1592\,\vec{a}_x - 2388\,\vec{a}_y)\,(\text{A/m})$$

Further,

$$\vec{H}_{t_2} = \vec{H}_{t_1} - (\vec{K}\times\vec{a}_n) \quad \text{or}$$

$$= [(1592\,\vec{a}_x - 2388\,\vec{a}_y) - (50\vec{a}_x \times \vec{a}_z)] \quad \text{or}$$

$$= (1592\,\vec{a}_x - 2438\,\vec{a}_y)$$

Also,

$$\vec{B}_{t_2} = \mu_2 \vec{H}_{t_2} \quad \text{or}$$

$$= 14\times 4\pi\times 10^{-7}(1592\,\vec{a}_x - 2438\,\vec{a}_y)$$

Finally, $\vec{B}_2$ can be obtained by using the relationship as given below:

$$\vec{B}_2 = \vec{B}_{n_2} + \vec{B}_{t_2}$$

6.25 There are two homogenous, linear and isotropic, media with interface at $x = 0$. $x < 0$ describes medium 1 $(\mu_{r1} = 4)$. $x > 0$ describes medium $2\,(\mu_{r2} = 10)$. Magnetic field medium 1 is $(30\vec{a}_x - 80\vec{a}_y + 70\vec{a}_z)$ (A/m). Find the magnetic field in medium 2. Also, find the magnetic flux density in medium 1.

Solution:

The magnetic field in medium 1 is obtained as

$$\vec{H}_1 = (30\vec{a}_x - 80\vec{a}_y + 70\vec{a}_z)(\text{A/m})$$

Further,

$$\vec{H}_1 = \vec{H}_{t_1} + \vec{H}_{n_1}$$

It is given that

$$\vec{H}_{t_1} = (-80\vec{a}_y + 70\vec{a}_z) \quad \text{and}$$

$$\vec{H}_{n_1} = 30\vec{a}_x$$

Applying boundary conditions, we get

$$\vec{H}_{t_1} = \vec{H}_{t_2} \quad \text{or}$$

$$\vec{H}_{t_2} = (-80\vec{a}_y + 70\vec{a}_z)\ (\text{A/m})$$

Now applying boundary condition on $\vec{B}$, we get

$$\vec{B}_{n_1} = \vec{B}_{n_2} \quad \text{or}$$

$$\mu_1 \vec{H}_{n_1} = \mu_2 \vec{H}_{n_2} \quad \text{or}$$

$$\vec{H}_{n_2} = \frac{\mu_1}{\mu_2}\vec{H}_{n_1} \quad \text{or}$$

$$= \frac{\mu_{r_1}\mu_0}{\mu_{r_2}\mu_0}\vec{H}_{n_1} \quad \text{or}$$

$$= \frac{4}{10} \times 30\vec{a}_x \quad \text{or}$$

$$= 12\vec{a}_x$$

$$\vec{H}_2 = \vec{H}_{t_2} + \vec{H}_{n_2} \quad \text{or}$$

$$= (+12\vec{a}_x - 80\vec{a}_y + 70\vec{a}_z)\ (\text{A/m})$$

Magnetic flux density $\vec{B} = \mu_1 \vec{H}_1$ or

$$= \mu_{r_1}\mu_0 \vec{H}_1 \quad \text{or}$$

$$= 4 \times 4\pi \times 10^{-7}(30\vec{a}_x - 80\vec{a}_y + 70\vec{a}_z) \quad \text{or}$$

$$= 16\pi(3\vec{a}_x - 8\vec{a}_y + 7\vec{a}_z)\ (\mu\text{Wb/m}^2)$$

6.26 Let $\vec{H}_1 = (-\vec{a}_x + 3\vec{a}_y + 2\vec{a}_z)$ (A/m) in region $(y - x - 3) \le 0$. Here, $\mu_{r_1} = 7$. Find $\vec{M}_1$ and $\vec{B}_1$. Also, find $\vec{H}_2$ and $\vec{B}_2$ in this region $(y - x - 3 \le 0)$. Consider $\mu_{r_2} = 4$.

Solution:

Plane is given as

$$y - x - 3 = 0$$

$(y - x) \le 3$ is region 1 as shown Figure 6.29. A point in this region may confirm it. Let us take origin (0, 0). It is in this region $(0 - 0 - 3) < 0$. Let the surface of the plane be $F(x, y) = y - x - 3$. A unit vector, which is normal to this plane, is given as

$$\vec{a}_n = \frac{\vec{F}}{|\vec{F}|}$$

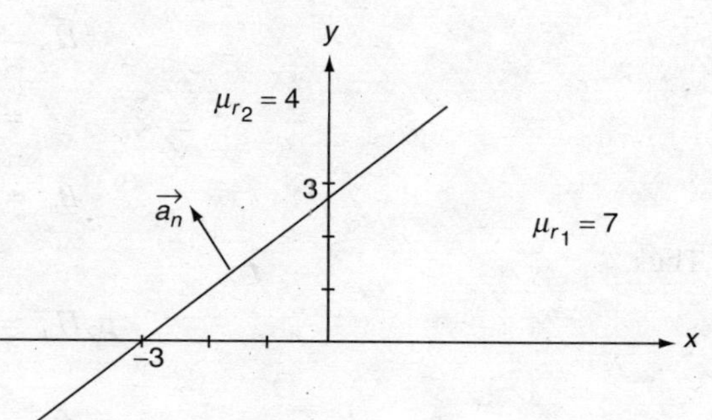

Figure 6.29 *Region* $(y - x) \le 3$

$$= \frac{(\vec{a}_y - \vec{a}_x)}{\sqrt{2}}$$

We know that

$$\vec{M}_1 = \chi_{m_1} \vec{H}_1$$

$$= (\mu_{r_1} - 1)\vec{H}_1$$

$$= (7-1)(-\vec{a}_x + 3\vec{a}_y + 2\vec{a}_z)$$

Also,

$$\vec{M}_1 = (-6\vec{a}_x + 18\vec{a}_y + 12\vec{a}_z)\ (\text{A/m}) \quad \text{and}$$

$$\vec{B}_1 = \mu_1 \vec{H}_1$$

$$= \mu_{r_1}\mu_0 \vec{H}_1 \quad \text{or}$$

$$= 7 \times 4\pi \times 10^{-7}(-\vec{a}_x + 3\vec{a}_y + 2\vec{a}_z)$$

Also,

$$\vec{H}_{n_1} = (\vec{H}_1 \cdot \vec{a}_n)\vec{a}_n \quad \text{or}$$

$$= \left[(-\vec{a}_x + 3\vec{a}_y + 2\vec{a}_z).\left(\frac{-\vec{a}_x + \vec{a}_y}{\sqrt{2}}\right)\right]\left(\frac{-\vec{a}_x + \vec{a}_y}{\sqrt{2}}\right) \quad \text{or}$$

$$= (-2\vec{a}_x + 2\vec{a}_y)$$

Further,

$$\vec{H}_1 = \vec{H}_{n_1} + \vec{H}_{t_1} \quad \text{or}$$

$$\vec{H}_{t_1} = \vec{H}_1 - \vec{H}_{n_1} \quad \text{or}$$

$$= (\vec{a}_x + \vec{a}_y + 2\vec{a}_z)$$

Applying boundary condition, we get

$$\vec{H}_{t_2} = \vec{H}_{t_1} \quad \text{or}$$

$$= (\vec{a}_x + \vec{a}_y + 2\vec{a}_z) \quad \text{and}$$

$$\vec{B}_{n_2} = \vec{B}_{n_1}$$

Thus,

$$\mu_2 \vec{H}_{n_2} = \mu_1 \vec{H}_{n_1} \quad \text{or}$$

$$\vec{H}_{n_2} = \frac{\mu_1}{\mu_2}\vec{H}_{n_1} \quad \text{or}$$

$$= \frac{\mu_{r_1}\mu_0}{\mu_{r_2}\mu_0}\vec{H}_{n_1} \quad \text{or}$$

$$= \frac{7}{4}(-2\vec{a}_x + 2\vec{a}_y) \quad \text{or}$$

$$= \left(-\frac{7}{2}\vec{a}_x + \frac{7}{2}\vec{a}_{y_z}\right)$$

Also,

$$\vec{H}_2 = \vec{H}_{n_2} + \vec{H}_{t_2} \quad \text{or}$$

$$= \left(-\frac{5}{2}\vec{a}_x + \frac{9}{2}\vec{a}_y + 2\vec{a}_z\right)\text{(A/m)}$$

$$\vec{B}_2 = \mu_2\vec{H}_2$$

$$= \mu_{r_2}\mu_0\vec{H}_2 \quad \text{or}$$

$$= 4 \times 4\pi \times 10^{-7}(-2.5\vec{a}_x + 4.5\vec{a}_y + 2\vec{a}_z) \quad \text{or}$$

$$\vec{B}_2 = (-12.56\vec{a}_x + 22.62\vec{a}_y + 10.0\vec{a}_z)\,(\mu\text{Wb/m}^2)$$

6.27 Magnetic flux lines are received at the iron air boundary. Following are incident with the normal of the boundary on the iron side:
(a) 30°, (b) 75° and (c) 45°.
Iron has a relative permeability of 270. What will be the angle with normal at which magnetic flux comes to axis?

Solution:
Let medium 1 be iron and medium 2 be air. It is shown in Figure 6.30.
Boundary conditions are?

$$\frac{\tan\alpha_1}{\tan\alpha_2} = \frac{\mu_{r_1}}{\mu_{r_2}} \quad \text{or}$$

$$\alpha_2 = \tan^{-1}\left[\frac{\mu_{r_2}}{\mu_{r_1}}\tan\alpha_1\right]$$

Thus, we get

(a) $$\alpha_2 = \tan^{-1}\left[\frac{1}{270}\tan 30°\right] = 0.122$$

(b) $$\alpha_2 = \tan^{-1}\left[\frac{1}{270}\tan 75°\right] = 0.792 \quad \text{and}$$

(c) $$\alpha_2 = \tan^{-1}\left[\frac{1}{270}\tan 45°\right] = 0.212$$

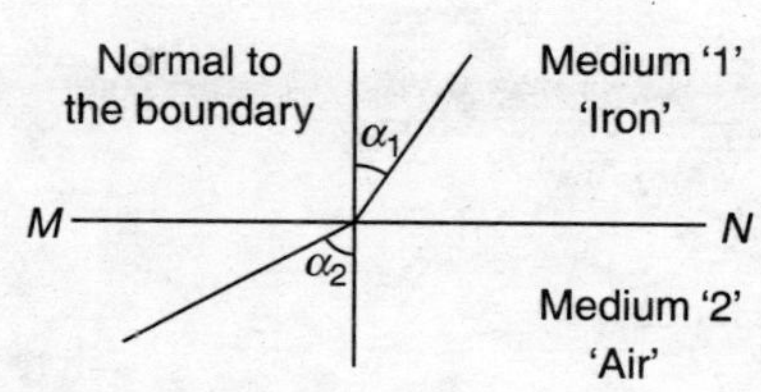

Figure 6.30 *Two mediums*

6.28 There are two ferromagnetic materials, which are separated by a plane boundary. Medium 1 has a relative permeability of 200. Medium 2 has a relative permeability of 2000. Magnetic field in medium 2 is directed at 750 with normal to the boundary. What is the angle between "field direction" and "normal" in medium 1? (Normal is with respect to the boundary.)

Solution:

Boundary conditions are obtained as

$$\frac{\tan\alpha_1}{\tan\alpha_2} = \frac{\mu_{r_1}}{\mu_{r_2}}$$

$$= \frac{200}{2000}$$

$$= \frac{1}{10}$$

The angle is obtained as

$$\alpha_1 = \tan^{-1}\left[\frac{1}{10}\tan\alpha_2\right] \quad \text{or}$$

$$= \tan^{-1}\left[\frac{1}{10}\times 3.73\right] \quad \text{or}$$

$$= \tan^{-1}[0.373] \quad \text{or}$$

$$= 20.46°$$

6.29 Find the torque (in N-m) acting on a circular current loop of radius 1 mm in the x–y plane. It is connected at the origin and with current 0.1 A flowing in the sense of increasing ϕ in a magnetic field $\hat{B} = 10^{-5}(2\vec{a}_x - 2\vec{a}_y - \vec{a}_z)$ Wb/m^2.

Solution:

We know that

$$\vec{T} = \overrightarrow{M}\times\vec{B}$$

$$= I\vec{A}\times\vec{B}$$

$$= I\pi r^2\vec{a}_z\times\vec{B}$$

$$= I\pi r^2\vec{a}_z\times 10^{-5}(2\vec{a}_x - 2\vec{a}_y - \vec{a}_z)$$

By putting the value of I and r, we get

$$\vec{T} = 0.1\times\pi\times 1\times 10^{-6}\,\vec{a}_z\times 10^{-5}(2\vec{a}_x - 2\vec{a}_y - \vec{a}_z)$$

$$= 10^{-12}\pi\,\vec{a}_z(2\vec{a}_x - 2\vec{a}_y - \vec{a}_z)$$

$$= 10^{-12}\pi\,(2\vec{a}_y + 2\vec{a}_x)$$

$$\vec{T} = 2\pi\times 10^{-12}(\vec{a}_y + \vec{a}_x)$$

Hence,

$$\vec{T} = 2\pi\times 10^{-12}(\vec{a}_y + \vec{a}_x)$$

6.30 For the vector $\vec{A} = (x+3y+2z)\vec{a}_x + (3az+y)\vec{a}_y + (3x^2+az+z)\vec{a}_z$ to be solenoidal, find the content "a."

Solution:

We know that $\nabla \cdot \vec{A} = 0$ for $\vec{A}$ to be solenoidal.

$$\Rightarrow \left[\frac{\partial A}{\partial_x}\vec{a}_x + \frac{\partial A}{\partial y}\vec{a}_y + \frac{\partial A}{\partial z}\vec{a}_z\right].[(x+3y+2z)\vec{a}_x + (3az+y)\vec{a}_y + (3x^2+az+z)\vec{a}_z]$$

$$\Rightarrow \frac{\partial}{\partial x}(x+3y+2z) + \frac{\partial}{\partial y}(3az+y) + \frac{\partial}{\partial z}(3x^2+az+z) = 0$$

$$\Rightarrow 1+1+a+1 = 0$$

$$\Rightarrow a = -3$$

6.31 Current density $(\vec{J})$ in a Cylindrical coordinate system is given as

$$J(r,\theta,z) = 0 \text{ to } 0 < r < a$$
$$= J_0(r/a^2)\vec{I}_Z \quad \text{for } a < r < b$$

where $\vec{I}_z$ is the unit vector along the z-coordinate axis. In the region $a < r < b$, what is the expression for the magnitude of the magnetic field intensity vector $(\vec{H})$?

Solution:

As it is given, $\vec{I}_z$ is the unit vector along the z-coordinate, so $r \times \phi$ component will be zero. We know that

$$\nabla \times \vec{H} = J$$

Taking only the z-coordinate, we get

$$\frac{1}{r}\left(\frac{\partial(rH_\theta)}{\partial r} - \frac{\partial H_r}{\partial \theta}\right)\vec{a}_z = J$$

$$\Rightarrow \frac{1}{r}\left(\frac{\partial(rH_\theta)}{\partial r} - \frac{\partial Hr}{\partial \theta}\right)\vec{a}_z = J_0\left(\frac{r}{a^2}\right)\left[\text{Given } J = J\cdot\left(\frac{r}{a^2}\right)\right]$$

After solving, we get

$$H = \frac{J_0(r^3 - a^3)}{3a^2 r}$$

6.32 Find the flux density at point (1, –3, 2) due to a uniform charge density ρ of 40 μC/m² at $x = 4$ (a plane).

Solution:

Electric field E due to an infinite plane of charge density ρ_s is given as

$$\vec{E} = \pm\frac{\rho_s}{2\varepsilon_o}\vec{a}_x \quad \text{and}$$

$$\vec{D} = \varepsilon_o\vec{E} \quad \text{or}$$

$$\pm\frac{\rho_s}{2\varepsilon_0}\vec{a}_x$$

The plane is at $x = 4$ (in this case), the point is at $x = 1$. This point is located below the plane. Here $\vec{a}_x = \vec{a}_x$. Flux density

$$\vec{D} = -\frac{\rho_s}{2}\vec{a}_x \quad \text{or}$$

$$= -\frac{-\rho_s}{2}\vec{a}_x = -\frac{-40}{2}\vec{a}_x$$

$$= -20\,\vec{a}_x \ (\mu\text{C/m}^2)$$

6.33 Two homogeneous isotropic dielectric mediums are separated at $z = 0$, ε_{r1} is 4 for region $z \le 0$, ε_{r2} is 6 for region $z > 0$. The electric field (E_1) in medium 1 is $\vec{E}_1 = (8\vec{a}_x + 9\vec{a}_y + 12\vec{a}_z)$. Find

(i) The normal and tangential component of $\vec{E}_1$ and
(ii) The angle between $\vec{E}_1$ and the normal (to surface).

Solution:
Electric field and its component are shown in Figure 6.31.
(i) Given that

$$\vec{E}_1 = 8\vec{a}_x + 9\vec{a}_y + 12\vec{a}_z$$

Unit vector $\vec{a}_z$ is normal to the boundary. The normal component of the electric field is given as

$$E_{1n} = 12\vec{a}_z \quad \text{or}$$

$$|\vec{E}_{1n}| = 12\left(\frac{V}{m}\right)$$

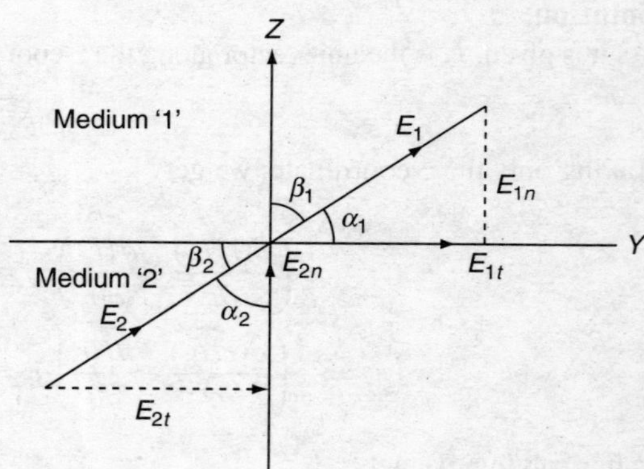

Figure 6.31 *Electric field and its component*

Further,

$$\vec{E}_1 = \vec{E}_{1t} + \vec{E}_{1n} \quad \text{or}$$

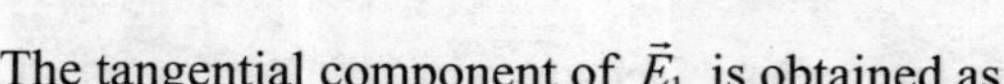

$$\vec{E}_{1t} = E_1 - E_{1n} = (8\vec{a}_x + 9\vec{a}_y)$$

The tangential component of $\vec{E}_1$ is obtained as

$$|E_{1t}| = \sqrt{8^2 + 9^2} = 12.04\left(\frac{V}{m}\right)$$

(ii) Let α_1 be the angle between $\vec{E}_1$ and normal to surface; we get

$$\tan\alpha_1 = \left|\frac{\vec{E}_{1n}}{E_{1t}}\right| \Rightarrow \alpha_1 = 44.90°$$

6.34 There is a rectangular loop of area A. It has uniform flux density B, which is normal to its plane (plane of loop). It is shown in Figure 6.32. Flux density varies harmonically with respect to time. Flux density is given as $B = B_o \sin \omega t$. What will be the total emf induced in the loop?

Figure 6.32 *Rectangular loop*

Solution:

It is a case of a stationary coil, where $\vec{B}$ is changing with respect to time. We know that

$$V_{\text{emf}} = \frac{-\partial \phi}{\partial t} \quad \text{or}$$

$$= -\int \frac{\partial B \cdot ds}{\partial t} \quad \text{or}$$

$$= -\int \frac{\partial (B_o \sin \omega t)}{\partial t} . ds \quad \text{or}$$

$$= \int_s -\omega B_o \cos \omega t . ds$$

or simply

$$= -A \omega B_o \cos \omega t$$

where A is the area of the loop.

6.35 Figure 6.33 shows a variable loop, which consists of two stationary parallel conductors. The loop is connected to a voltmeter at one end and to a moving bar at the other end. It is moving with uniform velocity of v (m/s). The loop is kept in uniform flux density $B\vec{a}_z$ (tesla). Obtain an expression for induced emf in this loop. Consider $L = 10$ cm, $B = 0.6\vec{a}_z$ and velocity $v = 20\sqrt{y}\vec{a}_y$ m/s. Let $y = 8$ cm at $t = 0$. At $t = 0.05$ sec, find

(a) The velocity of the bar,
(b) The position of the bar and
(c) The emf, which is induced in this loop and also find its polarity.

Solution:

y is the position of the moving bar. Flux (through the surface of the loop) at any time t is given as

$$\phi = ByL$$

Thus,

$$V_{\text{emf}} = \oint \vec{E} \cdot dl$$

$$= \frac{-\partial \phi}{\partial t}$$

$$= -BL \frac{\partial y}{\partial t}$$

$$= -BLv$$

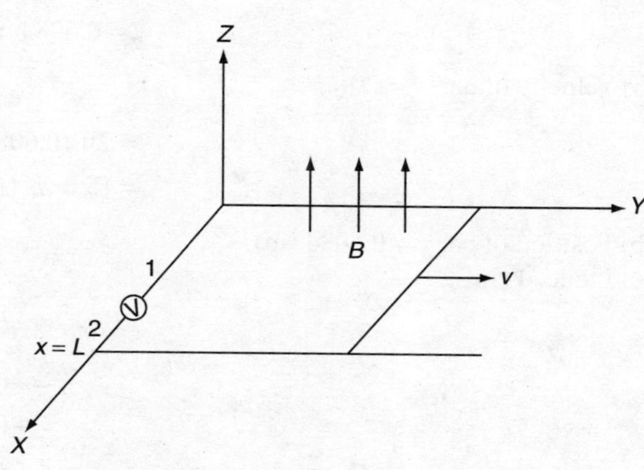

Figure 6.33 *Variable loop*

where v is the velocity of the bar. As per Lenz's Law, end 2 will have a higher potential than end 1 in the voltmeter. Voltmeter reading

$$V_{\text{emf}_{12}} = -BLv \text{ Volts}$$

It is given that

$$L = 10 \text{ (cm)}, \qquad \vec{B} = 0.6\vec{a}_z \text{ (T)}, \qquad v = 20\sqrt{y} \text{ (m/s)}$$

Also at $t = 0$, $y = 8$ cm.

Thus,

$$v = \frac{\partial y}{\partial t}$$

$$= 20\sqrt{y}$$

Further,

$$\int \frac{\partial y}{\sqrt{y}} = \int 20\,\text{dt} \quad \text{or}$$

$$2\sqrt{y} = 20t + c$$

where c is a constant. At $t = 0$, we get

$$2\sqrt{8 \times 10^{-2}} = c$$

$$c = 0.56$$

$$2\sqrt{y} = 20t + 0.56 \quad \text{or}$$

$$y = (10t + 0.28)^2$$

At $t = 0.05$, we get

$$y = (10 \times 0.05 + 0.28)^2 \quad \text{or}$$

$$y = 0.6084 \text{ m}$$

(a) Velocity of bar $v = 20\sqrt{y}$

$$= 20\sqrt{0.6084}$$

$$= 15.6\ \vec{a}_y \text{ (m/s)}$$

(b) Position of bar $y = 0.6084$ (m)

(c) Induced emf

$$V_{\text{emf12}} = -BLy \quad \text{or}$$

$$= -0.6 \times 10 \times 10^{-2} \times 15.6$$

$$= -0.936 \text{ (volts)}$$

Induced emf V_{emf12} is –ve, therefore end 2 will be at a higher voltage than end 1.

6.36 There is a straight conductor of length 0.6 m. It is lying on the x-axis. Its one end is at the origin. This conductor is kept in magnetic flux density $\vec{B} = 4.6\vec{a}_y$ (T) and velocity $v = 0.02 \sin 10^3\, t\, \vec{a}_z$ (m/s). Find the motional electric field intensity and the induced emf in this conductor.

Solution:

It is given that

$$\vec{B} = 4.5\vec{a}_y \text{ T}$$

$$\vec{v} = 0.02 \sin 10^3 t\, \vec{a}_z \text{ (m/s)}$$

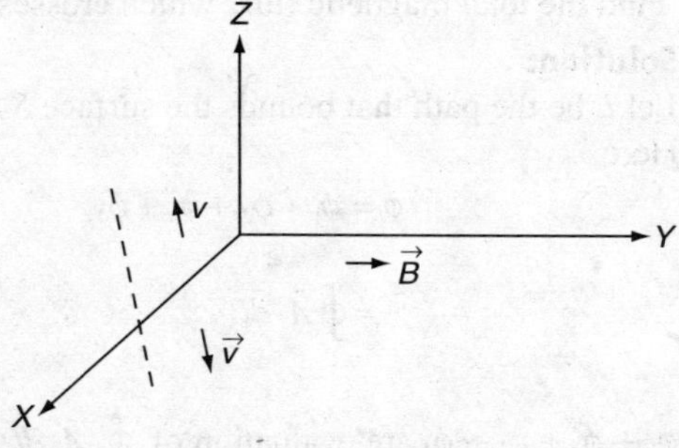

Figure 6.34 *Motional emf*

Motional emf producing field intensity is given as $\vec{E} = \vec{v} \times \vec{B}$

$$= 0.092 \sin 10^3 \text{t}(-\vec{a}_x) \text{ V/m}$$

Induced emf V_{emf} in a conductor is obtained as

$$V_{\text{emf}} = \int (\vec{v} \times \vec{B}).dl$$

$$= \int_0^{0.6} (0.092 \sin 10^3 t(-\vec{a}_z))dx\, \vec{a}_x$$

$$= -0.092 \sin 10^3 t \times 0.6$$

$$= -0.0552 \sin 10^3 t \text{ (V)}$$

6.37 A circular coil of 160 turns has a radius of 4.5 cm. The coil carries a current I of 0.6 Amp. Find the magnitude of the magnetic field.

Solution:

It is given that $I = 0.6$ Amp, $R = 4.5$ cm $= 0.045$ m, $h = R = 0.045$ m and $N = 160$.

(a) We know

$$B = \frac{\mu_0 NI}{2R}$$

$$= 10^{-7} \times 2 \times \frac{22}{7} \times \frac{160 \times 0.6}{0.045}$$

$$= 1.34 \times 10^{-3} \text{ T}$$

(b) Again

$$B = \frac{\mu_0}{4\pi} \cdot \frac{2\pi NI\, a^2}{(a^2 + h^2)^{3/2}}$$

$$= 10^{-7} \times 2 \times \frac{22}{7} \times \frac{160 \times 0.6 \times (0.45)^2}{[(0.045)^2 + (0.045)^2]^{3/2}}$$

$$= 4.73 \times 10^{-4} \text{ T}$$

6.38 A magnetic vector potential is

$$A = \frac{r^2}{12}\vec{a}_z \text{ (Wb/m)}$$

Find the total magnetic flux, which crosses surface $\phi = \pi/2$ $(3 \le r \le 6 \text{ m})$ and $(0 \le z \le 8 \text{ m})$.

Solution:

Let L be the path that bounds the surface S as shown in Figure 6.35.

Here,

$$\phi = \phi_1 + \phi_2 + \phi_3 + \phi_4$$

$$= \oint_L \vec{A}\cdot\vec{dl}$$

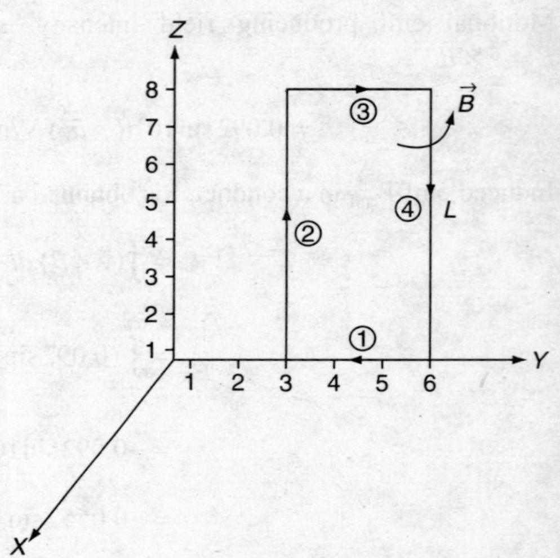

Figure 6.35 *Surface and magnetic field flux*

$\phi_1 + \phi_2 + \phi_3 + \phi_4$ are evaluation of $\oint_L \vec{A}\cdot dl$ along segments of path, L labelled as 1, 2, 3 and 4, as shown in Figure 6.35. Here, $\phi_1 = \phi_3 = 0$. Now,

$$\phi = \phi_2 + \phi_4$$

$$= -\frac{1}{9}\left[(3)^2\int_0^8 dz + (6)^2\int_0^8 dz\right] \quad \text{or}$$

$$= -\frac{1}{9}[\{(3)^2 - (6)^2\}\{\int_0^8 dz\}] \quad \text{or}$$

$$= -\frac{1}{9}[-3\times 9\times 8] \quad \text{or}$$

$$= 24 \text{ Wb}$$

6.39 A direct current I of 10 Amp flows through a long and straight round conductor. What will be the magnetic flux through half of the wire's cross-section per meter length?

Solution:

Half the portion is shown in Figure 6.36.

Assume B to be the magnetic induction at a distance r from the axis and R the radius of the conductor. By using Ampere's Circuital Law, we get

$$B\cdot 2\pi r = \mu_0\frac{i\pi r^2}{\pi R^2} \quad \text{or}$$

$$B = \frac{\mu_0}{2\pi}\frac{ir^2}{R^2}$$

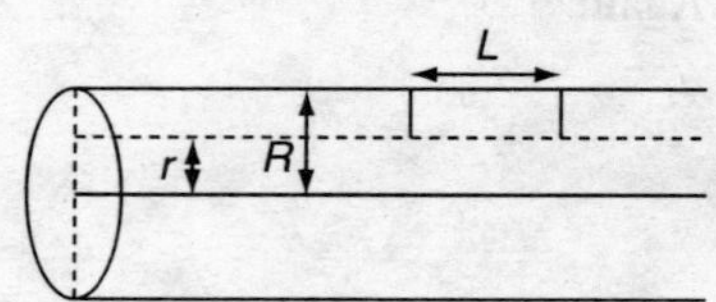

Figure 6.36 *Magnetic flux determination through half the wire's cross–section*

Magnetic flux ϕ through the shaded region is obtained as

$$\phi = \int_0^R B\cdot dr$$

$$= \int_0^R \frac{\mu_0}{2\pi}\frac{ir}{R^2}\,dr$$

$$= \frac{\mu_0}{4\pi} i$$

$$= \frac{\mu_0}{4\pi} \times 10$$

$$= 1\times 10^{-6}\,(\text{Wb/m}^2)$$

6.40 A magnetic vector potential '$\vec{A}$' is $-(\rho^2/10)\vec{a}_z$ (Wb/m). Find the magnetic flux that crosses surface $\phi = \pi/2\,(4 \le \rho \le 5m)$ and $(0 \le z \le 8m)$.

Solution:

We know that

$$\vec{B} = \nabla \times \vec{A}$$

$$= \frac{-\partial \cdot A_z}{\partial \rho}\vec{a}_\phi \quad \text{or}$$

$$\vec{B} = +\frac{\rho}{5}\vec{a}_\phi$$

$$\vec{d}_s = dp\ dz\ a\phi$$

Hence, total flux ϕ is given as

$$= \int B \cdot ds \quad \text{or}$$

$$= \frac{1}{5}\int_{z=0}^{z=8}\int_{\rho=4}^{\rho=5} \rho \cdot dl\ dz \quad \text{or}$$

$$= \frac{1}{5}\frac{\rho^2}{2}\Bigg|_4^5 \times 8 \quad \text{or}$$

$$= \frac{1}{5}\frac{(25-16)\times 8}{2} \quad \text{or}$$

$$= \frac{9\times 8}{10} = 7.2 \text{ Wb.}$$

Finally,

$$\phi = 7.2 \text{ Wb}$$

6.41 A magnetic vector potential $\vec{A}$ is given as $45\sin\theta\,\vec{a}_\theta$ (in Spherical coordinates). Obtain flux density at point $(3, \pi/3, \pi/4)$.

Solution:

We know that

$$\vec{B} = \nabla \times \vec{A} \quad \text{or}$$

$$= \begin{vmatrix} \dfrac{\vec{a}_r}{r^2 \sin\theta} & \dfrac{\vec{a}_\theta}{r \sin\theta} & \dfrac{\vec{a}_\phi}{r} \\ \dfrac{\partial}{\partial r} & \dfrac{\partial}{\partial \theta} & \dfrac{\partial}{\partial \phi} \\ 0 & 45\sin\theta & 0 \end{vmatrix} \quad \text{or}$$

$$\vec{B} = \frac{\partial}{\partial r}(45r\sin\theta)\cdot\frac{\vec{a}_\phi}{r}$$

$$\vec{B} = \frac{45\sin\theta}{r}\vec{a}_\phi$$

so,

$$B \text{ at point } \left(3, \frac{\pi}{3}, \frac{\pi}{4}\right) = \frac{45\sin(\pi/3)}{3}\vec{a}_\phi \quad \text{or}$$

$$\vec{B} = \frac{15\times\sqrt{3}}{2}\cdot\vec{a}_\phi$$

$$\vec{B} = 12.99\,\vec{a}_\phi$$

6.42 A magnetizing field H of 1400 (Am^{-1}) causes a magnetic flux ϕ of 4.2×10^{-5} (Weber) in an iron bar. The iron bar has a cross-sectional area S of 2 cm^2. Find the permeability and susceptibility of this iron bar.

Solution:

It is given that $H = 1400$ (A/m), $\phi = 4.2 \times 10^{-5}$ (Weber) and cross-sectional area = 2 cm^2 = 2×10^{-4} (m^2). We have to find μ and χ_m. We know that

$$B = \frac{\phi}{S}$$

$$= \frac{4.2\times10^{-5}}{2\times10^{-4}} = 0.06$$

$$= 0.21 \text{ (Wb/m}^2\text{)}$$

Also,

$$\mu = \frac{B}{H}$$

$$= \frac{4.2}{1400}$$

$$= 3\times10^{-4} \text{ (T/Am)}$$

Also,

$$\mu = \mu_0(1+\chi_m) \quad \text{or}$$

$$\chi_m = \left(\frac{\mu}{\mu_0} - 1\right) \quad \text{or}$$

$$\chi_m = \left(\frac{3\times10^{-4}}{4\pi\times10^{-7}} - 1\right)$$

$$\chi_m = (238.63 - 1)$$
$$\chi_m = 237.63$$

6.43 A magnetic material possesses $\mu_r = 2/\pi$. This magnetic material is put into a magnetic field having strength $H = 2\rho^2\vec{a}_\phi$ (A/m). Obtain magnetization.

Solution:
Magnetic flux density is given as

$$\vec{B} = \mu\vec{H}$$
$$= \mu_r\mu_0\vec{H}$$
$$= \frac{2}{\pi}\cdot 4\pi\times10^{-7}\times 2\rho^2\vec{a}_\phi \text{ (A/m)}$$

Finally,

$$\vec{B} = 16\times10^{-7}\rho^2\vec{a}_\phi \text{ (Wb/m}^2\text{)}$$

Also,

$$\vec{B} = \mu_o(\vec{H} + \vec{M})$$
$$= \mu_o\mu_r\vec{H} = \mu_o(\vec{H} + \vec{M}) \quad \text{or}$$
$$\vec{M} = \vec{H}(\mu_r - 1)$$
$$= \vec{H}\left(\frac{4}{\pi} - 1\right)$$

$$2\rho^2\vec{a}_\phi \times 0.27 \text{ (A/m)}$$

Finally,

$$\vec{M} = 0.546\rho^2\vec{a}_\phi \text{ (A/m)}$$

6.44 A toroid has 3000 turns. It has outer radii of 20 cm and 26 cm, respectively. Find the relative permeability of the core. A current of 2 Amp produces a magnetic field of intensity of 1.5 (T) in this core.

Solution:
It is given that the total number of turns = 3000, $I = 2$ Amp and $B = 1.5$ T.
Average radius

$$r = \frac{20 + 26}{2} = 23 \text{ cm}$$

$$\text{Number of turns/Length } (n) = \frac{3000}{2\pi r}$$

$$= \frac{3000}{2\pi \times 23 \times 10^{-2}}$$

$$= 20,488.6$$

Further,

$$\vec{B} = \mu\, n\, I$$

$$= \mu_o\, \mu_r\, nI \quad \text{or}$$

$$\mu_r = \frac{B_o}{\mu_o\, n\, I} \quad \text{or}$$

$$\mu_r = \frac{B}{\mu_o\, n\, I} \quad \text{or}$$

$$\mu_r = \frac{1.5}{4\pi \times 10^{-7} \times 20,488 \times 2}$$

$$\mu_r = 287.5$$

Finally,

$$\mu_r = 287.5$$

6.45 There is a boundary at $z = 0$ between two magnetic materials. Permittivities are $\mu_1 = z\mu_0$ (H/m) for region $1\,(z > 0)$ and $\mu_2 = 14\mu_o$(H/m) for region 2 ($z < 0$). Surface current density K is $60\,\vec{a}_x$(A/m) at the boundary ($z = 0$). There is a field $B_1 = (10\vec{a}_x - 16\vec{a}_y + 24\vec{a}_z)$ (mT) in region 1. What is the flux density $\vec{B}_2$ in region 2?

Solution:

Given that $\vec{B}_1 = (10\vec{a}_x - 16\vec{a}_y + 24\vec{a}_z)$ mT. It is shown in Figure 6.37.

Normal component of $\vec{B}_1$ is obtained as

$$\vec{B}_{n_1} = (\vec{B}_1 \cdot \vec{a}_n)\vec{a}_n$$

$$= [(10\vec{a}_x - 16\vec{a}_y + 24\vec{a}_z) \cdot \vec{a}_z]\vec{a}_z \quad \text{or}$$

$$= 24\vec{a}_z \text{ (mT)}$$

and also

$$\vec{B}_{n_2} = \vec{B}_{n_1}\, 24\vec{a}_z \text{ (mT)}$$

Tangential component of $\vec{B}_1$ is obtained as

$$\vec{B}_{t_1} = (\vec{B}_1 - \vec{B}_{n_1})$$

$$= (10\vec{a}_x - 16\vec{a}_y) \text{ (mT)}$$

$$= (3980\,\vec{a}_x - 6368\,\vec{a}_y) \text{ (A/m)}$$

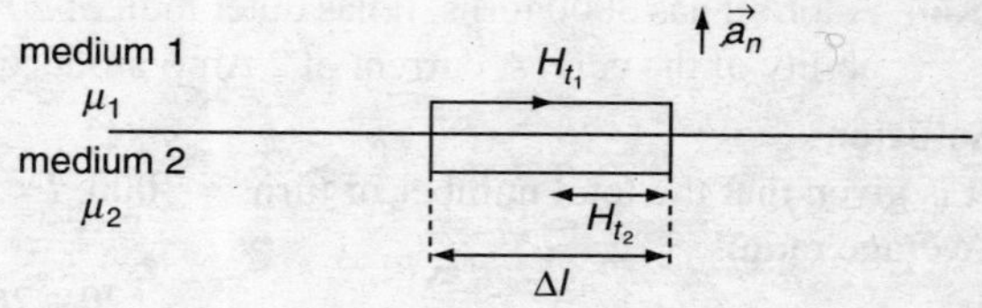

Figure 6.37 *Two magnetic materials*

Also,

$$\vec{H}_{t_1} = \frac{\vec{B}_{t_1}}{\mu_1} = \frac{(10\,\vec{a}_x - 16\,\vec{a}_y)\times 10^{-3}}{2\times 4\pi\times 10^{-7}}$$

$$= 3980\,\vec{a}_x - 6368\,\vec{a}_y (\text{A/m})$$

Further,

$$\vec{H}_{t_2} = \vec{H}_{t_1} - (K\times \vec{a}_n)$$

$$= 3980\,\vec{a}_x - 6368\,\vec{a}_y - (60\vec{a}_x \times \vec{a}_z)$$

$$= 3980\,\vec{a}_x - 6308\,\vec{a}_y$$

Also,

$$\vec{B}_{t_1} = \mu_2 \vec{H}_{t_2}$$

$$= 14\times 4\pi\times 10^{-7}(3980\,\vec{a}_x - 6308\,\vec{a}_y)$$

Finally, $\vec{B}_2$ can be obtained by using the relationship as given below:

$$\vec{B}_2 = \vec{B}_{n_2} + \vec{B}_{t_1}$$

6.46 There are two homogeneous, linear and isotropic, media with interface at $x = 0$. $x < 0$ describes medium 1 ($\mu_{r_1} = 14$). $x > 0$ describes medium 2 ($\mu_{r_2} = 12$). The magnetic field in medium 1 is $(80\vec{a}_x - 50\vec{a}_y + 45\vec{a}_z)$ (A/m). Find the magnetic field in medium 2. Also, find the magnetic flux density in medium 1.

Solution:

The magnetic field in medium 1 is obtained as

$$\vec{H}_{t_1} = (80\vec{a}_x - 50\vec{a}_y + 45\vec{a}_z)\,(\text{A/m})$$

Further,

$$\vec{H}_1 = \vec{H}_{t_1} + \vec{H}_{n_1}$$

It is given that

$$\vec{H}_{t_1} = -50\vec{a}_y + 45\vec{a}_z \quad \text{and}$$

$$\vec{H}_{n_1} = 80\vec{a}_x$$

Applying boundary conditions, we get

$$\vec{H}_{t_1} = \vec{H}_{t_2} \quad \text{or}$$

$$\vec{H}_{t_2} = (-50\vec{a}_y + 45\vec{a}_z)\,(\text{A/m})$$

Now applying the boundary condition on $\vec{B}$, we get

$$\vec{B}_{n_1} = \vec{B}_{n_2} \quad \text{or}$$

$$\mu_1 \vec{H}_{n_1} = \mu_2 \vec{H}_{n_2} \quad \text{or}$$

$$\vec{H}_{n_2} = \frac{\mu_1}{\mu_2} \vec{H}_{n_1}$$

$$= \frac{\mu_{r_1} \mu_0}{\mu_{r_2} \mu_0} \vec{H}_{n_1}$$

$$= \frac{14}{12} \times 80\, \vec{a}_x$$

$$= 93.3\, \vec{a}_x \quad \text{and}$$

$$\vec{H}_1 = 93.3\, \vec{a}_x - 50\, \vec{a}_y + 45\, \vec{a}_z \text{ (A/m)}$$

Magnetic flux density

$$\vec{B} = \mu_1 \vec{H}_1$$

$$= \mu_{r_1} \mu_0 \vec{H}_1$$

$$= 14 \times 4\pi \times 10^{-7} (80\, \vec{a}_x - 50\, \vec{a}_y + 45\, \vec{a}_z)$$

$$= 56\pi(16\, \vec{a}_x - 10\, \vec{a}_y + 9\, \vec{a}_z) \times 5 \times 10^{-7}$$

$$= 280\, \pi(1.6\, \vec{a}_x - \vec{a}_y + 0.9\, \vec{a}_z) \times 10^{-6}$$

$$= 280\pi(1.6\, \vec{a}_x - \vec{a}_y - 0.9\, \vec{a}_z)\ (\mu\text{Wb/m}^2)$$

6.47 Let $\vec{H}_1 = (-2\, \vec{a}_x + 5\, \vec{a}_y + 3\, \vec{a}_z)$ (A/m) in region $(y - x - 3) \leq 0$. Here, $\mu_{r_1} = 5$. Find $\vec{M}_1$ and $\vec{B}_1$. Also, find $\vec{H}_2$ and $\vec{B}_2$ in this region $(y - x - 3) \leq 0$. Consider $\mu_{r2} = 8$.

Solution:

Plane is given as $(y - x - 3) = 0$. $y - x \leq 3$ in region 1 as shown in Figure 6.38. Let us take origin (0, 0). It is in this region (0 –0 –3) < 0. Let the surface of the plane be $f(x - y) = y - x - 3$. A unit vector normal to the plane is

$$\vec{a}_n = \frac{\vec{f}}{|\vec{f}|}$$

$$= \frac{(\vec{a}_y - \vec{a}_x)}{\sqrt{2}}$$

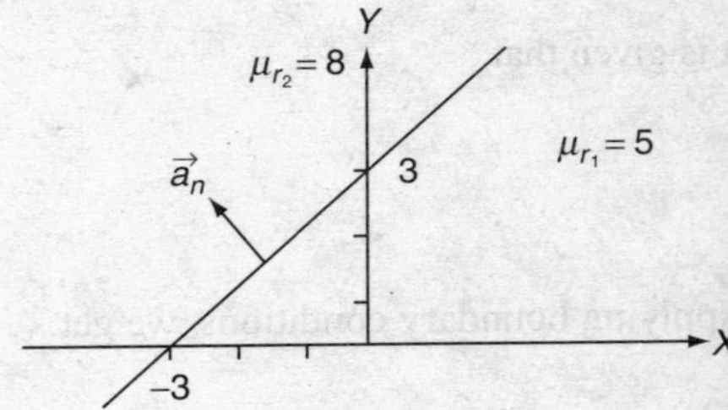

Figure 6.38 *Region* $y - x \leq 3$

As we know that

$$\vec{M}_1 = x_{m_1} \vec{H}_1$$

$$= (\mu_{r_1} - 1)\vec{H}_1$$

$$= (5 - 1)(-2\, \vec{a}_x + 5\, \vec{a}_y + 3\, \vec{a}_z)$$

$$= (-8\vec{a}_x + 20\vec{a}_y + 12\, \vec{a}_z)$$

$$\vec{B}_1 = \mu_1 \vec{H}_1$$

$$= \mu_{r_1} \mu_0 \vec{H}_1$$

$$= 5 \times 4\pi \times 10^{-7} (-2\,\vec{a}_x + 5\,\vec{a}_y + 3\,\vec{a}_z)$$

Also,

$$\vec{H}_{n_1} = (\vec{H}_1 \cdot \vec{a}_n)\vec{a}_n$$

$$= \left[(-2\,\vec{a}_x + 5\,\vec{a}_y + 3\,\vec{a}_z) \cdot \left(\frac{-\vec{a}_x + \vec{a}_y}{\sqrt{2}}\right)\right]\left[\frac{-\vec{a}_x + \vec{a}_y}{\sqrt{2}}\right]$$

$$= \left(\frac{2+5}{\sqrt{2}\sqrt{2}}\right)[-\vec{a}_x + \vec{a}_y]$$

$$= \frac{7}{2}[-\vec{a}_x + \vec{a}_y]$$

$$= -3.5\,\vec{a}_x + 3.5\,\vec{a}_y$$

Further,

$$\vec{H}_1 = \vec{H}_{n_1} + \vec{H}_{t_1}$$

$$H_{t_1} = \vec{H}_1 - \vec{H}_{n_1}$$

$$= (1.5\,\vec{a}_x + 1.5\,\vec{a}_y + 3\,\vec{a}_z)$$

Applying the boundary condition, we get

$$\vec{H}_{t_2} = \vec{H}_{t_1}$$

$$= (1.5\,\vec{a}_x + 1.5\,\vec{a}_y + 3\,\vec{a}_z) \quad \text{and}$$

$$\vec{B}_{n_2} = \vec{B}_{n_1}$$

$$\mu_2 \vec{H}_{n_2} = \mu_1 \vec{H}_{n_1}$$

$$\vec{H}_{n_2} = \frac{\mu_1}{\mu_2} \vec{H}_{n_1}$$

$$= \frac{\mu_{r_1}}{\mu_{r_2}} \cdot \frac{\mu_0}{\mu_0} \cdot \vec{H}_{n_1}$$

$$= \frac{5}{8} \cdot (-3.5\,\vec{a}_x + 3.5\,\vec{a}_y)$$

$$= (-2.18\,\vec{a}_x + 2.18\,\vec{a}_y)$$

Also,

$$\vec{H}_2 = \vec{H}_{n_2} + \vec{H}_{t_2}$$

$$= (-2.18\,\vec{a}_x + 2.18\,\vec{a}_y) + (1.5\,\vec{a}_x + 1.5\,\vec{a}_y + 3\,\vec{a}_z)$$

$$= (-0.68\,\vec{a}_x + 3.68\,\vec{a}_y + 3\,\vec{a}_z)\ (\text{A/m})$$

$$B_2 = \mu_2 \vec{H}_2$$

$$= \mu_{r_2} \cdot \mu_0 \times \vec{H}_2$$

$$= 8 \times 4\pi \times 10^{-7} (-0.68\,\vec{a}_x + 3.68\,\vec{a}_y + 3\,\vec{a}_z)$$

$$= 32\pi \times 10^{-6} (-0.068\,\vec{a}_x + 0.368\,\vec{a}_y + 0.3\,\vec{a}_z)$$

$$= (-6.834\,\vec{a}_x + 36.94\,\vec{a}_y + 30.15\,\vec{a}_z)\ (\mu\text{Wb/m}^2)$$

6.48 Magnetic flux lines are received at the iron air boundary. Incident angles with the normal of the boundary on the iron side are as follows:
(a) 15°, (b) 60° and (c)75°.
Iron has a relative permeability of 270. What will be the angle with the normal at which magnetic flux comes to the axis?

Solution:

Let medium 1 be iron and medium 2 be air. It is shown in Figure 6.39.
Boundary condition is given as

$$\frac{\tan \alpha_1}{\tan \alpha_2} = \frac{\mu_{r_1}}{\mu_{r_2}} \quad \text{or}$$

$$\alpha_2 = \tan^{-1}\left[\frac{\mu_{r_2}}{\mu_{r_1}} \tan \alpha_1\right]$$

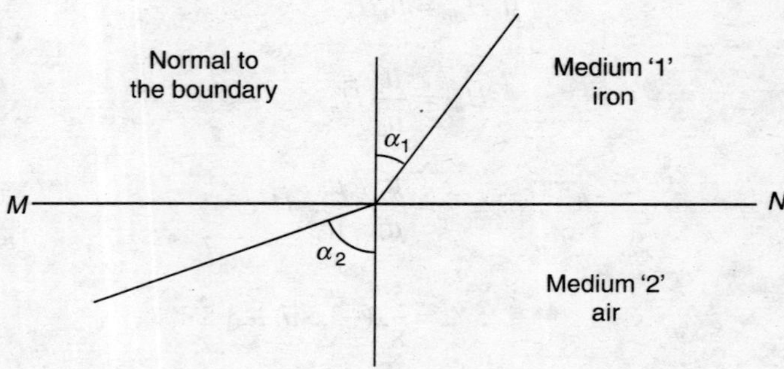

Figure 6.39 *Two mediums*

Thus, we get,

(a)
$$\alpha_2 = \tan^{-1}\left[\frac{1}{270}\tan 15°\right] = 0.056°$$

(b)
$$\alpha_2 = \tan^{-1}\left[\frac{1}{270}\tan 60°\right] = 0.367°$$

(c)
$$\alpha_2 = \tan^{-1}\left[\frac{1}{270}\tan 75°\right] = 0.792°$$

6.49 There are two ferromagnetic materials, which are separated by a plane boundary. Medium 1 has relative permeability of 250. Medium 2 has relative permeability of 1500. Magnetic field in medium 2 is directed at 60° with normal to boundary. What is the angle between "field direction" and "normal" in medium 1? (Normal is with respect to boundary.)

Solution:

Boundary condition is given as

$$\frac{\tan\alpha_1}{\tan\alpha_2} = \frac{\mu_{r_1}}{\mu_{r_2}}$$

$$= \frac{250}{1500} = 0.16$$

Angle is obtained as

$$\alpha_1 = \tan^{-1}[0.16\tan\alpha_2]$$

$$= \tan^{-1}[0.16 \times 1.732]$$

$$= 15.48°$$

6.50 Electric field density $\vec{D}$ in free space is $15\cos(\omega t + \beta z)\vec{a}_x$. Find an expression for $\vec{E}$, $\vec{B}$, and $\vec{H}$ in this case.

Solution:

It is given that $\vec{D} = 15\cos(\omega t + \beta z)\vec{a}_x$. Further,

$$\vec{E} = \frac{\vec{D}}{\varepsilon_0}$$

Thus,

$$\vec{E} = \frac{15\cos(\omega t - \beta z)}{\varepsilon_0}\vec{a}_x$$

Maxwell's equation gives

$$\nabla \times \vec{E} = \frac{-\partial B}{\partial t}$$

Thus,

$$\begin{vmatrix} \vec{a}_x & \vec{a}_y & \vec{a}_z \\ \frac{\partial}{\partial x} & \frac{\partial}{\partial y} & \frac{\partial}{\partial z} \\ E_x & E_y & E_z \end{vmatrix} = \frac{-\partial B}{\partial t}$$

However, $E_y = E_z = 0$. Thus,

$$\left|\frac{\partial E_x}{\partial z}\right| \vec{a}_y = \frac{-\partial B}{\partial t} \quad \text{or}$$

$$\frac{\partial}{\partial t}\frac{[15\cos(\omega t + \beta z)]}{\varepsilon_0}\vec{a}_y = \frac{-\partial \vec{B}}{\partial t}$$

$$\frac{-\partial B}{\partial t} = \frac{-15\sin(\omega t + \beta z)}{\varepsilon_0}\omega\,\vec{a}_y$$

$$B = \frac{15\,\omega}{\varepsilon_0}\int \sin(\omega t + \beta z)\,dt.\vec{a}_y \quad \text{or}$$

$$\vec{B} = \frac{-15\,\omega}{\varepsilon_0 \cdot \omega}\cos(\omega t + \beta z).\vec{a}_y \quad \text{or}$$

$$\vec{B} = \frac{-15}{\varepsilon_0}\cos(\omega t + \beta z)\,\vec{a}_y$$

Further, $\vec{H} = (\vec{B}/\mu_0)$. So,

$$\vec{H} = \frac{-15}{\varepsilon_0 \mu_0}\cos(\omega t + \beta z)\,\vec{a}_y \ \ (\text{Wb/m}^2)$$

6.51 In free space $\vec{B}$ is $10\,e^{j(\omega t+6z)}\,\vec{a}_y$. Prove that $\vec{E} = 1.6\,\omega e^{j(\omega t+6z)}\,\vec{a}_x$.

Solution:
It is given that

$$\vec{B} = 10\,e^{j(\omega t+6z)}\,\vec{a}_y$$

Further,

$$\vec{B} = \mu_0\,\vec{H}$$

Thus,

$$\vec{H} = \frac{10}{\mu_0}e^{j(\omega t+6z)}\,\vec{a}_y$$

Maxwell's equation gives

$$\nabla \times \vec{H} = \frac{\partial \vec{D}}{\partial t}$$

But, $\vec{H}_x = \vec{H}_z = 0.$

Thus,

$$\frac{\partial \vec{D}}{\partial t} = \begin{vmatrix} \vec{a}_x & \vec{a}_y & \vec{a}_z \\ \dfrac{\partial}{\partial x} & \dfrac{\partial}{\partial y} & \dfrac{\partial}{\partial z} \\ 0 & \dfrac{10\,e^{j(\omega t+6z)}}{\mu_0} & 0 \end{vmatrix} \frac{dy}{dx} \quad \text{or}$$

$$= \frac{\partial}{\partial z}\left(\frac{10\,e^{j(\omega t+6z)}}{\mu_0}\right)\vec{a}_x$$

$$\frac{\partial \vec{D}}{\partial t} = \frac{j\,60}{\mu_0}.e^{j(\omega t+6z)}\,\vec{a}_x$$

On integration, we get

$$\vec{D} = \frac{60}{\omega\mu_0}\,e^{j(\omega t+6z)}\,\vec{a}_x \quad \text{and}$$

$$\vec{E} = \frac{\vec{D}}{\varepsilon_0} = \frac{60}{\omega\mu_0\varepsilon_0}\,e^{j(\omega t+6z)}\,\vec{a}_x$$

However,

$$\frac{1}{\mu_0\varepsilon_0} = \left(\frac{\omega}{\beta}\right)^2$$

$$= \frac{\omega^2}{36}$$

Therefore,

$$\vec{E} = \frac{60}{\omega}\cdot\frac{\omega^2}{36}\,e^{j(\omega t+6z)}\vec{a}_x$$

$$\vec{E} = 1.6\,\omega e^{j(\omega t+6z)}\,\vec{a}_x$$

6.52 A medium is characterized by $\sigma = 0$, $\mu = \mu_0$, ε_0 and $\vec{E} = 20\sin(5^4 t - \beta z)\,\vec{a}_y$ (V/m). Find $\vec{H}$ and β in this case.

Solution:

Here Gauss's Law (for electric field) is getting satisfied. It is

$$\nabla\cdot\vec{E} = -\frac{\partial E_y}{\partial y} = 0$$

Faraday's Law gives

$$\nabla\times\vec{E} = -\frac{\mu\partial\vec{H}}{\partial t} \quad \text{or}$$

$$\vec{H} = \frac{1}{\mu}\int (\nabla \times \vec{E})\,dt$$

However,

$$\nabla \times \vec{E} = \begin{vmatrix} \vec{a}_x & \vec{a}_y & \vec{a}_z \\ \frac{\partial}{\partial x} & \frac{\partial}{\partial y} & \frac{\partial}{\partial z} \\ 0 & E_y & 0 \end{vmatrix} \quad \text{or}$$

$$= \frac{-\partial E_y}{\partial z} . \vec{a}_x \quad \text{or}$$

$$= 20\beta \cos(5^4 t - \beta z)\,\vec{a}_x \text{ (V/m)}$$

Thus,

$$\vec{H} = \frac{-20\beta}{\mu}\int \cos(5^4 t - \beta z)\,dt\,\vec{a}_x \text{ (V/m)} \quad \text{or}$$

$$= \frac{20\beta}{5^4 \mu}\sin(5^4 t - \beta z)\,\vec{a}_x$$

It is verified that

$$\nabla \cdot \vec{H} = \frac{\partial H_x}{\partial x} = 0$$

Thus, Gauss's Law for magnetic field is also getting satisfied. Ampere's Law gives

$$\nabla \times \vec{H} = \sigma \vec{E} + \varepsilon \frac{\partial E_x}{\partial t}$$

But here $\sigma = 0$. So,

$$\vec{E} = \frac{1}{\varepsilon}\int (\nabla \times \vec{H})\,dt$$

Now,

$$\nabla \times \vec{H} = \begin{vmatrix} \vec{a}_x & \vec{a}_y & \vec{a}_z \\ \frac{\partial}{\partial x} & \frac{\partial}{\partial y} & \frac{\partial}{\partial z} \\ H_x & 0 & 0 \end{vmatrix} \quad \text{or}$$

$$= \frac{\partial H_x}{\partial z} \cdot \vec{a}_y$$

$$= \frac{20\beta^2}{5^4 \mu}\cos(5^4 t - \beta z)\vec{a}_y$$

Thus,

$$\vec{E} = \frac{1}{\varepsilon}\int \frac{20\beta^2}{5^4\mu}\cos(5^4 t - \beta z)\vec{a}_y \quad \text{or}$$

$$\vec{E} = \frac{20\beta^2}{\mu\varepsilon 5^4}\cdot\frac{1}{5^4}\sin(5^4 t - \beta z)\vec{a}_y \quad \text{or}$$

$$\vec{E} = \frac{4\beta^2}{\mu\varepsilon 5^7}\sin(5^4 t - \beta z)\vec{a}_y$$

Comparing the above with the given expression of $\vec{E}$, we find

$$\frac{4\beta^2}{5^7\mu\varepsilon} = 20$$

$$\beta = \sqrt{\frac{20\times 5^7}{4}\cdot\mu\varepsilon}$$

$$\beta = \pm 5^4\sqrt{\mu_0\varepsilon_0}$$

$$= \frac{\pm 5^4}{3\times 10^8} = \pm 2.08\times 10^{-6} \quad \text{and}$$

$$\vec{H} = \pm\frac{20\times 2.08\times 10^{-6}}{4\pi\times 10^7\times 5^4}.\sin(5^4 t \pm 2.08\times 10^{-6} z)\vec{a}_x$$

$$= \frac{1.6\times 10^{-15}}{\pi}\sin(5^4 t \pm 2.08\times 10^{-6})\vec{a}_x \text{ (A/m)}$$

6.53 For a homogeneous medium, $\varepsilon_r = 36$ and $\mu_r = 1$. $\vec{E}$ and $\vec{B}$ are $5\pi e^{j(\omega t+\beta z)}\vec{a}_x$ (V/m) and $10\mu_0 H_0 e^{j(\omega t+\beta z)}\vec{a}_x$ (Tesla), respectively. Determine ω and H_0 if the wavelength λ of the signal is 0.4 m.

Solution:

Velocity of wave propagation in the medium $\varepsilon_r = 36$ is given as

$$v = \frac{C}{\sqrt{\varepsilon_r\mu_r}}$$

$$= \frac{3\times 10^8}{\sqrt{36\times 1}}$$

$$v = 0.5\times 10^8 \text{ m/s}$$

Wavelength (λ) = 0.4 m (given).

Thus,

$$\omega = 2\pi f$$

$$= 2\pi \frac{v}{\lambda} \quad \text{or}$$

$$\omega = 2\pi \times \frac{0.5\times 10^8}{0.4} \quad \text{or}$$

$$\omega = 7.8\times 10^8 \text{ (rad/s)}$$

Maxwell's equation gives

$$\nabla \times \vec{E} = \frac{-\partial \vec{B}}{\partial t}$$

Thus,

$$\begin{vmatrix} \vec{a}_x & \vec{a}_y & \vec{a}_z \\ \frac{\partial}{\partial x} & \frac{\partial}{\partial y} & \frac{\partial}{\partial z} \\ E_x & 0 & 0 \end{vmatrix} = \frac{-\partial \vec{B}}{\partial t} \quad \text{or}$$

$$\left|\frac{\partial E_x}{\partial z}\right| \vec{a}_y = \frac{-\partial \vec{B}}{\partial t} \quad \text{or}$$

$$\frac{\partial}{\partial z}(5\pi e^{j(\omega t+\beta z)})\vec{a}_y = \frac{-\partial \vec{B}}{\partial t}$$

$$\frac{\partial}{\partial z}(5\pi e^{j(\omega t+\beta z)})\vec{a}_y = \frac{-\partial}{\partial t}(10\mu_0 H_0 e^{j(\omega t+\beta z)})\vec{a}_y$$

$$j5\pi\beta e^{j(\omega t+\beta z)} = -j10\mu_0 H_0 \omega e^{j(\omega t+\beta z)}$$

Finally, $H_o = 4.16\times 10^{-3}$ (A/m) (considering a +ve sign for H_0).

6.54 In free space, $\vec{E} = 10\pi e^{j(50^6 t+\beta z)}\,\vec{a}_x$ (V/m) and $\vec{H} = H_o e^{j(50^6 t+\beta z)}\,\vec{a}_y$ (A/m). Find values of H_0 and $\beta\,(\beta > 0)$.

Solution:

For $\vec{E}$ and $\vec{H}$ representing a plane wave, we get

$$\frac{\omega}{\beta} = c = \frac{1}{\sqrt{\mu_0 \varepsilon_0}}$$

$$= 3\times 10^8 \text{ (m/s)}$$

It is given that $\vec{E} = 10\pi e^{j(50^6 t+\beta z)}\vec{a}_x$ (V/m). On comparison with the actual expression, we find

$$\omega = 50^6$$

Thus,

$$\beta = \frac{50^6}{3\times 10^8}$$

$$\beta = 52.08 \text{ (rad/m)}$$

Maxwell's equation gives

$$\nabla \times \vec{E} = \frac{-\partial B}{\partial t}$$

$$= \mu_0 \frac{\partial \vec{H}}{\partial t}$$

Thus,

$$\begin{vmatrix} \vec{a}_x & \vec{a}_y & \vec{a}_z \\ \frac{\partial}{\partial x} & \frac{\partial}{\partial y} & \frac{\partial}{\partial z} \\ E_x & E_y & E_z \end{vmatrix} = -\mu_0 \frac{\partial H_y}{\partial t}$$

However, $E_y = E_z = 0$. Therefore,

$$\left| \frac{\partial E_x}{\partial z} \right| \vec{a}_y = -\mu_0 \frac{\partial H_y}{\partial t} \quad \text{or}$$

$$\vec{a}_y \cdot \frac{\partial}{\partial z} \left| 10\pi e^{j(50^6 t)+\beta z)} \right| = -\mu_0 \frac{\partial}{\partial t} [H_0 e^{j(50^6 t+\beta z)}] \quad \text{or}$$

$$\beta 10\pi\, e^{j(50^6 t+\beta z)} \vec{a}_y = -4\pi \times 10^{-7} \times 50^6\, H_0\, e^{j(50^6 t+\beta z)}$$

$$52.08 \times 10\pi = -4\pi \times 10^{-7} \times 50^6 \times H_0 \quad \text{or}$$

$$H_0 = 0.083\ \text{(A/m) (considering +ve sign for } H_0)$$

6.55 A plane wave is travelling in a medium characterized by $\varepsilon_r = 9$ and $\mu_r = 4$. Electric field intensity is $40\sqrt{3\pi}$ (V/m). Find the energy density in the magnetic field. Also find the total energy density.

Solution:

Electric energy density W_E is obtained as

$$W_E = \frac{1}{2} \varepsilon E^2 \quad \text{or}$$

$$= \frac{1}{2} \times 8.854 \times 10^{-12} \times 9 \times (40)^2 \times 3\pi \quad \text{or}$$

$$= 601.2 \times 10^{-9} \quad \text{or}$$

$$= 601.2\ (\text{nJ/m}^3)$$

Total energy density W_T is

$$W_T = 1202.4\ (\text{nJ/m}^3)$$

6.56 In case of non-magnetic medium, $\vec{E} = 5\sin(4\pi\times10^6 t - 0.2x)\vec{a}_z$ (V/m). What are the values of ε_r, η and time average power? What is the total power that crosses to 9 cm^2 of plane $(4x + 2y) = 10$?

Solution:

It is given that $\beta = 0.2$:

$$\omega = 4\pi\times 10^6$$

$\mu = \mu_0$ (non-magnetic medium) and $\varepsilon = \varepsilon_0\,\varepsilon_r$. However,

$$\beta = \omega\sqrt{\mu\varepsilon} \quad \text{or}$$

$$= \omega\sqrt{\mu_0\varepsilon_0\mu_r\varepsilon_r} \quad \text{or}$$

$$= \frac{\omega\sqrt{\varepsilon_r}}{3\times10^8} \quad \text{or}$$

$$\varepsilon_r = \left(\frac{0.2\times3\times10^8}{4\pi\times10^6}\right)^2$$

$$\varepsilon_r = 22.79$$

Also,

$$\eta = \frac{120\pi}{\sqrt{\mu_r\varepsilon_r}}$$

$$\eta = 78.96\ (\Omega)$$

Time average power

$$P_{ave} = \vec{E}\times\vec{H} \quad \text{or}$$

$$= \frac{|\vec{E}|^2}{2\eta}\vec{a}_x$$

$$= \frac{25}{2\times78.96}$$

$$= 158\,\vec{a}_x\ (\text{mW/m}^2)$$

For the plane, $4x + 2y = 10$, we get

$$\vec{a}_x = \frac{4\vec{a}_x + 2\vec{a}_y}{\sqrt{10}}$$

Thus, total power

$$P = \int P_{avg}\, ds$$

$$= 158\times10^{-3}\times9\times10^{-4}\left[\frac{4\vec{a}_x + 2\vec{a}_y}{\sqrt{10}}\right]$$

$$= \frac{158\times9\times4\times10^{-7}}{\sqrt{10}} \quad (\because \vec{a}_x\cdot\vec{a}_y = 0)$$

$$= 1.798\times10^{-4}\ \text{W}$$

6.57 An electric vector $\vec{E}$ of an electromagnetic wave in free space in given by $E_x = E_z = 0$ and $E_y = Ae^{j(\omega t - z/c)}$. Find the H component in free space.

Solution:

$$\nabla \times \vec{E} = \begin{vmatrix} \vec{a}_x & \vec{a}_y & \vec{a}_z \\ \dfrac{\partial}{\partial x} & \dfrac{\partial}{\partial y} & \dfrac{\partial}{\partial z} \\ 0 & \mathrm{A}e^{j(\omega t - z/c)} & 0 \end{vmatrix} \quad \text{or}$$

$$= -\mu_0 \frac{\partial}{\partial t}\left(H_x \vec{a}_x + H_y \vec{a}_y + H_z \vec{a}_z\right)$$

$$-\mu_0 \frac{\partial H_x}{\partial t} = -\frac{\partial}{\partial z} Ae^{j(\omega t - z/c)} \quad \text{or}$$

$$\frac{\partial H_x}{\partial t} = \frac{A}{\mu_0} e^{j\left(\omega t - \frac{z}{c}\right)}(-j/c)$$

Therefore,

$$H_x = \frac{-A}{\mu_0}\frac{j}{c}\int e^{j\left(\omega t - \frac{z}{c}\right)} dt \quad \text{or}$$

$$= \frac{-Aj}{\mu_0 c(j\omega)} e^{j\left(\omega t - \frac{z}{c}\right)} \quad \text{or}$$

$$= \frac{-A}{c\mu_0 \omega} e^{j\left(\omega t - \frac{z}{c}\right)}$$

Putting

$$c = \frac{1}{\sqrt{\mu_0 \varepsilon_0}}$$

$$H_x = \frac{-\sqrt{\varepsilon_0/\mu_0}}{\omega} Ae^{j(\omega t - z/c)}$$

$$H_x = -\frac{1}{\omega} \cdot \sqrt{\frac{\varepsilon_0}{\mu_0}} \cdot E_y$$

$$H_y = H_z = 0$$

6.58 Using Ampere's Circuital Law in integral form, find $\vec{H}$ everywhere due to the following current density in Cylindrical coordinates:

$$\begin{aligned} J &= 0, \quad 0 < R < x \\ &= J_0 \vec{a}_z, \quad x < R < y \\ &= 0, \quad y < R < \infty \end{aligned}$$

Solution:

$$\oint H \cdot dl = I_{enclosed}$$

$$I = \oint_s J \cdot ds$$

As $J = 0 \quad \Rightarrow \quad I = 0$. For $R = 0$ to x; $J = 0, \quad \Rightarrow \quad I = 0$. For the $x < R < y$ region,

$$\oint H \cdot dl = J_0 \pi (R^2 - x^2)$$

$$\vec{H}_R = \frac{J_0 \pi (R^2 - x^2)}{2\pi R} . \vec{a}_\phi$$

For $y < R < \infty$,

$$\vec{H} \cdot 2\pi R = J_0 \pi (y^2 - x^2)$$

$$\vec{H} = \frac{J_0 \pi (y^2 - x^2)}{2\pi R} \vec{a}_\phi$$

For $R > b, \quad I = 0 \quad \Rightarrow \quad \vec{H} = 0$

6.59 Consider the volume current density distribution in Cylindrical coordinates as

$$J(r, \phi, z) = 0, \qquad 0 < R < x$$

$$= J_0 \left(\frac{R}{x} \right) \vec{a}_z, \qquad x < R < y$$

$$= 0, \qquad y < R < \infty$$

Solution:

$$0 < R < x, \qquad J = 0$$

$$\Rightarrow \oint \vec{H} . d\vec{l} = \int \vec{J} . ds = 0$$

$$\Rightarrow \vec{H} = 0$$

$$x < R < y,$$

$$\oint \vec{H} . d\vec{l} = \int \vec{J} \cdot ds$$

where $\qquad ds = R dR \, d\phi \quad$ or

$$= \int_0^{2\pi} \int_x^R J_0 \left(\frac{R}{x} \right) R \, dR \, d\phi \quad \text{or}$$

$$= \frac{J_0 2\pi}{x} \left(\frac{R^3}{3} \right)_x^R \quad \text{or}$$

$$= \frac{J_0 2\pi}{x} \left(\frac{R^3}{3} - \frac{x^3}{3} \right) \quad \text{or}$$

$$H \cdot 2\pi R = \frac{J_0 2\pi\left(R^3 - x^3\right)}{3x}$$

$$\vec{H} = \frac{J_0(R^3 - x^3)}{3xR}\vec{a}_\phi$$

At $R = y$,

$$\vec{H} = \frac{J_0(y^3 - x^3)}{3xy}$$

For $y < R < \infty$,

$$\vec{H} = \frac{J_0(y^3 - x^3)}{3xR}\vec{a}_\phi$$

6.60 Find the current distribution producing the following field distribution using Ampere's Law:

$$H = \begin{cases} J_0 R^2 \vec{a}_\phi, & 0 < R < a \\ J_0 \dfrac{a^3}{R}\vec{a}_\phi, & a < R < b \\ 0, & b < R < \infty \end{cases}$$

Solution:

By Ampere's Law,

$$\oint \vec{H} \cdot \vec{dl} = I$$

$$\nabla \times \vec{H} = \vec{J}$$

$\nabla \times \vec{H}$ in Cylindrical coordinates is obtained as

$$\left(\frac{1}{R}\frac{\partial H_z}{\partial \phi} - \frac{\partial H_\phi}{\partial z}\right)\vec{a}_R + \left(\frac{\partial H_R}{\partial z} - \frac{\partial H_z}{\partial R}\right)\vec{a}_\phi + \frac{1}{R}\left(\frac{\partial}{\partial R}(RH_\phi) - \frac{\partial H_R}{\partial \phi}\right)\vec{a}_z$$

H is a function of R and has ϕ component.

$$J = \frac{1}{R}\frac{\partial}{\partial R}(RH_\phi)$$

For $0 < R < a$,

$$J = \frac{1}{R}\frac{\partial}{\partial R}(J_0 R^3)$$

$$= \frac{J_0 3R^2}{R}$$

$$= J_0 3R\vec{a}_\phi$$

For $a < R < b$,

$$J = \frac{1}{R}\frac{\partial}{\partial R}\left(RJ_0 \frac{a^3}{R} \right) = 0$$

For $b < R < \infty$,

$$J = \frac{1}{R}\frac{\partial}{\partial R}(R \times 0) = 0$$

6.61 Calculate the flux density at the centre of the square loop having 110 turns with 12 m side carrying 2 Amp current. The loop is in air medium.

Solution:

$\mu_r = 1$, $\mu = \mu_0\ \mu_r$, $\mu_0 = 4\pi \times 10^{-7}$ H/m, $I = 2$ Amp, $N = 110$ and $a = 2$ m.

As

$$B = \frac{\mu I N 2\sqrt{2}}{\pi a} = \frac{4\pi \times 10^{-7} \times 110 \times 110 \times 2\sqrt{2}}{\pi \times 12}$$

$$B = 1.14 \times 10^{-3} \text{ (Tesla)}$$ or

6.62 A radial field $H = \left(\left(4.23 \times 10^6\right)/R\right) \cos\phi\ \vec{a}_\phi$ (A/m) exists in free space. Find the magnetic flux ϕ crossing the surface defined by

$$\frac{-\pi}{3} < \phi < \frac{\pi}{3} \quad \text{and} \quad 0 < z < 2\text{m}.$$

Solution:

Magnetic flux

$$\phi = \int_s \vec{B} \cdot d\vec{s}$$

$$\vec{ds}_\phi = d_\phi d_z \vec{a}_R$$

$$\vec{B} = \mu \vec{H}$$

$$\phi = \int_{\theta - \pi/3}^{\pi/3} \int_{z=0}^{2} \frac{\mu_0 \times 4.23 \times 10^6}{R} \times R\ d\phi\ dz \quad \text{or}$$

$$= 4\pi \times 10^{-7} \times 4.23 \times 10^6 \times \int_{\theta=\pi/3} d\phi\ z\Big|_0^2 \quad \text{or}$$

$$= 10.6 \times |\phi|_{-\pi/3}^{\pi/3}$$

$$= 22.26 \text{ Wb/m}^2$$

6.63 In Cylindrical coordinates, $\vec{J} = 10^4 \sin^2(2R)\vec{a}_z$. Obtain $\vec{H}$ from this current density and compare $\vec{J}$ with the answer.

Solution:

From Ampere's Circuital Law

$$\oint \vec{H}\cdot d\vec{l} = I$$

$$\int_s \nabla\times\vec{H}.ds = \int \vec{J}\cdot ds = \nabla\times\vec{H} = \vec{J}$$

$$\nabla\times\vec{H} = \frac{1}{R}\begin{vmatrix} \vec{a}_R & \overrightarrow{Ra_\phi} & \vec{a}_z \\ \dfrac{\partial}{\partial R} & \dfrac{\partial}{\partial\phi} & \dfrac{\partial}{\partial z} \\ H_R & RH_\phi & H_z \end{vmatrix}$$

$$= \frac{1}{R}\left(\frac{\partial H_z}{\partial\phi} - \frac{\partial H_\phi}{\partial z}\right)\vec{a}_R + \left(\frac{\partial H_R}{\partial z} - \frac{\partial H_z}{\partial R}\right)\vec{a}_\phi + \left(\frac{\partial}{\partial R}(RH_\phi) - \frac{\partial H_R}{\partial\phi}\right)\frac{\vec{a}_z}{R}$$

Therefore,

$$\vec{J} = \frac{1}{R}\frac{\partial H_z}{\partial\phi} - \frac{\partial H_\phi}{\partial z}$$

$$\vec{J}_\phi = \frac{\partial H_R}{\partial z} - \frac{\partial H_z}{\partial R} \quad \text{or}$$

$$\frac{\partial H_z}{\partial R} = \frac{\partial H_R}{\partial z} \quad \text{and}$$

$$\vec{J}_z = 10^4\sin^2(2R) = \frac{1}{R}\left[\frac{\partial(RH_\phi)}{\partial R} - \frac{\partial H_R}{\partial\phi}\right] \quad \text{or}$$

$$R10^4\left[1-\cos^2(2R)\right] = \left[\frac{\partial(RH_\phi)}{\partial R} - 0\right] \quad \text{or}$$

$$R10^4\left[1 - \frac{1-\cos 2(2R)}{2}\right] = \frac{\partial(RH_\phi)}{\partial R} \quad \text{or}$$

$$R10^4\left[\frac{1-\cos 4R}{2}\right] = \frac{\partial(RH_\phi)}{\partial R} \quad \text{or}$$

$$10^4\left[\frac{R}{2} - \frac{R\cos}{2}4R\right] = \frac{\partial(RH_\phi)}{\partial R} \quad \text{or}$$

$$RH_\phi = 10^4\left[\int \frac{R}{2}dR - \int \frac{R\cos(4R)}{2}dR\right]$$

Put

$$u = R,\ du = dR \text{ and } \cos uR = dv$$

$$v = \int \frac{\sin uR}{R}$$

Therefore,

$$RH_\phi = \frac{10^4}{2}\left[\frac{R^2}{2}\right] - 10^4\int \frac{R\cos}{2}4R\ dR \quad \text{or}$$

$$= \left[\frac{10^4}{2}\left(\frac{R^2}{2} - \frac{R\sin 4R}{4} + \int \frac{\sin 4R}{4}dR\right)\right] \quad \text{or}$$

$$= 10^4\left[\frac{R^2}{4} - \frac{R\sin 4R}{4} - \frac{\cos 4R}{16}\right] \quad \text{or}$$

$$H_\phi = 10^4\left[\frac{R}{4} - \frac{\sin 4R}{4} - \frac{\cos 4R}{16R}\right]$$

6.64 Across a parallel plate capacitor, an AC voltage source V_0 is connected, which has angular frequency ω. Show that the displacement current in the capacitor is the same as the current in the wire. Also, determine the magnetic field intensity at distance r from the wire.

Solution:
In a parallel plate capacitor, current is given as

$$i_c = C\frac{dv}{dt}$$

$$= CV_0\,\omega\cos\omega t$$

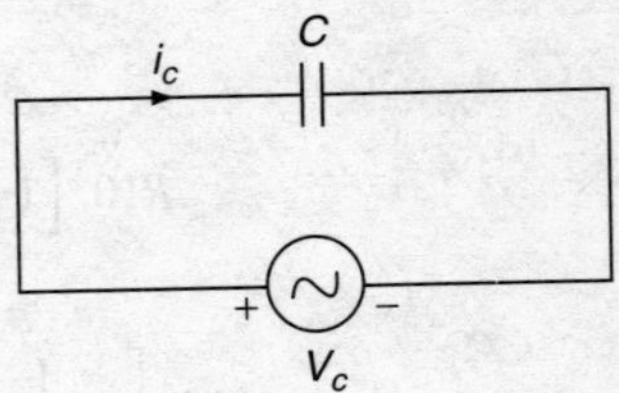

Figure 6.40 *Parallel plate capacitor with AC source*

For a parallel plate capacitor having an area A, plate distance d and having permittivity ε, capacitance C is given as

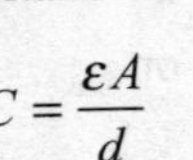

$$C = \frac{\varepsilon A}{d}$$

Further,

$$E = V_c/d$$

Therefore,

$$D = \varepsilon E = \varepsilon\frac{V_o}{d}\sin\omega t$$

Therefore, the displacement current is given as

$$i_D = \int \frac{\partial D}{\partial t} \cdot ds = \left(\frac{\varepsilon A}{d}\right) \cdot V_0 \cos \omega t$$

$$= C\, V_0\, \omega \cos \omega t$$

Hence, proved.

6.65 A circular loop of N turns of conducting wire lies in an x–y plane with its centre at the origin of the magnetic field given by

$$B = B_0 \sin\left(\frac{\pi r}{2b}\right)$$

where b is the radius of the loop and ω the angular frequency. Find the total induced emf in the loop.

Solution:

The loop is stationary in the varying magnetic field. Therefore, induced emf e may be given as

$$e = -N\frac{d\phi}{dt}$$

The magnetic flux is ϕ is given as

$$\phi = \int_s B \cdot ds \quad \text{or}$$

$$= \int_0^b \left(B_0 \sin\left(\frac{\pi r}{2b}\right) \cos \omega t \right) \vec{a}_z \cdot (2\pi\, r\, dr)\vec{a}_z \quad \text{or}$$

$$= \int_0^b 2\pi\; B_0 \cos \omega t . \sin\left(\frac{\pi r}{2b}\right) r dr \quad \text{or}$$

$$= 2\pi B_0 \cos \omega t \int_0^b r \sin\left(\frac{\pi r}{2b}\right) \cdot dr \quad \text{or}$$

$$= 2\pi B_0 \cos \omega t \left[\left(-r \cos\left(\frac{\pi r}{2b}\right) \cdot \left(\frac{2b}{\pi}\right)\right)_0^b - \int_0^b (-1) \cos\left(\frac{\pi r}{2b}\right)\left(\frac{2b}{\pi}\right) dr \right] \quad \text{or}$$

$$= 2\pi B_0 \cos \omega t \left[0 + \frac{2b}{\pi} \int_0^b \cos\left(\frac{\pi r}{2b}\right) dr \right] \quad \text{or}$$

$$= 2\pi B_0 \cos \omega t \left\{ \frac{2b}{\pi} \left(\sin\left(\frac{\pi r}{2b}\right) \cdot \left(\frac{2b}{\pi}\right)\right)_0^b \right\} \quad \text{or}$$

$$= \frac{8\pi b^2}{\pi^2} \cdot \cos \omega t\; B_0 \left(\sin \frac{\pi}{2} - \sin 0 \right) \quad \text{or}$$

$$\phi = \frac{8b^2}{\pi} B_0 \cos\omega t$$

Therefore,

$$e = -\frac{Nd\phi}{dt} = -N\frac{8b^2}{\pi} B_0 \cdot (-\sin\omega t)\cdot \omega \quad \text{or}$$

$$e = \frac{8N}{\pi} b^2 B_0 \, \omega \sin\omega t$$

6.66 A uniform plane wave with electric field intensity $E = E_x \vec{a}_x$ in a lossless dielectric medium propagates in the z-direction. The relative permittivity and permeability are 9 and 1, respectively. Assume E_x as sinusoidal with frequency 120 MHz having a maximum value of $+10^{-4}$ V/m at $t = 0$ and $z = 1/8$ m.

(a) Obtain an instantaneous expression for E for any time t and z.

(b) Obtain an instantaneous expression for H.

Solution:

$$\beta = \omega\sqrt{\mu\varepsilon} = \frac{\omega}{c}\sqrt{\mu_r \varepsilon_r}$$

$$= \frac{2\pi \times 120 \times 10^6}{3 \times 10^8}\sqrt{9}$$

$$= 24\pi \,(\text{rad/m})$$

(a) $E(z,t) = E_x \vec{a}_x = 10^{-4}\cos(2\pi \times 120 \times 10^6 t - \beta z + \phi)\vec{a}_x$

As $E_x = 10^{-4}$ the argument of cosine is zero, i.e.

$$2\pi \times 120 \times 10^6 t - \beta z - \phi = 0$$

Applying boundary condition at $t = 0$, $z = 118$

$$\phi = \beta z = (2.4\pi) \times \left(\frac{1}{8}\right) = 0.3\pi \text{ rad}$$

Therefore, $E(z,t) = 10^{-4}\cos(2\pi \times 120 \times 10^6 t - 2.4\pi z + 0.3\pi)$ V/m

(b) $H = H_y \, \vec{a}_y$

$$= \frac{F_x}{\eta}\vec{a}_y$$

$$\eta = \sqrt{\frac{\mu}{\in}} = \frac{\eta_o}{\sqrt{\in_o}} = \frac{120\pi}{3} = 40\pi \; (\Omega)$$

Therefore,

$$H(z,t) = \frac{10^{-4}}{40\pi}\cos(2\pi \times 120 \times 10^6 t - 2.4\pi z + 0.3\pi) \; (\text{A/m})$$

By applying Ampere's Circuital Law,

$$\oint H.dl = 2\pi r\, H_\phi$$

As no charge is deposited on the wire and $D = 0$,

$$\int J \cdot ds = i_c = CV_o\, \omega \cos \omega t$$

Now equating the above two integral terms

$$2\pi r\, H_\phi = \omega C V_o \cos \omega t \quad \text{or}$$

$$H_\phi = \frac{\omega\, C V_o\, \cos \omega t}{2\pi r} \text{ A/m}$$

6.67 A sinusoidal electric intensity of amplitude 100 V/m and frequency 2 GHz exists in a lossy dielectric medium having a relative permittivity of 2.5 and loss tangent of 0.005. Find the average power dissipated in the medium per cubic meter.

Solution:

The effective conductivity of the lossy medium is given by

$$\tan \delta_c = 0.005 = \frac{\sigma}{\omega \varepsilon_o \varepsilon_r} \quad \text{or}$$

$$\sigma = 0.005 \times 2\pi \times (2\times 10^9)\left(\frac{10^{-9}}{36\pi}\right)\times 2.5 \quad \text{or}$$

$$= 1.38 \times 10^{-3} \text{ (S/m)}$$

Therefore, the average power dissipated per unit volume is given by

$$P = \frac{1}{2} JE = \frac{1}{2}\sigma E^2 \quad \text{or}$$

$$= \frac{1}{2}\times (1.38\times 10^{-3}) \times (100)^2$$

$$= 6.94 \text{ W/m}^3$$

UNSOLVED QUESTIONS

6.1 Magnetic flux ϕ through a coil is perpendicular to its plane. Magnetic flux is directed into paper. It is varying according to $\phi = (6t^2 + 8t + 3)$ (mili Wb). Find the induced emf in this loop at time $t = 4$ sec.

6.2 Magnetic flux density $\vec{B}$ in a medium is

$$\vec{B} = \frac{1}{\rho}\cos\phi\, \vec{a}_\rho$$

Find the flux at a surface defined as $(-\pi/4 < \phi < \pi/4)$ and $(0 \le z \le 3 \text{ m})$.

6.3 A coil has 50 turns. A coil is pulled in 0.02 sec between the magnet. The coil's area includes a flux of 30×10^{-6} (weber/turn). In a place its flux is 10^{-5} (weber/turn). Find the average induced emf in the coil.

6.4 A magnetic field $\vec{B}$ in Cylindrical coordinates is

$$\vec{B} = \frac{B_0}{r}\vec{a}_\phi$$

B_0 is a constant. A rectangular plane as shown in Figure 6.41 is situated in the y–z plane. This plane is also parallel to the z-axis. The loop moves at a velocity $v = 10\,\vec{a}_y$ in the plane. Find the electric field around this loop.

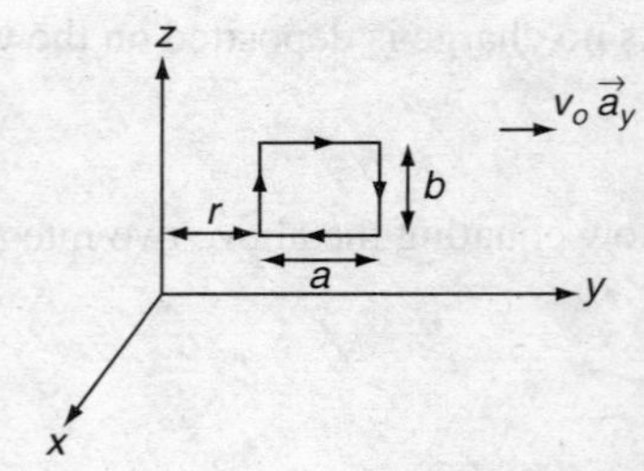

Figure 6.41 *A rectangular plane*

6.5 Find the magnetic field intensity. $\vec{H}$ is at the centre of a square current element. The length is 4 m. Current I is 3.0 Amp.

6.6 An infinitely long conductor is shown in Figure 6.42. It carries a current of 5 mA. The conductor makes segments, i.e. a semicircular loop on the $x-y$ plane, two straights, one parallel to the x-axis and the last one coinciding with the z-axis. Find the magnetic field induction at C (centre) of this semicircular loop.

6.7 A conducting triangular loop is shown in Figure 6.43. It carries a current of 8 Amp. What is $\vec{H}$ at a point (0, 0, 4) due to side 1 of this triangular loop?

6.8 A positive y-axis (semi-infinite line wrt origin) carries a current of 5 Amp. Current is in the $\vec{a}_y$ direction. Assuming it to be part of a large circuit, obtain $\vec{H}$ at a point A (4, 14,–2).

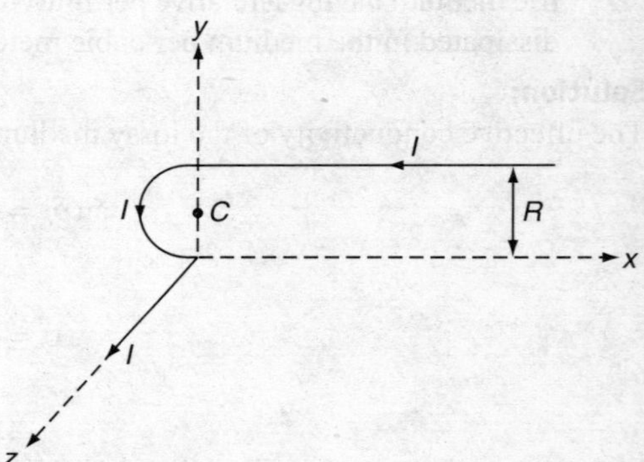

Figure 6.42 *Semicircular loop*

6.9 An electron in a Hydrogen atom circles around a proton at 2.5×10^6 (ms^{-1}) orbit, which has a radius of 3.5×10^{-11}(m). Find
(a) Equivalent current and
(b) Magnetic field, which will be eventually produced near the proton.

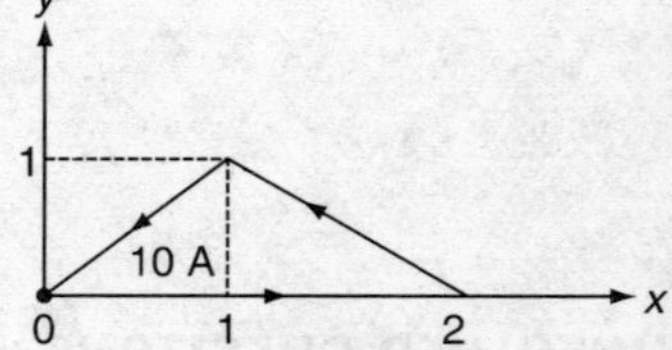

Figure 6.43 *A conducting triangular loop*

6.10 There is an infinitely long current element on the x-axis. It carries a 2.0 mA current. Current is in the $\vec{a}_x$ direction. Find $\vec{H}$ at a point A (2, 3, 1).

6.11 Volume current density distribution is given as

$$\vec{J}(r,\phi,z) = \text{(in Cylindrical coordinates)} \begin{cases} 0, & 0 < r < a \\ J_0(r/a)\vec{a}_z, & a \le r \le b \\ 0, & b < r < \infty \end{cases}$$

What will be the magnetic field intensity $\vec{H}$ in various regions?

6.12 Find the magnetic flux between conductors of a coaxial cable. Its length is 8 m. The inner radius of the conductor is 2 cm. The radius of the outer conductor is 4 m. Current enclosed is 5 A.

6.13 Find current density due to a magnetic field $\vec{H} = 20\sin x\ \vec{a}_y$.

6.14 Vector magnetic potential due to direct current flowing through (in free space) is $(x^2 + y^2)\vec{a}_z$ (μWb/m^2). Find the magnetic field due to current element at a point (3, 3, 4).

6.15 An electron moves at 2×10^8 m/sec in the positive x-axis direction. Find the magnetic field intensity at point (0, 0, 10^{-1}) m.

6.16 A plane $y = 1$ carries current per length $\vec{K} = 60\vec{a}_z$. Obtain $\vec{H}$ at a point (2, 4, −2).

6.17 A solenoid has radius 4 mm. Its length is 3 cm. There are 50 turns/m. It carries a current of 300 mA. Find (a) $|\vec{H}|$ at solenoid ends and (b) $|\vec{H}|$ at the center.

6.18 A circular toroid has its center at the origin. Its axis is the same as the z-axis. It has 100 turns. Here, $\rho_0 = 8$ cm and $a = 2$ cm. Find $|\vec{H}|$ at (5, 8, 2) cm if the toroid carries 40 mA current.

6.19 A current distribution gives a vector magnetic potential $A = (x^2 y\vec{a}_x + yx\vec{a}_y - 3xy\vec{a}_z)$ (Wb/m). Find the flux through the surface $z = 1$, $0 \le x \le 2$ and $-1 \le y \le 3$.

6.20 What will be the magnetic flux density at a point because of a long straight filamentary conductor, which carries a current of 5 mA in direction $\vec{a}_z$? (Hint: Use the concept of vector magnetic potential.)

6.21 Magnetic vector potential $\vec{A}$ is $(3e^{-r}\cos\phi\,\vec{a}_r - 3\cos\phi\,\vec{a}_z)$ (in Cylindrical coordinates). What will be the flux density at a point (3, $3\pi/4$, 0)?

6.22 The magnetizing field of 1200 (Am^{-1}) causes a magnetic flux of 1.6×10^{-5} (Weber) in an iron bar. The cross-sectional area of the bar is 0.5 m^2. Find the permeability and susceptibility of this iron bar.

6.23 A toroid with 1500 turns has inner core and outer core radii of 8 cm and 10 cm, respectively. Find the relative permeability of the core. A current of 0.8 Amp produces a magnetic field with intensity 4.5 (T) in this toroidal core.

6.24 In a material, flux density $\vec{B}$ is $= 0.05x\ \vec{a}_y$ (tesla) and magnetic susceptibility $\chi_m = 3.5$. Obtain the current density and also the volume current density.

6.25 In a case, two regions 1 and 2 are separated by a surface $2x - 4y + 3z = 0$. Region 1 permeability is $\mu_1 = 3\mu_0$. Region 2 permeability is $\mu_2 = 4\mu_0$. Point $P(1, 1, 1)$ is lying in region 2. If field $\vec{H}$ is $(3\vec{a}_x + 5\vec{a}_y - 5a_z)$ (A/m) then find $\vec{H}_2$.

6.26 Unit normal vector $\vec{a}_{n21}$ from region 2 ($\mu = 3\mu_0$) to region 1 ($\mu = 2\mu_0$) is given as $\vec{a}_{n21} = (0.57\vec{a}_x + 0.85\vec{a}_y - 0.28\vec{a}_z)$, $\vec{H}_1 = (8\vec{a}_x + 3\vec{a}_y - 4\vec{a}_z)$ (A/m) and $\vec{H}_2 = (M\vec{a}_x - 4\vec{a}_y + 5\vec{a}_z)$ (A/m). Find M and the two angles (normal to the interface).

SUMMARY

1. Magnetic field in free space is produced due to direct current.

2. The Lorentz force equation relates mechanical force to an electrical force. It is given as $\vec{F} = q(\vec{E} + \vec{u} \times \vec{B})$ (N).

3. There are two Faraday's Laws of electromagnetic induction. According to Faraday's Law induced emf

$$V_{emf} = -N\frac{d\phi}{dt} \text{ (V)}$$

4. According to Ampere's Law, current I is given as

$$I = \oint \vec{H} \cdot \vec{dl} \text{ (A)}$$

5. According to Lenz's Law, induced current has such a direction that it always opposes the change in magnetic flux.

6. As per the Biot–Savart Law, magnetic field intensity $d\vec{H}$ is given as

$$d\vec{H} = \frac{Id\vec{l} \times \vec{a}_r}{4\pi r^2}$$

7. Magnetomotance or Magnetomotive force (MMF) F is given as

$$F = NI = \oint \vec{H}.d\vec{l}$$

8. According to Ampere's Circuital Law, line integral $\vec{H}$ along a closed path is the same as the direct current, which is enclosed by that path.

9. Magnetic flux ϕ through a surface held in magnetic field $\vec{B}$ is given as

$$\phi = \int_s \vec{B}.d\vec{s}$$

   where ds is a small area element.

10. Magnetic flux density B is given as

$$B = \frac{\phi}{A} \text{(Webers)}$$

11. Magnetic scalar potential is denoted by V_m. Negative gradient of V_m gives magnetic field intensity $\vec{H}$.

12. Magnetic vector potential is denoted by $\vec{A}$. Magnetic flux density and magnetic vector potential are related to each other as $\vec{B} = \nabla \times \vec{A}$.

13. Magnetization M is given as

$$M = \frac{\text{dipole moment}}{\text{volume}}$$

14. Magnetization vector $\vec{J}_m$ is related to volume current density as

$$\vec{J}_m = \nabla + \vec{M}$$

15. Law of refraction is expressed as

$$\frac{\tan\theta_1}{\tan\theta_2} = \frac{\mu_1}{\mu_2}$$

16. Energy density W_H is given as

$$W_H = \frac{1}{2}\mu_o H^2 \text{ (Joules/m}^3\text{)} \quad \text{or}$$

$$W_H \frac{1}{2} = \vec{B}\cdot\vec{H} \text{ (Joules/m}^3\text{)} \quad \text{in vector form}$$

MULTIPLE-CHOICE QUESTIONS

1. Steady magnetic fields are governed by __________ law.
(a) Biot–Savart's (b) Ampere's Circuital
(c) Both (a) and (b) (d) None of these

2. There will be force of attraction between two current-carrying conductors if the currents are in __________ direction.
(a) Same (b) Opposite (c) None of these (d) Cannot say

3. Lorentz force equation comprises __________ and __________ forces.
(a) Electric, magnetic (b) Mechanical, chemical
(c) Both (a) and (b) (d) None of these

4. Magnetic dipole moment is a product of __________.
(a) Current and area (b) Area and its direction
(c) Current, area and its direction (d) None of these

5. According to Faraday's __________ Law, as long as changes happen in magnetic flux, induced emf persists.
(a) First (b) Second (c) Third (d) Fourth

6. According to __________ Law, induced current acts to produce an opposing flux.
(a) Bio–Savart's (b) Lenz's (c) Ampere's (d) Faraday's

7. In terms of current density, Biot–Savart's Law is expressed as __________.
(a) $\int \frac{\vec{J}\times\vec{a}_r}{2\pi r}.dv$ (b) $\int \frac{\vec{J}\times\vec{a}_r}{2\pi r^2}.dv$ (c) $\int \frac{\vec{J}\times\vec{a}_r}{4\pi r}.dv$ (d) $\int \frac{\vec{J}\times\vec{a}_r}{4\pi r^2}.dv$

8. Biot–Savart's Law can be applied to __________ length current-carrying conductors.
(a) Large (b) Medium (c) Small (d) Very small

9. According to Ampere's Circuital Law, field intensity $\vec{H}$ at a point at distance R from a very long straight filament conductor-carrying current I is given as

(a) $\vec{H} = \frac{I}{2\pi r^2}\vec{a}_\phi$ (b) $\vec{H} = \frac{I}{2\pi R}\vec{a}_\phi$ (c) $\vec{H} = \frac{I}{4\pi R^2}\vec{a}_\phi$ (d) $\vec{H} = \frac{I}{4\pi R}\vec{a}_\phi$

10. Ampere's Circuital Law is analogous to ________ Law in electrostatics.

(a) Lenz's (b) Gauss's (c) Biot–Savart's (d) Faraday's

11. Ampere's Circuital Law can be applied ________ the conductor.

(a) Inside (b) Outside (c) Both (a) and (b) (d) None of these

12. Magnetic flux density is the same as ________.

(a) Magnetic induction (b) Magnetic field strength
(c) Both (a) and (b) (d) None of these

13. Total flux passing through a closed surface held in a magnetic field is ________.

(a) Infinity (b) Zero (c) Unity (d) None of these

14. Which of the following is true for electrostatics?

(a) $\vec{E} = -\nabla V$ (b) $\nabla^2 V = 0$ (c) Both (a) and (b) (d) None of these

15. ________ gradient of magnetic scalar potential gives magnetic field intensity.

(a) Positive (b) Negative (c) Double (d) Integral

16. Magnetic vector potential for volume current is expressed as ________.

(a) $\int_s \frac{\mu_0\, Jdv}{4\pi r}$ (b) $\int_s \frac{\mu_0\, Jdv}{4\pi r^2}$ (c) $\int_s \frac{\mu_0\, Jdv}{2\pi r}$ (d) $\int_s \frac{\mu_0\, Jdv}{2\pi r^2}$

17. ________ can be obtained from vector magnetic potential $\vec{A}$.

(a) $\vec{B}$ (b) $\vec{H}$ (c) Both (a) and (b) (d) None of these

18. Magnetization is given as ________.

(a) $\frac{\text{Volume}}{\text{Dipole moment}}$ (b) $\frac{\text{Dipole moment}}{\text{Volume}}$

(c) Dipole moment × Volume (d) None of these

19. Magnetization volume current density in magnetic materials is due to ________.

(a) Applied non-uniform magnetic field (b) Non-uniform magnetic susceptibility of material
(c) Both (a) and (b) (d) None of these

20. According to the Lav' of refraction ________.

(a) $\tan\theta_1 = \frac{\mu_2}{\mu_1}\tan\theta_2$ (b) $\tan\theta_1 = \frac{\tan\theta_2}{\mu_1\mu_2}$

(c) $\tan\theta_1 = \frac{\mu_1\mu_2}{\tan\theta_2}$ (d) $\tan\theta_1 = \frac{\mu_1}{\mu_2}\tan\theta_2$

21. Energy density W_H is given as __________.

(a) $\frac{1}{2}HB$ (b) $\frac{1}{2}\mu_0 H^2$ (c) $\frac{1}{2}\vec{B}\cdot\vec{H}$ (d) All of these

ANSWERS TO MULTIPLE-CHOICE QUESTIONS

(1) c; (2) a; (3) a; (4) c; (5) a; (6) b; (7) d; (8) d; (9) b;
(10) b; (11) c; (12) c; (13) b; (14) c; (15) b; (16) a; (17) c; (18) b;
(19) c; (20) d; (21) d

7 Maxwell's Equations

7.1 INTRODUCTION

Electrostatic fields are denoted by $\vec{E}(x, y, z)$. Magnetostatic fields are represented by $\vec{H}(x, y, z)$. If electric and magnetic fields vary with time, then these fields are called "time-varying fields". In static electromagnetic (EM) fields, electric and magnetic fields are independent of one another. In dynamic EM fields, the two fields are interdependent. A time-varying electric field always involves a corresponding time-varying field. Time-varying EM fields are represented as $\vec{E}(x, y, z, t)$ and $\vec{H}(x, y, z, t)$. Time-varying EM fields are of more importance and are more practical than static EM fields.

Maxwell has given four equations for time-varying EM fields. With the help of Maxwell's equations, EM fields can be completely described. Maxwell's equations are also called "electromagnetic field equations". These equations shelter on fundamental laws of Faraday, Gauss and Ampere dealing with "displacement current". Electrostatic fields are produced by static electric charges. Magnetostatic fields are produced by the motion of electric charges having uniform velocity or magnetic charges (magnetic poles). Time-varying fields or waves are caused by accelerated charge. Time-varying fields and EM waves are analysed in communication systems, radar systems, biomedical engineering, etc. A "wave" is basically an energy carrier. A wave carries information. Wave is a function of time and space. It is as a physical phenomenon, which occurs at a place at a particular time. Waves can be reproduced at another place at a later time. Time delay is "propagational" owing to space separation between the first and second locations. This group of phenomena creates a "wave".

Maxwell contributed towards the analysis of EM wave through his Maxwell's equations. In EM waves, the time-varying magnetic field generates a time-varying electric field, which again generates a magnetic field. In this process, energy is propagated at light velocity through free space. Waves have very wide engineering applications in communication like TV, radio, mobile communications, radars, etc.

7.2 DISPLACEMENT CURRENT

Current through a capacitor is of a different nature than of that flowing through a resistor. In a capacitor, current flows when the applied voltage changes. Current through a capacitor is given as

$$i_c = \frac{dq}{dt} = C\frac{dv}{dt}$$

where q is the instantaneous charge, C the capacitance of a capacitor and V the voltage applied (changing). Instantaneous charge q in a capacitor is given as $q = CV$.

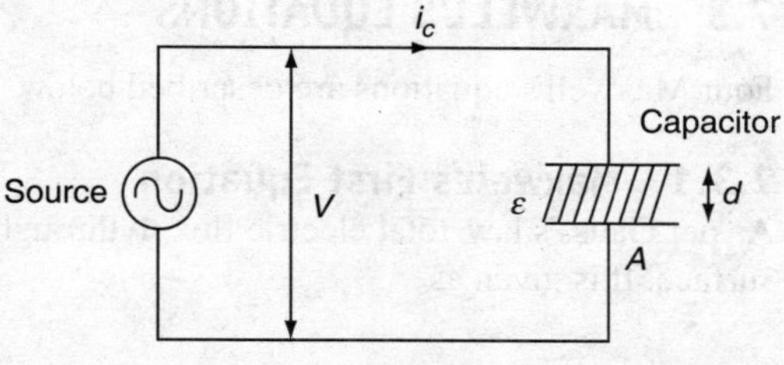

Figure 7.1 *Circuit*

Current i_c through a capacitor is known as the "displacement current". Actually no current flows through the capacitor. It is only an apparent current. In a parallel plate capacitor, as much current flows out of one plate as much as it flows into another plate. Figure 7.1 shows a circuit having a voltage source and a pure capacitor. Conduction current "i_c" in a wire is the "displacement current" in a capacitor. In the case of a capacitor

$$i_c = \frac{dq}{dt}$$

$$= C\frac{dv}{dt} \quad \text{and}$$

$$C = \frac{\varepsilon A}{d}$$

where v is the time-varying voltage, A the area of parallel plates, d the separation between two plates and ε the permittivity of the capacitor material.

Further,

$$i_c = \frac{\varepsilon A}{d}\frac{dv}{dt} \quad \text{or}$$

$$= \varepsilon A\frac{dE}{dt}$$

where $E = v/d$ is the "electric field intensity".

Also,

$$\frac{i_c}{A} = \varepsilon\frac{\partial E}{\partial t} \quad \text{or}$$

$$\vec{J}_d = \frac{dD}{dt}$$

Vector $\vec{D} = \varepsilon\vec{E}$ can vary with space.

Thus,

$$\vec{J}_d = \frac{d\vec{D}}{dt}$$

where $\vec{D}$ is the electric displacement and $\vec{J}_d$ the displacement current density.

The displacement current density $\vec{J}_d$ does not represent a current passing through a capacitor. $\vec{J}_d$ only represents an apparent current. $\vec{J}_d$ represents the rate at which charge flow takes place from one electrode to another electrode in the external electrical circuit. That is why it is called displacement current.

Concepts of "displacement current" and "displacement current density" were introduced by J. C. Maxwell. It deals with the production of magnetic fields in empty space. In empty space, the conduction current is "zero". In empty space, magnetic fields are due to "displacement currents".

7.3 MAXWELL'S EQUATIONS

Four Maxwell's equations are described below.

7.3.1 Maxwell's First Equation

As per Gauss's law, total electric flux Φ through a closed surface is equal to the charge enclosed by that surface. It is given as

$$\Phi = q_{enc}$$

However,

$$\Phi = \oint_s \vec{D} \cdot \vec{ds} \quad \text{and}$$

$$q_{enc} = \int_v \rho_v \, dv$$

Thus,

$$\oint_s \vec{D} \cdot \vec{ds} = \int_v \rho_v \, dv$$

By applying the Divergence Theorem,

we get

$$\oint_s \vec{D} \cdot \vec{ds} = \int_v \nabla \cdot \vec{D} \, dv \quad \text{or}$$

$$\int_v (\nabla \cdot \vec{D}) \, dv = \int_v \rho_v \, dv \quad \text{or}$$

$$\nabla \cdot \vec{D} = \rho_v$$

7.3.2 Maxwell's Second Equation

As per this equation/law, the total magnetic flux Φ_m emerging through a closed surface is always zero. It is given as

$$\Phi_m = \oint_s \vec{B} \cdot \vec{ds} = 0$$

Applying Gauss's Divergence Theorem to the above, we get

$$\oint_s \vec{B} \cdot \vec{ds} = \int_v (\nabla \cdot \vec{B}) \, dv \quad \text{or}$$

$$\nabla \cdot \vec{B} = 0$$

7.3.3 Maxwell's Third Equation

According to Ampere's circuit law, the line integral of the tangential component of $\vec{H}$ (around a closed path) is the same as the net current I_{enc} (which is enclosed by the path).

Thus, circulation $\vec{H}$ is equal to I_{enc}. It is given as

$$\oint \vec{H} \cdot \vec{dl} = I_{enc}$$

By applying Stoke's Theorem to the left-hand side, we get

$$\oint_L \vec{H} \cdot \vec{dl} = \int_v (\nabla \times \vec{H}) \cdot \vec{ds}$$

also
$$I_{enc} = \int_s \vec{J} \cdot \vec{ds}$$

Thus,

$$\nabla \times \vec{H} = \vec{J}$$

Maxwell gave the concept of displacement current $\vec{D}$, and thus the modified Ampere's law concept of displacement current density J_d in the case of time-varying fields is very important. For time-varying fields,

$$\nabla \times \vec{H} = \vec{J} + \vec{J}_d$$

By taking divergence of the above expression, we get

$$\nabla \cdot (\nabla \times \vec{H}) = \nabla \cdot \vec{J} + \nabla \cdot \vec{J}_d$$

Divergence of curl of a vector is zero. It gives

$$\nabla \cdot (\nabla \times \vec{H}) = 0 \quad \text{or}$$

$$\nabla \cdot \vec{J} = -\nabla \cdot \vec{J}_d$$

The Continuity equation for time-varying fields is given as

$$\nabla \cdot \vec{J} = -\frac{\partial \rho}{dt}$$

A comparison of the above two expressions for $\nabla \cdot \vec{J}$ yields

$$\nabla \cdot \vec{J}_d = \frac{\partial \rho}{dt}$$

As pe: Gauss's Law

$$\nabla \cdot \vec{D} = \rho \quad \text{or}$$

$$\nabla \cdot \frac{\partial \vec{D}}{\partial t} = \frac{\partial \rho}{dt}$$

Thus,

$$\nabla \cdot \vec{J}_d = \nabla \cdot \frac{\partial \rho}{dt}$$

Divergence of the two vectors is equal if their vectors are equal.

Therefore,
$$\vec{J}_d = \frac{\partial \rho}{dt}$$

On substitution, we get

$$\nabla \times \vec{H} = \vec{J}_d + \frac{\partial \rho}{dt}$$

where $\vec{J}$ is the conduction current density, $\vec{J}_d$ the displacement current density and $\vec{D}$ the displacement current.

7.3.4 Maxwell's Fourth Equation

The magnitude of induced emf is always equal to the rate of change of magnetic flux linking the circuit. It is given as

$$\text{Induced emf} = -\frac{d\phi_m}{dt}$$

Also,

$$\phi_m = \oint \vec{B} \cdot \vec{ds}$$

Thus,

$$\text{emf} = -\int_s \frac{d\vec{B}}{dt} \cdot \vec{ds}$$

A negative sign appears because of Lenz's Law. Lenz's Law states that induced emf sets up a current in a direction such that the magnetic effect produced opposes the very cause producing it.

The emf is a closed line integral of non-conservation electrical field, which is generated by a battery. It is given as

$$\text{emf} = \oint_L \vec{E} \cdot \vec{dL}$$

Thus,

$$\oint_L \vec{E} \cdot \vec{dL} = -\int_s \frac{d}{dt} \vec{B} \cdot \vec{ds}$$

Apply Stoke's Theorem in the LHS of the above expression, we get

$$\int_L \vec{E} \cdot \vec{dL} = \int_s (\nabla \times \vec{E}) \cdot \vec{ds}$$

On substitution, we get

$$\int_s (\nabla \times \vec{E}) \cdot \vec{ds} = -\int_s \frac{\partial}{\partial t} \vec{B} \cdot \vec{ds}$$

Thus,

$$\nabla \times \vec{E} = -\frac{\partial \vec{B}}{\partial t}$$

7.4 SUMMARY OF MAXWELL'S EQUATION FOR TIME-VARYING FIELDS

A summary of time-varying fields is given below:

Integral Form	Differential Form	Comments
$\int_s \vec{D}\cdot\vec{ds} = \int_v \rho\cdot dv$	$\nabla\cdot\vec{D} = \rho$	Gauss's Law
$\int_s \vec{B}\cdot\vec{ds} = 0$	$\nabla\cdot\vec{B} = 0$	Non-existence of magnetic monopole
$\oint_L \vec{H}\cdot\vec{dL} = \int_s \left(\vec{J} + \frac{\partial \vec{D}}{\partial t}\right)\cdot\vec{ds}$	$\nabla\times\vec{H} = \vec{J} + \frac{\partial \vec{D}}{\partial t}$	Ampere's Law
$\oint_L \vec{E}\cdot\vec{dL} = -\int_s \frac{\partial \vec{B}}{\partial t}\cdot\vec{ds}$	$\nabla\times\vec{E} = -\frac{\partial \vec{B}}{\partial t}$	"Conservativeness" of electrostatic fields

7.5 MAXWELL'S EQUATION FOR SINUSOIDAL TIME-VARYING FIELD

A periodic variation is analysable in terms of the sinusoidal variation having fundamental and harmonic frequencies. Such time-varying fields are expressed as

$$E = E_0 \cos \omega t \quad \text{or} \quad E = E_0 \sin \omega t$$

where E_0 is the maximum field strength. Time-varying fields $\tilde{E}(r,t)$ in terms of corresponding phasor quantity $E(r)$ are expressed as

$$\tilde{E}(r,t) = \mathrm{Re}\left[E(r)e^{j\omega t}\right]$$

Differential form

$$\nabla\cdot\vec{D} = \rho$$

$$\nabla\cdot\vec{B} = 0$$

$$\nabla\times\vec{E} = -j\omega\vec{B} \quad \text{and}$$

$$\nabla\times\vec{H} = \vec{J} + j\omega\vec{D}$$

Integral form

$$\oint \vec{D}\cdot\vec{ds} = \int \rho_v \cdot dv$$

$$\oint \vec{B}\cdot\vec{ds} = 0$$

$$\oint \vec{E}\cdot\vec{dL} = -j\omega\int \vec{B}\cdot\vec{ds} \quad \text{and}$$

$$\oint \vec{H}\cdot\vec{dL} = \int(\vec{J} + j\omega\vec{D})\cdot\vec{ds}$$

7.6 BOUNDARY CONDITIONS USING MAXWELL'S EQUATIONS

Maxwell's equations are applicable throughout space and time. Maxwell's equation used for solving EM fields must satisfy the boundary conditions that are applicable at the interfaces between different media. At the points of discontinuity, Maxwell's equations cannot give any information as they contain space derivatives. If Maxwell's equations have been evolved for an EM field in integral force, only then they can give information at points of discontinuity. The point of discontinuity is the boundary surface. Various relations between electric field E, magnetic field H, electric flux density D and magnetic flux density B in terms of Maxwell's equations are only valid for a point in the continuous medium. Solutions of Maxwell's equations inside the solution region are applicable and joinder to remainder universe through boundary conditions and initial conditions. Various boundary conditions at the boundary surface (point of discontinuity) are given below.

1. Boundary condition '1': $E_{\tan '1'} = E_{\tan '2'}$
   It means that the tangential component of an electric field at the boundary of medium '1' is the same as the tangential component of electric field at the boundary of medium '2'. Thus, at a point of discontinuity, the tangential component of an electric field is continuous.
2. Boundary condition '2': $H_{\tan '1'} - H_{\tan '2'} = J_s$
   Here, J_s is the surface current density expressed in A/m^2. It means that the tangential component of magnetic field H at the surface of a perfect conductor is discontinuous. Discontinuity is equal to J_s. Other than a perfect conductor, the tangential component of magnetic field H is continuous across the surface.
3. Boundary condition '3': $B_{n'1'} = B_{n'2'}$
   It means that at the boundary surface (point of discontinuity), the normal component of magnetic flux density B_n is continuous. '1' and '2' refer to the two media.
4. Boundary condition '4': $D_{n'1'} - D_{n'2'} = \rho_s$
   Here, ρ_s is the surface charge density expressed in C/m^2. It means that the normal component of electric flux density D_n is discontinuous for that surface, where surface charge density exists. Discontinuity is equal to ρ_s. For those surfaces, where the surface charge density is zero, the normal component of electric flux density D_n is continuous.

Proof of boundary condition '1'

$$E_{\tan '1'} = E_{\tan '2'}$$

Consider a rectangular loop as depicted in Figure 7.2.

The line integral of an electrical field E around a closed path is zero.

Thus,

$$\oint E.dl = 0$$

$$\Rightarrow E_{y1}\frac{\Delta y}{2} + E_{x1}\ \Delta x - E_{y3}\frac{\Delta y}{2} - E_{y4}\frac{\Delta y}{2} - E_{x2}\Delta x + E_{y2}\frac{\Delta y}{2} = 0$$

Put $\Delta y = 0$ for point of discontinuity.
Then,

$$E_{x1}\ \Delta x - E_{x2}\ \Delta x = 0 \quad \text{or}$$

$$E_{x1} = E_{x2}$$

Hence, two tangential components of the electric field are equal.

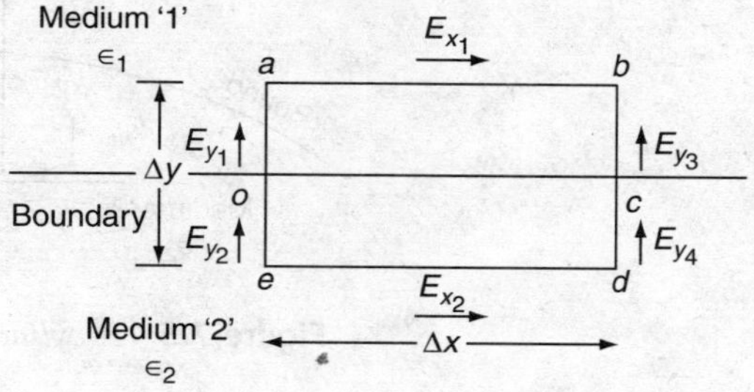

Figure 7.2 *A rectangular loop for proving boundary condition '1'*

Proof of boundary condition '2'

$$H_{\tan '1'} - H_{\tan '2'} = J_s$$

Consider a rectangular loop as depicted in Figure 7.3. Applying Ampere's circuital law

$$\oint H.dL = I$$

Here,

$$H_{x1}\ \Delta x + H_{y1}\frac{\Delta y}{2} + H_{y2}\frac{\Delta y}{2} - H_{y3}\frac{\Delta y}{2} - H_{y4}\frac{\Delta y}{2} - H_{x2}\Delta x = I$$

As $\Delta y = 0$ at the point of discontinuity, we get

$$H_{x1}\Delta x - H_{x2}\Delta x = I$$

$$H_{x1} - H_{x2} = \frac{I}{\Delta x} = J_s$$

where H_{x1} and H_{x2} are tangential components of magnetic field H.

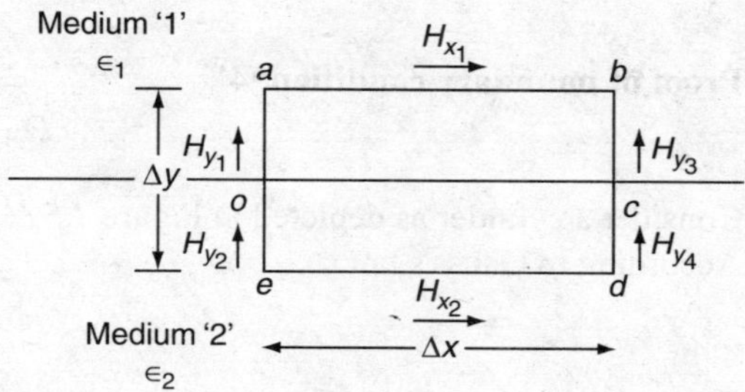

Figure 7.3 *A rectangular loop for proving boundary condition '2'*

Proof of boundary condition '3'

$$B_{n'1'} = B_{n'2'}$$

Consider a cylinder as depicted in Figure 7.4.

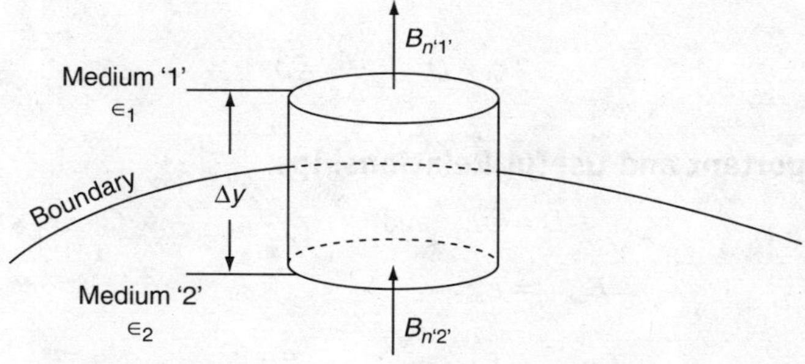

Figure 7.4 *A cylinder for proving boundary condition '3'*

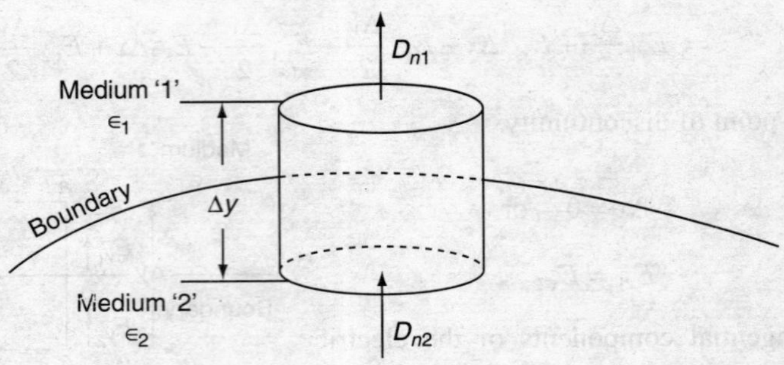

Figure 7.5 *A cylinder for proving boundary condition '4'*

As per Gauss's Law

$$\oint B.ds = 0$$

Here,

$$\oint B_{n'1'}a_y \ ds \ \vec{a}_y + \oint B_{n'2'}a_y \ ds(-\vec{a}_y) = 0$$

It yields

$$B_{n'1'}\Delta s - B_{n'2'}\Delta s = 0 \quad \text{or}$$

$$B_{n'1'} = B_{n'2'}$$

Proof of boundary condition '4'

$$D_{n1} - D_{n2} = \rho_s$$

Consider a cylinder as depicted in Figure 7.5.
According to Gauss's law

$$\oint D \cdot ds = Q$$

Thus,

$$D_{n1} \cdot \Delta s - D_{n2} \ \Delta s = Q$$

where Δs is the surface (m^2).
Thus,

$$D_{n1} - D_{n2} = \frac{Q}{\Delta s} = \rho_s$$

7.6.1 Other Important and Useful Relationships

$$D = \in E \quad \text{and}$$

$$E_{\tan'1'} = E_{\tan'2'}$$

so

$$D_{\tan'2'} = \frac{\in_2}{\in_1} D_{\tan'1'} \quad \text{or} \quad \frac{\in_{r2}}{\in_{r1}} D_{\tan'1'}$$

Thus, quantities pertaining to other mediums can be obtained at the boundary from the quantities given for another medium. Similarly,

$$D_{n'1'} - D_{n'2'} = \rho_s$$

It yields

$$\in_1 E_{n'1'} - \in_2 E_{n'2'} = \rho_s$$

Boundary condition $B_{n'1'} = B_{n'2'}$ gives

$$\mu_1 H_{n'1'} = \mu_2 H_{n'2'} \quad (\text{as } B = \mu H) \quad \text{or}$$

$$H_{n'1'} = \frac{\mu_2}{\mu_1} H_{n'2'} \quad \text{or} \quad \frac{\mu_{r2}}{\mu_{r1}} H_{n'2'}$$

Similarly,

$$H_{\tan'1'} - H_{\tan'2'} = J_s \text{ gives}$$

$$\frac{B_{\tan'1'}}{\mu_1} - \frac{B_{\tan'2'}}{\mu_2} = J_s$$

In scalar form, various boundary conditions and relationships are given as

$$\left.\begin{array}{l} E_{\tan'1'} - E_{\tan'2'} = 0 \\ H_{\tan'1'} - H_{\tan'2'} = J_s \\ B_{n'1'} = B_{n'2'} \\ D_{n'1'} - D_{n'2'} = \rho_s \end{array}\right\} \text{Boundary conditions}$$

$$\in_1 E_{n'1'} - \in_2 E_{n'2'} = \rho_s$$

$$\mu_{r1} H_{n'1'} - \mu_{r2} H_{n'2'} = 0$$

$$\frac{B_{\tan'1'}}{\mu_1} - \frac{B_{\tan'2'}}{\mu_2} = J_s$$

$$\frac{D_{\tan'1'}}{\in_{r1}} - \frac{D_{\tan'2'}}{\in_{r2}} = 0$$

$$J_{n'1'} = J_{n'2'}$$

$$J_{\tan'2'} = \frac{\sigma_2}{\sigma_1} J_{\tan'1'}$$

In vector form, various boundary conditions and their derived relationships are given as

$$\left.\begin{array}{l} \vec{a}_n \times (E_1 - E_2) = 0 \\ \vec{a}_n \times (H_1 - H_2) = J_s \\ \vec{a}_n.(B_1 - B_2) = 0 \\ \vec{a}_n.(D_1 - D_2) = 0 \end{array}\right\} \text{Boundary conditions}$$

$$\vec{a}_n.(\in_1 E_1 - \in_2 E_2) = \rho_s$$

$$\vec{a}_n.\left(\frac{H_1}{\mu_1} - \frac{H_2}{H_2}\right) = 0$$

$$\vec{a}_n \times \left(\frac{B_1}{\mu_1} - \frac{B_2}{\mu_2}\right) = J_s$$

$$\vec{a}_n.(J_1 - J_2) = 0$$

$$\vec{a}_n \times \left(\frac{J_1}{\sigma_1} - \frac{J_2}{\sigma_2}\right) = 0$$

SOLVED QUESTIONS

7.1 A parallel plate capacitor has a plate area of 15 cm^2. Plate separation is 6 mm. A voltage of 25 sin $10^2 t$ Volt is applied to its plates. Find the displacement current (assume $\varepsilon = 3\varepsilon_o$).

Solution:

It is given that $A = 15 \times 10^{-4} \text{m}^2$, $d = 6 \times 10^{-3}$ m, $V = 25 \sin 10^2 t$ Volt and $\varepsilon = 3\varepsilon_o$.

We know that

$$D = \varepsilon E = \varepsilon \frac{V}{d} \quad \text{and}$$

$$J_d = \frac{\partial D}{\partial t} = \frac{\varepsilon}{d} \cdot \frac{\partial v}{\partial t}$$

Thus,

$$I_d = J_d\, A = \frac{A \cdot \varepsilon}{d} \cdot \frac{\partial v}{\partial t} \quad \text{or}$$

$$= \frac{15 \times 10^{-4}}{6 \times 10^{-3}} \frac{3 \times 10^{-9}}{36\pi} \times 10^2 \times 25 \cos 10^2 t \quad \text{or}$$

$$= 16 \cdot 57 \cos 10^2 t \; (nA)$$

7.2 In free space, $\vec{H} = 50 \cos(\omega t - 5x)\vec{a}_z$ (A/m). What is the displacement current density $(\vec{J}_d)$?

Solution:

Maxwell's equation gives $\nabla \times \vec{H} = \dfrac{\partial \vec{D}}{\partial t}$ or

$$\begin{vmatrix} \vec{a}_x & \vec{a}_y & \vec{a}_z \\ \frac{\partial}{\partial x} & \frac{\partial}{\partial y} & \frac{\partial}{\partial z} \\ H_x & H_y & H_z \end{vmatrix} = \vec{J}_d$$

Here, in this case $\vec{H}$ is in the z-direction. Thus,

$$H_x = H_y = 0 \quad \text{or}$$

$$\vec{a}_x \left|\frac{\partial H_Z}{\partial y}\right| - \vec{a}_z \left|\frac{\partial H_Z}{\partial x}\right| + 0\vec{a}_z = \vec{J}_d$$

as

$$\vec{H} = 50\cos(\omega t - 15x)\vec{a}_z$$

$$\frac{\partial H_Z}{\partial y} = 0$$

Thus,

$$\vec{J}_d = \left|\frac{\partial \vec{H}_Z}{\partial x}\right| \vec{a}_y$$

$$\vec{J}_d = -250\sin(\omega t - 5x)\vec{a}_y \ (\text{A/m}^2)$$

7.3 Parallel plates in a capacitor have an area of 2 cm^2. These plates are separated by 0.2 cm. 20 sin $10^2 t$ Volts is applied to the capacitor. Find the displacement current in the case, when dielectric material (between plates) has a relative permittivity of 6.

Solution:
We know that displacement current density (J_d) is given as

$$J_d = \frac{\partial}{\partial t}\left(\frac{\varepsilon V}{d}\right) \quad \text{or}$$

$$= \frac{\varepsilon}{d}\frac{\partial V}{\partial t} \quad \text{or}$$

$$= \frac{\varepsilon_0 \varepsilon_r}{d}\frac{\partial V}{\partial t}$$

$$\text{Displacement current} = \frac{\varepsilon_o \varepsilon_r \cdot A}{d} \cdot \frac{\partial V}{\partial t}$$

$$= \frac{8.86\times10^{-12}\times6\times2\times10^{-4}}{0.2\times10^{-2}}\times20\times10^2\cos10^2 t$$

Thus,

$$J_d = 10632\times10^{-12}\cos10^2 t \ (\text{A}) \quad \text{or}$$

$$J_d = 10.632\cos 10^2 t \ (\text{nA})$$

7.4 Electric field density $\vec{D}$ in free space is 5 sin $(\omega t + \beta z)\vec{a}_x$. Find an expression for $\vec{E}$, $\vec{B}$ and $\vec{H}$ in this case.

Solution:
It is given that $\vec{D} = 5 \ \sin\ (\omega t + \beta z)\vec{a}_x$

Further,

$$\vec{E} = \frac{\vec{D}}{\varepsilon_o}$$

Thus,

$$\vec{E} = \frac{5\sin\left(\omega t - \beta z\right)}{\varepsilon_o}\vec{a}_x$$

Maxwell's equations gives

$$\nabla \times \vec{E} = -\frac{\partial \vec{B}}{\partial t}$$

Thus,

$$\begin{vmatrix} \vec{a}_x & \vec{a}_y & \vec{a}_z \\ \dfrac{\partial}{\partial x} & \dfrac{\partial}{\partial y} & \dfrac{\partial}{\partial z} \\ E_x & E_y & E_z \end{vmatrix} = -\frac{\partial \vec{B}}{\partial t}$$

However,

$$E_y = E_z = 0$$

Thus,

$$\left|\frac{\partial \vec{E}_x}{\partial z}\right|\vec{a}_y = -\frac{\partial \vec{B}}{\partial t} \quad \text{or}$$

$$\frac{\partial}{\partial t}\frac{\left[5\ \sin(\omega t + \beta z)\right]}{\varepsilon_o}\vec{a}_y = -\frac{\partial \vec{B}}{\partial t} \quad \text{or}$$

$$\frac{\partial \vec{B}}{\partial t} = -\frac{5\beta\cos(\omega t + \beta z)}{\varepsilon_o}\vec{a}_y \quad \text{or}$$

$$\vec{B} = -\frac{5\beta}{\varepsilon_o}\int \cos(\omega t + \beta z)\,dt \cdot \vec{a}_y$$

Therefore,

$$\vec{B} = -\frac{5\beta}{\omega\varepsilon_o}\sin(\omega t + \beta z)\,\vec{a}_y \text{ (Wb/m}^2\text{)}$$

Further,

$$\vec{H} = (\vec{B}/\mu_o)$$

so

$$\vec{H} = -\frac{5\beta}{\omega\varepsilon_o\mu_o}\sin(\omega t + \beta z)\ \vec{a}_y \text{ (Wb/m}^2\text{)}$$

7.5 In free space, $\vec{B}$ is given as $10e^{j(\omega t+5z)}\,\vec{a}_y$. Prove that $\vec{E} = -2\omega e^{j(\omega t+5z)}\vec{a}_x$.

Solution:

It is given that

$$\vec{B} = 10e^{j(\omega t+5z)}\vec{a}_y$$

Further,

$$\vec{B} = \mu_o \vec{H}$$

Thus,

$$\vec{H} = \frac{10}{\mu_o} e^{j(\omega t+5z)}\vec{a}_y$$

Maxwell's equation gives

$$\nabla \times \vec{H} = \frac{\partial \vec{D}}{\partial t}$$

However,

$$\vec{H}_x = \vec{H}_z = 0$$

Thus,

$$\frac{\partial \vec{D}}{\partial t} = \begin{vmatrix} \vec{a}_x & \vec{a}_y & \vec{a}_z \\ \dfrac{\partial}{\partial x} & \dfrac{\partial}{\partial y} & \dfrac{\partial}{\partial z} \\ 0 & \dfrac{10\, e^{j(\omega t+5z)}}{\mu_o} & 0 \end{vmatrix} \quad \text{or}$$

$$= \frac{\partial}{\partial z}\left(\frac{10\, e^{j(\omega t+5z)}}{\mu_o}\right)\vec{a}_x \quad \text{or}$$

$$\frac{\partial \vec{D}}{\partial t} = \frac{j50}{\mu_o} e^{j(\omega t+5z)}\vec{a}_x$$

On integration, we get

$$\vec{D} = \frac{50}{\omega\mu_o} e^{j(\omega t+5z)}\,\vec{a}_x \quad \text{and}$$

$$\vec{E} = \frac{\vec{D}}{\varepsilon_o} = \frac{50}{\omega\mu_o\varepsilon_o} e^{j(\omega t+5z)}\,\vec{a}_x$$

However,

$$\frac{1}{\mu_o\varepsilon_o} = \left(\frac{\omega}{\beta}\right)^2$$

$$= \frac{\omega^2}{25}$$

Therefore,

$$\vec{E} = -2\omega e^{j(\omega t + 5z)}\ \vec{a}_x$$

7.6 A medium is characterized by $\sigma = 0$, $\mu = \mu_o$, ε_o and $\vec{E} = 10\ \sin(10^4 t - \beta z)\ \vec{a}_y$ (V/m). Find $\vec{H}$ and β in this case.

Solution:

Here, Gauss's Law (for electric field) is getting satisfied. It is

$$\nabla \cdot \vec{E} = -\frac{\partial E_y}{\partial y} = 0$$

Faraday's Law gives

$$\nabla \times \vec{E} = -\mu \frac{\partial \vec{H}}{\partial t} \quad \text{or}$$

$$\vec{H} = -\frac{1}{\mu}\int (\nabla \times \vec{E})\, dt$$

However,

$$\nabla \times \vec{E} = \begin{vmatrix} \vec{a}_x & \vec{a}_y & \vec{a}_z \\ \frac{\partial}{\partial x} & \frac{\partial}{\partial y} & \frac{\partial}{\partial z} \\ 0 & E_y & 0 \end{vmatrix} \quad \text{or}$$

$$= -\frac{\partial E_y}{\partial z}\vec{a}_x \quad \text{or}$$

$$= 10\beta \cos(10^4 t - \beta z)\vec{a}_x (\text{V/m})$$

Thus,

$$\vec{H} = -\frac{10\beta}{\mu}\int \cos(10^4 t - \beta z) dt\ \vec{a}_x (\text{V/m}) \quad \text{or}$$

$$= -\frac{10\beta}{10^4 \mu}\sin(10^4 t - \beta z)\vec{a}_x$$

It is verified that

$$\nabla \cdot \vec{H} = \frac{\partial H_x}{\partial x} = 0$$

Thus, Gauss's Law for magnetic fields also gets satisfied. Ampere's Law gives

$$\nabla \times \vec{H} = \sigma \vec{E} + \varepsilon \frac{\partial \vec{E}_x}{\partial t}$$

But here,

$$\sigma = 0$$

Therefore,

$$\vec{E} = \frac{1}{\varepsilon}\int (\nabla\times\vec{H})\,dt$$

Now,

$$\nabla\times\vec{H} = \begin{vmatrix} \vec{a}_x & \vec{a}_y & \vec{a}_z \\ \frac{\partial}{\partial x} & \frac{\partial}{\partial y} & \frac{\partial}{\partial z} \\ H_x & 0 & 0 \end{vmatrix} \quad \text{or}$$

$$= \frac{\partial H_x}{\partial z}\vec{a}_y \quad \text{or}$$

$$= \frac{10\beta^2}{10^4\mu}\cos(10^4 t - \beta z)\vec{a}_y$$

Thus,

$$\vec{E} = \frac{1}{\varepsilon}\int \frac{10\beta^2}{10^4\mu}\cos(10^4 t - \beta z)\,\vec{a}_y \quad \text{or}$$

$$\vec{E} = \frac{10\beta^2}{\mu\varepsilon 10^4}\frac{1}{10^4}\sin(10^4 t - \beta z)\vec{a}_y \quad \text{or}$$

$$\vec{E} = \frac{\beta^2}{\mu\varepsilon 10^7}\sin(10^4 t - \beta z)\vec{a}_y$$

Comparing the above with the given expression of $\vec{E}$, we find

$$\frac{\beta^2}{\mu\varepsilon 10^7} = 10 \quad \text{or}$$

$$\beta = \pm\sqrt{10^8\mu\varepsilon} \quad \text{or}$$

$$= \pm 10^4\sqrt{\mu_o\varepsilon_o} \quad \text{or}$$

$$= \pm\frac{10^4}{3\times 10^8} \quad \text{or}$$

$$= \pm\frac{100}{3} \quad \text{and}$$

$$\vec{H} = \pm\frac{10\times(1000/3)}{4\pi\times 10^7(10^4)}\sin\left(10^4 t \pm \frac{1000z}{3}\right)\vec{a}_z \quad \text{or}$$

$$\vec{H} = \pm\frac{10^7}{12\pi}\sin\left(10^4 t \pm \frac{10^3 z}{3}\right)\vec{a}_x\,(\text{A/m})$$

7.7 Find magnetic field H_2 and its magnitude of medium '2' $(x > 0)$ if the magnetic field of medium '1' $(x < 0)$ is given as $H_1 = (2.0\vec{a}_x + 0.5\vec{a}_y - \vec{a}_z)$ (A/m). $\mu_{r1} = 2.0$ and $\mu_{r2} = 4.0$.

Solution:

Tangential and normal components of magnetic fields $H_{\tan'1'}$ and $H_{n'1'}$ are obtained as

$$H_{\tan'1'} = (0.5\vec{a}_y - \vec{a}_z)\,(\text{A/m}) \quad \text{and}$$

$$H_{n'1'} = 2.0\vec{a}_x(\text{A/m})$$

Further,

$$H_{\tan'1'} = H_{\tan'2'} \quad \text{and} \quad H_{n'2'} = \frac{\mu_{r1}}{\mu_{r2}} H_{n'1'}$$

Thus,

$$H_{\tan'2'} = (0.5\vec{a}_y - \vec{a}_z)\text{A/m} \quad \text{and}$$

$$H_{n'2'} = \frac{2}{4} \times 2.0\vec{a}_x = \vec{a}_x(\text{A/m})$$

Magnetic field H_2 is obtained as

$$H_2 = H_{n'2'} + H_{\tan'2'}$$
$$= (\vec{a}_x + 0.5\vec{a}_y - \vec{a}_z)\ \text{A/m}$$

The magnitude of magnetic field H_2 is obtained as

$$H_2 = \sqrt{(1)^2 + (0.5)^2 + (-1)^2}$$
$$= \sqrt{1 + 0.25 + 1} = \sqrt{2.25}$$
$$= 1.5\text{A/m}$$

7.8 Find E_2 and D_2 of medium '2' $(x > 0)$ if for medium '1' $(x < 0)$ $E_1 = (\vec{a}_x + 4\vec{a}_y + 2\vec{a}_z)$ (V/m). $\sigma_1 = \sigma_2 = 0$ and $\in_{r1} = 2$ and $\in_{r2} = 4$.

Solution: Tangential and normal components of electric field $E_{\tan'1'}$ and $E_{n'1'}$ are obtained as

$$E_{\tan'1'} = (4\vec{a}_y + 2\vec{a}_z)(\text{V/m}) \quad \text{and}$$

$$E_{n'1'} = \vec{a}_x(\text{V/m})$$

Further,

$$E_{\tan'1'} = E_{\tan'2'} \quad \text{and} \quad E_{n'2'} = \frac{\in r_1}{\in r_2} E_{n'1'}$$

Thus,

$$E_{\tan'2'} = (4\vec{a}_y + 2\vec{a}_z)(\text{V/m}) \quad \text{and}$$

$$E_{n'2'} = \frac{2}{4}(\vec{a}_x) = 0.5\vec{a}_x(\text{V/m})$$

Electric field E_2 is obtained as

$$E_2 = E_{n'2'} + E_{\tan'2'}$$
$$= (0.5\vec{a}_x + 4\vec{a}_y + 2\vec{a}_z)(\text{V/m})$$

Further,

$$D_2 = \epsilon_2 E_2$$
$$= \epsilon_0 \epsilon_{r2} E_2 = 4\epsilon_0 (0.5\vec{a}_x + 4\vec{a}_y + 2\vec{a}_z)$$
$$= \epsilon_0 (2\vec{a}_x + 16\vec{a}_y + 8\vec{a}_z)(\text{C/m}^2)$$

7.9 For the vector $\vec{A} = (2x+5y+az)\vec{a}_x + (3x+by-2z)\vec{a}_y + (2x+4y+cz)\vec{a}_z$, what will be the value of a, b and c for $\vec{A}$ to be "irrotational"

Solution:
For $\vec{A}$ to be "irrotational," $\nabla \times \vec{A} = 0$.

Now,
$$\nabla \times \vec{A} = \begin{vmatrix} \vec{a}_x & \vec{a}_y & \vec{a}_z \\ \partial/\partial x & \partial/\partial y & \partial/\partial z \\ (2x+5y+az) & (3x+by-2z) & (2x+4y+cz) \end{vmatrix}$$

$$= \vec{a}_x \left[\frac{\partial}{\partial y}(2x+4y+cz) - \frac{\partial}{\partial z}(3x+by-2z)\right]$$

$$-\vec{a}_y \left[\frac{\partial}{\partial x}(2x+4y+cz) - \frac{\partial}{\partial z}(2x+5y+az)\right]$$

$$+\vec{a}_z \left[\frac{\partial}{\partial x}(3x+by-2z) - \frac{\partial}{\partial y}(2x+5y+az)\right]$$

$$= \vec{a}_x(4+2) - \vec{a}_y(2-a) + \vec{a}_z(3-5)$$

However,

$$\nabla \times \vec{A} = 0 \quad \text{(given)}$$

$$-(2-a) = 0 \Rightarrow a = 2 \quad \text{and}$$

$$b = 0, c = 0$$

7.10 In a homogenous non-conducting region, $\mu_r = 4$, $\vec{E}$ $40\pi e^{j(\omega t-(7/3))}\vec{a}_y$ (V/m) and $\vec{H}$ is $5e^{j(\omega t-(7/3))}\vec{a}_y$.(V/m). Find values of ε_r and ω.

Solution:
It is given that $\mu_r = 4$, $|\vec{H}| = 5$, $|\vec{E}| = 40\pi$ and $\beta = 7/3$.
Now

$$\frac{|\vec{E}|}{|\vec{H}|} = \sqrt{\frac{\mu}{\varepsilon}} \quad \text{or}$$

$$\frac{|\vec{E}|}{|\vec{H}|} = \sqrt{\frac{\mu_0 \mu_r}{\varepsilon_0 \varepsilon_r}}$$

$$\frac{40\pi}{5} = 120\pi\sqrt{\frac{\mu_r}{\varepsilon_r}}$$

$$\sqrt{\varepsilon_r} = 30$$

$$\varepsilon_r = 900$$

Further,

$$\frac{\omega}{\beta} = v_p \quad \text{or}$$

$$\frac{\omega}{\beta} = \frac{1}{\sqrt{\mu\varepsilon}} \quad \text{or}$$

$$\frac{\omega}{\beta} = \frac{3\times 10^8}{\sqrt{\mu_r\varepsilon_r}} \quad \text{or}$$

$$\frac{\omega}{7/3} = \frac{3\times 10^8}{\sqrt{4\times 900}} \quad \text{or}$$

$$\omega = \frac{3\times 10^8 \times 7}{3\times 60}$$

or simply,

$$\omega = 1.16\times 10^7 \text{ rad/sec}$$

7.11 In a free space, it is given that $\vec{B} = B_m e^{j(\omega t - \beta z)}\ \vec{a}_y$.

Show that

$$\vec{E} = \frac{B_m\,\omega}{\beta} e^{j(\omega t - \beta z)}$$

Solution:
From Maxwell's equation, we know that

$$\nabla\times\vec{E} = \frac{-\partial B}{\partial t}$$

$$\frac{-\partial\vec{B}}{\partial t} = -B_m e^{j(\omega t - \beta z)}(j\omega)\vec{a}_y$$

Thus, it can be shown as given. It is also proved in problem 7.12.

7.12 In free space $\vec{D} = D_m\cos(\omega t + \beta z)\vec{a}_x$. Using Maxwell's equation, show that

$$\vec{B} = \frac{-\omega\mu_0 D_m}{\beta}\cos(\omega t + \beta z)\ \vec{a}_y$$

Solution:
We know that

$$\nabla\times\vec{E} = \frac{-\partial B}{\partial t} \quad \text{and} \quad \vec{D} = \varepsilon_0\vec{E}.$$

For free space,

$$\vec{E} = \frac{\vec{D}}{\varepsilon_0}$$

$$\nabla \times \vec{E} = \begin{vmatrix} \vec{a}_x & \vec{a}_y & \vec{a}_z \\ \dfrac{\partial}{\partial x} & \dfrac{\partial}{\partial y} & \dfrac{\partial}{\partial z} \\ \dfrac{Dm}{\varepsilon_0}\cos(\omega t - \beta z) & 0 & 0 \end{vmatrix} \quad \text{or}$$

$$= \vec{a}_x(0-0) - \vec{a}_y\left(0 - \frac{\partial}{\partial z}\left(\frac{D_m}{\varepsilon_0}\right)\cos(\omega t + \beta z)\right) + \vec{a}_z(0-0) \quad \text{or}$$

$$= -\frac{D_m}{\varepsilon_0}\sin(\omega t + \beta z)\cdot(+\beta)\,\vec{a}_y \quad \text{or}$$

$$= \frac{-D_m\beta}{\varepsilon_0}\sin(\omega t + \beta z)\;\vec{a}_y$$

As

$$\nabla \times \vec{E} = \frac{-\partial \vec{B}}{\partial t}$$

$$\vec{B} = -\int (\nabla \times \vec{E})\; dt \quad \text{or}$$

$$= \int \frac{D_m\beta}{\varepsilon_0}\sin(\omega t + \beta z)\,\vec{a}_y\, dt \quad \text{or}$$

$$\vec{B} = -\frac{D_m\beta}{\varepsilon_0\omega}\cos(\omega t + \beta z)\cdot\vec{a}_y$$

As we know that in free space

$$\frac{\omega}{\beta} = v_p = c = \frac{1}{\sqrt{\mu_0\varepsilon_0}}$$

$$\frac{\omega^2}{\beta^2} = \frac{1}{\mu_0\varepsilon_0}$$

$$\vec{B} = \frac{-D_m\beta\mu_0}{\mu_0\varepsilon_0}\frac{\cos(\omega t + \beta z)}{\omega}\vec{a}_y$$

Replacing

$$\frac{1}{\mu_0\varepsilon_0} \quad \text{with} \quad \frac{\omega^2}{\beta^2}$$

we get

$$\vec{B} = \frac{-\omega\mu_0 D_m}{\beta}(\cos\omega t + \beta z)\ \vec{a}_y$$

Now,

$$\nabla \times E = \begin{vmatrix} \vec{a}_x & \vec{a}_y & \vec{a}_z \\ \frac{\partial}{\partial x} & \frac{\partial}{\partial y} & \frac{\partial}{\partial z} \\ E_x & E_y & E_z \end{vmatrix} \quad \text{or}$$

$$= \left(\frac{\partial E_z}{\partial y} - \frac{\partial E_y}{\partial z}\right)\vec{a}_x - \left(\frac{\partial E_z}{\partial x} - \frac{\partial E_x}{\partial z}\right)\vec{a}_y + \left(\frac{\partial E_y}{\partial x} - \frac{\partial E_x}{\partial y}\right)\vec{a}_z$$

By comparing and equating the like coefficient of j on both sides, we get

$$-B_m(j\omega)e^{j(\omega t-\beta z)} = -\left(\frac{\partial E_z}{\partial x} - \frac{\partial E_x}{\partial z}\right) = \frac{\partial E_x}{\partial z}$$

(As E_z is not function of x, $\partial E_z/\partial x = 0$),

$$\vec{E}_x = -B_m(j\omega)\int e^{j(\omega t-\beta z)}dz$$

$$= \frac{-B_m(j\omega)}{-(j\beta)}e^{j(\omega t-\beta z)}$$

$$= \frac{B_m\omega}{\beta}e^{j(\omega t-\beta z)}$$

7.13 A cylindrical conductor of radius 2×10^{-2} m has an internal magnetic field

$$\vec{H} = (3\times10^4)\left(\frac{R}{2} - \frac{R^2}{6\times10^{-2}}\right)\vec{a}_\phi(\text{A/m})$$

Find the total current in the conductor.

Solution:

We know that from Ampere's circuital law

$$\oint \vec{H}\cdot\vec{dl} = I$$

It is given that $R = 2\times10^{-2}$ (m).

So,

$$\oint \vec{H}\cdot d\vec{l} = (3\times10^4)\left[\frac{2\times10^{-2}}{2} - \frac{(2\times10^{-2})^2}{6\times10^{-2}}\right]\int_0^{2\pi} d\phi$$

$$= 3\times10^4\left[10^{-2} - \frac{4\times10^{-4}}{6\times10^{-2}}\right]\times 2\pi$$

$$= 6\pi\times10^4[0.01 - 0.006]$$

$$= 0.0753\times10^4$$

$$= 753 \text{ A}$$

7.14 Compute the total magnetic flux ϕ crossing the $z=0$ plane in Cylindrical coordinates for $R \le 4\times10^{-2}$ m if

$$\vec{B} = \frac{0.5}{R}\sin^2\phi\vec{a}_z \text{ (Tesla)}$$

Solution:

Magnetic flux ϕ is given as

$$\phi = \int_s \vec{B}\cdot\vec{d}_s$$

$$\overrightarrow{ds} = RdR\ d\phi\ \vec{a}_r$$

$$\phi = \int_s \vec{B}\cdot\overrightarrow{ds} = [0.5\times R]_0^{4\times10^{-2}} \int_0^{2\pi}\left(\frac{1-\cos 2\phi}{2}\right) d\phi$$

$$= 0.02\times\left(\frac{\phi}{2} - \frac{\sin 2\phi}{4}\right)_0^{2\pi}$$

$$= 0.02\times\pi$$

$$= 0.062 \text{ Wb/m}^2$$

7.15 The electric field intensity of a linearly polarized uniform plane wave propagating is +*z* direction in seawater is $E = 200\cos(10^6\pi t)\vec{a}_x$ (V/m) at $z = 0$. The constructive parameters of sea water are $\varepsilon_r = 72$, $\mu_r = 1$ and $\sigma = 4$ (S/m). Determine (i) Attenuation constant, (ii) Phase constant, (iii) Intrinsic impedance and (iv) Phase velocity and skin depth.

Solution:

As

$$\omega = 10^6\pi \text{ rad/s}$$

$$f = \frac{\omega}{2\pi} = 5\times10^5 \text{ Hz}$$

$$\frac{\sigma}{\omega\varepsilon} = \frac{\sigma}{\omega\varepsilon_o\varepsilon_r} = \frac{4}{10^6\pi\left(1/36\pi\times10^{-9}\right)\times72} = 20\ (>>1)$$

Therefore, we can use the formula for a good conductor.

(a) Attenuation constant

$$\alpha = \sqrt{\pi f\mu\sigma} = \sqrt{(5\times10^5\pi\times4\pi\times10^{-7})\times4}$$

$$= 2.80 \text{ Np/m}$$

(ii) Phase constant

$$\beta = \sqrt{\pi f\mu\sigma} = \alpha = 2.80 \text{ Np/m}$$

(iii) Intrinsic impedance

$$\eta = (1+j)\sqrt{\frac{\pi f \mu}{\sigma}}$$

$$= (1+j)\sqrt{\frac{\pi(5\times10^5)\times(4\pi\times10^{-7})}{4}} = 0.56\,e^{\pi/4}$$

(iv) Phase velocity

$$v_p = \frac{\omega}{\beta} = \frac{10^6\,\pi}{2.80} = 357.1\times10^3\ \text{(m/s)}$$

Skin depth

$$\delta = \frac{1}{\alpha} = 357.14\times10^{-3}\,\text{(m)}$$

7.16 Find the angle by which direction of $\vec{E}$ charges, as it crosses the boundary between two dielectrics if $\varepsilon_{r1} = 9$ and $\varepsilon_{r2} = 8$.

Solution:

$\varepsilon_{r1} = 9$, $\varepsilon_1 = 9\varepsilon_0$ and $\varepsilon_2 = 8\varepsilon_0$ or

$$\theta_{ic} = \sin^{-1}\sqrt{\frac{\varepsilon_2}{\varepsilon_1}} = \sin^{-1}\sqrt{\left(\frac{8}{9}\right)}$$

$$\theta_{ic} = 68.4°$$

7.17 In a system, five thousand lines of forces enter a volume of space. One thousand lines of forces leave this volume. Find the charge contained in it.

Solution:

Let q be the charge contained in this volume of space. As per Gauss's Law, the net electric flux $\phi = \dfrac{q_{in}}{\varepsilon_o}$ or

$$= \frac{q}{\varepsilon_o}$$

$$q = \varepsilon_o \phi$$

The number of lines of forces passing normally through the surface are given as

$$= (5000 - 1000)$$

$$= 4000$$

Thus,

$$\phi = 4000$$

Now charge $q = 8.85 \times 10^{-12} \times 4000$ or $q = 3.54 \times 10^{-8}$ C.

7.18 In the above question, find the total dipole-moment of the slab.

Solution:

Total dipole moment

$$= p \times \text{volume}$$
$$= 2.5 \times 0.09$$
$$= 0.225$$

7.19 For a homogeneous medium $\varepsilon_r = 36$ and $\mu_r = 1$. $\vec{E}$ and $\vec{B}$ are $10\pi\, e^{j(\omega t+\beta z)}\vec{a}_x(\text{V/m})$ and $20\mu_o H_o\, e^{j(\omega t+\beta z)}\vec{a}_x(\text{Tesla})$, respectively. Determine ω and H_o if the wavelength λ of the signal is 0.9 m.

Solution:

Velocity of wave propagation in the medium $\varepsilon_r = 36$ is given as

$$v = \frac{C}{\sqrt{\varepsilon_r \mu_r}}$$
$$= \frac{3\times10^8}{\sqrt{36\times1}}$$
$$= 0\cdot5\times10^8\ (\text{m/s})$$

Wave length $(\lambda) = 0.9$ m (given).

$$\omega = 2\pi f$$

Thus,

$$= 2\pi \times \frac{v}{\lambda} \quad \text{or}$$
$$\omega = 2\pi \times \frac{0\cdot5\times10^8}{0\cdot9} \quad \text{or}$$
$$\omega = 3\cdot4\times10^8\,(\text{rad/s})$$

Maxwell's equation gives

$$\nabla\times\vec{E} = -\frac{\partial\vec{B}}{\partial t}$$

Thus,

$$\begin{vmatrix} \vec{a}_x & \vec{a}_y & \vec{a}_z \\ \dfrac{\partial}{\partial x} & \dfrac{\partial}{\partial y} & \dfrac{\partial}{\partial z} \\ E_x & 0 & 0 \end{vmatrix} = -\frac{\partial\vec{B}}{\partial t} \quad \text{or}$$

$$\left|\frac{\partial E_x}{\partial z}\right|\vec{a}_y = -\frac{\partial\vec{B}_y}{\partial t} \quad \text{or}$$

$$\frac{\partial}{\partial z}(10\pi\, e^{j(\omega t+\beta z)})\vec{a}_y = -\frac{\partial}{\partial t}20\mu_o H_o e^{j(\omega t+\beta z)}\,\vec{a}_y \quad \text{or}$$

$$j10\pi\beta\, e^{j(\omega t+\beta z)} = -j2\mu_o\omega H_o e^{j(\omega t+\beta z)}$$

Finally,

$$H_o = 0\cdot 417(\text{A/m}) \quad (\text{considering +ve sign for } H_o)$$

7.20 In a free space, $\vec{E} = 5\pi e^{j(10^6+\beta z)}\,\vec{a}_x$ (V/m) and $\vec{H} = H_o\, e^{j(10^6+\beta z)}\,\vec{a}_y$ (A/m). Find values of H_o and $\beta\,(\beta > 0)$.

Solution:

For $\vec{E}$ and $\vec{H}$ representing a plane wave, we get

$$\frac{\omega}{\beta} = c$$

$$= \frac{1}{\sqrt{\mu_o\varepsilon_o}}$$

$$= 3\times 10^8\,(\text{m/s})$$

It is given that $\vec{E} = 5\pi e^{j(10^6+\beta Z)}\,\vec{a}_x$ (V/m). On comparison with the actual expression, we find

$$\omega = 10^6$$

Thus,

$$\beta = \frac{10^6}{3\times 10^8}$$

$$= \frac{1}{300}(\text{rad/m})$$

Maxwell's equation gives

$$\nabla\times\vec{E} = -\frac{\partial\vec{B}}{\partial t}$$

$$= -\mu_o\frac{\partial\vec{H}}{\partial t}$$

Thus,

$$\begin{vmatrix} \vec{a}_x & \vec{a}_y & \vec{a}_z \\ \frac{\partial}{\partial x} & \frac{\partial}{\partial y} & \frac{\partial}{\partial z} \\ E_x & E_y & E_z \end{vmatrix} = -\mu_o\frac{\partial H_y}{\partial t}$$

However,

$$E_y = E_z = 0$$

Therefore,

$$\left|\frac{\partial E_x}{\partial z}\right|\vec{a}_y = -\mu_o\frac{\partial H_y}{\partial t} \quad \text{or}$$

$$\vec{a}_y \frac{\partial\left[5\pi e^{j(10^6 t+\beta z)}\right]}{\partial z} = -\mu_o \frac{\partial\left[H_o\, e^{j(10^6 t+\beta z)}\right]}{\partial t} \quad \text{or}$$

$$\beta 5\pi\, e^{j(10^6 t+\beta z)}\, \vec{a}_y = -(4\pi\times10^{-7})10^6\, H_o e^{j(10^6 t+\beta z)} \quad \text{or}$$

$$\frac{1}{300}\times5\pi = -4\pi\times10^{-7}\times10^6\times H_o$$

Thus,

$$H_o = -\frac{5}{120}(\text{A/m}) \quad \text{or}$$

$$H_o = 5/120\ (\text{A/m}) \qquad (H_o \text{ must be +ve})$$

UNSOLVED QUESTIONS

7.1 In a free space, $\vec{E} = 8\cos(\omega t - 40x)\,\vec{a}_y$ (V/m). Find, $\vec{J}_d$.

7.2 In air,

$$\vec{E} = \frac{\sin\theta}{r}\cos(4\times10^7 t - \beta r)\,\vec{a}_\phi (V/m)$$

What is the value of β?

7.3 A medium is characterized by $\sigma = 0, \mu = \mu_0$ and $\varepsilon = 5\varepsilon_o$.

$$\vec{E} = 10\sin(10^8 t - 2\beta z)\,\vec{a}_y (\text{V/m})$$

Find $\vec{H}$.

7.4 In a free space,

$$\vec{E} = 20\pi\, e^{j(10^9 t+3\beta z)}\, \vec{a}_z (\text{V/m}) \quad \text{and}$$

$$\vec{H} = H_m\, e^{j(10^9 t+3\beta z)}\, \vec{a}_y (\text{A/m})$$

Find the values of β and H_m.

7.5 In a free space, magnetic field $\vec{H}$ of an EM wave is $0.6\omega\varepsilon_o\cos(\omega t - 20x)\vec{a}_z$ (A/m). Find electric field E and displacement current density $\vec{J}_d$.

7.6 An electric field $\vec{E}$ in a source-free medium is $2{\cdot}5\cos(10^8 - 2\beta z)\vec{a}_x$ (V/m). Find $\vec{B}$ and $\vec{H}$. (Assume that $\varepsilon = \varepsilon_o$, $\mu = \mu_o$ and $\sigma = 0$).

7.7 In a free space, $\vec{H}$ is $0{\cdot}5\cos(4\times10^8 t - kx)\vec{a}_y$ (A/m). What are the values of k, λ, and T?

7.8 A plane wave is travelling in a lossless unbounded medium in the +ve x-direction. It possesses the same permeability as free space. Permittivity is seven times that of free space. Find the phase velocity of the wave. If the electric field intensity E has only the y component of amplitude a of 12 (volt/metre), find the amplitude and directions of magnetic field intensity.

7.9 A plane wave of 15 GHz is travelling in free space. Its amplitude E_x is 15 (V/m). Find
(a) Phase velocity, wavelength and propagation constant,
(b) Amplitude and direction of magnetic field intensity and
(c) Characteristic impedance of medium.

7.10 Electric field intensity of a uniform plane wave travelling in air is 8000 (V/m). It is in the $\vec{a}_y$ direction. The wave is propagating in the $\vec{a}_x$ direction. The wave frequency is 10^7 (rad/s). What are the length, frequency, time period and amplitude of $\vec{H}$?

7.11 An electric field $\vec{E}_y$ is $10\cos(4\pi 10^9 t - \beta x)\vec{a}_y$. It is propagating through a lossless medium characterized by $\mu_r = 1$ and $\varepsilon_r = 90$. The wave frequency is 500 MHz. Find the wavelength and its corresponding magnetic field $\vec{H}$.

7.12 What will be the depth of penetration for Cu at 2 MHz? The conductivity of Cu is 58 (MΩ/m) and the permeability is 1·26 (μH/m).

7.13 What is the propagation constant γ for a material characterized by $\mu_r = 1$, $\varepsilon_r = 10$ and $\sigma = 0{\cdot}15$ (pS/m)? The wave frequency is 2.6 MHz.

7.14 A lossy dielectric material is characterized as $\mu_r = 1$, $\varepsilon_r = 50$ and $\sigma = 30$(S/m). Find the values of attenuation constant and phase constant.

7.15 A non-magnetic material has $\varepsilon_r = 1{\cdot}25$ and $\sigma = 10^{-6}$ (S/m). What are the loss tangent and intrinsic impedance for a 4·5 MHz wave?

7.16 A lossless medium has $\eta = 40\pi$ and $\mu_r = 1$, $\vec{H} = -0{\cdot}2\cos(\omega t - z)\vec{a}_z + 0{\cdot}2\sin(\omega t - z)\vec{a}_z$ (A/m). Find ε_r, ε and $\vec{E}$ in this case.

7.17 A plane wave is travelling in a lossy medium in the $+y$ direction. The lossy medium is characterized by $\varepsilon_r = 3, \mu_r = 1$ and $\sigma = 10^{-3}$ (Ω/m). $\vec{E}$ is $60\cos(10^8 \pi t + \pi/4)\vec{a}_z$ (V/m) at $y = 0$. Find
(a) $\vec{E}$ at ($y = 2$ m, $t = 3$ ns) and
(b) $\vec{H}$ at ($y = 3$ m, $t = 3$ ns)

7.18 A plane of 2-GHZ wave is travelling in air. The peak electric field intensity is 1.5 (Volt/m). It is incident normally on a large sheet of Cu. What will be the average power absorbed per square meter of the Cu sheet?

7.19 List and prove various boundary conditions using Maxwell's equations.

7.20 Find the magnetic field H_2 and its magnitude ($x > 0$) of medium '2' if the magnetic field of medium '1' ($x < 0$) is $H_1 = (2.0\vec{a}_x + 0.5\vec{a}_y - \vec{a}_z)$ (A/m), $\mu_{r1} = 1.0$ and $\mu_{r2} = 2.0$.

(**Answer:** $H_2 = (\vec{a}_x + 0.5\vec{a}_y - \vec{a}_z)$ (A/m) and 1.5 (A/m))

7.21 Find E_2 of medium '2' ($x > 0$), if for medium '1' ($x < 0$) $E_1 = (2\vec{a}_x + 2\vec{a}_y + \vec{a}_z)$(V/m). $\sigma_1 = \sigma_2 = 0$ and $\epsilon_{r1} = 1$ and $\epsilon_{r2} = 2$.

(**Answer:** $E_2 = (\vec{a}_x + 2\vec{a}_y + \vec{a}_z)$(V/m)

SUMMARY

1. Electrostatic fields are denoted by $\vec{E}(x,y,z)$. Magnetic fields are denoted by $\vec{H}(x,y,z)$. Hence, varying *EM* fields are represented as $\vec{E}(x,y,z,t)$ and $\vec{H}(x,y,z,t)$.

2. Maxwell's equations completely describe electromagnetic fields. That is why sometimes these equations are also called electromagnetic field equations.

3. Electrostatic fields are produced by static electric charges.

4. Magnetostatic fields are produced by moving electric charges.

5. A wave is an energy carrier. It is a function of time and space.

6. In *EM* waves, time-varying magnetic field generates a time-varying electric field, which again generates time-varying magnetic field and so on.

7. Waves are used in radio, TV, mobiles, radar, etc.

8. Concepts of displacement current and displacement current density are used in Maxwell's equations.

9. In empty space, the conduction current in zero and magnetic fields are due to displacement currents, only.

10. As per Maxwell's first equation, $\nabla, \vec{D} = \rho v$.

11. As per Maxwell's second equation, $\nabla, \vec{B} = 0$.

12. As per Maxwell's third equation,

$$\nabla \cdot \vec{H} = \vec{J} = \vec{J}_d + \frac{\partial \rho}{\partial t}$$

13. As per Maxwell's fourth equation,

$$\nabla \times \vec{E} - \frac{\partial \vec{B}}{\partial t}$$

14. Maxwell's equations for sinusoidal time-varying field in differential form is given as

$$\nabla \times \vec{H} = \vec{J} + j\omega\vec{D}, \nabla \times \vec{E} = -j\omega\vec{B},\ \nabla \cdot\ \vec{B} = 0 \text{ and } \nabla \cdot \vec{D} = \rho.$$

15. Maxwell's equations for sinusoidal time-varying field in integral form is given as

$$\oint \vec{D} \cdot \vec{ds} = \int \rho v \cdot dv, \oint \vec{\mathrm{B}} \cdot \vec{ds} = 0, \oint \vec{E} \cdot \vec{dL} = -j\omega \int \vec{B} \cdot \vec{ds} \text{ and } \oint \vec{\mathrm{H}} \cdot \vec{dL} = \int (\vec{j} + j\omega\vec{D}) \cdot \vec{ds}.$$

16. Boundary condition '1' using Maxwell's equations is given as

$$E_{\tan\text{'1'}} = E_{\tan\text{'2'}}$$

17. Boundary condition '2' using Maxwell's equations is given as

$$H_{\tan\text{'1'}} - H_{\tan\text{'2'}} = J_s$$

18. Boundary condition '3' using Maxwell's equations is given as

$$B_{n'1'} = B_{n'2'}$$

19. Boundary condition '4' using Maxwell's equation is given as

$$D_{n'1'} - D_{n'2'} = \rho_s$$

MULTIPLE-CHOICE QUESTIONS

1. Maxwell's equations shelter on __________ law(s).
(a) Faraday's (b) Gauss's (c) Ampere's (d) All of these

2. Conduction current through a wire is __________ displacement current in capacitor.
(a) Same as (b) Different from (c) Twice of (d) None of these

3. In empty space, conduction current is __________.
(a) Infinity (b) Unity (c) Zero (d) None of these

4. As per Maxwell's first equation __________.
(a) $\nabla\times\vec{D}=\rho_v$ (b) $\nabla\cdot\vec{D}=\rho_v$ (c) $\nabla\times\vec{D}=\rho_s$ (d) $\nabla\cdot\vec{D}=\rho_s$

5. Maxwell's second equation gives __________.
(a) $\nabla\times\vec{B}=0$ (b) $\nabla\cdot\vec{B}=0$ (c) $\nabla\times\vec{B}=1$ (d) $\nabla\cdot\vec{B}=1$

6. According to Ampere's Circuital Law __________.
(a) $\oint_v \vec{H}\cdot\overrightarrow{dv}=I_{enc}$ (b) $\oint_s \vec{H}\cdot\overrightarrow{ds}=I_{enc}$
(c) $\oint_L \vec{H}\cdot\overrightarrow{dl}=I_{enc}$ (d) None of these

7. Maxwell's third equation gives __________.
(a) $\nabla\times\vec{H}=\vec{J}_d+\frac{\partial\rho}{\partial t}$ (b) $\nabla\cdot\vec{H}=J_d\frac{\partial\rho}{\partial t}$
(c) $\nabla\cdot\vec{H}=\vec{J}_d+\frac{\partial\rho}{\partial t}$ (d) $\nabla\times\vec{H}=J_d\frac{\partial\rho}{\partial t}$

8. Maxwell's fourth equation gives __________.
(a) $\nabla\cdot\vec{E}=-\frac{\partial\vec{B}}{\partial t}$ (b) $\nabla\times\vec{E}=-\frac{\partial\vec{B}}{\partial t}$ (c) $\nabla\cdot\vec{E}=\frac{\partial\vec{B}}{\partial t}$ (d) $\nabla\times\vec{E}=\frac{\partial\vec{B}}{\partial t}$

9. emf is closed __________ integral of non-conservational electric field that is generated by battery.
(a) Line (b) Surface (c) Volume (d) None of these

10. Maxwell's equations in __________ form give unformation at points of discontinuity in electro-magnetic fields.
(a) Differential (b) Integral (c) Algebraic (d) None of these

11. Which of the following is a boundary condition?

(a) $E_{\tan'1'} = \varepsilon_0 E_{\tan'2'}$ (b) $E_{\tan'1'} = \frac{d}{dt} E_{\tan'2'}$ (c) $E_{\tan'1'} = E_{\tan'2'}$ (d) None of these

12. Another boundary condition using Maxwell's equations is given as ________.

(a) $H_{\tan'1'} + H_{\tan'2'} = 0$ (b) $H_{\tan'1'} - H_{\tan'2'} = 0$
(c) $H_{\tan'1'} + H_{\tan'2'} = J_s$ (d) $H_{\tan'1'} - H_{\tan'2'} = J_s$

13. At the point of discontinuity, ________ component of magnetic flux density is continuous.

(a) Tangential (b) Normal (c) None of these (d) Cannot say

14. For those surfaces where surface charge density is ________ normal component of electric flux density is continuous.

(a) Infinity (b) Unity (c) Zero (d) None of these

15. Which of the following vector form of boundary condition is incorrect?

(a) $\vec{a}_n (E_1 - E_2) = 0$ (b) $\vec{a}_n (B_1 - B_2) = 0$ (c) $\vec{a}_n (D_1 - D_2) = 0$ (d) None of these

16. Waves are used in ________.

(a) TV (b) Radio (c) Radar (d) All of these

17. Current through a capacitor is expressed as ________.

(a) $\frac{\varepsilon A}{d} \frac{dv}{dt}$ (b) $\frac{d}{\varepsilon A} \frac{dv}{dt}$ (c) $\frac{\varepsilon}{Ad} \frac{dv}{dt}$ (d) $\frac{Ad}{\varepsilon} \frac{dv}{dt}$

18. Displacement current density $\vec{J}_d$ ________ current passing through a capacitor.

(a) Represents (b) Does not represent (c) Is the same as (d) None of these

19. Line integral of an electric field around a closed path is ________.

(a) infinity (b) unity (c) zero (d) None of these

20. ________ are caused by accelerated charges

(a) Time-varying fields (b) Waves
(c) Both (a) and (b) (d) None of these

ANSWERS TO MULTIPLE-CHOICE QUESTIONS

(1) d; (2) a; (3) c; (4) b; (5) b; (6) c; (7) a; (8) b; (9) a;
(10) b; (11) c; (12) d; (13) b; (14) c; (15) a; (16) d; (17) a; (18) b;
(19) c; (20) c

Electromagnetic Waves

8

8.1 INTRODUCTION

Space variations of electric and magnetic field components are associated time variations of magnetic and electric field components, respectively. This interdependence gives rise to "electromagnetic wave propagation".

8.2 WAVE EQUATIONS FOR WAVES TRAVELLING IN SPACE

General wave equations using Maxwell's equations are derived. In electromagnetics (EM), the basic and fundamental relations that need to be satisfied are

$$\nabla \times \vec{H} = \vec{J} + \frac{\partial \vec{D}}{\partial t}$$

$$\nabla \times \vec{E} = -\frac{\partial \vec{B}}{\partial t}$$

$$\nabla \cdot \vec{D} = \rho_v$$

$$\nabla \cdot \vec{B} = 0$$

$$\vec{D} = \varepsilon \vec{E}$$

$$\vec{B} = \mu \vec{H} \quad \text{and}$$

$$\vec{J} = \sigma \vec{E}$$

Maxwell's equations if solved are simultaneous equations, which give the laws that must be obeyed by both $\vec{E}$ and $\vec{H}$.

8.3 CHARACTERISTICS OF WAVES

These characteristics are as follows:

(1) A amplitude of wave. A has the same unit as E_y. A is basically a time harmonic.
(2) $(\omega t - \beta x)$ is the phase (in radians) of the wave.
(3) ω is angular frequency.
(4) β is phase constant. It is also called as "wave number" (in radians/meter).

E varies with both, viz. time t and space variable z. E can be plotted as a function of t, while keeping z constant one time and vice versa the next time. Plots of E (z, t = constant) and E (t, z = constant) are depicted in Figures 8.1 and 8.2, respectively. From Figure 8.1, we find that wave takes distance λ to repeat itself. λ is called the "wavelength" (meters). From Figure 8.2, we find that wave takes time T to repeat itself. T is called "period" (seconds). Thus, time T is taken by the wave to travel distance λ at speed u. Then

$$\lambda = uT$$

Also, $T = (1/f)$. f is the frequency. f denotes the number of cycles per second of the wave. f is expressed in Hertz (Hz). Thus,

$$u = f\lambda$$

Because of the relationship between wavelength and frequency, the position of a radio station can be found very easily within its band either with the help of frequency or wavelength. Usually, "frequency" is preferred over wavelength. It is so because

$$\omega = 2\pi f \quad \text{and}$$

$$\beta = \frac{\omega}{u}$$

Also,

$$T = \frac{1}{f} = \frac{2\pi}{\omega}$$

Thus,

$$\beta = \frac{2\pi}{\lambda}$$

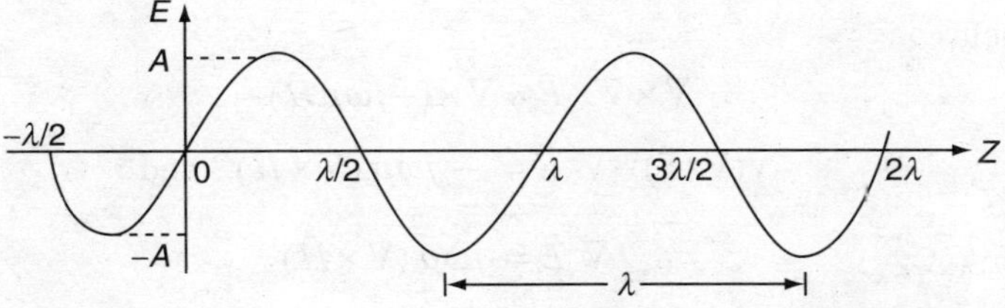

Figure 8.1 *Variation of E (z, t = constant)*

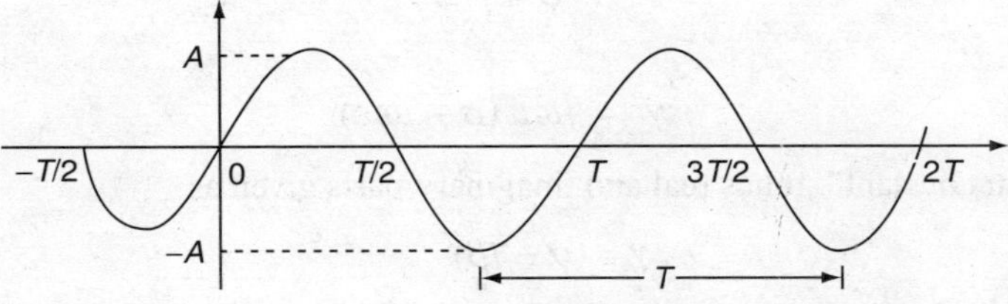

Figure 8.2 *Variation of E (t, z = constant)*

Wave $E = A\sin(\omega t - \beta z)$ travels with u in the z-direction. Consider a fixed point P on the wave. We depict this equation at times $t = 0$, $T/4$ and $T/2$ as shown in Figure 8.2. As can be seen in Figure 8.2, the wave advances with time. Point P moves in the $+z$ direction. Point P has constant phase.
Thus,

$$\omega t - \beta z = \text{Constant} \quad \text{or}$$

$$\frac{\partial z}{\partial t} = \frac{\omega}{\beta} = u$$

As expressed above, wave travels with velocity u in the $+z$ direction. Similarly, one can show that $B\sin(\omega t + \beta z)$ in the equation $E = B\,e^{j(\omega t + \beta z)}$ travels at velocity u in the $-z$ direction.

8.4 WAVE PROPAGATION IN LOSSY DIELECTRIC MEDIUM

Wave propagation in lossy dielectric is considered as a general case. From it, wave propagation in other media types can be derived (special cases). In a lossy dielectric medium, an EM wave loses power during its propagation. Losses mainly happen due to poor conduction.

A lossy dielectric medium is a partially conducting medium (imperfect conductor or imperfect dielectric medium). It has $\sigma \neq 0$. In the case of a lossless dielectric medium (perfect or good dielectric medium), $\sigma = 0$.

Consider a linear, isotropic and homogeneous lossy dielectric medium. It is charge free $(\rho_v = 0)$. Maxwell's equation in this case is

$$\nabla \times \vec{H} = (\sigma + j\omega\varepsilon)\vec{E} \qquad \text{(Time factor } e^{j\omega t} \text{ is suppressed.)}$$

$$\nabla \times \vec{E} = -j\omega\mu\vec{H}$$

$$\nabla \cdot \vec{E} = 0 \quad \text{and}$$

$$\nabla \cdot \vec{H} = 0$$

Curl of both side yields

$$\nabla \times \nabla \times \vec{E} = \nabla \times (-j\omega\mu\vec{H})$$

$$\nabla(\nabla \cdot \vec{E}) - \nabla^2\vec{E} = -j\omega\mu\,(\nabla \times \vec{H}) \quad \text{and}$$

$$\nabla^2\vec{E} = j\omega\mu\,(\nabla \times \vec{H})$$

Similarly,

$$\nabla^2\vec{E} = j\omega\mu\,(\sigma + j\omega\varepsilon)\vec{E} \quad \text{or}$$

$$\nabla^2\vec{E} = \gamma^2\vec{E}$$

Here,

$$\gamma^2 = j\omega\mu\,(\sigma + j\omega\varepsilon)$$

γ is the "propagation constant". It has real and imaginary parts given as

$$\gamma = (\alpha + j\beta)$$

γ is called the "attenuation constant". It has units (neper/m). β is the phase constant (rad/m).

The solution of a wave equation is

$$E_y = E_m \; e^{j\omega t} \; e^{\pm\gamma x} \qquad \text{(considering only } E_y \text{ component)} \quad \text{or}$$

$$E_y = E_m \; e^{-\alpha x} \; e^{j(\omega t-\beta x)}$$

Here,

$$-\gamma x = (-\alpha x - j\beta x)$$

Factor e^{-ax} is "attenuation" of wave. Phase constant β (for a lossy dielectric medium) is quite different from "phase constant". In the case of a perfect dielectric medium having the same dielectric constant and permeability, β increases as conductivity increases. Wavelength corresponding to the given frequency becomes small. Also, velocity of propagation is less. Thus,

$$\gamma^2 = j\omega\mu(\sigma + j\omega\varepsilon) \quad \text{or}$$

$$\gamma = \sqrt{j\omega\mu(\sigma + j\omega\varepsilon)} \quad \text{or}$$

$$= j\omega\sqrt{\mu\varepsilon}\sqrt{1 - j\frac{\sigma}{\omega\varepsilon}}$$

Y can be separated into α and β as

$$(\alpha + j\beta)^2 = j\omega\mu(\sigma + j\omega\varepsilon)$$

$$(\alpha^2 - \beta^2) = \omega^2\mu\varepsilon \quad \text{and}$$

$$2\alpha\beta = \omega\mu\sigma$$

On solving, we get

$$\alpha = \omega\sqrt{\frac{\mu\varepsilon}{2}\left[\sqrt{1+\left(\frac{\sigma}{\omega\varepsilon}\right)^2} - 1\right]}$$

$$\beta = \omega\sqrt{\frac{\mu\varepsilon}{2}\left[\sqrt{1+\left(\frac{\sigma}{\omega\varepsilon}\right)^2} + 1\right]}$$

Phase velocity

$$(v_P) = \left(\frac{\omega}{\beta}\right)$$

Thus,

$$v_P = \frac{\omega}{\omega\sqrt{\frac{\mu\varepsilon}{2}\left[1+\left(\frac{\sigma}{\omega\varepsilon}\right)^2 + 1\right]}} \quad \text{or}$$

$$v_P = \frac{1}{\sqrt{\frac{\mu\varepsilon}{2}\left[\sqrt{1+\left(\frac{\sigma}{\omega\varepsilon}\right)^2}+1\right]}}$$

Wavelength $\lambda = (2\pi/\beta)$. It is

$$\lambda = \frac{1}{f\sqrt{\frac{\mu\varepsilon}{2}\left[\sqrt{1+\left(\frac{\sigma}{\omega\varepsilon}\right)^2}+1\right]}}$$

Here,

$$f = \frac{\omega}{2\pi}$$

E_y component of $\vec{E}$ (varying wrt x) is given as

$$\nabla \times \vec{E} = \frac{\partial E_y}{\partial x} \cdot \vec{a}_z \quad \text{or}$$

$$\nabla \times \vec{E} = -\gamma E_m \, e^{-\gamma x} \vec{a}_z$$

Putting $\Delta \times \vec{E}$ from Maxwell's equation, we get

$$j\omega\mu H = -\gamma E_m \cdot e^{-\gamma x} \vec{a}_z$$

$\vec{H}$ has H_z component corresponding to E_y. Thus,

$$H_z = \frac{\gamma}{j\omega\mu} E_y$$

Intrinsic impedance η is

$$\eta = \frac{E_y}{H_z}$$

$$= \frac{j\omega\mu}{\gamma}$$

Thus,

$$\eta = \sqrt{\frac{j\omega\mu}{\sigma + j\omega\varepsilon}}$$

8.5 WAVE PROPAGATION IN A GOOD CONDUCTOR

For a good conductor

$$\sigma >> \omega\varepsilon \quad \text{or}$$

$$\frac{\sigma}{\omega\varepsilon} >> 1$$

Propagation constant γ is

$$\gamma^2 = j\omega\mu\ (\sigma + j\omega\varepsilon) \quad \text{or}$$

$$\gamma^2 = j\omega\mu\sigma\left(1 + j\frac{\omega\varepsilon}{\sigma}\right) \quad \text{or}$$

$$\gamma^2 \approx j\omega\mu\sigma \quad \text{or}$$

$$\gamma = \sqrt{j\omega\mu\sigma} \quad \text{or}$$

$$= \sqrt{\omega\mu\sigma}\ \ \angle 45^\circ \quad \text{or}$$

$$= \sqrt{(\omega\mu\sigma)/2} + j\sqrt{(\omega\mu\sigma)/2}$$

From here, we get

$$\alpha = \beta = \sqrt{\frac{\omega\mu\sigma}{2}}$$

$$= \sqrt{\pi f\,\mu\sigma}$$

Velocity of propagation

$$v_p = \left(\frac{\omega}{\beta}\right) \quad \text{or}$$

$$v_p = \sqrt{\left(\frac{2\omega}{\mu\sigma}\right)}$$

Intrinsic impedance is

$$\eta = \sqrt{\frac{j\omega\mu}{\sigma + j\omega\varepsilon}}$$

$$\approx \sqrt{\frac{j\omega\mu}{\sigma}}$$

$$\eta = \sqrt{\frac{\omega\mu}{\sigma}} \quad \angle 45^\circ$$

8.6 WAVE PROPAGATION IN GOOD DIELECTRICS

For a good dielectric medium

$$\sigma << \omega\varepsilon \quad \text{or}$$

$$\frac{\sigma}{\omega\varepsilon} << 1$$

As per the Binominal theorem, we get

$$(1+x)^n \quad = 1+nx+\frac{n(n-1)}{2!}x^2+\frac{n(n-1)(n-2)}{2!}x^3+\cdots$$

Alternatively, we may also write

$$\sqrt{\left(1+\frac{\sigma^2}{\omega^2\varepsilon^2}\right)} \quad \cong 1+\frac{\sigma^2}{2\omega^2\varepsilon^2}$$

Further,

$$\alpha = \omega\sqrt{\frac{\mu\varepsilon}{2}\left[\sqrt{1+\left(\frac{\sigma}{\omega\varepsilon}\right)^2}-1\right]} \quad \text{or}$$

$$\approx \omega\sqrt{\frac{\mu\varepsilon}{2}\left(1+\frac{\sigma^2}{2\omega^2\varepsilon^2}-1\right)}$$

Thus,

$$\alpha = \frac{\sigma}{2}\sqrt{\frac{\mu}{\varepsilon}} \quad \text{and}$$

$$\beta = \omega\sqrt{\frac{\mu\varepsilon}{2}\left[\sqrt{1+\left(\frac{\sigma}{\omega\varepsilon}\right)^2}+1\right]} \quad \text{or}$$

$$\beta \approx \omega\sqrt{\frac{\mu\varepsilon}{2}\left[1+\frac{\sigma^2}{2\omega^2\varepsilon^2}+1\right]} \quad \text{or}$$

$$\beta \approx \omega\sqrt{\mu\varepsilon}\sqrt{\left[1+\frac{\sigma^2}{4\omega^2\varepsilon^2}\right]} \quad \text{or}$$

$$\beta \approx \omega\sqrt{\mu\varepsilon}\left[1+\frac{\sigma^2}{8\omega^2\varepsilon^2}\right]$$

Velocity of wave

$$v = \frac{\omega}{\beta} \quad \text{or}$$

$$v = \frac{\omega}{\omega\sqrt{\mu\varepsilon}\left[1+\frac{\sigma^2}{8\omega^2\varepsilon^2}\right]} \quad \text{or}$$

$$v = \frac{1}{\sqrt{\mu\varepsilon}}\left[1 - \frac{\sigma^2}{8\omega^2\varepsilon^2}\right]$$

The effect of a small amount of loss slightly reduces wave propagation velocity.
Intrinsic impedance

$$\eta = \sqrt{\frac{j\omega\mu}{(\sigma + j\omega\varepsilon)}} \quad \text{or}$$

$$= \sqrt{\frac{\mu}{\varepsilon}\left(\frac{1}{1+\dfrac{\sigma}{j\omega\varepsilon}}\right)} \quad \text{or}$$

$$= \sqrt{\frac{\mu}{\varepsilon}}\left(1+\frac{\sigma}{j\omega\varepsilon}\right)^{-1/2} \quad \text{or}$$

$$= \sqrt{\frac{\mu}{\varepsilon}}\left(1-\frac{\sigma}{2j\omega\varepsilon}\right) \quad \text{or}$$

$$\eta = \sqrt{\frac{\mu}{\varepsilon}}\left(1+\frac{j\sigma}{2\omega\varepsilon}\right)$$

Loss adds a small reactive component to intrinsic impedance η. The above approximations are valid in the case $(\sigma/\omega\varepsilon) < 0{\cdot}1$ and for a perfect dielectric medium $(\sigma/\omega\varepsilon) \le 0.01$.

8.7 WAVE PROPAGATION IN FREE SPACE

For free space, we know that

$$\sigma = 0$$

$$\varepsilon = \varepsilon_o \quad \text{and}$$

$$\mu = \mu_o$$

Thus,

$$\alpha = 0$$

$$\beta = \omega\sqrt{\mu_o\varepsilon_o}$$

$$u = \frac{1}{\sqrt{\mu_o\varepsilon_o}} = 3\times10^8 \text{ (m/s)} \quad \text{and}$$

$$\lambda = \frac{2\pi}{\beta} = \frac{2\pi}{\omega\sqrt{\mu_o\varepsilon_o}}$$

8.8 DEPTH OF PENETRATION—SKIN DEPTH

In a high conductivity medium, wave is attenuated as it progresses. It happens due to the losses that occur in the propagation medium. In good conductors, the "rate-of-attenuation" is quite large. Wave penetrates a very short distance before reducing to zero "skin depth", or "depth-of-penetration" is a measure of depth till which an EM wave can penetrate in a given medium. Skin depth is denoted by δ. Its unit is meter. Let the wave propagate in the $+x$ direction. Wave attenuation is given as

$$E = E_o\, e^{-\alpha x}$$

where α is the attenuation constant and E_o the amplitude of the wave at the surface of the conducting medium. Further, at

$$x = \delta$$

$$E = E_o\, e^{-\infty\delta}$$

On comparison, we get

$$E_o\, e^{-\infty\delta} = E_o\, e^{-1} \quad \text{or}$$

$$\alpha\,\delta = 1 \quad \text{or}$$

$$\delta = (1/\alpha)$$

Thus,

$$\delta = \frac{1}{\omega\sqrt{\dfrac{\mu\varepsilon}{2}\left[\sqrt{1+\dfrac{\sigma^2}{\omega^2\varepsilon^2}}-1\right]}}$$

In the case of a good conductor,

$$\frac{\sigma}{\omega\varepsilon} >> 1$$

Thus,

$$\delta = \frac{1}{\alpha} \approx \sqrt{\frac{2}{\omega\mu\sigma}}$$

$$= \sqrt{\frac{1}{\pi f \mu\sigma}}$$

[Please note that at 1 MHz, δ is 25 cm into seawater and 7.1 m into fresh water.]

8.9 THE POYNTING VECTOR AND POYNTING THEOREM

EM waves transport energy from one point to another point (Maxwell's equation). The rate of energy transportation is obtained as

$$\nabla \times \vec{E} = -\mu \frac{\partial \vec{H}}{\partial t} \quad \text{and}$$

$$\nabla \times \vec{H} = \sigma \vec{E} + \frac{\varepsilon \partial \vec{E}}{\partial t}$$

Further,

$$\vec{E}\cdot(\nabla\times\vec{H}) = \sigma\vec{E}\cdot\vec{E} + \varepsilon E\cdot\frac{\partial\vec{E}}{\partial t}$$

We know that

$$\nabla\cdot(\vec{A}\times\vec{B}) = \vec{B}\cdot(\nabla\times\vec{A}) - \vec{A}\cdot(\nabla\times\vec{B})$$

Put

$$\vec{A} = \vec{H} \quad \text{and}$$

$$\vec{B} = \vec{E}$$

Thus,

$$\nabla(\vec{H}\times\vec{E}) = \vec{E}\cdot(\nabla\times\vec{H}) - \vec{H}\cdot(\nabla\times\vec{E}) \quad \text{or}$$

$$\vec{E}\cdot(\nabla\times\vec{H}) = \nabla\cdot(\vec{H}\times\vec{E}) + \vec{H}\cdot(\nabla\times\vec{E})$$

Putting $\vec{E}\cdot(\nabla\times\vec{H})$ in Maxwell's equation, we find

$$\nabla\cdot(\vec{H}\times\vec{E}) + \vec{H}\cdot(\nabla\times\vec{E}) = \sigma E^2 + \vec{E}\cdot\varepsilon\frac{\partial\vec{E}}{\partial t}$$

However,

$$\vec{H}\cdot(\nabla\times\vec{E}) = \vec{H}\cdot\left(-\mu\frac{\partial\vec{H}}{\partial t}\right) = -\frac{\mu}{2}\cdot\frac{\partial}{\partial t}(\vec{H}\cdot\vec{H})$$

Thus,

$$\nabla\cdot(\vec{H}\times\vec{E}) = -\vec{H}\cdot\left(\mu\frac{\partial\vec{H}}{\partial t}\right) = \sigma E^2 + \varepsilon\frac{\partial E^2}{\partial t}$$

$$\nabla\cdot(\vec{H}\times\vec{E}) = \frac{\mu}{2}\frac{\partial H^2}{\partial t} + \sigma E^2 + \frac{\varepsilon\partial E^2}{\partial t} \quad \text{or}$$

$$\nabla\cdot(\vec{E}\times\vec{H}) = -\frac{\mu}{2}\frac{\partial H^2}{\partial t} - \frac{\varepsilon\partial E^2}{\partial t} - \sigma E^2$$

Taking volume integral, we get

$$\int_v \nabla\cdot(\vec{E}\times\vec{H})\,dv = -\frac{\partial}{\partial t}\int_v\left[\frac{1}{2}\varepsilon E^2 + \frac{1}{2}\mu H^2\right]dv - \int_v \sigma E^2\,dv$$

As per the Divergence Theorem, we get

$$\oint_s \cdot(\vec{E}\times\vec{H})\cdot d\vec{s} = -\frac{\partial}{\partial t}\int_v\left[\frac{1}{2}\varepsilon E^2 + \frac{1}{2}\mu H^2\right]dv - \int_v \sigma E^2\,dv$$

where $\oint_s \nabla\cdot(\vec{E}\times\vec{H})\cdot d\vec{s}$ is the total power leaving the volume.

$-\frac{\partial}{\partial t}\int_v\left[\frac{1}{2}\varepsilon E^2 + \frac{1}{2}\mu H^2\right]dv$ is the rate of decrease in energy (this energy is stored in electric and magnetic fields.)

$\int_v \sigma E^2 \, dv$ is the Ohmic dissipated power.

The above expression is called "Poynting Theorem". Poynting vector $\vec{P}\,(=\vec{E}\times\vec{H})$ is instantaneous power density (W/m^2). Power flows in the direction perpendicular to $\vec{E}$ and $\vec{H}$. It is in the direction of vector $\vec{E}\times\vec{H}$.

According to the Poynting Theorem, the net power flowing out of a given volume (v) is equal to the time rate of decrease in energy stored in v minus losses (mainly conduction). Thus,

$$\int_v (\sigma E^2)\, dv = -\frac{\partial}{\partial t}\int_v \left[\frac{1}{2}\mu H^2 + \frac{1}{2}\varepsilon E^2\right] dv - \oint_s (\vec{E}\times\vec{H})\cdot d\vec{s}$$

Thus, the rate of energy dissipation in volume v is equal to the rate at which stored EM energy is decreasing in v plus inward rate of flow of energy through the surface of the given volume. Consider a plane wave, which is traveling in the x-direction. Thus,

$$\vec{E} = E_y \vec{a}_y \quad \text{and}$$

$$\vec{H} = H_z \vec{a}_z$$

Further,

$$\vec{P} = \vec{E}\times\vec{H}$$

$$= E_y \vec{a}_y \times H_z \vec{a}_z \quad \text{or}$$

$$\vec{P} = P\,\vec{a}_x$$

where P is the magnitude of the "Poynting Vector". In perfect dielectric materials, $\vec{E}$ and $\vec{H}$ fields are described as

$$E_y = E_m \cos(\omega t - \beta x) \quad \text{and}$$

$$H_z = \frac{E_m}{\eta}\cos(\omega t - \beta x)$$

Thus,

$$P_x = \frac{E_m^2}{\eta}\cos^2(\omega t - \beta x)$$

Integrating over a time period and dividing by T as

$$P_{x_{ave}} = \frac{1}{T}\int_o^T \frac{E_m^2}{\eta}\cos^2(\omega t - \beta x)\, dt$$

It gives us the time average power density. Further, putting $T = (1/f)$ above, we get

$$P_{x_{ave}} = f\int_o^{(1/f)} \frac{E_m^2}{\eta}\cos^2(\omega t - \beta x)\, dt$$

$$= \frac{f E_m^2}{\eta} \int_0^{(1/f)} [1+\cos 2\,(\omega t - \beta x)]\, dt$$

$$= \frac{f E_m^2}{\eta} \left[t + \frac{1}{2\omega} \sin\,(2\omega t - \beta x) \right]_0^{(1/f)}$$

Thus,

$$P_{x_{ave}} = \frac{1}{2} \frac{E_m^2}{\eta}\ (\text{W/m}^2)$$

As a general case, we get

$$P_{x_{ave}} = \frac{1}{2} E_m H_m$$

Cross product defines the Poynting Vector. Fields are supposed to be in real form. If $\vec{E}$ and $\vec{H}$ are in complex form with a common time dependence $e^{j\omega}$, then the average value of $\vec{P}$ is

$$\vec{P}_{ave} = \frac{1}{2} \text{Re}\ (\vec{E} \times \vec{H}^*)$$

where $\vec{H}^*$ is the complex conjugate of $\vec{H}$. Assume that

$$\vec{E}\,(x,t) = E_o\, e^{-\alpha z} \cos\,(\omega t - \beta x)\vec{a}_y$$

and also $\eta = |\eta| \angle \theta_\eta$ (for lossy dielectric medium). Then we get

$$\vec{H}\,(x,t) = \frac{E_0^2}{|\eta|} e^{-\alpha z} \cos\,(\omega t - \beta x - \theta_n)\vec{a}_z \quad \text{or}$$

$$\vec{P}\,(x,t) = \frac{E_0^2}{|\eta|} e^{-2\alpha x} \cos\,(\omega t - \beta x) \cdot \cos\,(\omega t - \beta x - \theta_n)\vec{a}_x \quad \text{or}$$

$$\vec{P}\,(x,t) = \frac{E_0^2}{2|\eta|} e^{-2\alpha x} \left[\cos\theta_\eta + \cos\,(2\omega t - 2\beta x - \theta_\eta)\right]\vec{a}_x$$

Here,

$$\cos A\ \cos B = \frac{1}{2} [\cos\,(A - B) + \cos(A + B)]$$

Power density has a DC component. Power density also has a second harmonic component. Thus, the time average value of Poynting Vector is given as

$$P_{ave}(x) = \frac{1}{2} \frac{E_o^2}{|\eta|} e^{-2\alpha x} \cos\theta_\eta$$

8.10 REFLECTION OF A PLANE WAVE AT NORMAL INCIDENCE

Consider a plane EM wave incident normally on the surface of a perfect dielectric as shown in Figure 8.3. Part of this energy is transmitted and the other part of energy is reflected.

A perfect dielectric medium has zero conductivity. Therefore, there is no loss of power during propagation through a dielectric medium. Subscripts i, r and t in Figure 8.3 refer to incident, reflected and transmitted waves, respectively. Let a plane wave propagating in the $+x$ direction be incident normally on boundary $x = 0$ between medium 1 ($x < 0$; characterized by $\sigma_1, \varepsilon_1, \mu_1$) and medium 2 ($x > 0$; characterized by $\sigma_2, \varepsilon_2, \mu_2$). It is shown in Figure 8.4.

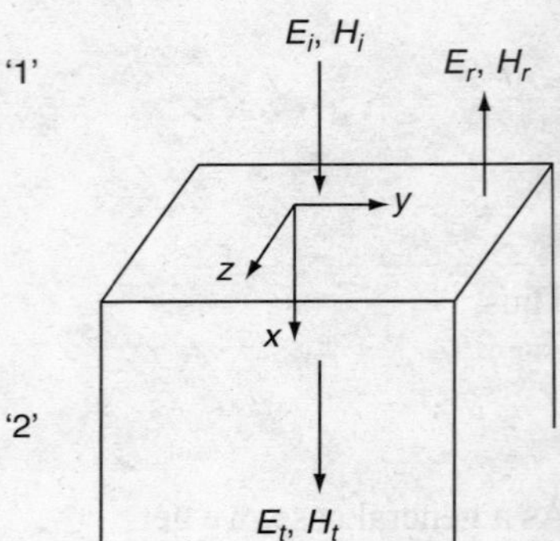

Figure 8.3 *Plane EM wave incident on a surface*

$\vec{E}_i, \vec{H}_i$ travels along $+\vec{a}_x$ in medium 1. Thus,

$$\vec{E}_i(x,t) = E_{im}\, e^{-\gamma_1 x}\vec{a}_y \qquad \text{(time factor } e^{j\omega t} \text{ suppressed) and}$$

$$\vec{H}_i(x,t) = \frac{E_{im}}{\eta_1} e^{-\gamma_1 x}\vec{a}_z$$

$(\vec{E}_r, \vec{H}_r)$ travels along $-\vec{a}_x$ in medium 1. Thus,

$$\vec{E}_r(x, t) = E_{rm}\, e^{+\gamma_1 x}\vec{a}_y \quad \text{and}$$

$$\vec{H}_r(x, t) = -\frac{E_{rm}}{\eta_1} e^{+\gamma_1 x}\vec{a}_z$$

where $\vec{E}_r(x, t)$ is along the $\vec{a}_y$ direction. For normal incident $\vec{E}_i$, $\vec{E}_r$ and $\vec{E}_t$ have the same polarization.

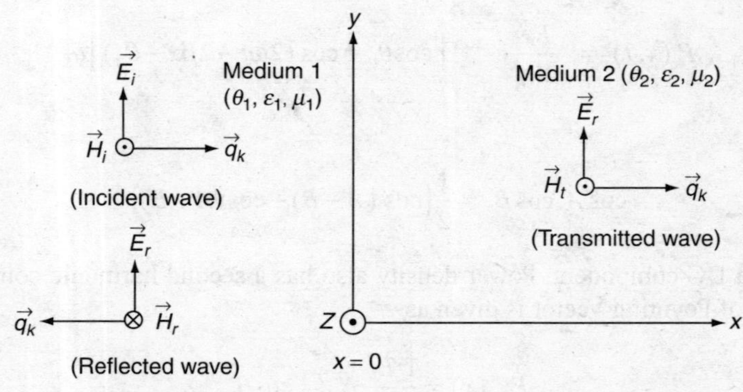

Figure 8.4 *Plane wave propagation in two mediums*

$(\vec{E}_t,\ \vec{H}_t)$ travels along $+\vec{a}_x$ in medium 2. Thus,

$$\vec{E}_t(x,t) = E_{tm}\, e^{-\gamma_2 x}\vec{a}_y \quad \text{and}$$

$$\vec{H}_t(x,t) = \frac{E_{tm}}{\eta} e^{-\gamma_2 x}\vec{a}_z$$

E_{im}, E_{rm} and E_{tm} are magnitudes of incident, reflected and transmitted electric fields at $x = 0$, respectively. Figure 8.4 shows that total field in medium 1 comprises incident and reflected fields. Medium 2 comprises transmitted field only. Thus, we get

$$\vec{E}_1 = \vec{E}_i + \vec{E}_r \quad \text{and}$$

$$\vec{H}_1 = \vec{H}_i + \vec{H}_r$$

Also,

$$\vec{E}_2 = \vec{E}_t \quad \text{and}$$

$$\vec{H}_2 = \vec{H}_t$$

At interface $x = 0$, boundary conditions require that tangential components of both fields $\vec{E}$ and $\vec{H}$ should be continuous. The waves are transverse, and also $\vec{E}$ and $\vec{H}$ fields are tangential to the interface. Thus, at $z = 0$, we find

$$\vec{E}_{1\tan} = \vec{E}_{2\tan}$$

It gives

$$\vec{E}_i(0) + \vec{E}_r(0) = \vec{E}_t(0) \quad \text{or}$$

$$E_{im} = E_{rm} + E_{to}$$

Also,

$$\vec{H}_i(0) + \vec{H}_r(0) = \vec{H}_t(0)$$

Thus,

$$\frac{(E_{im} - E_{rm})}{\eta_1} = \frac{E_{to}}{\eta_2}$$

On solving, we get

$$E_{rm} = \frac{(\eta_2 - \eta_1)}{(\eta_2 + \eta_1)} E_{im} \quad \text{and}$$

$$E_{tm} = \frac{2\eta_2}{(\eta_2 + \eta_1)} E_{im}$$

Reflection coefficient Γ is

$$\Gamma = \frac{E_{rm}}{E_{im}}$$

$$= \frac{(\eta_2 - \eta_1)}{(\eta_2 + \eta_1)}$$

Transmission coefficient τ is

$$\tau = \frac{E_{tm}}{E_{im}}$$

$$= \frac{2\eta_2}{(\eta_2 + \eta_1)}$$

Please note that

(i) $(1+\Gamma) = \tau$.
(ii) Both Γ and τ are dimensionless. They can be complex also.
(iii) Also, $0 \le |\Gamma| \le 1$.

The above is a general case. Now, let us consider the following two special cases:

Special Case 1: Medium 1 is a perfect dielectric medium (lossless, $\sigma_1 = 0$). Medium 2 is a perfect conductor ($\sigma_2 \simeq \infty$). Here,

$$\eta_2 = 0$$

Therefore,

$$\Gamma = -1 \quad \text{and}$$

$$\tau = 0 \qquad \text{(Wave is totally reflected.)}$$

Fields in a perfect conductor vanish. Therefore, there is no transmitted wave $(\vec{E}_2 = 0)$. This totally reflected wave combines with incident waves and becomes a "standing wave". A standing wave stands. It does not travel. A standing wave consists of two travelling waves, viz. $\vec{E}_i$ and $\vec{E}_r$. These two travelling waves are of equal amplitudes but are opposite in directions.
Standing wave in medium 1 is given as

$$\vec{E}_{1s} = \vec{E}_{is} + \vec{E}_{rs} \quad \text{or}$$

$$(E_{im}\, e^{-\gamma_1 x} + E_{rm} e^{+\gamma_1 x})\, \vec{a}_y$$

However,

$$\Gamma = -1$$

$$\sigma_1 = 0$$

$$\alpha_1 = 0 \quad \text{and}$$

$$\gamma_1 = j\beta_i$$

Thus,

$$\vec{E}_{1s} = -\vec{E}_{im}\, (e^{i\beta_1 x} - e^{-J\beta_1 x})\vec{a}_y \quad \text{or}$$

$$\vec{E}_{1s} = -2j\vec{E}_{im} \sin \beta_1 x \cdot \vec{a}_y$$

Also,

$$\vec{E}_1 = \text{Re}(\vec{E}_{1s}\, e^{j\omega t}) \quad \text{or}$$

$$\vec{E}_1 = 2\vec{E}_{im} \sin \beta_1 x \sin \omega t \ \vec{a}_y$$

The magnetic field component $\vec{H}_1$ of the wave is given as

$$\vec{H}_1 = \frac{2E_{im}}{\eta_1} \cos \beta_1 x \cdot \cos \omega t \, \vec{a}_y$$

Special Case 2: Mediums 1 and 2 are lossless ($\sigma_1 = \sigma_2 = 0$). Here η_1 and η_2 will be real. Γ and τ will also be real. Γ can be positive or negative. Let us consider the following cases:

(a) $\Gamma > 0\ (\eta_2 > \eta_1)$: There is a standing wave in medium 1. There is also a transmitted wave in medium 2. Amplitudes of incident and reflected waves are not equal. Maximum and minimum values of $\vec{E}_1\ (= \vec{E}_i + \vec{E}_r)$ will appear at different points.

(b) $\Gamma < 0\ (\eta_2 < \eta_1)$: There is a standing wave in medium 1. Locations of maximum and minimum values of $\vec{E}_1\ (= \vec{E}_i + \vec{E}_r)$ will interchange for the case $\Gamma > 0$.

Minimum $|\vec{H}_1|$ occurs, when $|\vec{E}_1|$ is maximum and vice versa. The transmitted wave in medium 2 is purely a travelling wave. There are no maxima and minima in this region. The ratio of the maximum value to the minimum value of electric field intensity in the case of a standing wave is known as "standing-wave ratio (SWR)". It is given as

$$\text{SWR } S = \frac{|\vec{E}_1|_{\max}}{|\vec{E}_1|_{\min}}$$

$$= \frac{|\vec{H}_1|_{\max}}{|\vec{H}_1|_{\min}}$$

$$= \frac{1+|\Gamma|}{1-|\Gamma|} \quad \text{or}$$

$$|\Gamma| = \frac{(S-1)}{(S+1)}$$

Γ ranges from -1 to $+1$. S ranges from 1 to ∞. S is expressed on a logarithmic scale by convention. Also, S (in dB) $= 20 \log_{10} S$

8.11 REFLECTION OF A PLANE WAVE AT OBLIQUE INCIDENCE

Consider a plane wave incident upon a boundary surface. It is not parallel to the plane containing $\vec{E}$ and $\vec{H}$. The boundary conditions will be more complex. Some part of this wave will be transmitted, whereas the other part of the wave will be reflected. Here, in the above case, the transmitted wave will be refracted. The direction of wave propagation will be altered as shown in Figure 8.5.

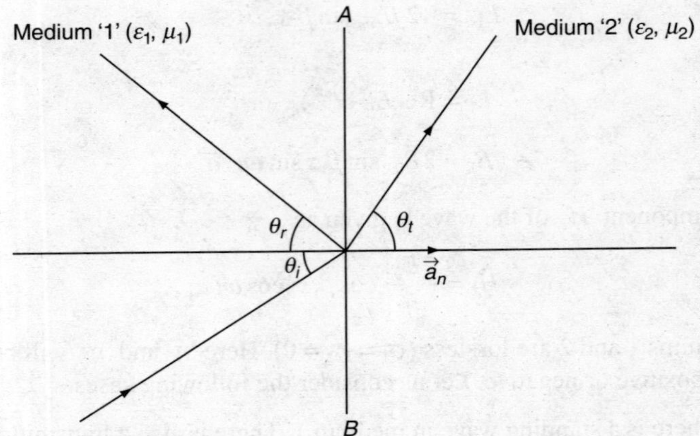

Figure 8.5 *Wave propagation*

Here, θ_r is the angle of reflection and θ_t the angle of transmission. In the above case, $\theta_r = \theta_i$. By Snell's Law, we get

$$\eta_1 \sin\theta_i = \eta_2 \sin\theta_t$$

where $\eta_1 = C\sqrt{\mu_1\varepsilon_1}$ and $\eta_2 = C\sqrt{\mu_2\varepsilon_2}$ are the refractive indices of the media obtained as

$$\frac{\sin\theta_1}{\sin\theta_t} = \frac{\sqrt{\mu_2\varepsilon_2}}{\sqrt{\mu_1\varepsilon_1}}$$

Power transmitted (per square meter) is given by $\vec{E}\times\vec{H}$. $\vec{E}$ and $\vec{H}$ both are at right angles to each other. Therefore, the power transmitted (per square meter) is E^2/η. The power of the incident wave striking AB will be $(1/\eta_1)E_{im}^2\cos\theta_i$. The power of the reflected wave will be proportional to $(1/\eta_1)E_{rm}^2\cos\theta_i$. The power of the transmitted wave (through the boundary) will be proportional $(1/\eta_2)E_{tm}^2\cos\theta_t$. By the Law of Conservation of Energy, we get

$$\frac{1}{\eta_1}E_{im}^2\cos\theta_i = \frac{1}{\eta_1}E_{rm}^2\cos\theta_i + \frac{1}{\eta_2}E_{tm}^2\cos\theta_t \quad \text{or}$$

$$\frac{E_{rm}^2}{E_{im}^2} = 1 - \frac{\eta_1 E_{tm}^2\cos\theta_t}{\eta_2 E_{im}^2\cos\theta_i} \quad \text{or}$$

$$\frac{E_{rm}^2}{E_{im}^2} = 1 - \frac{\sqrt{\mu_1/\varepsilon_1}\,E_{tm}^2\cos\theta_t}{\sqrt{\mu_2/\varepsilon_2}\,E_{im}^2\cos\theta_i}$$

For oblique incidences, let us consider these two special cases:

Case I: Field $\vec{E}$ is perpendicular to the plane of incidence.
Case II: Field $\vec{E}$ is parallel to the plane of incidence.

Any other case of polarization can be considered as a linear combination of these two cases as mentioned, above.

8.11.1 Parallel Polarization

Field $\vec{E}$ is parallel to the plane of incidence $\vec{H}$, which is parallel to the reflecting surface. Applying the boundary condition considering that the tangential component field of $\vec{E}$ is continuous across the boundary (Figure 8.6), we get

$$(E_{im} - E_{rm})\cos\theta_i = E_{tm}\cos\theta_t \quad \text{or}$$

$$\frac{E_{tm}}{E_{im}} = \left[1 - \frac{E_{rm}}{E_{im}}\right]\frac{\cos\theta_i}{\cos\theta_t}$$

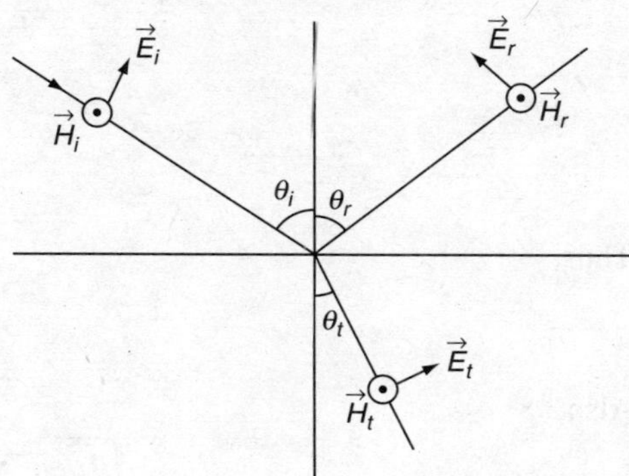

Figure 8.6 *Parallel polarization*

One solving, we find

$$\left(\frac{E_{rm}}{E_{im}}\right)^2 = 1 - \frac{\sqrt{\mu_1/\varepsilon_1}}{\sqrt{\mu_2/\varepsilon_2}}\left(1 - \frac{E_{rm}}{E_{im}}\right)^2 \frac{\cos\theta_i}{\cos\theta_t} \quad \text{or}$$

$$\frac{\sqrt{\mu_1/\varepsilon_1}}{\sqrt{\mu_2/\varepsilon_2}}\left(1 - \frac{E_{rm}}{E_{im}}\right)^2 \frac{\cos\theta_i}{\cos\theta_t} = 1 - \left(\frac{E_{rm}}{E_{im}}\right)^2 \quad \text{or}$$

$$\frac{\sqrt{\mu_1/\varepsilon_1}}{\sqrt{\mu_2/\varepsilon_2}}\left(1 - \frac{E_{rm}}{E_{im}}\right)\frac{\cos\theta_i}{\cos\theta_t} = \left(1 + \frac{E_{rm}}{E_{im}}\right) \quad \text{or}$$

$$\frac{E_{rm}}{E_{im}}\left[1 + \frac{\sqrt{\mu_1/\varepsilon_1}}{\sqrt{\mu_2/\varepsilon_2}}\frac{\cos\theta_i}{\cos\theta_t}\right] = \frac{\sqrt{\mu_1/\varepsilon_1}}{\sqrt{\mu_2/\varepsilon_2}}\frac{\cos\theta_i}{\cos\theta_t} - 1$$

Thus,

$$\Gamma = \frac{E_{rm}}{E_{im}} = \left[\frac{\sqrt{\mu_1/\varepsilon_1}\cos\theta_i - \sqrt{\mu_2/\varepsilon_2}\cos\theta_t}{\sqrt{\mu_1/\varepsilon_1}\cos\theta_i + \sqrt{\mu_2/\varepsilon_2}\cos\theta_t}\right]$$

Similarly, we find the transmission coefficient (τ) as

$$\tau = \frac{E_{tm}}{E_{im}} = \frac{2\sqrt{\mu_2/\varepsilon_2}\cos\theta_i}{\sqrt{\mu_1/\varepsilon_1}\cos\theta_i + \sqrt{\mu_2/\varepsilon_2}\cos\theta_t}$$

Expressing Γ and τ in terms of as η_1 and η_2, we get

$$\Gamma = \frac{(\eta_2\cos\theta_t - \eta_1\cos\theta_i)}{(\eta_2\cos\theta_t + \eta_1\cos\theta_i)} \quad \text{and}$$

$$\tau = \frac{2\eta_2\cos\theta_i}{(\eta_2\cos\theta_t + \eta_1\cos\theta_i)}$$

A case of interest is finding the possibility of obtaining no reflection at an angle. For this, put

$$\Gamma = 0$$

Thus, we get

$$\sqrt{\mu_1/\varepsilon_1}\cos\theta_i - \sqrt{\mu_2/\varepsilon_2}\cos\theta_t = 0 \quad \text{or}$$

$$\sqrt{\mu_1/\varepsilon_1}\cos\theta_i = \sqrt{\mu_2/\varepsilon_2}\cos\theta_t$$

However, we know

$$\frac{\sin\theta_i}{\sin\theta_t} = \frac{\eta_2}{\eta_1} \quad \text{or}$$

$$= \frac{C\sqrt{\mu_2/\varepsilon_2}}{C\sqrt{\mu_i/\varepsilon_i}}$$

Thus,

$$\sin\theta_t = \sqrt{\frac{\mu_1\varepsilon_1}{\mu_2\varepsilon_2}}\sin\theta_i$$

Also,

$$\sqrt{\frac{\mu_2}{\varepsilon_2}}\sqrt{1-\sin^2\theta_t} = \sqrt{\frac{\mu_1}{\varepsilon_1}}\cos\theta_i \quad \text{or}$$

$$\frac{\mu_2}{\varepsilon_2}(1-\sin^2\theta_t) = \frac{\mu_1}{\varepsilon_1}\cos^2\theta_i \quad \text{or}$$

$$\frac{\mu_2}{\varepsilon_2}\left[1-\frac{\mu_1\varepsilon_1}{\mu_2\varepsilon_2}\sin^2\theta_i\right] = \frac{\mu_1}{\varepsilon_1}(1-\sin^2\theta_i) \quad \text{or}$$

$$\left(1-\frac{\mu_1\varepsilon_2}{\mu_2\varepsilon_1}\right) = \sin^2\theta_i\left[\frac{\mu_1\varepsilon_1}{\mu_2\varepsilon_2}-\frac{\mu_1\varepsilon_2}{\mu_2\varepsilon_1}\right] \quad \text{or}$$

$$\left[\frac{\mu_2\varepsilon_1}{\mu_1\varepsilon_2}-1\right] = \sin^2\theta_i\left[\left(\frac{\varepsilon_1}{\varepsilon_2}\right)^2-1\right]$$

Finally,

$$\sin^2\theta_i = \left[\frac{1-(\mu_2\varepsilon_1/\mu_1\varepsilon_2)}{1-(\varepsilon_1/\varepsilon_2)^2}\right]$$

Consider dielectric media as non-magnetic, then

$$\mu_1 = \mu_2 = \mu_0$$

Thus,

$$\sin^2\theta_i = \frac{1}{1+(\varepsilon_1/\varepsilon_2)}$$

where θ_i is the Brewster angle (or θ_B). θ_B for any combination of ε_1 and ε_2 is given as

$$\tan\theta_B = \sqrt{\frac{\varepsilon_2}{\varepsilon_1}}$$

8.11.2 Perpendicular Polarization

Field $\vec{E}$ is perpendicular to the plane of incidence. It is parallel to the reflecting surface. Let E_{im} be in the positive x-direction (outward as shown in Figure 8.7). Also, E_{rm} and E_{tm} have been considered positive and in the positive x-direction. Applying boundary conditions that the tangential component of field $\vec{E}$ is continuous across the boundary, we get

$$E_{im} + E_{rm} = E_{tm}$$

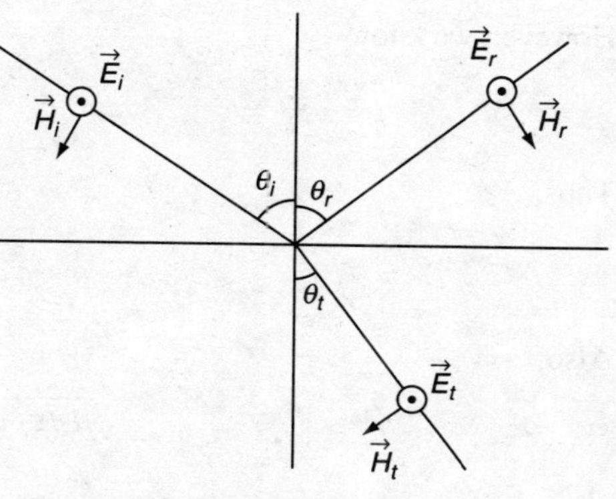

Figure 8.7 *Perpendicular polarization*

Also,

$$\frac{E_{tm}}{E_{im}} = 1 + \left(\frac{E_{rm}}{E_{im}}\right)$$

However,

$$\left(\frac{E_{rm}}{E_{im}}\right)^2 = 1 - \sqrt{\frac{\mu_1/\varepsilon_1}{\mu_2/\varepsilon_2}} \frac{E_{tm}^2 \cos\theta_t}{E_{im}^2 \cos\theta_i} \quad \text{or}$$

$$\sqrt{\frac{\mu_1/\varepsilon_1}{\mu_2/\varepsilon_2}} \left(1 + \frac{E_{rm}}{E_{im}}\right)^2 \frac{\cos\theta_t}{\cos\theta_i} = 1 - \left(\frac{E_{rm}}{E_{im}}\right)^2 \quad \text{or}$$

$$\sqrt{\frac{\mu_1/\varepsilon_1}{\mu_2/\varepsilon_2}} \left(1 + \frac{E_{rm}}{E_{im}}\right) \frac{\cos\theta_t}{\cos\theta_i} = 1 - \left(\frac{E_{rm}}{E_{im}}\right) \quad \text{or}$$

$$\sqrt{\frac{(\mu_1/\varepsilon_1)\cos\theta_t}{(\mu_1/\varepsilon_2)\cos\theta_i}} - 1 = \left[-\frac{\sqrt{\mu_1/\varepsilon_1}}{\sqrt{\mu_2/\varepsilon_2}} \frac{\cos\theta_t}{\cos\theta_i} \frac{E_{rm}}{E_{im}} - \frac{E_{rm}}{E_{im}}\right]$$

Finally,

$$\Gamma = \frac{E_{rm}}{E_{im}} = \frac{\sqrt{(\mu_2/\varepsilon_2)}\cos\theta_i - \sqrt{(\mu_1/\varepsilon_1)}\cos\theta_t}{\sqrt{(\mu_2/\varepsilon_2)}\cos\theta_i + \sqrt{(\mu_1/\varepsilon_1)}\cos\theta_i}$$

The above expression is for "reflection coefficient" in the case of horizontal polarization. Similarly, we can find "transmission coefficient" (τ) as

$$\tau = \frac{E_{tm}}{E_{im}}$$

$$= \left[\frac{2\left(\sqrt{\mu_2/\varepsilon_2}\right)\cos\theta_i}{\left(\sqrt{\mu_2/\varepsilon_2}\right)\cos\theta_i + \left(\sqrt{\mu_1/\varepsilon_1}\right)\cos\theta_t}\right]$$

Expressing values of Γ and τ in terms of η_1 and η_2, we get

$$\Gamma = \frac{(\eta_2 \cos\theta_i - \eta_1 \cos\theta_t)}{(\eta_2 \cos\theta_i + \eta_1 \cos\theta_t)}$$

$$\tau = \frac{2\eta_2 \cos\theta_i}{(\eta_2 \cos\theta_i + \eta_1 \cos\theta_t)}$$

A case of interest is finding the possibility of obtaining no reflection at an angle. It will occur when

$$\Gamma = 0$$

Thus,

$$\sqrt{\mu_2/\varepsilon_2}\cos\theta_i = \sqrt{\mu_1/\varepsilon_1}\cos\theta_t$$

However, we know

$$\frac{\sin\theta_i}{\sin\theta_t} = \frac{\sqrt{\mu_2\varepsilon_2}}{\sqrt{\mu_1\varepsilon_2}}$$

Thus,

$$\sin\theta_t = \frac{\sqrt{\mu_1\varepsilon_1}}{\sqrt{\mu_2\varepsilon_2}}\sin\theta_i$$

Also,

$$\sqrt{\mu_2/\varepsilon_2}\cos\theta_i = \sqrt{\mu_1/\varepsilon_1}\sqrt{1-\sin^2\theta_t} \quad \text{or}$$

$$(\mu_2/\varepsilon_2)\cos^2\theta_i = (\mu_1/\varepsilon_1)(1-\sin^2\theta_t) \quad \text{or}$$

$$\left(\frac{\mu_2}{\varepsilon_2}\right)(1-\sin^2\theta_i) = \left(\frac{\mu_1}{\varepsilon_1}\right)\left(1-\frac{\mu_1\varepsilon_1}{\mu_2\varepsilon_2}\sin^2\theta_i\right) \quad \text{or}$$

$$\left(1-\frac{(\mu_1/\varepsilon_1)}{(\mu_2/\varepsilon_2)}\right) = \sin^2\theta_i\left(1-\frac{\mu_1^2}{\mu_2^2}\right)$$

Finally,

$$\sin^2\theta_i = \frac{(1-(\mu_1\varepsilon_2/\mu_2\varepsilon_1))}{1-(\mu_1/\mu_2)^2}$$

where θ_i is the Brewster angle. It is denoted as θ_B. In the case of non-magnetic media, we get

$$\mu_1 = \mu_2 = \mu_0 \quad \text{and}$$

$$\sin^2\theta_i \to \infty$$

Therefore, θ_B does not exist. The sine of an angle can never be greater than unity. Consider

$$\mu_1 \neq \mu_2 \qquad \text{and} \qquad \varepsilon_1 = \varepsilon_2$$

Then we get

$$\sin^2\theta_i = \frac{\mu_2}{(\mu_1+\mu_2)} \quad \text{or}$$

$$\sin\theta_i = \sqrt{\frac{\mu_2}{(\mu_1+\mu_2)}}$$

or simply

$$\tan\theta_i = \sqrt{\mu_2/\mu_1}$$

The above case is only theoretically possible. It is very rare in practice.

8.12 BREWSTER'S ANGLE

It is also called the polarization angle. It was found by Scottish physicist, Sir David Brewster (1781–1868). Brewster angle is denoted by θ_B. Normally, some part of light is reflected at the boundary of two media when light moves between two media having different refractive indices. However, if an angle of incidence is so chosen that the line having one particular polarization is not reflected, then this angle of incidence is called the Brewster angle (θ_B). At the Brewster angle, polarization cannot be reflected. At the Brewster angle, light's electric field lies in same plane as that of incident ray. Light with polarization parallel to the plane is called "p-polarized." Light with perpendicular polarization is called "s-polarized." At the Brewster angle, an incident unpolarized light's reflected light is always s-polarized. Thus, precisely Brewster's angle is that angle of incidence at which there is zero reflection.

In the case of parallel polarization, Brewster angle θ_B is given as

$$\theta_B = \tan^{-1}\sqrt{\frac{\epsilon_{r2}}{\epsilon_{r1}}}$$

In the case of perpendicular polarization, for zero reflection

$$\epsilon_{r1} = \epsilon_{r2}$$

where ϵ_{r1} and ϵ_{r2} are refractive indices of two media. Polarized sunglasses are made on the concept of Brewster's angle. These sunglasses reduce glare from the sun. Photographers use this principle to remove reflections due to water. Polarized windows/glasses are provided so as to keep the house cool.

SOLVED QUESTIONS

8.1 An electric field in a free space is $\vec{E} = 15\cos(10^6 t + \beta x)\vec{a}_y$ (Volt/m). What will be the direction of wave propagation? Also, find the values of β and the time that the wave takes to travel a distance of $\lambda/2$.

Solution:

Here, $(\omega t + \beta x)$ contains a +ve sign. Thus, wave propagates in the $-\vec{a}_x$ direction. In free space

$$u = c = 3\times10^8 \text{ (m/s)}$$

Thus,

$$\beta = \frac{\omega}{c}$$

$$= \frac{10^6}{3\times10^8}$$

$$= \frac{1}{300}$$

$$\beta = \frac{1}{300} \text{ (rad/m)}$$

Let T be the period wave. It takes T seconds to travel a distance λ at a speed of 3×10^8 (m/s). The time taken to travel distance $\lambda/2$ is

$$t = T/2$$

$$= \frac{\pi}{\omega}$$

$$= 3.142 \text{ µsec}$$

Wave is travelling at speed of light c; thus,

$$\frac{\lambda}{2} = ct$$

$$t = \frac{\lambda}{2c}$$

However,

$$\lambda = \frac{2\pi}{\beta}$$

$$= 600\ \pi$$

Thus,

$$t = \frac{600\pi}{2\times 3\times 10^8}$$

$$t = 3.142\ \mu\text{sec}$$

8.2 Let $\vec{H}\,(z,t) = 20\cos\,(10^5 t + 30z)\,\vec{a}_y$ (A/m). Find the amplitude, frequency, phase constant and wavelength in this case.

Solution:

It is given that $\vec{H}(z,t) = 20\cos\,(10^5 t + 30z)\vec{a}_y$. Standard equation is

$$\vec{H}(z,t) = H_o \cos\,(\omega t + \beta z)\vec{a}_y$$

On comparison, we find the amplitude of magnetic field

$$= 20\ (\text{A/m})$$

Also,

$$\omega = 10^5$$

Therefore,

$$f = \frac{10^5}{2\pi}$$

$$= 15.92\ (\text{kHz})$$

Phase constant $\beta = 30$ (rad/m) and wavelength

$$\lambda = (2\pi/\beta)$$

$$= \frac{2}{30}$$

$$= 0.209\ (\text{m})$$

8.3 Electric field intensity of a uniform plane wave (in air) is 3750 (V/m). It is in the $\vec{a}_y$ direction. Wave is propagating in the $\vec{a}_x$ direction having a frequency of 10^6 (rad/sec). Find, the wavelength time period and amplitude of $\vec{H}$ in this case.

Solution:

It is given that

$$\vec{E}_y = 3750\cos(10^6 t - \beta x)\,\vec{a}_y$$

$$\varepsilon_0 = 8.86 \times 10^{-12} \text{ (F/m)} \quad \text{and}$$

$$\mu_0 = 4\pi \times 10^{-7} \text{ (H/m)}$$

Wavelength

$$\lambda = \frac{v}{f}$$

$$= \frac{3 \times 10^8}{(10^6/2\pi)} \quad \text{or}$$

$$= 1886 \text{ (m)}$$

Time period

$$T = \frac{1}{f} = \frac{1}{(10^6/2\pi)} \quad \text{or}$$

$$= 6.284 \text{ (μs)}$$

Further,

$$\frac{\vec{E}}{\vec{H}} = \eta$$

$$= \sqrt{\frac{\mu_0}{\varepsilon_0}}$$

$$\approx 377\ \Omega$$

Also,

$$|\vec{H}| = \frac{|\vec{E}|}{\eta} \quad \text{or}$$

$$= \frac{3750}{377} \quad \text{or}$$

$$= 9.95 \text{ (A/m)}$$

Thus,

$$\vec{H} = 9.95 \cos(10^6 t - \beta x)\vec{a}_z \text{ (A/m)}$$

8.4 An electric field $\vec{E}_y = 20\cos(3\pi 10^8 t - \beta x)\vec{a}_y$ travels through a lossless medium characterized by $\mu_r = 2$ and $\varepsilon_r = 39$. It has a frequency of 400 MHz. Find values of various parameters, which are associated with this wave.

Solution:

Phase constant β is obtained as

$$\beta = \frac{\omega}{v_P} \quad \text{or}$$

$$= \omega\sqrt{\mu\varepsilon} \quad \text{or}$$

$$= 3\pi \times 10^8 \sqrt{\mu_o \varepsilon_0 \mu_r \varepsilon_r} \quad \text{or}$$

$$= 3\pi \times 10^8 \sqrt{2 \times 39}\sqrt{\mu_0 \varepsilon_0} \quad \text{or}$$

$$= 3\pi \times 10^8 \frac{\sqrt{78}}{3 \times 10^8}$$

Thus,

$$\beta = 27.75 \text{ (rad/m)}$$

Wavelength

$$\lambda = \frac{2\pi}{\beta}$$

$$= 0.226 \text{ (m)}$$

Propagation velocity

$$v_p = \frac{\omega}{\beta} \quad \text{or}$$

$$= \frac{1}{\sqrt{78}\sqrt{\mu_0 \vec{\varepsilon}_0}} \quad \text{or}$$

$$= 0.33 \times 10^8 \text{ (m/s)}$$

and intrinsic impedance

$$\eta = \sqrt{\frac{\mu}{\varepsilon}} \quad \text{or}$$

$$= \sqrt{\frac{\mu_0 \mu_r}{\varepsilon_0 \varepsilon_r}} \quad \text{or}$$

$$= 377\sqrt{\frac{\mu_r}{\varepsilon_r}}$$

$$= 377\sqrt{\frac{2}{39}}$$

or simply

$$= 85.38\ (\Omega)$$

8.5 A medium has $\varepsilon_r = 4$ and $\mu_r = 9$. A wave having frequency $f = 0.3$ MHz propagates through this medium. What are the values of propagation constant and intrinsic impedance of this medium for $\sigma = 0$.

Solution:

We know that propagation constant

$$\gamma = \sqrt{j\omega\mu(\sigma + j\omega\varepsilon)}$$

It is given that

$$\mu_r = 9, \quad \varepsilon_r = 4, \quad \sigma = 0 \quad \text{and} \quad f = 0.3 \times 10^6 \text{ Hz}$$

Thus,

$$\gamma = \sqrt{j\omega\mu\,(0 + j\omega\varepsilon)} \quad \text{or}$$

$$= j\omega\sqrt{\mu\varepsilon} \quad \text{or}$$

$$= j \times 2\pi f \times \sqrt{\mu_0 \varepsilon_0 \mu_r \varepsilon_r} \quad \text{or}$$

$$= j \times 2\pi f \times \sqrt{4 \times 9}\sqrt{\mu_0 \varepsilon_0} \quad \text{or}$$

$$= j \times 2\pi \times 0.3 \times 10^6 \times \frac{6}{3 \times 10^8}$$

Finally,

$$\gamma = j\,0.03768\,(\text{m}^{-1})$$

Intrinsic impedance η is obtained as

$$\eta = \sqrt{\frac{j\omega\mu}{(\sigma + \omega\varepsilon)}} \quad \text{or}$$

$$= \sqrt{\frac{\mu}{\varepsilon}} \quad \text{or}$$

$$= \sqrt{\frac{\mu_0}{\varepsilon_0}}\sqrt{\frac{\mu_r}{\varepsilon_r}} \quad \text{or}$$

$$= \sqrt{\frac{9}{4}} \times 120\pi$$

Finally,

$$\eta = 180\pi\,(\Omega)$$

8.6 A uniform plane wave $\vec{E}(= |\vec{E}_Y|\vec{a}_y)$ propagates in the x-direction in a simple lossless medium characterized by $\varepsilon_r = 9, \mu_r = 4$ and $\sigma = 0$. E_y is sinusoidal and has a frequency of $100\,\text{MHz}$. Its maximum value is $+10^{-2}$ (Volt/m) at $t = 0$ and at $x = 1/6$ (m). Obtain an instantaneous expression of $\vec{E}$ in terms of t and x.

Solution:

It is given that

$$\vec{E}(x, t) = |\vec{E}_Y|\vec{a}_y$$

$$= 10^{-2} \cos\,(2\pi 10^8 t - \beta x + \phi)\vec{a}_y \qquad (\cos \omega t \text{ taken as reference})$$

β is obtained as

$$\beta = \omega\sqrt{\mu\varepsilon} \quad \text{or}$$

$$= \frac{\omega\sqrt{\mu_r \varepsilon_r}}{c} \quad \text{or}$$

$$= \frac{2\pi \times 10^8}{3 \times 10^8} \sqrt{4 \times 9} \quad \text{or}$$

$$= 4\,\pi \text{ (rad/sec)}$$

$\vec{E}_y$ is $+10^{-2}$, when the cosine function is zero. Thus,

$$(2\pi \times 10^{8t} - \beta x + \phi) = 0 \text{ at } t = 0 \text{ and at } x = 1/6 \text{ (m)}$$

Thus,

$$\phi = \beta x \quad \text{or}$$

$$= 4\pi \times \frac{1}{6} \quad \text{or}$$

$$= \frac{2\,\pi}{3} \text{ (rad)}$$

Finally, we get

$$\vec{E}(x,t) = 10^{-2} \cos\left(2\pi 10^8 t - 4\pi x + \frac{2\pi}{3}\right) \vec{a}_y \quad \text{or}$$

$$\vec{E}(x,t) = 10^{-2} \cos\left(2\pi 10^8 t - 4\pi\left(x - \frac{1}{6}\right)\right) \vec{a}_y \text{ (V/m)}$$

8.7 A uniform plane wave having frequency 10^3 Hz travels characterized by $\sigma = 10^{-2}$ (S/m), $\varepsilon_r = 64$ and $\mu = \mu_0$. Find values of different parameters of this plane wave.

Solution:

It is given that $\varepsilon_r = 64$, $\mu_r = 1$ and $\sigma = 10^{-2}$ (S/m). Now finding

$$\frac{\sigma}{\omega\varepsilon} = \frac{10^{-2}}{2\pi \times 1000 \times 64 \times 8 \cdot 854 \times 10^{-12}} \quad \text{or}$$

$$= 2808.75 \qquad \text{(It is greater than unity)}$$

This medium is a very good conductor. Attenuation constant

$$\alpha = \sqrt{\pi + \mu\sigma} \quad \text{or}$$

$$= \sqrt{\pi \times 1000 \times 4\pi 10^{-7} \times 10^{-2}} \quad \text{or}$$

$$= 2\pi \times 10^{-3} \text{ (Np/m)}$$

Phase constant

$$\beta = \alpha = 2\pi \times 10^{-3} \text{ (rad/m)}$$

Intrinsic impedance

$$\eta = \sqrt{\frac{\omega\mu}{\sigma}} \angle 45° \quad \text{or}$$

$$= \sqrt{\frac{2\pi \times 1000 \times 4\pi \times 10^{-7}}{10^{-2}}} \angle 45° \quad \text{or}$$

$$= 0.2\sqrt{2\pi} \angle 45° \quad \text{or}$$

$$= 0.2\pi \ (1 + j)\,(\Omega)$$

Wave velocity

$$v = \frac{\omega}{\beta} \quad \text{or}$$

$$= \frac{2\pi \times 1000}{2\pi \times 10^{-3}} \quad \text{or}$$

$$= 10^6\ (\text{m/s})$$

Wavelength

$$\lambda = \frac{2\pi}{\beta} \quad \text{or}$$

$$= \frac{2\pi}{2\pi \times 10^{-3}} \quad \text{or}$$

$$= 1000\ (\text{m})$$

8.8 What is skin depth δ at 6.4 MHz in Al, which is characterized by $\sigma = 38.2$ (MS/m) and $\mu_r = 1$. Also, find the values of propagation constant and wave velocity.

Solution:

We know that

$$\delta = \frac{1}{\sqrt{\pi f \mu \sigma}} \quad \text{or}$$

$$= \frac{1}{\sqrt{\pi \times 6.4 \times 10^6 \times 4\pi \times 10^{-7} \times 38.2 \times 10^6}}$$

$$= 0.03219\ (\text{mm})$$

However,

$$\alpha = \beta = \delta^{-1}$$

Now,

$$\gamma = (\alpha + j\beta) \quad \text{or}$$

$$\gamma = (31.06 \times 10^3 + j31.06 \times 10^3)\,(\text{m}^{-1})$$

Also,

$$v = \frac{\omega}{\beta}$$

$$= 2\pi \times 6.4 \times 10^6 \times 0.03219 \times 10^{-3}$$

$$= 1294.4\ (\text{m/sec})$$

8.9 In a homogenous non-conducting region, $\mu_r = 4$. $\vec{E}$ is $60\pi\, e^{j(\omega t(-9/4)y)}\vec{a}_y$ (V/m) and $\vec{H}$ is $5\, e^{j(\omega t-(9/4)y)}\vec{a}_y$ (V/m). Find the values of ε_r and ω.

Solution:

It is given that $\mu_r = 4, |\vec{H}| = 5, |\vec{E}| = 60\pi$ and $\beta = (9/4)$.

Now,

$$\frac{|\vec{E}|}{|\vec{H}|} = \sqrt{\frac{\mu}{\varepsilon}} \quad \text{or}$$

$$\left|\frac{\vec{E}}{\vec{H}}\right| = \sqrt{\frac{\mu_o \mu_r}{\varepsilon_o \varepsilon_r}} \quad \text{or}$$

$$\frac{60\pi}{5} = 120\pi\sqrt{\frac{\mu_r}{\varepsilon_r}} \quad \text{or}$$

$$\frac{60\pi}{5} = 120\pi\sqrt{\frac{4}{\varepsilon_r}} \quad \text{or}$$

$$\sqrt{\varepsilon_r} = 20 \quad \text{or}$$

$$\varepsilon_r = 400$$

Further,

$$\frac{\omega}{\beta} = v_p \quad \text{or}$$

$$\frac{\omega}{\beta} = \frac{1}{\sqrt{\mu\varepsilon}} \quad \text{or}$$

$$\frac{\omega}{\beta} = \frac{3\times 10^8}{\sqrt{\mu_r \varepsilon_r}} \quad \text{or}$$

$$\frac{\omega}{(9/4)} = \frac{3\times 10^8}{\sqrt{4\times 400}} \quad \text{or}$$

$$\omega = \frac{3\times 10^8}{4\times 10}\times\frac{9}{4}$$

or simply

$$1.68\times 10^7 \text{ (rad/sec)}$$

8.10 Intrinsic impedance of a lossy dielectric medium is $100\angle 30°$ (Ω) at a specified frequency. The plane wave propagating through this dielectric medium has magnetic field component $\vec{H} = 5\, e^{-\alpha x}\cos^{(\omega t-(3/4)x)}\vec{a}_y$ (A/m) at that frequency. Find the expression for $\vec{E}$ and the value of α, in this case. Also, find wave polarization.

Solution:

Plane wave travels along $\vec{a}_x$. Magnetic field component is along $\vec{a}_y$. Hence, the electric field component will be along $-\vec{a}_z$ as $(-\vec{a}_z)\times\vec{a}_y = \vec{a}_x$. Here,

$$|\vec{H}| = 5$$

Now

$$\frac{|\vec{E}|}{|\vec{H}|} = \eta$$

$$= 100\ \angle 30^\circ \quad \text{or}$$

$$\frac{|\vec{E}|}{|\vec{H}|} = 100\ e^{j\pi/6}$$

Finally,

$$\vec{E} = -500\ e^{-\alpha x}\cos\left(\omega t - \frac{3}{4}x + \frac{\pi}{6}\right)\vec{a}_z\ \text{(V/m)}$$

Further,

$$\alpha = \omega\sqrt{\frac{\mu\varepsilon}{2}\left[\sqrt{1+\left(\frac{\sigma}{\omega\varepsilon}\right)^2} - 1\right]} \quad \text{and}$$

$$\beta = \omega\sqrt{\frac{\mu\varepsilon}{2}\left[\sqrt{1+\left(\frac{\sigma}{\omega\varepsilon}\right)^2} + 1\right]}$$

Thus,

$$\frac{\alpha}{\beta} = \omega\left[\frac{\sqrt{1+\left(\frac{\sigma}{\omega\varepsilon}\right)^2} - 1}{\sqrt{1+\left(\frac{\sigma}{\omega\varepsilon}\right)^2} + 1}\right]^{1/2}$$

However,

$$\frac{\sigma}{\varepsilon} = \tan 2\theta$$

$$= \tan 60^\circ$$

$$= \sqrt{3}$$

Thus,

$$\frac{\alpha}{\beta} = \left[\frac{\sqrt{1+3} - 1}{\sqrt{1+3} + 1}\right]^{1/2} \quad \text{or}$$

$$\frac{\alpha}{\beta} = \left[\frac{1}{3}\right]^{1/2} \quad \text{or}$$

$$\alpha = \frac{3}{4\sqrt{3}} \quad \text{or}$$

$$\alpha = 0.433 \text{ (Np/m)}$$

8.11 In a lossless medium $\eta = 40\pi$, $\mu_r = 1$ and $\vec{H} = 2\cos(\omega t - z)\vec{a}_x + 5\sin(\omega t - z)\vec{a}_y$ (A/m). Find ε_r, ω and $\vec{E}$ in this case.

Solution:

It is given that $\sigma = 0$, $\alpha = 1$ and $\beta = 1$.

$$\eta = \sqrt{\frac{\mu}{\varepsilon}}$$

$$= \frac{120\pi}{\sqrt{\varepsilon_r}} \quad \text{or}$$

$$40\pi = \frac{120\pi}{\sqrt{\varepsilon_r}}$$

Thus,

$$\varepsilon_r = 9$$

Now,

$$\beta = \omega\sqrt{\mu\varepsilon} \quad \text{or}$$

$$= \omega\sqrt{\mu_o\varepsilon_o}\sqrt{\mu_r\varepsilon_r} \quad \text{or}$$

$$\beta = \frac{\omega\sqrt{1\times 4}}{3\times 10^8} \quad \text{or}$$

$$\omega = \frac{\beta\times 3\times 10^8}{2} \quad \text{or}$$

$$\omega = 1.5\times 10^8 \text{ (rad/sec)}$$

Further,

$$\vec{H} = \vec{H}_1 + \vec{H}_2$$

Here,

$$\vec{H}_1 = -2\cos(\omega t - z)\vec{a}_x \text{ (A/m)} \quad \text{and}$$

$$\vec{H}_2 = 5\sin(\omega t - z)\vec{a}_y \text{ (A/m)}$$

Electric field is given as

$$\vec{E} = \vec{E}_1 + \vec{E}_2$$

Wave propagates along the +z-axis, $\vec{H}_1$ propagates along the $-x$-axis. $\vec{E}_1$ component will propagate along the +y-axis as $\vec{a}_y \times (-\vec{a}_x) = \vec{a}_z$. Magnitude of $\vec{E}_1$ is obtained as

$$\left|\vec{E}_1\right| = \eta\left|\vec{H}_1\right| \quad \text{or}$$
$$= 40\,\pi\,(2) \quad \text{or}$$
$$= 80\,\pi$$

Thus,

$$\vec{E}_1 = 80\pi\ \cos(\omega t - z)\vec{a}_y$$

Similarly, $\vec{E}_2$ component propagates along the $\vec{a}_x$ axis as

$$\vec{a}_x \times \vec{a}_y = \vec{a}_z$$

The magnitude of $\vec{E}_2$ is obtained as

$$\left|\vec{E}_2\right| = \eta\left|\vec{H}_2\right| \quad \text{or}$$
$$= 40\,\pi\,(5) \quad \text{or}$$
$$= 200\,\pi$$

Thus,

$$\vec{E}_2 = 200\,\pi \sin(\omega t - z)\vec{a}_x$$

Finally, $\vec{E}$ is obtained by adding $\vec{E}_1$ and $\vec{E}_2$as

$$\vec{E} = 80\pi \cos(\omega t - z)\vec{a}_y + 200\pi \sin(\omega t - z)\vec{a}_x$$

8.12 A uniform plane propagating in a medium has $\vec{E} = 3\,e^{-\alpha z}\sin(8\times 10^7 t - \beta z)\vec{a}_y$ (V/m). Medium is characterized by $\varepsilon_r = 9, \mu_r = 30$ and $\sigma = 12(\Omega^{-1}/\text{m})$. Find values of α and β and expression for $\vec{H}$.

Solution:
Finding $\sigma/\omega\varepsilon$ as

$$\frac{\sigma}{\omega\varepsilon} = \frac{12}{8\times 10^7 \times 9 \times (10^{-9}/36\pi)}$$
$$= 1885$$

Further,

$$\frac{\sigma}{\omega\varepsilon} >> 1$$

Thus, medium is a good conductor.

$$\alpha = \beta = \sqrt{\frac{\mu\omega\sigma}{2}} \quad \text{or}$$

$$= \sqrt{\frac{4\pi\times 10^{-7} \times 30 \times 8 \times 10^7 \times 12}{2}} \quad \text{or}$$

$$= 134.52$$

$$\alpha = 134.52\ (\text{Np/m}) \quad \text{and} \quad \beta = 134.52\ (\text{rad/m})$$

Now,

$$\eta = \sqrt{\frac{\mu\omega}{\sigma}}$$

$$= \sqrt{\frac{4\pi 10^{-7} \times 30 \times 8 \times 10^{7}}{12}}$$

$$= 15.85$$

Further,

$$\tan 2\theta_\eta = \frac{\sigma}{\omega\varepsilon} = 1885 \text{ or}$$

$$\theta_\eta = (\pi/4)$$

Wave is propagating in direction of the +z-axis. $\vec{E}$ is in direction of $\vec{a}_y$. Therefore, $\vec{H}$ will propagate in the $-\vec{a}_x$ direction. The amplitude of $\vec{H}$ is obtained as

$$\left|\vec{H}\right| = \frac{\left|\vec{E}\right|}{\eta}$$

$$= \frac{3}{15.85}$$

$$= 0.189$$

Finally,

$$\vec{H} = -0.189 e^{-134.52z} \sin\left(8 \times 10^{-7} t - 134.52z - \frac{\pi}{4}\right)\vec{a}_x \text{ (A/m)}$$

8.13 In a free space, $\vec{E}$ is given as 15 sin$(\omega t - \beta z)\vec{a}_x$ (V/m). What is the average power P that crosses a circular disk of radius 4 meters in the "z = constant" plane?

Solution:

It is given that

$$\vec{E} = 15 \sin\,(\omega t - \beta z)\vec{a}_x$$

$$= 15\, e^{j(\omega t - \beta z)}\vec{a}_x$$

Wave is propagating in the +z-axis. $\vec{E}$ is in the +x-axis. $\vec{H}$ will propagate in the +y-axis as $((\vec{a}_x) \times (\vec{a}_y)) = \vec{a}_z$. Finding $\left|\vec{H}\right|$ as

$$\left|\vec{H}\right| = \frac{\left|\vec{E}\right|}{\eta} \text{ or}$$

$$= \frac{15}{120\pi} \text{ or}$$

$$\vec{H} = \frac{15}{120\pi} e^{j(\omega t - \beta z)}\vec{a}_y$$

The complex conjugate of $\vec{H}$ is obtained as

$$\vec{H}^* = \frac{15}{120\pi} e^{j(\omega t - \beta z)} \vec{a}_y$$

Finding P_{av} as

$$P_{av} = \frac{1}{2}\text{Re}(\vec{E} \times \vec{H}^*) \quad \text{or}$$

$$= \frac{1}{2} \times 15 \times \frac{15}{120\pi} e^{j(\omega t - \beta z)} \cdot e^{-j(\omega t - \beta z)} \vec{a}_x \times \vec{a}_y \quad \text{or}$$

$$P_{av} = 0.298 \vec{a}_z \ (\text{W/m}^2)$$

The average power crossing a circular disk is obtained as

$$P = 0.298 \ \times \pi \times (4)^2 \text{ Watts}$$

$$P = 15 \text{ Watts}$$

8.14 A plane wave is travelling in a medium characterized by $\varepsilon_r = 4$ and $\mu_r = 3$. Electric field intensity is $20\sqrt{3\pi}$ (V/m). Find the energy density in the magnetic field. Also, find the total energy density.

Solution:

Electric energy density W_E is obtained as

$$W_E = \frac{1}{2}\varepsilon E^2 \quad \text{or}$$

$$= \frac{1}{2} \times 8.854 \times 10^{-12} \times 4 \times (20)^2 \times 3\pi \quad \text{or}$$

$$= 6.67 \times 10^{-8} \quad \text{or}$$

$$= 66.7 \times 10^{-9}$$

$$W_E = 6.67 \ (\text{nJ/m}^3)$$

In the case of a travelling wave, we get

$$W_H = W_E = 66.7 \ (\text{nJ/m}^3)$$

Total energy density W_T is

$$W_T = 133.4 \ (\text{nJ/m}^3)$$

8.15 In case of a non-magnetic medium, $\vec{E} = 6\sin(2\pi \times 10^6 t - 0.4x)\vec{a}_z$ (V/m). What are the values of ε_r, η and the time average power? What is the total power that crosses to 25 cm^2 of the plane $(3x + y) = 10$?

Solution:

It is given that $\beta = 0.4$, $\omega = 2\pi \times 10^6$, $\mu = \mu_0$ (non-magnetic medium) and $\varepsilon = \varepsilon_0 \varepsilon_r$.

However,

$$\beta = \omega\sqrt{\mu\varepsilon} \quad \text{or}$$

$$= \omega\sqrt{\mu_0 \mu_r \varepsilon_0 \varepsilon_r} \quad \text{or}$$

$$= \frac{\omega\sqrt{\varepsilon_r}}{3\times10^8} \quad \text{or}$$

$$\varepsilon_r = \left(\frac{0.4\times3\times10^8}{2\pi\times10^6}\right)^2 \quad \text{or}$$

$$\varepsilon_r = 364.7$$

Also,

$$\eta = \frac{120\pi}{\sqrt{\mu_r\varepsilon_r}} \quad \text{or}$$

$$\eta = 19.74\ (\Omega)$$

Time average power

$$P_{ave} = \vec{E}\times\vec{H} \quad \text{or}$$

$$= \frac{|\vec{E}|^2}{2\eta}\vec{a}_x \quad \text{or}$$

$$= \frac{36}{2\times19.74}\vec{a}_x \quad \text{or}$$

$$= 912\ \vec{a}_x\ (\text{mW/m}^2)$$

For the plane $(3x+y)=10$, we get

$$\vec{a}_x = \frac{(3\vec{a}_x+\vec{a}_y)}{\sqrt{10}}$$

Thus, total power

$$P = \int P_{ave}ds \quad \text{or}$$

$$= (912\times10^{-3}\vec{a}_x)\cdot(25\times10^{-4})\left[\frac{(3\vec{a}_x+\vec{a}_y)}{\sqrt{10}}\right] \quad \text{or}$$

$$= \frac{912\times10^{-3}\times25\times10^{-4}\times3}{\sqrt{10}}((\vec{a}_x\cdot\vec{a}_y)=0) \quad \text{or}$$

$$= 2162.9\ (\mu\text{W})$$

8.16 In a free space $z\le0$, a plane wave has $\vec{H} = 20\cos(8\times10^7t-\beta z)\vec{a}_x\,(\text{mA/m})$. The wave is incident normally on a lossless medium characterized by $\varepsilon_r = 4$ and $\mu_r = 9$ in region $z\ge0$. Obtain reflected wave $(\vec{H}_r, \vec{E}_r)$ and transmitted wave $(\vec{H}_t, \vec{E}_t)$.

Solution:

Figure 8.8 shows the region $(z\ge0)$ and free space $(z\le0)$.

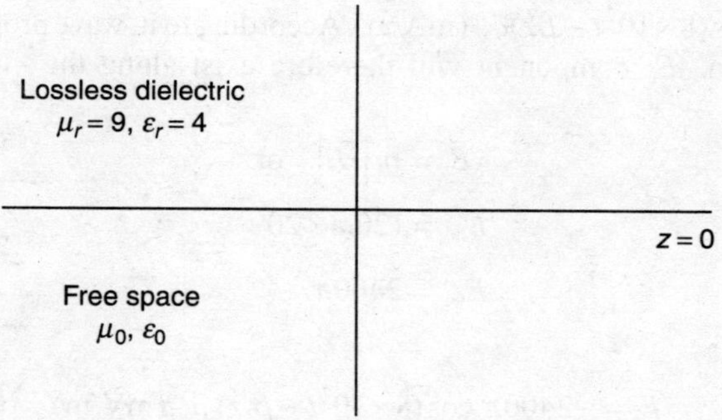

Figure 8.8 *Region and free space*

In free space,

$$\beta_1 = \frac{\omega}{c} = \frac{8\times10^7}{3\times10^8} = \frac{8}{30} = \frac{4}{15}$$

Also,

$$\eta_1 = \eta_0 = 120\pi$$

In the case of a lossless dielectric medium, we get

$$\beta_2 = \omega\sqrt{\mu\varepsilon} \quad \text{or}$$
$$= \omega\sqrt{\mu_o\varepsilon_o\mu_r\varepsilon_r} \quad \text{or}$$
$$= \frac{\omega}{c}\sqrt{9\times4} \quad \text{or}$$
$$= \frac{4}{15}\times3\times2 \quad \text{or}$$
$$= \frac{8}{5}$$

Further,

$$\eta_2 = \sqrt{\mu/\varepsilon} = \sqrt{\frac{\mu_o\mu_r}{\varepsilon_o\varepsilon_r}} = \frac{3}{2}\eta_o = 180\,\pi$$

It's given that $\vec{H}_i = 20\cos(8\times10^7 t - \beta z)\vec{a}_x$ (mA/m). According to it, wave propagates in the +z direction. $\vec{H}_i$ is in the +x direction. $\vec{E}_i$ component will therefore exist along the –y-axis as $(-\vec{a}_y)\times\vec{a}_x = \vec{a}_z$. Magnitude of

$$\vec{E}_i = \eta_1 \left|\vec{H}_i\right| \quad \text{or}$$

$$E_{im} = 120\pi \times 20 \quad \text{or}$$

$$E_{im} = 2400\pi$$

Also,

$$\vec{E}_i = -2400\pi\cos(8\times10^7 t - \beta_1 z)\,\vec{a}_y \text{ (mV/m)} \quad \text{or}$$

$$\vec{E}_i = -2400\pi\cos(8\times10^7 t - (4/15)z)\,\vec{a}_y \text{ (mA/m)}$$

Finding

$$\frac{E_{rm}}{E_{im}} = \frac{(\eta_2 - \eta_1)}{(\eta_2 + \eta_1)}$$

$$= \frac{1}{5} \quad \text{or}$$

$$E_{rm} = \frac{E_{im}}{5} \quad \text{or}$$

$$= 480\,\pi$$

Thus, we find

$$\vec{E}_r = -480\pi\cos\left(8\times10^7 t + \frac{4z}{15}\right)\vec{a}_y \text{ (mV/m)} \quad \text{and}$$

$$\vec{H}_r = \frac{-480\pi}{\eta_1}\cos\left(8\times10^7 t + \frac{4z}{15}\right)\vec{a}_x \text{ (mA/m)} \quad \text{or}$$

$$\vec{H}_r = -4\cos\left(8\times10^7 t + \frac{4z}{15}\right)\vec{a}_x \text{ (mA/m)}$$

It is therefore evident that the reflected wave travels along the –z-axis. For a transmitted wave, we get

$$\frac{E_{tm}}{E_{im}} = \tau$$

$$= 1 + \Gamma$$

$$= \frac{6}{5} \quad \text{or}$$

$$E_{tm} = \frac{6\,E_{im}}{5} \quad \text{or}$$

$$E_{tm} = \frac{6\times2400\pi}{5}$$

$$= 2880\,\pi$$

The direction of $\vec{E}_t$ will remain the same as $\vec{E}_i$. Thus,

$$\vec{E}_t = -E_{tm}\cos(8\times10^7 t - \beta_2 z)\,\vec{a}_y \text{ (mV/m)} \quad \text{or}$$

$$\vec{E}_t = -2880\pi\cos\left(8\times10^7 t - \frac{8z}{5}\right)\vec{a}_y \text{ (mV/m)} \quad \text{and}$$

$$\vec{H}_t = -\frac{2880\pi}{\eta_2}\cos\left(8\times10^7 t - \frac{8z}{5}\right)\vec{a}_y \text{ (mV/m)} \quad \text{or}$$

$$\vec{H}_t = 16\cos\left(8\times10^7 t - \frac{8z}{5}\right)\vec{a}_x \text{ (mV/m)}$$

8.17 A perpendicularly polarized wave travels from a region characterized by $\varepsilon_r = 8$, $\mu_r = 1$, $\sigma = 0$ to a free space at an incidence angle of 10°. The incident field is 5.0 (V/m). Find the reflected and transmitted electric fields. Also, find the reflected and transmitted magnetic fields.

Solution:

For medium 1, we have $\varepsilon_{r_1} = 8$ and $\mu_{r_1} = 1$. Thus,

$$\eta_1 = \sqrt{\frac{\mu_1}{\varepsilon_1}}$$

$$= 42.42\pi\ (\Omega)$$

For medium 2 (free space), we get

$$\eta_2 = \sqrt{\frac{\mu_o}{\varepsilon_o}}$$

$$= 120\pi\ (\Omega)$$

As per Snell's Law, we get

$$\frac{\sin\theta_i}{\sin\theta_t} = \sqrt{\frac{\varepsilon_2}{\varepsilon_1}} \quad \text{or}$$

$$\frac{\sin 10^\circ}{\sin\theta_t} = \sqrt{\frac{\varepsilon_o}{\varepsilon_o\times 8}} \quad \text{or}$$

$$\theta_t = 29.42^\circ$$

Reflection coefficient for electric field is obtained as

$$\frac{E_{rm}}{E_{im}} = \frac{\sqrt{\varepsilon_1}\cos\theta_i - \sqrt{\varepsilon_2}\cos\theta_t}{\sqrt{\varepsilon_1}\cos\theta_i + \sqrt{\varepsilon_2}\cos\theta_t} \quad \text{or}$$

$$= \frac{\sqrt{\varepsilon_0\varepsilon_{r_1}}\cos 10^\circ - \sqrt{\varepsilon_0}\cos 29.42^\circ}{\sqrt{\varepsilon_0\varepsilon_{r_1}}\cos 10^\circ + \sqrt{\varepsilon_0}\cos 29.42^\circ} \quad \text{or}$$

$$= \frac{(2.79 - 0.871)}{(2.79 + 0.871)} \quad \text{or}$$

$$\frac{E_{rm}}{E_{im}} = 0.524 \quad \text{or}$$

$$\vec{E}_{rm} = 2.62 \text{ (V/m)}$$

Transmitted electric field

$$E_{tm} = (E_{im} + E_{rm})$$
$$= (5 + 2.62)$$
$$= 7.62 \text{ V/m}$$

Incident magnetic field

$$H_{im} = \frac{E_{im}}{\eta_1}$$
$$= \frac{5}{42.42}$$
$$= 0.1178 \text{ (A/m)}$$

Reflected magnetic field H_{rm} is obtained as

$$H_{rm} = \frac{E_{rm}}{\eta_1} \quad \text{or}$$
$$= \frac{2.62}{42.42} \quad \text{or}$$
$$= 0.0617 \text{ (A/m)}$$

The transmitted magnetic field

$$H_{tm} = \frac{E_{tm}}{\eta_2} \quad \text{or}$$
$$= \frac{7.62}{120\pi} \quad \text{or}$$
$$= 0.0202 \text{ (A/m)}$$

8.18 A parallel polarized wave is incident from "air" into "paraffin". Find the Brewster angle (ε_r for paraffin is 3).

Solution:

According to the formula, we have

$$\tan\theta_B = \sqrt{\frac{\varepsilon_2}{\varepsilon_1}} \quad \text{or}$$
$$= \frac{\sqrt{\varepsilon_0 \varepsilon_{r_2}}}{\sqrt{\varepsilon_o \varepsilon_{r_1}}} \quad \text{or}$$

$$= \sqrt{\frac{3}{1}}$$

Thus,

$$\theta_B = 60°$$

8.19 What are the reflection and transmission coefficients of an electric field wave that is travelling in air and is incident normally on a boundary that is between air and a dielectric medium? (Assume $\mu_r = 8$ and $\varepsilon_r = 2$.)

Solution:

Mediums are air and dielectric medium. We can write

$$\eta_1 = \sqrt{\frac{\mu_o}{\varepsilon_o}}$$

$$= 377\ \Omega \quad \text{and}$$

$$\eta_2 = \sqrt{\frac{\mu}{\varepsilon}}$$

$$\eta_2 = \sqrt{\frac{\mu_o}{\varepsilon_o}}\sqrt{\frac{\mu_r}{\varepsilon_r}}$$

$$\eta_2 = 377 \times 2$$

$$= 754\ \Omega$$

The reflection coefficient for normal incidence is obtained as

$$\frac{E_{rm}}{E_{im}} = \frac{(\eta_2 - \eta_1)}{(\eta_2 + \eta_1)} \quad \text{or}$$

$$\frac{E_{rm}}{E_{im}} = \frac{(754 - 377)}{(754 + 377)}$$

Thus,

$$\Gamma = \frac{1}{3}$$

The transmission coefficient for normal incidence is obtained as

$$\frac{E_{tm}}{E_{im}} = \frac{2\eta_2}{(\eta_2 + \eta_1)} \quad \text{or}$$

$$= \frac{(2 \times 754)}{(754 + 377)}$$

Thus, transmission coefficient

$$\tau = \frac{4}{3}$$

8.20 An electric field in free space is $\vec{E} = 10\cos(20^6 t + \beta x)\vec{a}_v$ (V/m). What will be the direction of wave propagation? Also, find the values of β and the time that the wave takes to travel a distance of $\lambda/2$.

Solution:
Here, (ωt + βx) contains a +ve sign. Thus, the wave propagates in the $-\vec{a}_x$ direction. In free space, $v = c = 3 \times 10^8$ (m/s). Thus,

$$\beta = \frac{\omega}{c}$$
$$= \frac{20^6}{3\times10^8}$$
$$= 0.213 \text{ (rad/m)}$$

Let T be the period of the wave. It takes T seconds to travel a distance λ at a speed of 3×10^8 (m/s). The time taken to travel distance $\lambda/2$ is obtained as

$$t = T/2$$
$$= \pi/\omega$$
$$= 0.049 \text{ µsec}$$

Wave is travelling at speed of light c; thus,

$$\lambda/2 = c\,t$$
$$t = \frac{\lambda}{2c}$$

However,

$$\lambda = \frac{2\pi}{\beta} = \frac{2\pi}{0.213} = 29.49$$

Thus,

$$t = \frac{29.49}{2\times3\times10^8}$$
$$t = 0.049 \text{ µsec}$$

8.21 Let $\vec{H}(z,t) = 10\sin(10^5 t + 20z)\vec{a}_y$ (A/m). Find the amplitude, frequency phase constant and wavelength in this case.

Solution:
It is given that $H(z,t) = 10\sin(10^5 t + 20z)\vec{a}_y$. The standard equation is

$$H(z,t) = H_0\cos(\omega t + \beta z)\vec{a}_y$$

On comparison, we find the amplitude of magnetic field as

$$=10 \text{ (A/m)}$$

Also, $\omega = 10^5$. Therefore,

$$f = \frac{10^5}{2\pi}$$
$$= 15.92 \text{ kHz}$$

Phase constant $\beta = 20$ (rad/m) and wavelength $\lambda = (2\pi/\beta)$

$$= \frac{2\pi}{20}$$

$$= 0.314 \text{ (m)}$$

8.22 Electric field intensity of a uniform plane wave (in air) is 4000 (V/m). It is in the $\vec{a}_y$ direction. Wave is propagating in the $\vec{a}_x$ direction having a frequency of 10^5 rad/sec. Find the wavelength, time period and amplitude of $\vec{H}$ in this case.

Solution:

It is given that

$$\vec{E}_y = 4000\cos(10^5 t - \beta x)\vec{a}_y$$

$$\varepsilon_0 = 8.86\times 10^{-12} \text{ (F/m)} \quad \text{and}$$

$$\mu_0 = 4\pi\times 10^{-7} \text{ (H/m)}$$

Wavelength

$$\lambda = v/f$$

$$= \frac{3\times 10^8}{(10^5/2\pi)} \quad \text{or}$$

$$\lambda = 18849.5 \text{ (m)}$$

Time period

$$T = \frac{1}{f} = \frac{1}{(10^5/2\pi)}$$

$$= 0.628 \ \mu\text{sec}$$

Further,

$$\frac{\vec{E}}{\vec{H}} = \eta$$

$$= \sqrt{\frac{\mu_0}{\varepsilon_0}}$$

$$= 377\ \Omega$$

Also,

$$|\vec{H}| = \frac{|\vec{E}|}{\eta}$$

$$= \frac{4000}{377}$$

$$= 10.6 \text{ (A/m)}$$

Thus,

$$H = 10.6\cos(10^5 t - \beta x)\vec{a}_z$$

8.23 An electric field $\vec{E}_y = 40\cos(5\pi 10^8 t - \beta x)\vec{a}_y$ travels through a lossless medium characterized by $\mu_r = 12$ and $\varepsilon_r = 36$. It has a frequency of 200 MHz. Find values of various parameters, which are associated with this wave.

Solution:

Phase constant β is obtained as

$$\beta = \frac{\omega}{v_p} \quad \text{or}$$

$$= \omega\sqrt{\mu\varepsilon} \quad \text{or}$$

$$= 5\pi\times 10^8 \sqrt{\mu_0\varepsilon_0\mu_r\varepsilon_r} \quad \text{or}$$

$$= 5\pi\times 10^8 \times \frac{\sqrt{12\times 36}}{3\times 10^8} \quad \text{or}$$

$$= 108.8$$

Thus,

$$\beta = 108.8\ (\text{rad/m})$$

Wavelength

$$\lambda = \frac{2\pi}{\beta}$$

$$= 0.057\ (\text{m})$$

Propagation velocity

$$v_p = \frac{\omega}{\beta}$$

$$= \frac{1}{\sqrt{12\times 36}}\cdot\frac{1}{\sqrt{\mu_0\varepsilon_0}}$$

$$= \frac{1}{20.78}\times 3\times 10^8$$

$$= 0.14\times 10^8\ (\text{m/s})$$

Intrinsic impedance

$$\eta = \sqrt{\mu/\varepsilon} \quad \text{or}$$

$$= \sqrt{\frac{\mu_0\mu_r}{\varepsilon_0\varepsilon_r}} \quad \text{or}$$

$$= 377\sqrt{\frac{\mu_r}{\varepsilon_r}} \quad \text{or}$$

$$= 377\sqrt{\frac{12}{36}} \quad \text{or}$$

$$= 217.6\ \Omega$$

or simply

$$\eta = 217.6\ \Omega$$

8.24 A medium has $\varepsilon_r = 6$ and $\mu_r = 8$. A wave having frequency $f = 0.5$ MHz propagates through this medium. What are the values of propagation constant and intrinsic impedance of this medium for $\sigma = 0$?

Solution:

We know that propagation constant

$$r = \sqrt{j\omega\mu(\sigma + j\omega\varepsilon)}$$

It is given that

$$\mu_r = 8, \qquad \varepsilon_r = 6, \qquad \sigma = 0, \qquad \text{and} \quad f = 0.5 \text{ MHz}$$

Thus,

$$r = \sqrt{j\omega\mu(0 + j\omega\varepsilon)} \quad \text{or}$$

$$= j\omega\sqrt{\mu\varepsilon} \quad \text{or}$$

$$= j \times 2\pi f \times \sqrt{\mu_0\varepsilon_0\mu_r\varepsilon_r} \quad \text{or}$$

$$= j \times 2\pi f \times \sqrt{6 \times 8} \times \sqrt{\mu_0\varepsilon_0} \quad \text{or}$$

$$= j \times 2\pi \times 0.5 \times 10^6 \times \frac{6.92}{3 \times 10^8}$$

Finally,

$$r = j0.072\ (\text{m}^{-1})$$

Intrinsic impedance η is obtained as

$$\eta = \sqrt{\frac{j\omega\mu}{(\sigma + \omega\varepsilon)}} \quad \text{or}$$

$$= \sqrt{\frac{\mu}{\varepsilon}} \quad \text{or}$$

$$= \sqrt{\frac{\mu_0}{\varepsilon_0}} \cdot \sqrt{\frac{\mu_r}{\varepsilon_r}} \quad \text{or}$$

$$= \sqrt{\frac{8}{6}} \times 120\pi$$

Finally,

$$\eta = 435.31\ \Omega.$$

8.25 A uniform plane wave $\vec{E} = \left(\left|\vec{E}_y\right| \vec{a}_y\right)$ propagates in the x-direction in a simple lossless medium characterized by $\varepsilon_r = 10$, $\mu_r = 5$ and $\varepsilon = 0$. $\vec{E}_y$ is sinusoidal and has a frequency of 100 MHz. Its maximum value is $+ 20^{-2}$ (Volt/m) at $t = 0$ and at $x = 1/6$ (m). Obtain the instantaneous expression of $\vec{E}$ in terms of t and x.

Solution:
It is given that

$$\vec{E}(x, t) = \left|E_y\right| \vec{a}_y$$

$$= 20^{-2} \cos(2\pi\, 10^8\, t - \beta x + \phi)\, \vec{a}_y \quad \text{(cos } \omega t \text{ is taken as reference)}$$

β is obtained as

$$\beta = \omega\sqrt{\mu\varepsilon} \quad \text{or}$$

$$= \frac{\omega\sqrt{\mu_r\, \varepsilon_r}}{C} \quad \text{or}$$

$$= \frac{2\pi \times 10^8}{3 \times 10^8}\sqrt{10 \times 5} \quad \text{or}$$

$$\beta = 14.8 \text{ rad/ sec}$$

$\vec{E}_y$ is 20^{-2}, when the cosine function is 0. Thus, $(2\pi \times 10^8 - \beta x + \phi) = 0$ at $t = 0$ and at $x = 1/6$ m. Thus,

$$\phi = \beta x$$

$$= 14.8 \times \frac{1}{6}$$

$$\phi = 2.46 \text{ rad}$$

Finally, we get

$$E(x, t) = 20^{-2} \cos(2\pi 10^8 t - 14.8x + 2.46)\, \vec{a}_y \text{ (V/m)}$$

8.26 A uniform plane wave having frequency 100 Hz travels through a medium characterized by $\sigma = 20^{-2}$ (S/m), $\varepsilon_r = 36$ and $\mu = \mu_0$. Find values of different parameters of this plane wave.

Solution:
It is given that $\varepsilon_r = 36$, $\mu_r = 1$ and $\sigma = 20^{-2}$ (S/m).

Now finding

$$\frac{\sigma}{\omega\varepsilon} = \frac{20^{-2}}{2\pi \times 100 \times 36 \times 8.84 \times 10^{-12}}$$

$$= 12{,}482.7 \quad \text{(It is much greater than unity.)}$$

This medium is a very good conductor.

Attenuation constant

$$\alpha = \sqrt{\pi f \mu \sigma} \quad \text{or}$$

$$= \sqrt{\pi \times 100 \times 4\pi \times 10^{-7} \times 20^{-2}}$$

$$= 9.93 \times 10^{-4} \text{ (Np/m)}$$

Phase constant

$$\beta = \alpha = 9.93\times10^{-4} \text{ (Np/m)}$$

Intrinsic impedance

$$\eta = \sqrt{\frac{\omega\mu}{\sigma}}\angle 45^\circ$$

$$= \sqrt{\frac{2\pi\times100\times4\pi\times10^{-7}}{20^{-2}}}\angle 45^\circ$$

$$= 0.56\angle 45^\circ$$

$$= 0.39(1+j)\ \Omega$$

Wave velocity

$$v = \frac{\omega}{\beta}$$

$$= \frac{2\pi\times100}{9.93\times10^{-4}}$$

$$= 632\ 747.7 \text{ m/s}$$

Wavelength

$$\lambda = \frac{2\pi}{\beta} = \frac{2\pi}{9.93\times10^{-4}}$$

$$= 6327.4 \text{ (m)}$$

8.27 What is the skin depth δ at 5 MHz in a medium characterized by σ = 88.2 (MS/m) and μ_r = 1? Also, find the values of propagation constant and wave velocity.

Solution:

We know that

$$\delta = \frac{1}{\sqrt{\mu f \pi\sigma}} \quad \text{or}$$

$$= \frac{1}{\sqrt{\pi\times5\times10^{6}\times4\pi\times10^{-7}\times88.2\times10^{6}}} \quad \text{or}$$

$$= 2.39\times10^{-5}$$

$$\alpha = \beta = \delta^{-1}$$

Now,

$$\gamma = (\alpha + j\beta)$$

$$\gamma = (41841 + j41841)\ (\text{m}^{-1})$$

Also,

$$v = \frac{\omega}{\beta}$$
$$= \frac{2\pi \times 5 \times 10^6}{4184}$$
$$v = 750.84 \text{ m/s}$$

8.28 Intrinsic impedance of a lossy dielectric medium is $400\angle 30°$ (Ω) at a specified frequency. The plane wave propagating through this dielectric medium has the magnetic field component $\vec{H} = 8e^{-\alpha x}\cos^{(\omega t-(3/4)x)}\vec{a}_y$ (A/m) at that frequency. Find the expression for $\vec{E}$ and the value of σ in this case. Also, find the wave polarization.

Solution:

Plane wave travels along $\vec{a}_x$. The magnetic field component is along $\vec{a}_y$. Hence, the electric field component will be along $-\vec{a}_z$ as $(-\vec{a}_z)\times\vec{a}_y = \vec{a}_x$. Here,

$$\left|\vec{H}\right| = 8$$
$$\frac{\left|\vec{E}\right|}{\left|\vec{H}\right|} = \text{‘}\eta\text{’} \quad \text{or}$$
$$= 400\angle 30°$$
$$= 400e^{j\pi/4}$$

Finally,

$$\vec{E} = -3200\,e^{-\alpha x}\cos\left(\omega t - \left(\frac{3}{4}\right)x + \pi/4\right)\vec{a}_z \text{ (V/m)}$$

Further,

$$\alpha = \omega\sqrt{\frac{\mu\varepsilon}{2}\left[\sqrt{1+\left(\frac{\sigma}{\omega\varepsilon}\right)^2} - 1\right]} \quad \text{and}$$

$$\beta = \omega\sqrt{\frac{\mu\varepsilon}{2}\left[\sqrt{1+\left(\frac{\sigma}{\omega\varepsilon}\right)^2} + 1\right]}$$

Thus,

$$\frac{\alpha}{\beta} = \omega\left[\frac{\sqrt{1+\left(\frac{\sigma}{\omega\varepsilon}\right)^2} - 1}{\sqrt{1+\left(\frac{\sigma}{\omega\varepsilon}\right)^2} + 1}\right]^{1/2}$$

However,

$$\frac{\sigma}{\omega\varepsilon} = \tan 2\theta$$
$$= \tan 60°$$
$$= \sqrt{3}$$

Thus,

$$\frac{\alpha}{\beta} = \left[\frac{\sqrt{1+3}-1}{\sqrt{1+3}+1}\right]^{1/2}$$

$$\frac{\alpha}{\beta} = \left[\frac{1}{3}\right]^{1/2}$$

$$\alpha = \frac{3}{4\sqrt{3}}$$

$$\alpha = 0.433 \text{ (Np/m)}$$

8.29 In a lossless medium, $\eta = 60\pi$, $\mu_r = 1$ and $\vec{H} = 12\cos(\omega t - z)\vec{a}_x + 15\sin(\omega t - z)\vec{a}_y$ (A/m). Find ε_r, ω and $\vec{E}$ in this case.

Solution:
It is given that $\sigma = 0$, $\alpha = 1$ and $\beta = 1$. Also,

$$\eta = \sqrt{\frac{\mu}{\varepsilon}}$$

$$= \frac{120\pi}{\sqrt{\varepsilon_r}}$$

$$60\pi = \frac{120\pi}{\sqrt{\varepsilon_r}}$$

$$\varepsilon_r = 4$$

Now,

$$\beta = \omega\sqrt{\mu\varepsilon} \quad \text{or}$$

$$= \omega\sqrt{\mu_0\varepsilon_0}\sqrt{\mu_r\varepsilon_r} \quad \text{or}$$

$$\beta = \frac{\omega\sqrt{1\times 4}}{3\times 10^8} \quad \text{or}$$

$$\omega = \frac{\beta\times 3\times 10^8}{2} \quad \text{or}$$

$$\omega = 1.5\times 10^8 \text{ (rad/sec)}$$

Further,

$$\vec{H} = \vec{H}_1 + \vec{H}_2$$

Here,

$$\vec{H}_1 = -12\cos(\omega t - z)\,\vec{a}_x \text{ (A/m)}$$

$$H_2 = 15\sin(\omega t - z)\,\vec{a}_y \text{ (A/m)}$$

Electric field is given as

$$\vec{E} = \vec{E}_1 + \vec{E}_2$$

Wave propagates along the +z-axis. $\vec{H}_1$ propagates along the $-x$-axis. $\vec{E}_0$ component will propagate along the $+y$-axis as $\vec{a}_y \times (-\vec{a}_x) = \vec{a}_z$. The magnitude of $\vec{E}_1$ is obtained as

$$|\vec{E}_1| = \eta |\vec{H}_1| \quad \text{or}$$
$$= 60\pi(12) \quad \text{or}$$
$$= 720\pi$$

Thus,

$$\vec{E}_1 = 720\pi \cos(\omega t - z)\vec{a}_y$$

Similarly, E_2 component propagates along the $\vec{a}_x$ -axis as

$$\vec{a}_x \times \vec{a}_y = \vec{a}_z$$

The magnitude of $\vec{E}_2$ is obtained as

$$|\vec{E}_2| = \eta |\vec{H}_2| \quad \text{or}$$
$$= 60\pi(15) \quad \text{or}$$
$$= 900\,\pi$$

Thus,

$$\vec{E}_2 = 900\pi \sin(\omega t - z)\vec{a}_x$$

Finally, E is obtained by adding $\vec{E}_1$ and $\vec{E}_2$ as $\vec{E} = 720\pi \cos(\omega t - z)\vec{a}_y + 900 \sin(\omega t - z)\vec{a}_x$.

8.30 A uniform plane wave propagating in a medium having $\vec{E} = 8e^{-\alpha z} \sin(18\times 10^7 t - \beta z)\,\vec{a}_y$ (V/m) medium is characterized by $\varepsilon_r = 4$, $\mu_r = 20$ and $\sigma = 2\,(\Omega^{-1}/\text{m})$. Find the values of α and β.

Solution:

Finding $\sigma/\omega t$ as

$$\frac{\sigma}{\omega t} = \frac{2}{18\times 10^7 \times 4 \times (10^{-9}/36\pi)}$$
$$= 314.15$$
$$\frac{\sigma}{\omega t} >> 1$$

Thus, the medium is good conductor.

Thus,

$$\alpha = \beta = \sqrt{\frac{\mu\omega\sigma}{2}} \quad \text{or}$$
$$= \sqrt{\frac{4\pi\times 10^{-7} \times 20 \times 18\times 10^7 \times 2}{2}} \quad \text{or}$$
$$= 67.25$$

$$\alpha = 67.25 \text{ (Np/m) and also } \beta = 67.25 \text{ (rad/m)}$$

$$\eta = \sqrt{\frac{\mu\omega}{\sigma}}$$

$$= \sqrt{\frac{4\pi \times 10^{-7} \times 20 \times 18 \times 10^{7}}{2}}$$

$$\eta = 47.55$$

Further,

$$\tan 2\theta_{\eta} = \frac{\sigma}{\omega\varepsilon} = 314.15 \quad \text{or}$$

$$\theta = 44.9°$$

The wave is propagating in the direction of the $+z$ -axis. $\vec{E}$ is in the direction of $\vec{a}_y$. Therefore, $\vec{H}$ will propagate in the $-\vec{a}_x$ direction. The amplitude of $\vec{H}$ is obtained as

$$|\vec{H}| = \frac{|\vec{E}|}{\eta}$$

$$= \frac{8}{47.55}$$

$$= 0.168$$

Finally,
$$\vec{H} = -0.168 e^{-67.25z} \sin(18 \times 10^{-7} t - 67.25 - 44.9°)\vec{a}_x \text{ (A/m)}$$

8.31 In a free space, $\vec{E}$ is given as 45 sin $(\omega t - \beta z)\vec{a}_x$ (V/m). What is the average power p_{av}, which crosses a circular disc of radius 2 meters in the z = constant plane?

Solution:

It is given that

$$\vec{E} = 45 \sin(\omega t - \beta z)\vec{a}_x$$

$$= 45 e^{j(\omega t - \beta z)}\vec{a}_x$$

The wave is propagating in the +ve z-axis. $\vec{E}$ is in the +x-axis. $\vec{H}$ will propagate in the +y-axis as $\vec{a}_x \times \vec{a}_y = \vec{a}_z$. Finding $|\vec{H}|$ as

$$|\vec{H}| = \frac{|\vec{E}|}{\eta} \quad \text{or}$$

$$= \frac{45}{120\pi} e^{j(\omega t - \beta z)}\vec{a}_y$$

The complex conjugate of $\vec{H}$ is obtained as

$$\vec{H}^* = \frac{45}{120\pi} e^{j(\omega t - \beta z)}\vec{a}_y$$

Finding P_{av} as

$$P_{av} = \frac{1}{2}\text{Re}(\vec{E} \times \vec{H})$$

$$= \frac{1}{2} \times 45 \times \frac{45}{120\pi} e^{j(\omega t - \beta z)} \cdot e^{j(\omega t - \beta z)} \vec{a}_x \times \vec{a}_y \quad \text{or}$$

$$P_{\text{av}} = 2.68\, \vec{a}_z \ (\text{W/m}^2)$$

The average power crossing a circular disc is obtained as

$$= 2.68 \times \pi \times (2)^2 \text{ Watts}$$

$$= 33.75 \text{ Watts}$$

8.32 In free space, $z \leq 0$, a plane wave has $\vec{H} = 12\cos(18 \times 10^7 t - \beta z)\vec{a}_x$ (mA/m). Wave is incident normally on a lossless medium characterized by $\varepsilon_r = 9$ and $\mu_r = 4$ in region $z \geq 0$. Obtain reflected wave $(\vec{H}_r, \vec{E}_r)$ and transmitted wave $(\vec{H}_t, \vec{E}_t)$.

Solution:

Figure 8.9 shows the region $(z \geq 0)$ and free space $(z \leq 0)$.

In free space,

$$\beta_1 = \frac{\omega}{c} \quad \text{or}$$

$$= \frac{18 \times 10^7}{3 \times 10^8} \quad \text{or}$$

$$= 0.6$$

Also,

$$\eta_1 = \eta_0 = 120\pi$$

In the case of a lossless dielectric medium, we get

$$\beta_2 = \omega\sqrt{\mu\varepsilon}$$

$$= \omega\sqrt{\mu_0\varepsilon_0\mu_r\varepsilon_r}$$

$$= \frac{\omega}{c}\sqrt{9 \times 4}$$

$$= 0.6 \times \sqrt{36}$$

$$\beta_2 = 3.6$$

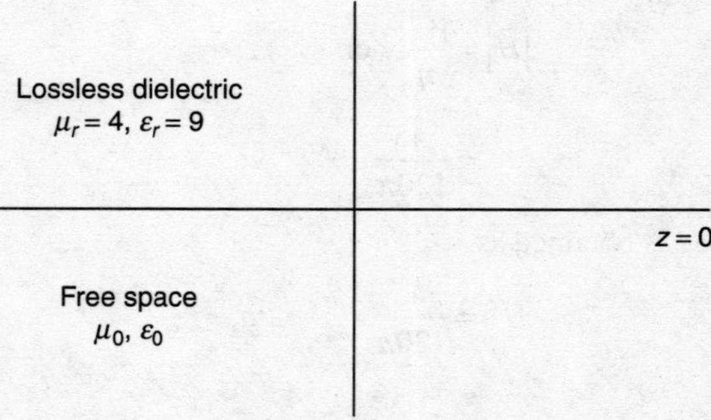

Figure 8.9 *Region and free space*

Further,

$$\eta_2 = \sqrt{\mu/\varepsilon}$$

$$= \sqrt{\frac{\mu_0\mu_r}{\varepsilon_0\varepsilon_r}}$$

$$= \sqrt{\frac{4}{9}}\eta_0$$

$$= \frac{2}{3}\times 120\pi$$

$$= 80\pi$$

It is given that $\vec{H}_i = 12\cos(18\times 10^7 t - \beta z)\,\vec{a}_x$ (mA/m). According to it, wave propagates in the +z-direction. $\vec{H}_i$ is in the +x-direction. E_i component will therefore exist along the −y-axis as $(-\vec{a}_y)\times\vec{a}_x = \vec{a}_z$. The magnitude of

$$\vec{E}_i = \eta_1 \left|\vec{H}_i\right| \quad \text{or}$$

$$E_{\text{im}} = 120\pi\times 12 \quad \text{or}$$

$$E_{\text{im}} = 1440\pi$$

Also,

$$\vec{E}_i = -1440\pi\cos(18\times 10^7 t - \beta_1 z)\,\vec{a}_y \text{ (mV/m)} \quad \text{or}$$

$$\vec{E}_i = -1440\pi\cos(18\times 10^7 t - 0.6z)\,\vec{a}_y \text{ (mV/m)}$$

Finding

$$\frac{E_{\text{rm}}}{E_{\text{im}}} = \frac{(\eta_2 - \eta_1)}{(\eta_2 + \eta_1)}$$

$$= -0.1$$

$$E_{\text{rm}} = -144\pi$$

Thus, we find

$$\vec{E}_r = -144\pi\cos(18\times 10^7 t + 0.6z)\,\vec{a}_y \text{ (mV/m)} \quad \text{and}$$

$$\vec{H}_r = \frac{-144\pi}{\eta_1}\cos(18\times 10^7 t + 0.6z)\vec{a}_x \text{ (mA/m)} \quad \text{or}$$

$$\vec{H}_r = 1.2\cos(18\times 10^7 t + 0.6z)\vec{a}_x \text{ (mA/m)}$$

It is therefore evident that the reflected wave travels along the −z-axis. For a transmitted wave, we get

$$\frac{E_{\text{tm}}}{E_{\text{im}}} = \tau$$

$$= 1 + \Gamma$$

$$= 1.1$$

$$\begin{aligned} E_{tm} &= 1.1 E_{im} \\ &= 1.1 \times 1440\pi \\ &= 1584\pi \end{aligned}$$

The direction of $\vec{E}_t$ will be the same as $\vec{E}_i$.

Thus,
$$\vec{E}_t = -E_m \cos(18 \times 10^7 t - \beta_2 z)\vec{a}_y \text{ (mV/m)}$$

$$\vec{E}_t = -1584\pi \cos(18 \times 10^7 - 3.6z)\vec{a}_y \text{ (mV/m)} \quad \text{and}$$

$$\vec{H}_t = \frac{-1584\pi}{\eta_2} \cos(18 \times 10^7 - 3.6z)\vec{a}_y \text{ (mV/m)}$$

$$\vec{H}_t = 19.8 \cos(18 \times 10^7 - 3.6z)\vec{a}_y \text{ (mV/m)}$$

8.33 A perpendicular polarized wave travels from a region characterized by $\varepsilon_r = 9$, $\mu_r = 1$, $\sigma = 0$ to a free space at an incidence angle of 10°. The incident field is 6.0 (V/m). Find reflected and transmitted electric fields. Also, find reflected and transmitted magnetic fields.

Solution:

For medium 1, we have $\varepsilon_{r_1} = 9$ and $\mu_1 = 1$. Thus,

$$\begin{aligned} \eta_1 &= \sqrt{\frac{\mu_1}{\varepsilon_{r_1}}} \\ &= 120\pi\sqrt{\frac{1}{9}} \\ &= 40\pi \ (\Omega) \end{aligned}$$

For medium 2 (free space), we get

$$\begin{aligned} \eta_2 &= \sqrt{\frac{\mu_0}{\varepsilon_0}} \\ &= 120\pi \ (\Omega) \end{aligned}$$

As per Snell's Law, we get

$$\frac{\sin\theta_i}{\sin\theta_t} = \sqrt{\frac{\varepsilon_2}{\varepsilon_1}} \quad \text{or}$$

$$\frac{\sin 10°}{\sin\theta_t} = \sqrt{\frac{\varepsilon_0}{\varepsilon_0 \times 9}}$$

$$\theta_t = 31.33°$$

The reflection coefficient for an electric field is obtained as

$$\frac{E_{\text{rm}}}{E_{\text{im}}} = \frac{\sqrt{\varepsilon_1}\cos\theta_i - \sqrt{\varepsilon_2}\cos\theta_t}{\sqrt{\varepsilon_1}\cos\theta_i + \sqrt{\varepsilon_2}\cos\theta_t} \quad \text{or}$$

$$= \frac{\sqrt{\varepsilon_0 \varepsilon_{r1}} \cos 10° - \sqrt{\varepsilon_0} \cos 31.33°}{\sqrt{\varepsilon_0 \varepsilon_{r1}} \cos 10° + \sqrt{\varepsilon_0} \cos 31.33°}$$

$$= \frac{2.95 - 0.85}{2.95 + 0.85} \quad \text{or}$$

$$= \frac{2.1}{3.80} \quad \text{or}$$

$$\frac{E_{\text{rm}}}{E_{\text{im}}} = 0.55 \quad \text{or}$$

$$E_{rm} = 3.3 (\text{V/m})$$

Transmitted electric field

$$E_{\text{tm}} = E_{\text{im}} + E_{\text{rm}}$$
$$= 6 + 3.3$$
$$= 9.3 \text{ V/m}$$

Incident magnetic field

$$H_{\text{im}} = \frac{E_{\text{im}}}{\eta_1} \quad \text{or}$$

$$= \frac{6}{40\pi} \quad \text{or}$$

$$= 0.04 \text{ (A/m)}$$

The reflected magnetic field H_{rm} is obtained as

$$H_{\text{rm}} = \frac{E_{\text{rm}}}{\eta_1} \quad \text{or}$$

$$= \frac{3.3}{40\pi} \quad \text{or}$$

$$= 0.02 \text{ (A/m)}$$

Transmitted magnetic field

$$H_{\text{tm}} = \frac{E_{\text{tm}}}{\eta_2} \quad \text{or}$$

$$= \frac{9.3}{120\pi} \quad \text{or}$$

$$= 0.024 \text{ A/m}$$

8.34 A parallel polarized wave is incident from the air into paraffin. Find the Brewster angle (ε_r for paraffin is 3).

Solution:

According to the formula, we have

$$\tan\theta_B = \sqrt{\frac{\varepsilon_2}{\varepsilon_1}}$$

$$= \sqrt{\frac{\varepsilon_0\varepsilon_{r2}}{\varepsilon_0\varepsilon_{r1}}}$$

$$= \sqrt{\frac{3}{1}}$$

Thus,

$$\theta_B = 60°.$$

8.35 What are reflection and transmission coefficients of an electric field wave that is travelling in air and is incident normally on boundary that is between air and a dielectric medium? (Assume $\mu_r = 9$ and $\varepsilon_r = 4$.)

Solution:

Mediums are air and dielectric medium. We can write

$$\eta_1 = \sqrt{\frac{\mu_0}{\varepsilon_0}}$$

$$= 377\ \Omega$$

In addition,

$$\eta_2 = \sqrt{\frac{\mu_0}{\varepsilon_0}}\sqrt{\frac{\mu_r}{\varepsilon_r}}$$

$$\eta_2 = 377 \cdot \frac{3}{2}$$

$$\eta_2 = 565.5\ \Omega$$

The reflection coefficient for normal incidence is obtained as

$$\frac{E_{\text{rm}}}{E_{\text{im}}} = \frac{\eta_2 - \eta_1}{\eta_2 + \eta_1} \quad \text{or}$$

$$\frac{E_{\text{rm}}}{E_{\text{im}}} = \frac{565.5 - 377}{565.5 - 377}$$

Thus,

$$\Gamma = \frac{188.5}{942.5}$$

$$= 0.2$$

Transmission coefficient for normal incidence is obtained as

$$\frac{E_{\text{rm}}}{E_{\text{im}}} = \frac{2\eta_2}{(\eta_2 + \eta_1)}$$

$$= \frac{2\times 565.5}{(565.5+377)} = 1.2$$

Thus, transmission coefficient $\tau = 1.2$.

8.36 The measured phase velocity of a dielectric medium is 150×10^6 m/s at t_1 and 220×10^6 m/s at t_2. Find the refractive index at t_1 and t_2.

Solution:

$$\text{Refractive index of medium} = \frac{\text{Velocity of light in free space}}{\text{Velocity of light in dieletric media}}$$

$$= \frac{c}{v_\rho}$$

However,

$$v_\rho = \frac{c}{\sqrt{\mu_0 t_0}}$$

$$= \frac{c}{\frac{c}{\sqrt{\varepsilon_r}}} = \sqrt{\varepsilon_r}$$

$$\text{Refractive index at } t_1 = \frac{c}{v_\rho} = \frac{3\times 10^8}{150\times 10^6} = 2$$

$$\text{Refractive index at } t_2 = \frac{3\times 10^8}{220\times 10^6} = 1.36$$

8.37 In free space, $\vec{E} = 60\sin(\omega t - 40x)\,\vec{a}_y$ V/m. Find I_d, $\vec{H}$ and ω?

Solution:

$$J_a = \frac{\partial \vec{D}}{\partial t} \quad \text{where} \quad \vec{D} = \varepsilon_0 \vec{E}$$

$$\vec{D} = \varepsilon_0\ 60\sin(\omega t - 40x)\vec{a}_y$$

$$\frac{\partial D}{\partial t} = 60\ \varepsilon_0\ \cos(\omega t - 40x)\cdot \omega\ \vec{a}_y$$

Therefore,

$$J_d = 60\omega\ \varepsilon_0\ \cos(\omega t - 40x)\ \vec{a}_y$$

As we know,

$$\frac{\omega}{\beta} = v_p = c = 3\times 10^8$$

In free space, $\beta = 50$ and $\omega = 1.5\times10^{10}$ rad/sec.

$$\nabla\times\vec{E} = \begin{vmatrix} \vec{a}_x & \vec{a}_y & \vec{a}_z \\ \dfrac{\partial}{\partial x} & \dfrac{\partial}{\partial y} & \dfrac{\partial}{\partial z} \\ 0 & E_y & 0 \end{vmatrix} \quad \text{or}$$

$$= \left(0 - \frac{\partial E_y}{\partial z}\right)\vec{a}_x - (0)\,\vec{a}_y + \left(\frac{\partial E_y}{\partial x} - 0\right)\vec{a}_z = i_c\left(\frac{\partial E_y}{\partial x}\right)$$

$$\frac{\partial E_y}{\partial x} = 60\cos(\omega t - 40x)(-40)$$

$$= -2400\cos(\omega t - 40x)\,\vec{a}_z$$

$$= \nabla\times\vec{E}$$

$$\nabla\times\vec{E} = \frac{-\partial B}{\partial t} = \frac{-\partial}{\partial t}(\mu_0\,\vec{H}) \qquad (\text{As } \vec{B} = \mu\vec{H}, \text{ in free space } \mu = \mu_0)$$

Therefore,

$$\mu_0\vec{H} = -\int\left(\nabla\times\vec{E}\right)dt = -2400\int\cos(\omega t - 40x)dt$$

$$\vec{H} = \frac{-2400(\sin\omega t - 40x)}{\omega\mu_0}$$

$$\vec{H} = \frac{-2400\times10^7}{1.5\times10^{10}\times4\pi}\sin(\omega t - 40x)\,\vec{a}_z$$

$$\vec{H} = -0.127\sin(\omega t - 40x)\,\vec{a}_z$$

Phase velocity

$$v_p = \frac{c}{\sqrt{\mu_r\varepsilon_r}}$$

$$= \frac{3\times10^8}{\sqrt{50}} \quad \text{or}$$

$$v_p = 4.243\times10^7 \text{ m/s}$$

$$\frac{\omega}{\beta} = v_p \quad \text{so}$$

$$\omega = \beta\times v_p \quad \text{or}$$

$$\omega = 2.43\times10^8 \text{ rad/sec}$$

Now,

$$\nabla\times\vec{E} = \begin{vmatrix} \vec{a}_x & \vec{a}_y & \vec{a}_z \\ \dfrac{\partial}{\partial x} & \dfrac{\partial}{\partial y} & \dfrac{\partial}{\partial z} \\ E_x & E_y & E_z \end{vmatrix}$$

As $E_y = E_z = 0$, for this case

$$\nabla \times \vec{E} = 0\,\vec{a}_x - \left(\frac{-\partial E_x}{\partial z}\right)\vec{a}_y + (0-0)\,\vec{a}_x$$
$$= \frac{\partial E_x}{\partial z}\,\vec{a}_y$$

Therefore,

$$\nabla \times \vec{E} = \frac{\partial}{\partial z}(40\pi\, e^{j(\omega t+\beta z)})\,\vec{a}_y$$
$$= 40\pi e^{j(\omega t+\beta 2)}(j\beta)\vec{a}_y$$

Also,

$$\frac{-\partial B}{\partial t} = -4\mu_0 H_m e^{j(\omega t+\beta z)}(j\omega)$$

From the above two equations,

$$-4\mu_0 H_m e^{j(\omega t+\beta z)}(j\omega) = 40\pi e^{j(\omega t+\beta z)}(j\beta)$$
$$H_m = \frac{-40\pi}{40\mu_0\omega\beta}$$
$$= \frac{-10\pi}{\mu_0\omega\beta}$$
$$|H_m| = \frac{10\pi}{\mu_0\omega\beta} = 0.143\ (\text{A/m})$$

8.38 On a long straight conducting wire surface, find out the Poynting vector. The wire has radius r, conductivity σ and direct current I. Also verify the Poynting theorem.

Solution:

It is given here that DC current is there. Therefore, it will be uniformly distributed over its cross-sectional area. Assume that the axis of the wire coincides with the z-axis. Therefore,

$$J = \frac{I}{\pi r^2}\vec{a}_z \quad \text{and}$$
$$E = \frac{J}{\sigma} = \frac{I}{\sigma\pi r^2}\vec{a}_z$$

Also, on the surface of the wire

$$H = \frac{1}{2\pi r}\,\vec{a}_\phi$$

Therefore, on the surface of the wire, the Poynting vector is given as

$$\vec{P} = \vec{E} \times \vec{H} \quad \text{or}$$
$$= \frac{I^2}{2\sigma\pi^2 r^3}\,(\vec{a}_z \times \vec{a}_\phi) \quad \text{or}$$
$$= \frac{I^2}{2\sigma\pi^2 r^3}(-\vec{a}_R)$$

For verification of the Poynting Theorem, integrating $\vec{P}$ over the wall of the wire segment is

$$\oint_s \vec{p}\cdot\vec{ds} = \oint_s \vec{p}\cdot\vec{a}_R\, ds \quad \text{or}$$

$$= \left(\frac{I^2}{2\sigma\pi^{2r^3}}\right)(2\pi bl) \quad \text{or}$$

$$= I^2\left(\frac{l}{\sigma\pi b^2}\right) \quad \text{or}$$

$$= I^2 R \ \text{(ohmic power loss)}$$

where, R is the resistance of the wire.

8.39 A uniform plane wave is a lossless medium with intrinsic impedance η_1 incident normally onto another lossless medium with intrinsic impedance η_2 through a plane boundary. Obtain an expression for the time average power densities in both media.

Solution:

Time average Poynting vector is given by

$$\vec{P}_{\text{avg}} = \frac{1}{2} Re(\vec{E}\times\vec{H})$$

For medium 1,

$$(\vec{P}_{\text{avg}})_1 = \frac{E_{i0}^2}{2\eta_1}\text{Re}[(1+\Gamma_L e^{j2\beta_1 z})(1-\Gamma e^{-j2\beta_1 z})]\vec{a}_z$$

$$= \frac{E_{i0}^2}{2\eta_1}\text{Re}[(1-\Gamma_L^2)+\Gamma(e^{j2\beta_1 z}-e^{-j2\beta_1 z})]\vec{a}_z$$

$$= \frac{E_{i0}^2}{2\eta_1}\text{Re}[(1-\Gamma_L^2)+j2\Gamma\sin 2\beta_1 z]\ \vec{a}_z$$

$$= \frac{E_{i0}^2}{2\eta_1}(1-\Gamma_L^2)\ \vec{a}_z$$

For medium 2,

$$\vec{P}_{\text{avg2}} = \frac{E_{i0}^2}{2\eta_2}\tau^2$$

As the given media is lossless, we have

$$\left(\vec{P}_{\text{avg}}\right)_1 = \left(\vec{P}_{\text{avg}}\right)_2$$

Therefore,

$$1-\Gamma_L^2 = \frac{\eta_1}{\eta_2}\tau^2$$

8.40 A wave propagates from a dielectric medium to the interface with free space. If the angle is the critical angle of 25°, find the relative permittivity.

Solution:

Critical angle is the angle of incidence for which the reflection angle is 90°.

$$\frac{\sin i}{\sin r} = \sqrt{\frac{\varepsilon_2}{\varepsilon_1}}$$

$$\frac{\sin 25°}{\sin 90°} = \sqrt{\frac{8.82 \times 10^{-12}}{\varepsilon_1}}$$

Therefore,

$$\frac{(0.42)^2}{1} = \frac{8.85 \times 10^{-12}}{\varepsilon_1} \quad \text{or}$$

$$\varepsilon_1 = 5.6\ \varepsilon_0 \quad \text{or}$$

$$\varepsilon_{r1} = 5.6$$

8.41 A travelling electric field in free space of amplitude 500 V/m strikes a perfect dielectric medium as shown in Figure 8.10. Find E_f.

Solution:

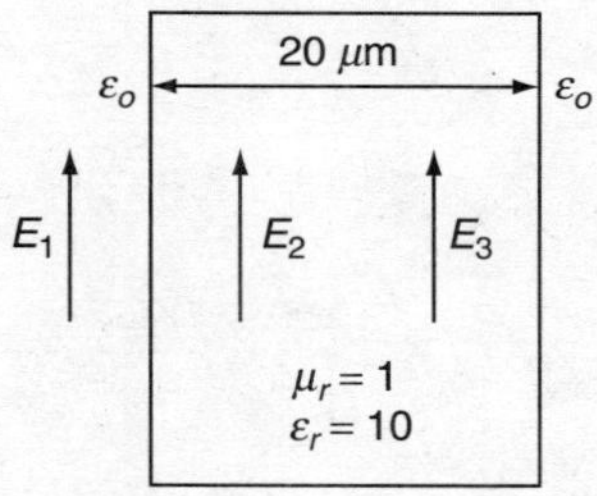

$$\eta_1 = 377\,\Omega$$

$$\eta_2 = 377\sqrt{\frac{1}{10}} = 119.2\,\Omega$$

$$E_1 = 500\text{ V/m}$$

$$\frac{E_2}{E_1} = \frac{2\eta_2}{\eta_1 + \eta_2} = \frac{2 \times 119.2}{377 + 119.2} = 0.48$$

Figure 8.10 *Dielectric*

Therefore,

$$E_2 = 240\text{ V/m}$$

$$E_3 = E_2 e^{-\alpha z} \qquad \text{(for perfect dielectric } \alpha = 0\text{)}$$

$$E_2 = E_3 = 240\text{ V/m}$$

$$\frac{E_4}{E_3} = \frac{2\eta_2}{\eta_2 + \eta_1} = \frac{2 \times 377}{377 + 119.2} = 1.519$$

$$\frac{H_r}{H_i} = \frac{-E_r}{E_i} = \frac{(\eta_1 - \eta_2)}{(\eta_1 + \eta_2)} = 0.39$$

$$\frac{H_\varepsilon}{H_i} = \frac{2\eta_1}{\eta_1 + \eta_2} = \frac{2 \times \sqrt{\frac{\mu_0}{\varepsilon_0}}}{\sqrt{\frac{\mu_0}{\varepsilon_0}} + \sqrt{\frac{\mu_0}{5\mu_0}}} = \frac{2}{1 + \frac{1}{\sqrt{5}}} = 0.81$$

8.42 A 150-MHz plane wave penetrates though 1 A of $S = 10^4$ mho S/m $\varepsilon_2 = \mu_r = 1$. Calculate the skin depth and also the depth at which the wave amplitude decreases to 18.1% of its initial value.

Solution:

$$\text{Skin depth} = \text{loss tangential} = \frac{\sigma}{\omega t}$$

$$= \frac{10^4}{2\pi \times 150 \times 10^6 \times 8.86 \times 10^{-12}}$$

$$= 119755.4\ (>> 1)$$

$$\delta = \sqrt{\frac{2}{\omega\mu\sigma}} = \sqrt{\frac{2}{2\pi f \mu\sigma}} \quad \text{or}$$

$$= \frac{1}{\sqrt{\mu + \mu\sigma}} = \frac{1}{\sqrt{150 \times 10^6 \times 4\pi \times 10^{-7} \times 10^4}} \quad \text{or}$$

$$= 7.28 \times 10^{-4}\ \text{m}$$

Now,

$$e^{-\alpha x} = 0.181 \quad \text{or}$$

$$e^{+\alpha x} = \frac{1}{0.181} = 5.52 \quad \text{or}$$

$$\alpha x = \ln(5.52) \quad \text{or}$$

$$\alpha x = 1.708 \quad \text{or}$$

$$x = \frac{1.708}{\alpha}$$

Further,

$$\sigma = \sqrt{\frac{\omega\mu\sigma}{2}}$$

$$= \sqrt{\frac{2\pi f \mu\sigma}{2}}$$

$$\alpha = \sqrt{\pi f \mu\sigma}$$

$$= \sqrt{150 \times 10^6 \times 4\pi \times 10^{-7} \times 10^4 \times \pi}$$

$$= 2433.4$$

$$\Rightarrow \quad x = \frac{1.708}{2433.4} = 7.01 \times 10^{-4}\ \text{m}$$

8.43 Given $\vec{E}_x = 120\pi e^{j(20^8 t + \beta z)}$ and $\vec{H}_y = H_0 e^{j(20^8 t + \beta_z)}$. Find values of H_0 and β.

Solution:

$$\beta = \frac{\omega}{v_0} = \frac{20^8}{3 \times 10^8} = 85.83\ \text{rad/m}$$

$$H_0 = \frac{E_0}{\eta} = \frac{120\pi}{120\pi} = 1 \text{ (Amp./m)}$$

8.44 Given two dielectric media, medium 1 is a free space and medium 2 is a dielectric medium having $\varepsilon_2 = 2\varepsilon_0$ and $\mu = \mu_0$. Find the reflection coefficient for oblique incidence $\angle i = 60°$ for parallel polarization.

Solution:

$$\varepsilon_1 = \varepsilon_0 \quad \text{and} \quad \mu_1 = \mu_0$$

$$\varepsilon_2 = 2\varepsilon_0$$

$$\angle i = 60°$$

$$\frac{\sin i}{\sin r} = \sqrt{\frac{\varepsilon_2}{\varepsilon_1}} = \sqrt{\frac{2\varepsilon_0}{\varepsilon_0}} = \sqrt{2}$$

$$\sin r = \frac{\sin i}{\sqrt{2}} = 0.61$$

$$r = 37.58°$$

$$\cos r = 0.79$$

For parallel polarization,

$$R = \frac{\sqrt{\varepsilon_2}\cos\theta_1 - \sqrt{\varepsilon_1}\cos\theta_2}{\sqrt{\varepsilon_2}\cos\theta_1 + \sqrt{\varepsilon_2}\cos\theta_2}$$

where $\theta_1 = \angle i$ and $\theta_2 = r$. Therefore,

$$R = \frac{\sqrt{2\varepsilon_0}\,(1/2) - \sqrt{\varepsilon_0} \times 0.79}{\sqrt{2\varepsilon_0}\,(1/2) + \sqrt{\varepsilon_0} \times 0.79}$$

$$= \frac{(0.707 - 0.79)}{(0.707 + 0.79)}$$

$$R = -0.05$$

8.45 A dielectric slab of the flat surface with relative permittivity 6 is disposed with its surface normal to a uniform field with flux density of 3 Coulombs per m^3. The slab occupies a volume of 0.09 m^3. It is uniformly polarized. Determine polarization in the slab.

Solution:

We know that

$$D = \varepsilon\, E$$

$$= \varepsilon_o\, \varepsilon_r\, E$$

$$= 3 \text{ C/m}^3$$

Now,

$$E = \frac{D}{\varepsilon_o \varepsilon_r}$$

$$= \frac{3}{\varepsilon_o \times 6}$$

$$\Rightarrow \quad \varepsilon_o E = \frac{3}{6} = 0.5$$

Also, polarization

$$p = (D - \varepsilon_o E)$$

$$= 3 - 0.5$$

Thus,

$$p = 2.5 \text{ C/m}^3$$

UNSOLVED QUESTIONS

8.1 A sinusoidal plane wave travelling through a medium having a breakdown strength of medium is 70 (kV/m) rms. The relative permittivity is 3. What will be the mean power flow density?

8.2 A transmission is characterized by $\gamma = (450 + j1448 \cdot 5)$ (per m) and $\eta = 40\angle 18$ (Ω). The wave frequency is 200 (MHz). Electric field $\vec{E}$ is $50\,e^{-\alpha z}$ cos $(3\pi\, 10^9 t - \beta z)\vec{a}_x$. Find the average power/m^2 of the electromagnetic wave at $z = 1.2$ mm.

8.3 The loss tangent of ice at 2000 MHz is 0.007. ε_r is 2.20. A uniform plane wave travelling in it has an amplitude of 200 (V/m). Find the time average power density at $x = 7$ m.

8.4 A parallel polarized electromagnetic wave is incident from air into flint glass. Flint glass is characterized by $\mu_r = 1.8$ and $\varepsilon_r = 11.2$. What will be the angle?

8.5 In a case, medium 1 is a free space. Medium 2 is characterized by $\varepsilon_r = 3$ and $\mu_r = 1.2$. Find the values of coefficients for oblique incidence $\theta_i = 30°$ in case of
(a) Perpendicular polarization and
(b) Parallel polarization

8.6 Define the Brewster angle. Explain its phenomenon.

8.7 What are the applications of the Brewster angle in real life?

SUMMARY

1. An electromagnetic wave is characterized by its amplitude, phase, angular frequency and wave number.

2. Speed of wave is given as $f\lambda$. Here, f is the frequency (Hz) and λ is the wavelength (meters).

3. β is the symbol for phase constant. It is also called wave number $\beta = \dfrac{2\pi}{\lambda}$

4. Wave equation in a lossy dielectric medium is given as $E_y = E_m e^{-\alpha x} e^{j(\omega t-\beta x)}$; $e^{-\alpha x}$ is called wave attenuation.
Phase velocity v_p is given as ω/β.

Intrinsic impedance η is given as $\sqrt{\dfrac{j\omega\mu}{\sigma + j\omega\varepsilon}}$.

5. For a good conductor, $\sigma >> \omega\varepsilon$. Here, intrinsic impedance η is given as $\sqrt{\dfrac{\omega\mu}{\sigma}}\angle 45°$.

6. For a good dielectric medium, $\sigma << \omega\varepsilon$. Wave velocity is given as $\dfrac{1}{\sqrt{\mu\varepsilon}}\left[1-\dfrac{\sigma^2}{8\omega^2\varepsilon^2}\right]$. Intrinsic impedance η is given as $\sqrt{\dfrac{\mu}{\varepsilon}}\left(1+\dfrac{j\sigma}{2\omega\varepsilon}\right)$.

7. For free space, $\sigma = 0$, $\varepsilon = \varepsilon_o$ and $\mu = \mu_o$. Wavelength is given as $\dfrac{2\pi}{\omega\sqrt{\mu_o\varepsilon_o}}$.

8. Depth of penetration–skin depth δ is given as $\sqrt{1/\pi f \mu\sigma}$. It is 25 cm into sea water and 7.1 m into fresh water.

9. The Poynting Vector $\vec{P}$ is given as $\vec{E}\times\vec{H}$. It is instantaneous power density and is expressed in w/m^2.

10. According to the Poynting Theorem, the net power flowing out of a given volume is equal to the time rate of decrease in energy stored minus conduction losses and others.

11. Reflection coefficient Γ is given as $(\eta_2 - \eta_1)/(\eta_2 + \eta_1)$. It ranges from –1 to +1.

12. Transmission coefficient ∂ is given as $2\eta_2/(\eta_2 + \eta_1)$.

13. Reflection coefficient Γ and transmission coefficient ∂ are related as $(1 + \Gamma) = \tau$.

14. Standing wave ratio (SWR) is given as $(1+|\Gamma|)/(1-|\Gamma|)$. It ranges from 1 to 10.

15. According to Snell's Law, $\eta_1 \sin\theta_i = \eta_2 \sin\theta_t$.

16. Brewster angle θ_B is given as $\tan\theta_B = \sqrt{\varepsilon_2/\varepsilon_1}$.

17. Light with polarization parallel to the plane is called "p-polarized" and with perpendicular polarization is called "s-polarized".

18. At the Brewster angle, an incident unpolarized light's reflected light is always "s-polarized".

19. For perpendicular polarization (zero reflection), $\varepsilon_2 = \varepsilon_1$.

MULTIPLE-CHOICE QUESTIONS

1. At the Brewster angle, polarization ______________.
   (a) Cannot be reflected
   (b) Is reflected at 30°
   (c) Is reflected at 90°
   (d) None of these

2. Wave number has units of ________.
(a) Radians (b) Meter (c) Radians/meter (d) None of these

3. Wave speed in terms of frequency f and wavelength λ is expressed as ________.
(a) f/λ (b) λ/f (c) λf (d) $(\lambda + f)$

4. For a lossy dielectric medium, ________.
(a) $\sigma = 0$ (b) $\sigma \neq 0$ (c) None of these (d) Cannot say

5. Wave attenuation is given as ________.
(a) $e^{+\beta x}$ (b) $e^{-\beta x}$ (c) $e^{+\alpha x}$ (d) $e^{-\alpha x}$

6. In the case of a perfect dielectric medium, phase constant ________ as conductivity increases.
(a) Increases (b) Decreases
(c) Remains unchanged (d) None of these

7. Phase velocity is given as ________.
(a) $\omega\beta$ (b) β/ω (c) ω/β (d) None of these

8. For a good conductor ________.
(a) $\dfrac{\sigma}{\omega\varepsilon} = 0$ (b) $\dfrac{\sigma}{\omega\varepsilon} << 1$ (c) $\dfrac{\sigma}{\omega\varepsilon} >> 1$ (d) $\dfrac{\sigma}{\omega\varepsilon} = \infty$

9. For a good dielectric medium ________.
(a) $\dfrac{\sigma}{\omega\varepsilon} = 0$ (b) $\dfrac{\sigma}{\omega\varepsilon} << 1$ (c) $\dfrac{\sigma}{\omega\varepsilon} >> 1$ (d) $\dfrac{\sigma}{\omega\varepsilon} = \infty$

10. In good conductors, rate of attenuation is ________.
(a) Small (b) Large (c) Infinity (d) Zero

11. Poynting Vector $\vec{P}$ is obtained as ________.
(a) $\vec{E} \cdot \vec{H}$ (b) $\vec{E} + \vec{H}$ (c) $\vec{E} - \vec{H}$ (d) $\vec{E} \times \vec{H}$

12. Power density has ________.
(a) A DC component (b) A second harmonic component
(c) Both (a) and (b) (d) None of these

13. Reflection coefficient Γ is ________.
(a) ≥ 100 (b) $= 10$ (c) ≤ 1 (d) None of these

14. Which of the following expressions is correct?
(a) $(1 + \Gamma) = \tau$ (b) $(1 + \tau) = \Gamma$ (c) $(1 + \Gamma)\,\tau = 0$ (d) $(1 + \tau)\Gamma = 0$

15. Transmission coefficient is given as ________.
(a) $\dfrac{2\eta_1}{(\eta_2 + \eta_1)}$ (b) $\dfrac{2\eta_2}{(\eta_2 + \eta_1)}$ (c) $\dfrac{(\eta_2 + \eta_1)}{2\eta_1}$ (d) $\dfrac{(\eta_2 + \eta_1)}{2\eta_2}$

16. Standing wave consists of two travelling waves of __________ amplitudes and __________ is direction.
(a) Unequal, same
(b) Unequal, opposite
(c) Equal, same
(d) Equal, opposite

17. Standing wave ratio S is given as __________.
(a) $\frac{1+|\Gamma|}{1-|\Gamma|}$
(b) $\frac{1-|\Gamma|}{1+|\Gamma|}$
(c) $\frac{|\Gamma|}{1+|\Gamma|}$
(d) $\frac{|\Gamma|}{1-|\Gamma|}$

18. SNR ranges from ______.
(a) 0 to 1
(b) 1 to 10
(c) 10 to 100
(d) 1 to ∞

19. As per Snell's Law _______.
(a) $\frac{\sin\theta_i}{\sin\theta_t} = \frac{\eta_1}{\eta_2}$
(b) $\frac{\sin\theta_i}{\sin\theta_t} = \frac{\eta_2}{\eta_1}$
(c) $\frac{\tan\theta_i}{\tan\theta_t} = \frac{\eta_1}{\eta_2}$
(d) $\frac{\tan\theta_i}{\tan\theta_t} = \frac{\eta_2}{\eta_1}$

20. Brewster's angle is given as $\tan\theta_B =$ __________.
(a) $\sqrt{\frac{\varepsilon_1}{\varepsilon_2}}$
(b) $\sqrt{\frac{\varepsilon_2}{\varepsilon_1}}$
(c) $\sqrt{\varepsilon_1 \varepsilon_2}$
(d) None of these

ANSWERS TO MULTIPLE-CHOICE QUESTIONS

(1) a; (2) c; (3) c; (4) b; (5) d; (6) a; (7) c; (8) c; (9) b;
(10) c; (11) d; (12) c; (13) c; (14) a; (15) b; (16) d; (17) a; (18) d;
(19) b; (20) b

9

Transmission Lines

9.1 INTRODUCTION

The previous chapter dealt with wave propagation in unbounded media (media of infinite extent). Such wave propagation is "unguided". In unguided propagation, a uniform plane wave exists throughout the space. EM energy associated with such a wave is spread across a very wide area. Wave propagation in "unbounded-media" finds its use/application in TV or radio broadcasting. In such applications, information is being transmitted for anybody who is interested. Such wave propagation is not desirable in telephonic applications. This is because information is received privately by one person only.

Another very important means of transmitting power or information is "guided structures". Guided structures guide (or direct) propagation of energy from a source to a load. Examples of such structures are transmission lines, wave guides, etc.

In power systems, transmission lines are very commonly used for power transmission and distribution, especially at low frequencies. Transmission lines are used to connect a source to a load. A source can be an electric generator, an oscillator or a transmitter. The load can be an antenna, a furnace or an oscilloscope.

Transmission lines such as twisted pair and coaxial cables (thin net and thick net) are used in Ethernet and Internet (computer networks). Transmission line problems can be very easily solved using the electric circuit theory and the EM field theory.

Transmission line analysis involves derivation of transmission line equations and characteristic quantities. Smith charts are used. Various practical applications of transmission lines, knowledge and information on transients in transmission lines help their analysis.

9.2 PARAMETERS OF TRANSMISSION LINES

A transmission line can be described by its line parameters. Line parameters include resistance per unit length R, conductance per unit length G, inductance per unit length L and capacitance per unit length C. These transmission line parameters depend on dimensions of the line and also on properties of conductors and the dielectric materials used. We will first consider simple transmission lines.

The conductivity of the transmission line conductors is usually high so that the effect of series resistance on the propagation constant is made negligible.

It is given as

$$\gamma = \sqrt{(R + j\omega L)(G + j\omega C)}$$

When $R = 0$, we get

$$\gamma = j\omega\sqrt{LC}\left(1+\frac{G}{j\omega C}\right)^{1/2}$$

Propagation constant γ for a TEM wave in a medium is given as

$$\gamma = j\omega\sqrt{\mu\varepsilon}\left(1+\frac{\sigma}{j\omega\varepsilon}\right)^{1/2}$$

where μ, ε and σ are known as constitutive parameters.
Further,

$$\frac{G}{C} = \frac{\sigma}{\varepsilon}$$

Thus,

$$LC = \mu\varepsilon$$

This is valid in the case of a lossless transmission line. The velocity of propagation (v) in the case of a lossless transmission line is

$$= \frac{1}{\sqrt{LC}}$$

It is the same as the velocity of propagation ($1/\sqrt{\mu\varepsilon}$) of a guided plane wave in the dielectric medium of a transmission line.

9.3 TYPES OF TRANSMISSION LINES

A transmission line consists of two conductors. Common transmission lines allowing transverse electromagnetic (TEM) wave propagation is given below.

9.3.1 Coaxial Transmission Lines

It has an inner conductor and a coaxial outer conducting sheath. Both are separated by a dielectric medium. Both electric and magnetic fields are confined to the dielectric region here. Coaxial transmission lines are very popular as cables, telephone cables, TV cables, probes and leads of measuring instruments. Power cables are generally coaxial transmission lines. The cross-section of a coaxial transmission line is shown in Figure 9.1.

Capacitance C per unit length of the coaxial transmission line is given as

$$C = \frac{2\pi\varepsilon}{\ln(b/a)}$$

The inductance of a coaxial transmission line is given as

$$L = \frac{\mu_o}{2\pi}\ln(b/a)$$

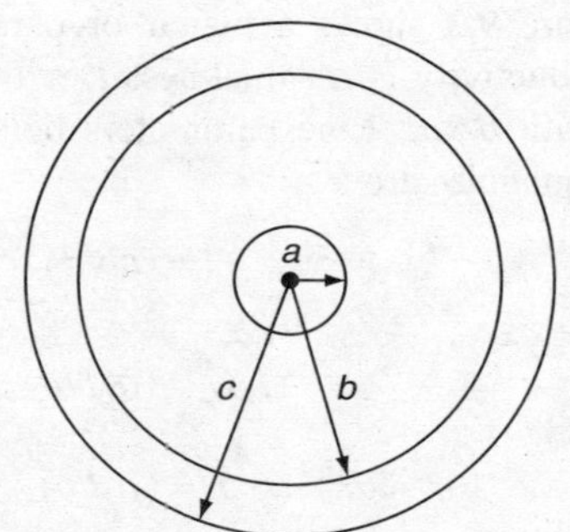

Figure 9.1 *Coaxial transmission line*

It is only the external inductance of a conductor. The flux inside the conductor has not been considered.

The conductance G per unit length is simply the reciprocal of the "leakage insulation resistance" of the coaxial transmission line. As shown in Figure 9.2, let the radius be x, dx cylindrical insulation thickness, then the insulation resistance dR_i (of cylindrical strip) of the dielectric medium is given as

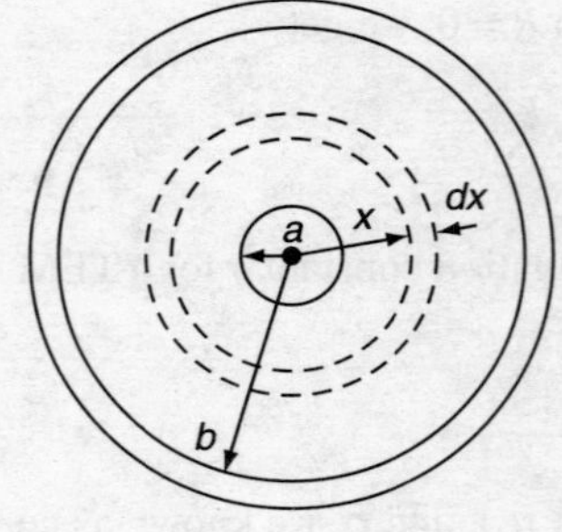

Figure 9.2 *Coaxial transmission line*

$$dR_i = \frac{1}{\sigma}\frac{dx}{2\pi x \cdot L} \qquad \text{(where } \sigma \text{ is the conductivity)}$$

The total insulation resistance per meter length is obtained as

$$R_i = \frac{1}{2\pi\sigma}\int_a^b \frac{dx}{x} \qquad (L = 1;\ \text{per unit length})\ \text{or}$$

$$= \frac{1}{2\pi\sigma}\ln(b/a)$$

The conductance per meter length of a coaxial transmission line is

$$G = \frac{1}{R_i} \quad \text{or}$$

$$= \frac{2\pi\sigma}{\ln(b/a)}$$

The resistance per meter length of a coaxial transmission line can be found by adding resistances of the inner conductor and the outer conductor. It is

$$R = \frac{1}{\sigma_C \pi}\left[\frac{1}{a^2} + \frac{1}{(c^2 - b^2)}\right]$$

where σ_C is the conductivity at low frequencies of the coaxial transmission line conductor.

9.3.2 Planar Transmission Lines

Figure 9.3 shows a planar or parallel plane transmission line. It has two conducting planes. Say, conductivity is σ_c, thickness t, width b and separation d. Dielectric medium parameters are μ, σ and ε with $b >> d$. Line parameters per unit length at high frequencies are

$$C = (\varepsilon b/d)$$

$$G = (\sigma_c b/d)$$

$$L = (\mu d/b) \quad \text{and}$$

$$R = (2/\sigma_c \delta b)$$

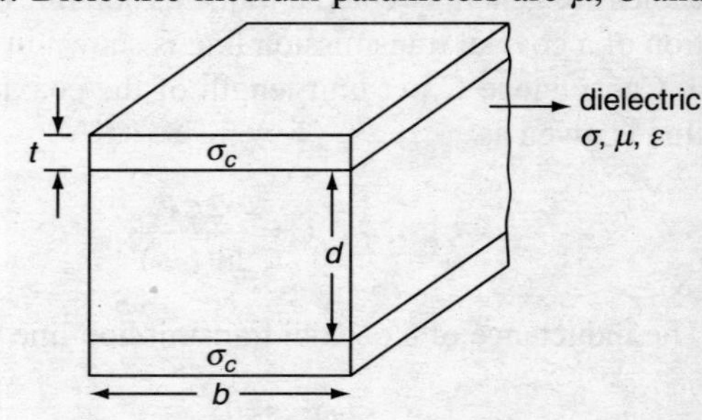

Figure 9.3 *Planar transmission line*

9.3.3 Two-wire Transmission Lines

This transmission line has a pair of parallel conductivity wires. The parallel conducting wires are separated. A two-wire transmission line is shown in Figure 9.4. These transmission lines are used in power system, telephone lines, etc. Let in two-wire transmission line conductor have radius a. Conductivity is σ_c. The distance between these two conductors is d. Medium parameters are μ, ε and σ. The capacitance per unit length is

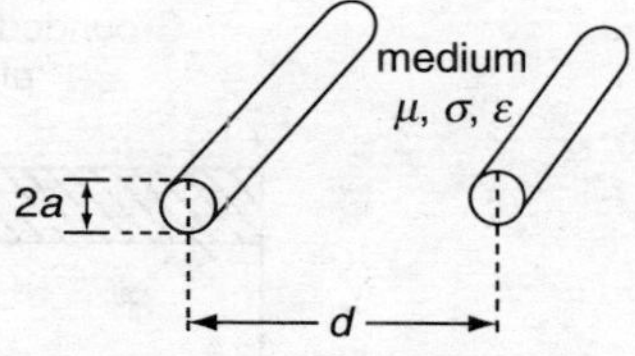

Figure 9.4 *Two-wire transmission line*

$$C = \frac{\pi\varepsilon}{\cosh^{-1}(d/2a)} \quad \text{(Valid for both high and low frequencies) or}$$

$$C \simeq \frac{\pi\varepsilon}{\ln(d/a)}; \quad a << d$$

Conductance per unit length is

$$G = \frac{\pi\sigma}{\cosh^{-1}(d/2a)} \quad \text{(Valid for both high and low frequencies)}$$

The inductance per unit length of a two-wire transmission line at low frequencies is a little larger than that at higher frequencies. It is so because at low frequencies, the internal inductance of the conductor is also considered. The inductance per unit length is

$$L = \frac{\mu}{\pi}\cosh^{-1}(d/2a) \quad \text{or}$$

$$L = \frac{\mu}{\pi}\ln(d/a); \quad a << d \qquad \text{(at high frequencies) and}$$

$$L = \frac{\mu}{\pi}\left[\frac{1}{4} + \cosh^{-1}(d/2a)\right] \qquad \text{(at low frequencies)}$$

Resistance per unit length is

$$R = \frac{1}{\pi\, a\, \delta\, \sigma_c} \quad \text{(at high frequencies) and}$$

$$R = \frac{2}{\sigma_c\, \pi\, a^2} \quad \text{(at low frequencies)}$$

9.4 MICROSTRIP LINE

A microstrip line has a dielectric substrate over a grounding conducting plane. There is a thin narrow metal strip over the substrate. Strip lines are fabricated using printed circuit technologies. Strip lines can be integrated very easily along with other circuit components. Stray fields of string lines interface with neighbouring circuits. Triplate lines are also used. In triplate lines, conducting planes (on both sides of the dielectric substrate) and a thin metal strip are put in the middle as shown in Figure 9.5.

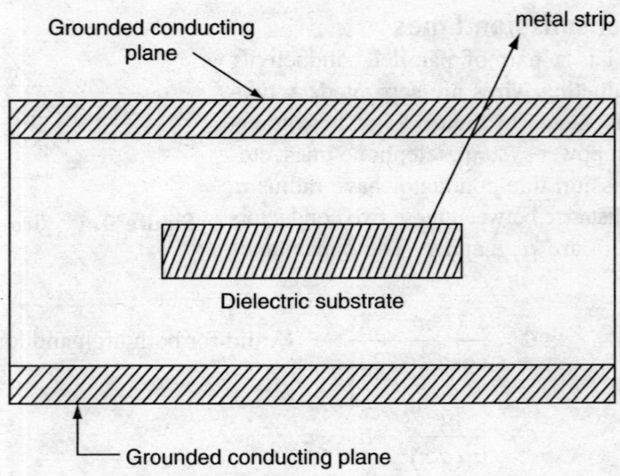

Figure 9.5 *Triplate line*

9.5 TRANSMISSION LINE EQUATION

A two-conductor transmission line can carry a TEM wave. The electric and magnetic fields on this line will be transverse to the wave propagation direction. In a TEM wave, field $\vec{E}$ is related to voltage V and field $\vec{H}$ is related to current I. It is given as

$$V = -\int \vec{E} \cdot d\vec{l} \quad \text{and}$$

$$I = \oint \vec{H} \cdot d\vec{l}$$

where V and I are used to solve transmission line problems and we need not to obtain field quantities. $\vec{E}$ and $\vec{H}$ (i.e. solving Maxwell's equation and the associated boundary conditions). Thus, the model is quite simple and convenient to analyse.

Consider an incremental length Δx of a conductor transmission line. The equivalent circuit of a portion of this transmission line is shown in Figure 9.6. This equivalent circuit comprises various line parameters R, L, G and C. Figure 9.6 is called an "L – type" equivalent circuit. Here, it is assumed that wave propagates in the $+x$ direction, i.e. generator to load.

By applying Kirchhoff 's voltage law, we get

$$V(x,t) = R\Delta x\, I(x,t) \; + L\Delta x \frac{\partial I}{\partial t}(x,t) \; + V(x+\Delta x, t) \quad \text{or}$$

$$\frac{-V(x+\Delta x, t) - V(x,t\,)}{\Delta x} = R\, I(x,t) + L\frac{\partial I(x,t)}{\partial t}$$

Taking the limit as $\Delta x \to 0$, we get

$$-\frac{\partial V(x,t)}{\partial t} = RI(x,t) + L\frac{\partial I(x,t)}{\partial t}$$

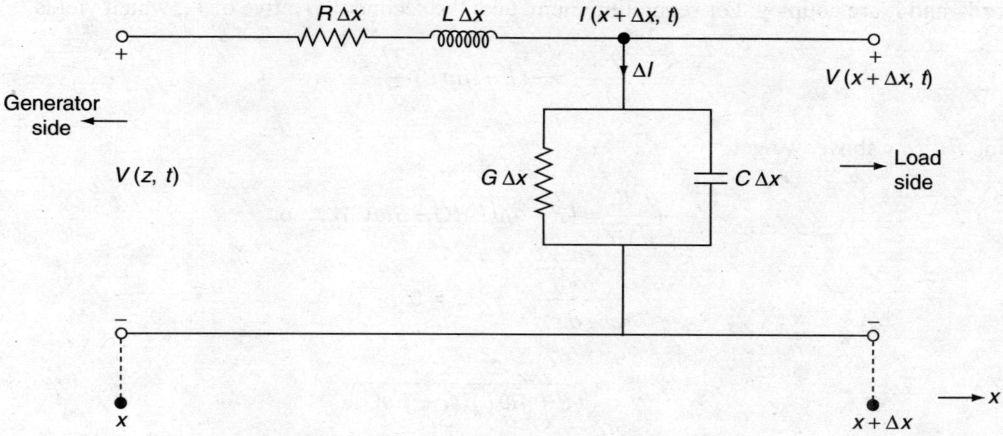

Figure 9.6 *"L-type" equivalent circuit model*

By applying Kirchhoff's current law, we get

$$I(x, t) = I(x + \Delta x, t) + \Delta I$$

$$= I(x + \Delta x, t) \; + G\Delta x \; V(x + \Delta x, t) \; + C\Delta x \frac{\partial V}{\partial t}(x + \Delta x, t) \quad \text{or}$$

$$\frac{-I(x + \Delta x, t) + I(x, t)}{\Delta x} = GV(x + \Delta x, t) \; + C\frac{\partial V}{\partial t}(x + \Delta x, t)$$

Putting $\Delta x \to 0$, we get

$$-\frac{\partial I(x, t)}{\partial t} = GV(x, t) \; + C\frac{\partial V(x, t)}{\partial t}$$

Assuming harmonic time as

$$V(x, t) = \text{Re}[V_s(x) \cdot e^{j\omega t}] \quad \text{and}$$

$$I(x, t) = \text{Re}[I_s(x) \cdot e^{j\omega t}]$$

where $V_s(x)$ and $I_s(x)$ are phasor forms of $V(x, t)$ and $I(x, t)$.
Thus,

$$-\frac{\partial V_s}{\partial x} = (R + j\omega L) I_s \quad \text{and}$$

$$-\frac{\partial I_s}{\partial x} = (G + j\omega C) V_s$$

where V_s and I_s are coupled. For separating them, take the second derivative of V_s, which yields

$$\frac{\partial^2 V_s}{\partial x^2} = -(R + j\omega L)\frac{\partial I_s}{\partial x}$$

Putting $\partial I_s/\partial x$ above, we get

$$-\frac{\partial^2 V_s}{\partial x^2} = (R + j\omega L)(G + j\omega C)V_s \quad \text{or}$$

$$\frac{\partial^2 V_s}{\partial x^2} - \gamma^2 V_s = 0$$

where

$$\gamma = \sqrt{(R + j\omega L)(G + j\omega C)}$$
$$= (\alpha + j\beta)$$

The second derivative of I_s gives

$$\frac{\partial^2 I_s}{\partial x^2} = (R + j\omega L)(G + j\omega C) I_s \quad \text{or}$$

$$\frac{\partial^2 I_s}{\partial x^2} - \gamma^2 I_s = 0$$

where γ is $(\alpha + j\beta)$ (known as "propagation constant"), α the attenuation constant (neper/meter) and β the phase constant (radians/meter). The above expressions are called "wave equations" for voltage and current. Wavelength λ and wave velocity u are given as

$$\lambda = (2\pi/\beta) \quad \text{and}$$

$$u = \frac{\omega}{\beta} \quad \text{or}$$
$$= f\lambda$$

The solution of the above linear homogenous differential wave equations is obtained as

$$V_s(x) = V_0^+ e^{-\gamma x} + V_0^- e^{+\gamma z} \quad \text{and}$$

$$I_s(x) = I_0^+ e^{-\gamma x} + I_0^- e^{+\gamma z}$$

where V_0^+, V_0^-, I_0^+ and I_0^- are wave amplitudes. The + sign denotes a wave travelling in the $+z$ direction and the – sign denotes a wave travelling in the $-z$ direction.

Instantaneous voltage expression is obtained as

$$V(x,t) = \text{Re}[V_s(x) \cdot e^{j\omega t}] \quad \text{or}$$
$$= V_0^+ e^{-\alpha x} \cos(\omega t - \beta x) + V_0^- e^{+\alpha x} \cos(\omega t + \beta x)$$

9.5.1 Characteristic Impedance (Z_o)

Characteristic impedance Z_o is the ratio of square root of series impedance per unit length (Z) to square root of the shunt admittance per unit length (Y). Characteristic impedance is

$$Z_o = \sqrt{\frac{Z}{Y}}$$

$$= \sqrt{\frac{(R + j\omega L)}{(G + j\omega C)}}$$

Alternatively, Z_o is a ratio of voltage wave to current wave at a point on the transmission line, which is travelling positively. Z_o is similar to η. η is the intrinsic impedance of the wave propagation medium.

Further,

$$Z_o = \frac{V_0^+}{I_0^+}$$

$$= -\frac{V_0^-}{I_0^-} \quad \text{or}$$

$$= \frac{(R + j\omega L)}{\gamma} \quad \text{or}$$

$$= \frac{\gamma}{(G + j\omega C)} \quad \text{or}$$

$$= \sqrt{\frac{(R + j\omega L)}{(G + j\omega C)}} \quad \text{or}$$

$$= (R_o + jX_o)$$

where R_o is the real part of Z_o and Y_o the imaginary part of Z_o.

9.6 LOSSLESS AND LOSSY TRANSMISSION LINES

The transmission lines discussed so far were of lossy type. Conductors of a lossy line are imperfect ($\sigma_c \neq \infty$). Also, the dielectric medium in which these imperfect conductors are embedded is lossy ($\sigma_c \neq 0$). Two special cases are considered here.

9.6.1 Lossless Transmission Line

A transmission line is "lossless" if conductors of a line are perfect ($\sigma_c \approx \infty$). Also, the dielectric medium separating them should be lossless ($\sigma \neq 0$). Thus, if $\sigma_c \approx \infty$ and $\sigma \approx 0$, we get $R = G = 0$ (a necessary condition). This is a necessary condition for a lossless transmission line.

Further, if $\alpha = 0$, $\quad \beta = \omega\sqrt{LC}$ and

$$\gamma = j\omega\sqrt{LC}$$

we get

$$v = \frac{\omega}{\beta} \quad \text{or}$$

$$= \frac{1}{\sqrt{LC}}$$

Characteristic impedance

$$Z_0 = (R_0 + jX_0) \quad \text{or}$$

$$= \sqrt{\frac{L}{C}}$$

where $$R_0 = \sqrt{\frac{L}{C}} \quad \text{and}$$

$$X_0 = 0$$

9.6.2 Distortionless Transmission Line

In a distortionless line, the attenuation constant α is frequency independent. The phase constant β of a distortionless line is linearly dependent on the frequency. For a distortionless line, the line parameters must hold the below-mentioned relationship:

$$\frac{R}{L} = \frac{G}{C}$$

Thus,

$$\gamma = \sqrt{(R + j\omega l)\cdot(G + j\omega C)} \quad \text{or}$$

$$= \sqrt{R\left(1 + \frac{j\omega L}{R}\right)G\cdot\left(1 + \frac{j\omega C}{G}\right)} \quad \text{or}$$

$$= \sqrt{RG}\left(1 + \frac{j\omega C}{G}\right) \quad \text{or}$$

$$= (\alpha + j\beta)$$

where $\alpha = \sqrt{RG}$ and $\beta = \omega\sqrt{LC}$.

Thus, α does not depend on the frequency. β is a linear function of frequency.

Phase velocity $$v = \frac{\omega}{\beta} \quad \text{or}$$

$$= \frac{1}{\sqrt{LC}} \quad \text{or}$$

$$= f\lambda$$

Characteristic impedance Z_o is obtained as

$$Z_o = (R_0 + jX_o) \quad \text{or}$$

$$= \sqrt{\frac{(R + j\omega L)}{(G + j\omega C)}} \quad \text{or}$$

$$= \sqrt{\frac{R(1 + j\omega L/R)}{G(1 + j\omega C/G)}}$$

Also,

$$Z_o = \sqrt{\frac{R}{G}} \quad \text{or}$$

$$Z_o = \sqrt{\frac{L}{C}}$$

Thus,

$$R_o = \sqrt{\frac{R}{G}} = \sqrt{\frac{L}{C}} \quad \text{and} \quad X_o = 0$$

The distortionless line has a non-vanishing attenuation constant. The distortionless line has a constant velocity. Also, it has a constant real characteristics impedance (similar to that of a lossless line). A lossless line is a distortionless line. A distortionless line is not necessarily a lossless line. In power transmission, lossless lines are desired. Telephone lines should be distortionless.

9.7 INPUT IMPEDANCE

A transmission line of length l connected to load Z_L is shown in Figure 9.7. The generator takes the line as a load. Let this input impedance be Z_{in}. Input impedance, standing wave ratio (SWR) and power flow through the line are obtained below.

Let the transmission line extend from $z = 0$ (at the generator) to $z = l$ (at load). Voltage and current waves are given as

$$V_s(z) = V_0^+ e^{-\gamma z} + V_0^- e^{+\gamma z} \quad \text{and}$$

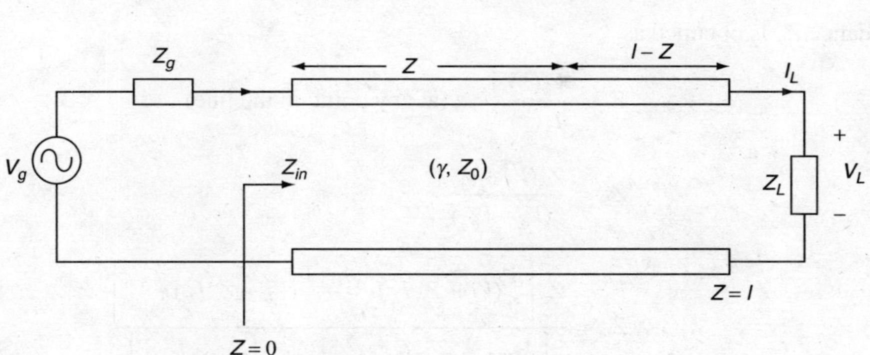

Figure 9.7 *Transmission line*

$$I_s(z) = \frac{V_0^+}{Z_o}e^{-\gamma z} - \frac{V_0^-}{Z_o}e^{+\gamma z}$$

V_0^+ and V_0^- are known as "terminal conditions".

Let

$$V_o = V \quad \text{at} \quad z = 0 \quad \text{and}$$

$$I_o = I \quad \text{at} \quad z = 0$$

On substitution, we get

$$V_o^+ = \frac{1}{2}(V_o + Z_o I_o)$$

$$V_o^- = \frac{1}{2}(V_o - Z_o I_o)$$

From Figure 9.7, we get

$$V_o = \frac{Z_{in}}{(Z_{in} + Z_g)}V_g \quad \text{and}$$

$$I_o = \frac{V_g}{(Z_g + Z_{in})}$$

Let the conditions at the load be

$$V_L = V(z = l) \quad \text{and}$$

$$I_L = I(z = l)$$

Solving the above, we get

$$V_o^+ = \frac{1}{2}(V_L + Z_o I_L)e^{\gamma l} \quad \text{and}$$

$$V_o^- = \frac{1}{2}(V_L - Z_o I_L)e^{-\gamma l}$$

Input impedance Z_{in} is obtained as

$$Z_{in} = \frac{V_s(z)}{I_s(z)} \qquad \text{(at any point on the line)}$$

$$Z_{in} = \frac{Z_o(V_o^+ + V_o^-)}{(V_o^+ - V_o^-)}$$

$$Z_{in} = \frac{Z_o\left[\frac{1}{2}(V_L + Z_o I_L)e^{\gamma l} + \frac{1}{2}(V_L - Z_o I_L)e^{-\gamma l}\right]}{\left[\frac{1}{2}(V_L + Z_o I_L)e^{\gamma l} - \frac{1}{2}(V_L - Z_o I_L)e^{-\gamma l}\right]}$$

$$Z_{in} = \frac{Z_o \frac{1}{2}\left[V_L(e^{\gamma l} + e^{-\gamma l}) + Z_o I_L(e^{\gamma l} - e^{-\gamma l})\right]}{\frac{1}{2}\left[V_L(e^{\gamma l} + e^{-\gamma l}) + Z_o I_L(e^{\gamma l} - e^{-\gamma l})\right]}$$

Also,

$$(e^{\gamma l} + e^{-\gamma l}) = 2\cosh \gamma l \quad \text{and}$$

$$e^{\gamma l} - e^{-\gamma l} = 2\sinh \gamma l$$

Thus,

$$\tanh \gamma l = \frac{(e^{\gamma l} - e^{-\gamma l})}{(e^{\gamma l} + e^{-\gamma l})}$$

Further,

$$Z_{in} = \frac{Z_o\left[(V_L/I_L) + Z_o\,(e^{\gamma l} - e^{-\gamma l})/(e^{\gamma l} + e^{-\gamma l})\right]}{\left[(V_L/I_L) + \frac{(e^{\gamma l} - e^{-\gamma l})}{(e^{\gamma l} + e^{-\gamma l})} + Z_o\right]} \quad \text{or}$$

$$Z_{in} = Z_o\left[\frac{Z_L + (Z_o \tanh \gamma l)}{Z_o + (Z_L \tanh \gamma l)}\right]$$

In the case of a lossless line, $\gamma = j\beta$.

Thus,

$$\tanh \gamma l = \tanh j\beta l \quad \text{or}$$

$$= j\tan \beta l \quad \text{and}$$

$$Z_{in} = Z_o\left[\frac{Z_L + jZ_o \tan \beta l}{Z_o + jZ_L \tan \beta l}\right]$$

The average power delivered by the generator to the line is

$$(P_{av})_i = \frac{1}{2}\mathrm{Re}\left[V_i I_i^*\right]$$

The average power delivered to the load is

$$(P_{av})_L = \frac{1}{2}\mathrm{Re}\left[V_L I_L^*\right] \quad \text{or}$$

$$= \frac{1}{2}\left|\frac{V_L}{Z_L}\right|^2 R_L \quad \text{or}$$

$$= \frac{1}{2}\left|I_L\right|^2 R_L$$

In the case of a lossless line, load power is equal to input power. A special case may be when the line is terminated with its characteristics impedance, i.e. $Z_L = Z_o$.

9.8 REFLECTION COEFFICIENT AND STANDING WAVE RATIO

Load impedance is different from characteristic impedance Z_o. If the transmission line terminates at a load, both the incident wave (from generator) and the reflected wave (from load) exist. Voltage at any point on transmission line is expressed as

$$V(x) = V_0^+ e^{-\gamma x} + V_0^- e^{\gamma x}$$

Conventionally, Γ_L is the voltage reflection coefficient. Γ_L is the ratio of reflected wave to incident wave (at load). It is given as

$$\Gamma_L = \frac{V_0^- e^{-\gamma l}}{V_0^+ e^{\gamma l}}$$

The voltage at any point on the transmission line is given as

$$V(x) = \frac{1}{2}(V_L + Z_o I_L)e^{\gamma(l-x)} + \frac{1}{2}(V_L - Z_o I_L)e^{-\gamma(l-x)}$$

Putting V_0^- and V_0^+, we get

$$\Gamma_L = \frac{(Z_L - Z_o)}{(Z_L + Z_o)}$$

where $V_L = Z_L I_L$. Voltage reflection coefficient at any point on a transmission line is the ratio of magnitude of reflected voltage wave to magnitude of incident wave. It is given as

$$\Gamma(z) = \frac{V_0^- e^{\gamma Z}}{V_0^+ e^{-\gamma Z}} \quad \text{or}$$

$$= \frac{V_0^-}{V_0^+} e^{2\gamma Z}$$

The current reflection coefficient at any point on the transmission line is given as current reflection coefficient = $-\Gamma_L$ (negative of voltage reflection coefficient at the same point). The ratio of the maximum voltage to the minimum voltage along a finite terminated line is known as the standing wave ratio (SWR) S:

$$\text{SWR } S = \frac{|V_{\max}|}{|V_{\min}|}$$

$$= \frac{1+|\Gamma_L|}{1-|\Gamma_L|} \quad \text{or}$$

$$|\Gamma_L| = \frac{(S-1)}{(S+1)}$$

The maximum value of input impedance is given as

$$|Z_{in}|_{\max} = \frac{V_{\max}}{I_{\min}} \quad \text{or}$$

$$= SZ_o$$

The minimum value of input impedance is given as

$$|Z_{in}|_{\min} = \frac{V_{\min}}{I_{\max}} \quad \text{or}$$

$$= \frac{Z_0}{S}$$

Special cases when the transmission line is connected to load Z_L are discussed below.

9.8.1 Shorted Line

In this case, $Z_L = 0$. Input impedance is given as

$$Z_{SC} = Z_{in}\big|_{Z_L=0}$$

$$= jZ_o \tan \beta_1$$

The tan β_1 value varies between $-\infty$ and $+\infty$. The input impedance of a lossless line can be purely inductive or purely capacitive. It depends on the value of β_1.

If the length of a short-circuited line is very small, then $\beta l \ll 1$. The input impedance in this case is given as

$$Z_{SC} = jZ_o \tan \beta l \quad \text{or}$$

$$= jZ_o \beta l \quad \text{or}$$

$$= j\sqrt{\frac{L}{C}}\omega\sqrt{LC}\, l \quad \text{or}$$

$$= j\omega Ll$$

It is simply the impedance of an inductance of Ll Henries.

Further,

$$\Gamma_L = -1 \quad \text{and}$$

$$S = \infty$$

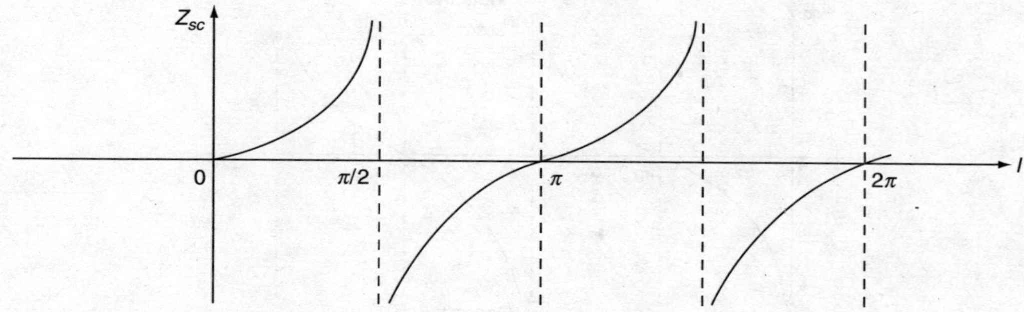

Figure 9.8 *Input impedance in a shorted line*

9.8.2 Open-circuited Line

In this case, $Z_L \to \infty$. Input impedance is given as

$$Z_{OC} = \lim_{Z_L \to \infty} Z_{in} \quad \text{or}$$

$$= \frac{Z_o}{j \tan \beta_1} \quad \text{or}$$

$$= -j Z_o \cot \beta_1$$

The input impedance in the case of an open-circuited lossless transmission line is purely reactive. The transmission line can be either capacitive or inductive. This is because cot β_1 can be either positive or negative, cot β_1 depends on the value of β_1.

If the transmission line is very short, then $\beta l << 1$ and $\tan \beta l = \beta l$.

Thus, we find $$Z_{OC} = -j \frac{Z_o}{\tan \beta l} \quad \text{or}$$

$$= -\frac{jZ_o}{\beta l} \quad \text{or}$$

$$= -j \frac{\sqrt{(L/C)}}{\omega \sqrt{LC}\, l} \quad \text{or}$$

$$= -\frac{j}{\omega C\, l}$$

It is basically the impedance of a capacitor. Practically, infinite impedance at higher frequencies is not possible at the load end. This is because of coupling to nearby objects and radiations coming from the open end.

Here, $\Gamma_L = 1$ and $S = \infty$. Also,

$$Z_{SC} Z_{OC} = Z_o^2$$

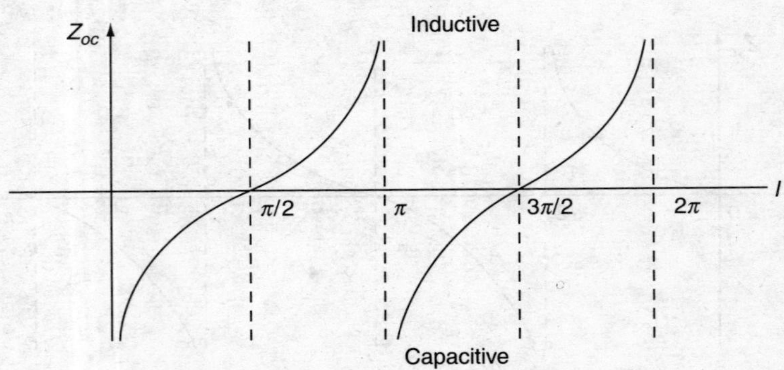

Figure 9.9 *Input impedance of an open-circuited line*

9.8.3 Matched Line

In this case, $Z_L = Z_o$. The input impedance is given as

$$Z_{in} = Z_o \quad \text{and}$$

$$\Gamma_L = 0$$

Also, $S = 1$.

Thus, in this case, the whole wave is transmitted. Due to complete transmission of the wave, there is no reflection. The incident power is completely absorbed in the load. The maximum power transfer that takes place at the transmission line is totally matched to the load (Maximum Power Transfer Theorem).

9.8.4 Quarter-wave Transmission Line

In this case, the length of the length is equal to quarter-wave length, i.e. $l = \lambda/4$. Further,

$$\beta l = (\pi/2)\left(\beta l = \frac{2\pi}{\lambda} \times \frac{\lambda}{4} = \frac{\pi}{2}\right)$$

If the length of the line is an odd multiple, $\lambda/4$, i.e. $l = (2n-1)\lambda/4$ with $n = 1, 2, 3 \ldots$ then βl is given as

$$\beta l = \frac{2\pi}{\lambda}(2n-1)\frac{\lambda}{4} \quad \text{or}$$

$$\beta l = \frac{\pi\,(2n-1)}{2}$$

Thus,

$$\tan \beta l = \tan\left[\frac{(2n-1)\pi}{2}\right]$$

$$\tan \beta l \to \pm\infty$$

Here, input impedance is

$$Z_{in} = \frac{Z_o^2}{Z_L}$$

where Z_o is the characteristics impedance and Z_L the load impedance. This expression is called "quarter-wave transformation". This transmission line is called the "quarter-wave transformer". An open-circuited quarter-wave transmission line appears as a short circuit at the input terminals. Conversely, a short-circuited quarter-wave line appears as an open circuit at the input terminals.

9.8.5 Half-wave Transmission Line

In this case, the length of the line is equal to the half-wave length, i.e. $l = \lambda/2$.

Further,

$$\beta l = \pi\left(\beta l = \frac{2\pi}{\lambda}, \frac{\lambda}{2} = \pi\right)$$

If the length of the line is an integer multiple of $\lambda/2$, i.e. $l = n\lambda/2$ with $n = 1, 2, 3 \ldots$ then βl is given as

$$\beta l = \frac{2\pi}{\lambda} \cdot \left(\frac{n\lambda}{2}\right)$$

$$= n\pi$$

Thus,

$$\tan \beta l = 0$$

Here, the input impedance is

$$Z_{in} = Z_L$$

where Z_L is the load impedance. A half-wave lossless transmission line as discussed above transforms the load impedance without any change to the input terminals. Characteristic impedance and the propagation constant are obtained from input impedance under short-circuited conditions.

When $\quad Z_L \to \infty; \quad Z_{oc} = Z_o \coth \gamma l$

and if $\quad Z_L \to 0; \quad Z_{oc} = Z_o \coth \gamma l$

From above, we get

$$Z_o = \sqrt{Z_{oc} \cdot Z_{sc}} \quad \text{and}$$

$$\gamma = \frac{1}{l} \tanh^{-1} \sqrt{\frac{Z_{sc}}{Z_{oc}}} \text{ per meter}$$

The above expressions are valid for both lossless and lossy half-wave transmission lines.

9.9 TRANSMISSION LINE IMPEDANCE MATCHING

Transmission lines are widely used for power and information transmission. In the case of radio frequency power transmission, maximum power should be transmitted to the load with minimum losses. For this, the load should be matched with the characteristic impedance of the transmission line. In matched loads, the SWR of the transmission line is close to unity. Mismatched loads and junctions cause echoes and eventually distort the signal carrying information. Various methods of impedance matching are described below.

9.9.1 Quarter-wave Transformer (Matching)

If, $Z_o \neq Z_L$, the load is mismatched. It causes a reflected wave on the line. For maximum power transfer, the load should be matched with the transmission line ($Z_o = Z_L$). In this case, there is no reflection ($|\Gamma_L| = 0$ or $S = 1$). Matching can be very easily achieved by using shorted sections of transmission lines.

Let,

$$l = \lambda/4$$

Further,

$$\beta l = \left(\frac{2\pi}{\lambda} \times \frac{\lambda}{4}\right) \quad \text{or}$$

$$= (\pi/2)$$

We get

$$Z_{in} = Z_o \left[\frac{Z_L + j\,Z_o \tan \pi/2}{Z_o + j\,Z_L \tan \pi/2} \right] \quad \text{or}$$

$$= \frac{Z_o^2}{Z_L}$$

A mismatched load Z_L can be matched with a transmission line having characteristic impedance Z_o by putting a transmission line before the load, which is $\lambda/4$ long and has characteristic impedance Z_o. It is shown in Figure 9.10. This $\lambda/4$ long section of the transmission line is called "quarter-wave transformer". Z_o is chosen such that

$$Z_{in} = Z_o$$

Figure 9.10 *Matching load Z_L to a transmission line*

Thus,

$$Z_o' = \sqrt{Z_o\,Z_L}$$

where Z_o, Z_o and Z_L are real.

As an example, if a 120 Ω (Z_L) load is to be matched to a 30 Ω (Z_o) transmission line, then the characteristic impedance of a quarter-wave transformer should be $\sqrt{(30)(120)} = 60\ \Omega$. Conversely, a 60 Ω quarter-wave transformer is needed to match a 30 Ω load to a 120 Ω line.

A standing wave will exist between the transformer and the load. There will not be any standing wave because of matching. The reflected wave or standing wave will only be eliminated at the desired wavelength λ (or frequency f). Reflection will take place at a slightly different wavelength. The main disadvantage of a quarter-wave transformer is that it is a frequency-sensitive device and operates on a very narrow band.

9.9.2 Single-stub Tuner

The drawbacks of a quarter-wave transformer as a line-matching device are eliminated by using a single-stub tuner. The single-stub tuner has an open (or shorted) section of the transmission line (say, of length d), which is connected in parallel to the main line at a distance (say, l) from the load. It is shown in Figure 9.11. It is quite difficult to use a series stub. An open-circuited tuner will radiate energy at high frequencies. Shunt short-circuited parallel stubs are preferred over a series stack.

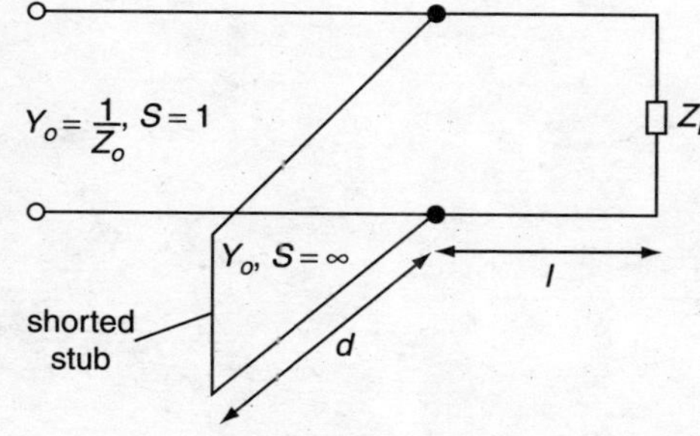

Figure 9.11 *Single-stub tuner*

For "stub matching", find a location that is nearest to the load. At this location, the real part of the line admittance (reciprocal of line impedance) must be equal to the line characteristic admittance Y_o (reciprocal of the line characteristic impedance of Z_o). The imaginary part of the line admittance gets cancelled due to the parallel placement of the short-circuited stub of a suitable length with the line. The imaginary part of the line admittance gets cancelled due to the condition that susceptance must be equal to negative of the imaginary part of the line admittance at the right of the stub. Line admittance as seen from the left of the stub is given as

$$(Y_o + jB) + (-jB) = Y_o$$

Line impedance as seen from the left stub into the line–stub junction is Z_o. Thus, "stub-matching" is achieved.

9.10 SMITH CHART

The Smith chart is a graphical calculator for high-frequency circuit applications. It was invented by Phillip H. Smith (1905–1987). The Smith chart replaces complex algebra and numbers with geometrical constructs.

The Smith chart is a four-dimensional representation of complex impedances with respect to coordinates. Impedance at a point along the transmission line depends on its properties and what is connected to it. The load is normally an antenna. Smith charts provide information on transmission line behaviour for a very wide range of impedances. It is a polar plot of complex reflection coefficients (Γ). It considers normalized complex load impedance. It is a dimensionless quantity. Normalized complex load impedance is obtained as

$$\text{Complex load impedance} = \frac{\text{Actual load impedance } (Z_L)}{\text{Characteristic impedance } (Z_o)}$$

Characteristic impedance Z_0 is a real quantity for lossless transmission lines.

Thus,

$$\frac{Z_L}{Z_o} = \frac{1+\Gamma(d)}{1-\Gamma(d)}$$

where $\Gamma(d)$ is the phasor by which Smith charts are represented. In terms of admittances, we find

$$\frac{Y_L}{Y_O} = \frac{1-\Gamma(d)}{1+\Gamma(d)}$$

$$Y_L = \frac{1}{Z_L} \quad \text{and}$$

$$Y_0 = \frac{1}{Z_0} \quad \text{and}$$

$$\frac{Z_L}{Z_0} = z_n \quad \text{(normalized complex load impedance)}$$

$$z_n = (r + jx)$$

In a lossless transmission line, the reflection coefficient (Γ) domain is a circle of unitary radius as depicted in Figure 9.12.

For lossless transmission lines, same domain (circle of unitary radius) of Smith chart. In the case of lossy transmission lines, the reflection coefficient (Γ) might be more than unity; therefore, the domain of

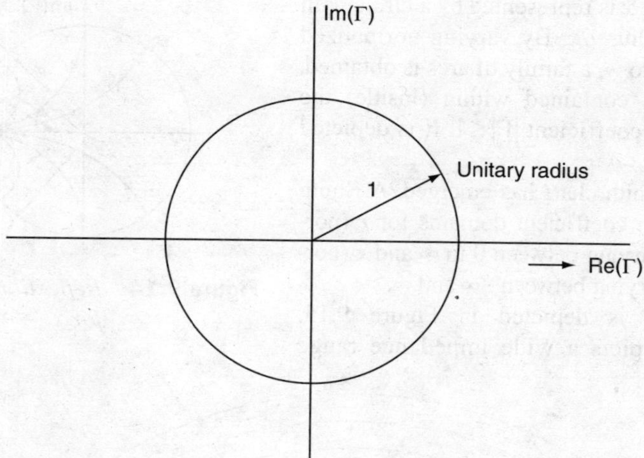

Figure 9.12 *Reflection coefficient (Γ) domain for a lossless transmission line*

the Smith chart is extended. It is because of complex characteristic impedance. It terms of the reflection coefficient (Γ), normalized complex impedance can be written as

$$(r+jx)=\frac{1+\mathrm{Re}(\Gamma)+j\,\mathrm{Im}(\Gamma)}{1-\mathrm{Re}(\Gamma)-j\,\mathrm{Im}(\Gamma)}$$

$$(r+jx)=\frac{1-\mathrm{Re}^2(\Gamma)-\mathrm{Im}^2(\Gamma)+j2\,\mathrm{Im}(\Gamma)}{(1-\mathrm{Re}(\Gamma))^2+\mathrm{Im}^2(\Gamma)}$$

From here,

$$r=\frac{1-\mathrm{Re}^2(\Gamma)-\mathrm{Im}^2(\Gamma)}{(1-\mathrm{Re}(\Gamma))^2+\mathrm{Im}^2(\Gamma)}\quad \text{and}$$

$$x=\frac{2\,\mathrm{Im}(\Gamma)}{(1-\mathrm{Re}(\Gamma))^2+\mathrm{Im}^2(\Gamma)}$$

The real part eventually gives $(1/(1+r))^2$. It is the equation of a circle. The imaginary part eventually gives $1/x^2$. It is also an equation of a circle. The real part gives all possible impedances on a complex plane with coordinates (Re(Γ), Im (Γ)). Normalized resistance r is represented by a circle with centre $(r/(1+r), 0)$ and radius $1/(1+r)$. By varying normalized resistance r from 0 to ∞, a family of circles is obtained, which is completely contained within (inside) the domain of reflection coefficient $|\Gamma|\le 1$. It is depicted in Figure 9.13.

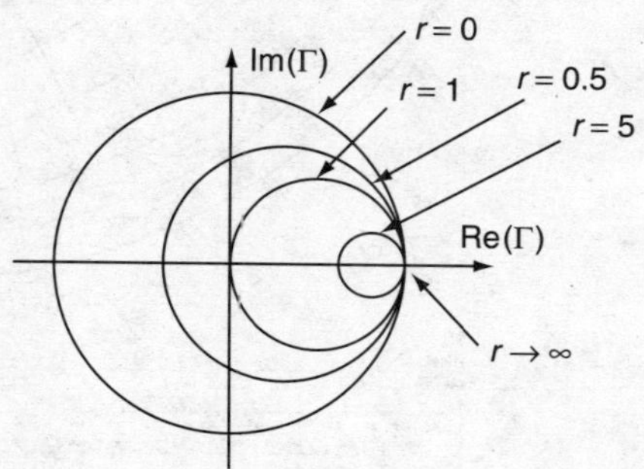

Figure 9.13 *Reflection coefficient domains for* $r \to 0$ *to* ∞

The imaginary part gives all possible impedances on a complex plane with coordinates (Re(Γ), Im(Γ)).

Normalized reactance x is represented by a circle with centre $(1, 1/x)$ and radius $1/x$. By varying normalized reactance x from $-\infty$ to ∞, a family of arcs is obtained, which is completely contained within (inside) the domain of reflection coefficient $|\Gamma| \leq 1$. It is depicted in Figure 9.14.

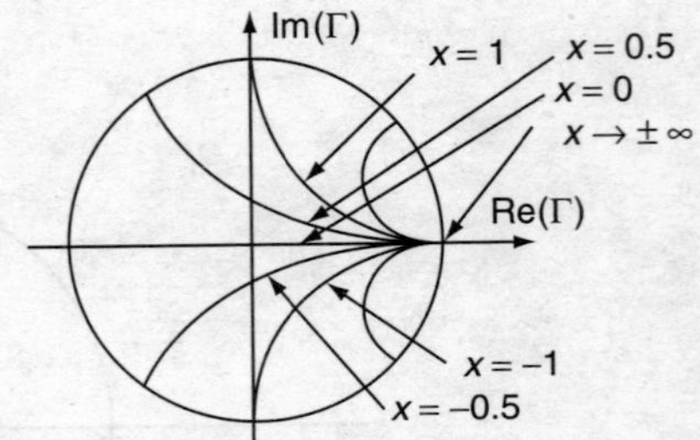

Figure 9.14 *Reflection coefficient domains for* $x \rightarrow -\infty$ to ∞

That is how the Smith chart has emerged. A Smith chart shows reflection coefficient domains for r (normalized resistance) varying between 0 to ∞ and x (normalized reactance) varying between $-\infty$ and ∞.

The Smith chart is depicted in Figure 9.15. The Smith chart depicts a wide impedance range

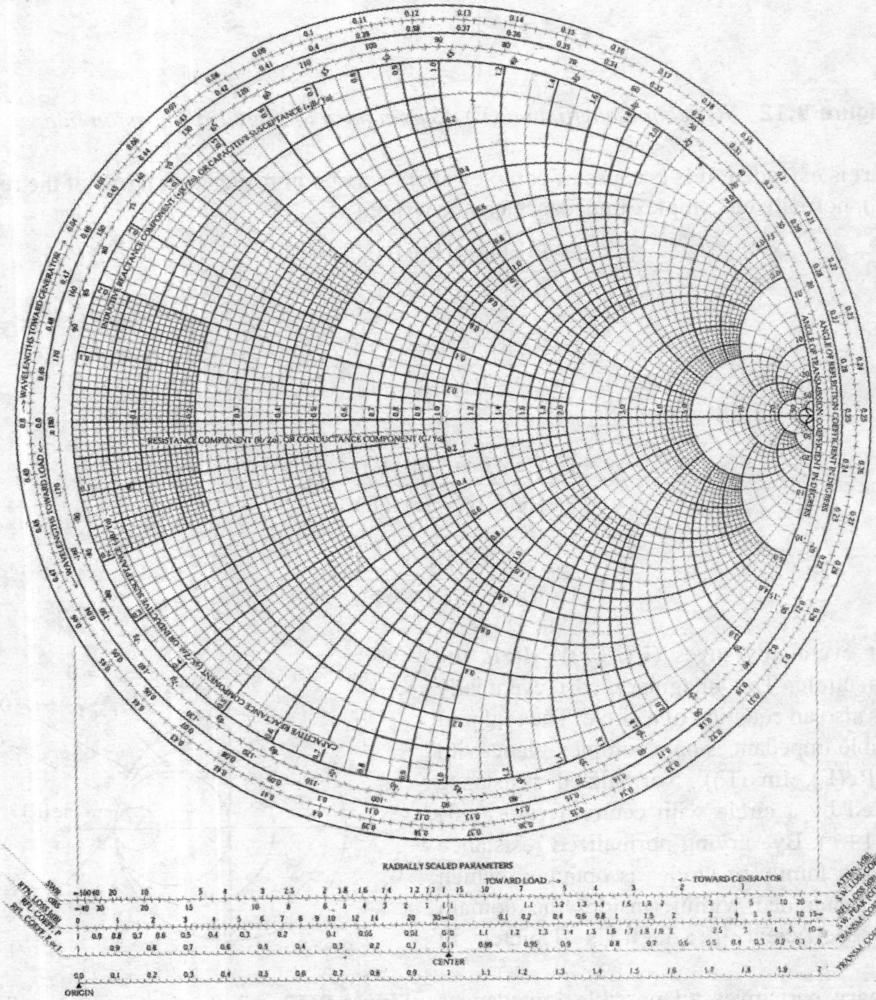

Figure 9.15 *Smith chart*

that a transmitter may encounter for quite high voltage standing wave ratio (VSWR). High VSWR refers to high mismatch.

9.10.1 Steps for Using Smith Chart for Obtaining Reflection Coefficient

Let us say that $Z(d)$ is given and $\Gamma(d)$ is to be obtained. Then the following steps are to be followed:

Step 1: Normalize the given impedance. Normalized impedance $z_n(d)$ is obtained as

$$z_n(d) = \frac{Z_d(d)}{Z_0} = (r + jx)$$

Step 2: Locate the circle of constant normalized resistance r as obtained in Step 1.
Step 3: Locate the arc of constant normalized reactance x as obtained in Step 1.
Step 4: Locate the intersection of two curves as located in Steps 2 and 3. This intersection indicates the reflection coefficient (Γ) in complex plane (Re(Γ), Im(Γ)). It directly gives the magnitude and phase angle for $\Gamma(d)$.

9.10.2 Steps for Using Smith Chart for Obtaining Line Impedance

Let us say that $\Gamma(d)$ is given and $z(d)$ is to be obtained. Then the following steps are to be followed:

Step 1: Locate the complex point representing $\Gamma(d)$ on the Smith chart.
Step 2: Corresponding to the reflection coefficient point on the Smith chart, observe and note down values of normalized resistance r and normalized reactance x.
Step 3: Normalized impedance $z_n(d)$ is given as

$$z_n(d) = (r + jx)$$

and actual impedance

$$Z(d) = Z_o z_n(d) = Z_o(r + jx) = (Z_o r + jZ_o x)$$

9.10.3 Steps for Using Smith Chart for Obtaining the Reflection Coefficient and Line Impedance

Let us say that the load reflection coefficient Γ_R and normalized load impedance Z_R are given, and $\Gamma(d)$ and $Z(d)$ are to be obtained. Then the following steps are to be followed:

Step 1: Locate Γ_R and Z_R on the Smith chart. In a lossless transmission line, $|\Gamma_R| = |\Gamma(d)|$.
Step 2: Draw a circle of $|\Gamma_R|$. Here, $|\Gamma_R|$ is the constant amplitude of reflection coefficient.
Step 3: Travel on the drawn circle by an angle θ is in the clockwise direction from a starting point representing the load.

$$\begin{aligned}\Gamma(d) &= \Gamma_R \exp(-j2\beta d) \\ \theta &= 2\beta d \\ &= 2\times\left(\frac{2\pi}{\lambda}\right)\times d\end{aligned}$$

Step 4: The location corresponding to θ on the Smith chart as travelled in Step 3 corresponds to new location d on the transmission line. Observe and record values of $\Gamma(d)$ and $Z(d)$ from the Smith chart.

9.10.4 Steps for Using the Smith Chart for Obtaining d_{max} and d_{min}

Let us say that Γ_R and Z_R are given and d_{max} and d_{min} are to be obtained. Then the following steps are to be followed:

Step 1: Locate the load reflection coefficient Γ_R and normalized load impedance z_{nR} on the Smith chart.

Step 2: Draw a circle of $|\Gamma_R|$. This circle will intersect the reflection coefficient's real axis at two points—at the first point, where $\Gamma(d)$ is a real positive, which corresponds to d_{max} and the other point, where $\Gamma(d)$ is a real negative, which corresponds to d_{min}.

Step 3: From the outer graduation of the Smith chart, observe and record distances normalized to the wavelength. Alternatively, from the angle between the real axis and the Γ_R vector, d_{max} and d_{min} can be easily obtained.

9.10.5 Steps for Using the Smith chart for Obtaining Voltage Standing Wave Ratio

VSWR is an important index in electromagnetic fields and especially transmission lines. VSWR is given as

$$\text{VSWR} = \frac{V_{max}}{V_{min}}$$

$$= \frac{1+|\Gamma_R|}{1-|\Gamma_R|}$$

From the Smith chart, VSWR can be very easily obtained by observing the value of real normalized impedance at maximum location d_{max}, where Γ is real and positive. $z_n(d_{max})$ is always either equal to or more than one.

Thus,

$$z_n(d_{max}) = \text{VSWR} = \frac{1+\Gamma(d_{max})}{1-\Gamma(d_{max})}$$

Let us say that Γ_R and Z_R are given, VSWR is to be obtained. Then the following steps are to be followed:

Step 1: Locate the load reflection coefficient Γ_R and normalized load impedance z_{nR} on the Smith chart.

Step 2: Draw a circle of $|\Gamma_R|$.

Step 3: Identify the intersection of the circle with the real axis for the value of d_{max}.

Step 4: Carefully see that another circle of constant normalized resistance intersects the point corresponding to d_{max}. From the normalized resistance, obtain the value of VSWR.

9.10.6 Step for Using the Smith Chart for Obtaining Characteristic Admittance (Y_o)

Normalized impedance $z_n(d)$ is given as

$$z_n(d) = \frac{1+\Gamma(d)}{1-\Gamma(d)}$$

Similarly, normalized admittance $y_n(d)$ is obtained as

$$Y_n(d) = \frac{1-\Gamma(d)}{1+\Gamma(d)} \quad \text{and}$$

$$Y(d) = Y_o.y_n(d)$$
$$= \frac{y_n(d)}{z_o}$$

Let us say that z_r and z_d are given and Y_R is to be obtained. Then the following steps are to be followed:

Step1: Locate the load reflection coefficient Γ_R and the normalized load impedance z_{nR} on the Smith chart.

Step 2: Draw a circle of $|\Gamma_R|$. Further, $|\Gamma_R| = |\Gamma(d)|$.

Step 3: Observe and record a point corresponding to normalized admittance on the circle as drawn in Step 2, which is exactly diametrically opposite (turning it through 180°) to the point corresponding to normalized impedance.

9.10.7 Advantages of the Smith Chart

The main advantages of the Smith chart have been given below:

(1) It is an efficient graphical representation in the complex plane.
(2) It gives both impedance and admittance.
(3) It also gives information on SNR, noise figure and stability regions.
(4) It is a Reimann surface. In numbers of half wavelengths along the transmission line, it is cyclical, etc.

9.11 SOME APPLICATIONS OF TRANSMISSION LINES

Any cable or wire is treated as a transmission line if its length is greater that 1/10 th of the wavelength of the signal travelling through it. Various transmission line types are the coaxial cable, strip line, microstrip, twisted pair, balanced line, star quad, twin lead, lecher lines, wave guides, single-wire line, optical fiber, acoustic transmission lines, etc. Some important applications of transmission lines are as follows:

(1) Transmission of high-frequency signals with minimum power loss.
(2) Transmission lines function as a filter. For this filter, an open-circuited or short-circuited transmission line is wired in the filter. These are called stub filters.
(3) Transmission lines are used as pulse generators. For example, Blumlein transmission line, etc.

SOLVED QUESTIONS

9.1 A lossy cable has $R = 6.75$ (Ω/m), $L = 3.0$ (μH/m), $C = 2$ (pF/m) and $G = 0$. It operates at $f = 0.5$ GHz. What are the values of the propagation constant, attenuation constant and phase constant of the line?

Solution:

It is given that $R = 6.75(\Omega/\text{m})$, $L = 3.0$ (μH/m), $C = 2\times10^{-12}$ (F/m), $G = 0$ and $f = 500\times10^6$ (Hz). Propagation constant

$$\gamma = \sqrt{(R + j\omega L)(G + j\omega C)} \quad \text{or}$$

$$\gamma = \sqrt{(6.75 + j2\pi\times500\times10^6\times3\times10^{-6})(0 + j2\pi\times500\times10^6\times2\times10^{-12})} \quad \text{or}$$

$$\gamma = (3.06\times10^{-3} + j7.69)\ (\text{m}^{-1})$$

Also,

$$\gamma = (\alpha + j\beta)$$

By comparison, we get attenuation constant

$$\alpha = 1.125 \times 10^{-3}\,(\text{dB/m})$$

and phase constant

$$\beta = 3.142\,(\text{rad/m})$$

9.2 An air line has a characteristic impedance of 50 Ω. Its phase constant is 4 rad/m at a frequency of 10 MHz. Find the inductance per meter. Also, find the capacitance per meter of this air line.

Solution:

An air line is a lossless line. Thus,

$$\sigma = 0$$

Also,

$$R = G = 0 \quad \text{and}$$

$$\alpha = 0$$

$$Z_o = R_o = \sqrt{\frac{L}{C}} \quad \text{and}$$

$$\beta = \omega\sqrt{LC}$$

From the above expressions, we get

$$\frac{R_o}{\beta} = \frac{1}{\omega_c} \quad \text{or}$$

$$C = \left(\frac{\beta}{\omega\, R_o}\right) \quad \text{or}$$

$$= \left(\frac{4}{2\pi \times 10 \times 10^6 \times 50}\right) \quad \text{or}$$

$$= 1273.06\,(\text{pF/m}) \quad \text{and}$$

$$L = R_0^2\, C$$

$$= (50)^2\,(1273.06 \times 10^{-12})$$

$$= 3.18\,(\mu\text{H/m})$$

9.3 A distortion line has $Z_o = 70\ (\Omega)$, $\alpha = 15(\text{mNp/m})$ and $u = 0.5\ c$. $c = 3\times10^8$ (m/s). Find the value of R, L, G, C and λ at 100 MHz.

Solution:

In the case of a distortionless line, we know that

$$RC = GL \quad \text{or}$$

$$G = \left(\frac{RC}{L}\right)$$

Further,

$$Z_o = \sqrt{\frac{L}{C}}$$

Thus,

$$\alpha = \sqrt{RG} \quad \text{or}$$

$$R = \sqrt{\frac{C}{L}} \quad \text{or}$$

$$= \frac{R}{Z_o} \quad \text{or}$$

$$R = \alpha Z_o$$

However, we know that

$$u = \frac{\omega}{\beta} \quad \text{or}$$

$$= \frac{1}{\sqrt{LC}}$$

Thus,

$$R = \alpha Z_o = (15 \times 10^{-3})(70)$$

$$= 1.05\ (\Omega/\text{m})$$

Also,

$$L = \left(\frac{Z_o}{u}\right) \quad \text{or}$$

$$= \left(\frac{70}{0.5\times3\times10^8}\right) \quad \text{or}$$

$$= 466.2\ (\text{nH/m}) \quad \text{and}$$

$$G = \frac{\alpha^2}{R} \quad \text{or}$$

$$= \left(\frac{225 \times 10^{-6}}{1.05}\right) \quad \text{or}$$

$$= 214.07\ (\mu\text{S/m})$$

Now,

$$u Z_o = \frac{1}{C} \quad \text{or}$$

$$C = \frac{1}{(0.5 \times 3 \times 10^8 \times 70)}$$

$$= 95.23\ (\text{pF/m}) \quad \text{and}$$

$$\lambda = \frac{u}{f} \quad \text{or}$$

$$= \left(\frac{0.5 \times 3 \times 10^8}{100 \times 10^6}\right) \quad \text{or}$$

$$= 1.5\ (\text{m})$$

9.4 In the case of a transmission line, series inductance is 0.4 mH and capacitance 0.2 mF/km. Losses are negligible. What are the values of characteristic impedance Z_o and phase velocity v_p?

Solution:

We know that

$$Z_o = \sqrt{\frac{L}{C}} \quad \text{(in the case of lossless lines)}$$

Also,

$$R_o = G_o = 0$$

Thus,

$$Z_o = \sqrt{\frac{0.4 \times 10^{-3}}{0.2 \times 10^{-3}}} \quad \text{or}$$

$$Z_o = 1.414\ (\text{ohms})$$

Phase velocity

$$v_p = (1/\sqrt{LC}) \quad \text{or}$$

$$= \frac{1}{\sqrt{0.4\times10^{-3}\times0.2\times10^{-3}}} \quad \text{or}$$

$$= \frac{10^4}{2\sqrt{2}} \quad \text{or}$$

$$v_p = 3\cdot 53\times10^3\,(\text{Km/sec})$$

9.5 A lossless 100 Ω air-spaced transmission line of 5 m length terminates at an impedance $Z_R = (120 + j40)\ \Omega$. It is operating at 150 MHz. Find the input impedance.

Solution:
It is given that the transmission line is lossless, and its parameters are $Z_o = 100\ \Omega$, $f = 150$ (MHz), $l = 5.0$ (m) and $Z_R = (120 + j40)\ \Omega$. We know that

$$v_p = \left(\frac{\omega}{\beta}\right) \quad \text{or}$$

$$\beta = \left(\frac{\omega}{v_p}\right) \quad \text{or}$$

$$= \left(\frac{2\pi f}{v_p}\right) \quad \text{or}$$

$$= \left(\frac{2\pi\times150\times10^6}{3\times10^8}\right) \quad \text{or}$$

$$= \pi\,(\text{rad/m})$$

Further,

$$\beta l = 5\pi$$

It is given that the transmission line is lossless. Thus,

$$Z_i = Z_o\left[\frac{Z_R + jZ_o\tan\beta l}{Z_o + jZ_R\tan\beta l}\right] \quad \text{or}$$

$$= 100\left[\frac{(120 + j40) + j\times100\tan5\pi}{100 + j(120 + j40)\tan5\pi}\right] \quad \text{or}$$

$$= 100\left[\frac{(120 + j40) + j \times 100 \times 0}{100 + j(120 + j40) \times 0}\right] \quad \text{or}$$

$$= (120 + j40)\ (\Omega)$$

9.6 In the case of a transmission line, open- and short-circuit impedances are 80 $\angle -20°\ \Omega$ and 20 $\angle -30°\ \Omega$, respectively. It is operating at 1.5 KHz. Find Z_o of this transmission line.

Solution:
It is given that

$$(Z_i)_{OC} = 80\angle -20° \quad \text{and}$$

$$(Z_i)_{SC} = 20\angle -30°$$

Characteristic impedance Z_o of a transmission line is obtained as

$$Z_o = \sqrt{(Z_i)_{OC} \cdot (Z_i)_{SC}}$$

In this case,

$$Z_o = \sqrt{80\angle -20° \times 20°\angle -30} \quad \text{or}$$

$$= \sqrt{80 \times 20}\,\frac{\angle -20° - 30°}{2} \quad \text{or}$$

$$= 40\angle -25.5°\ \Omega$$

9.7 A transmission line has its characteristic impedance $Z_o = R_o = 30\ \Omega$. The transmission line is terminated in an impedance $Z_R = (30 + j30)\ \Omega$. What is impedance Z_i at a point that is $\lambda/8$ (m) away from the load.

Solution:
It is given that $l = \lambda/8$ m and $\beta = (2\pi/\lambda)$. We know that

$$Z_i = Z_o\left[\frac{Z_R + jZ_o \tan \beta l}{Z_o + jZ_R \tan \beta l}\right]$$

Electrical length

$$(\beta l) = \left(\frac{2\pi}{\lambda} \times \frac{\lambda}{8}\right)$$

$$= \frac{\pi}{4}$$

Thus,

$$Z_i = Z_o\left[\frac{Z_R + jZ_o \tan \pi/4}{Z_o + jZ_R \tan \pi/4}\right] \quad \text{or}$$

$$Z_i = 30\left[\frac{(30+j30)+j30\times1}{30+j(30+j30)}\right] \quad \text{or}$$

$$Z_i = 30\left[\frac{30+j60}{j30}\right] \quad \text{or}$$

$$Z_i = (-30j+60)\ \Omega$$

Impedance at $l = \lambda/8$ is $(60-30j)\ \Omega$.

9.8 A transmission line operating at a very high frequency has a capacitance of 0.40 μF per km. Its inductance is 96 mH per km. What are the values of characteristic impedance Z_o and propagation constant γ at $f = 100$ MHz?

Solution:

It is given that

$$L = 96 \text{ mH/km} \quad \text{or}$$

$$= 96\times10^{-3} \text{ H/km}$$

$$C = 0.04\ \mu\text{F/km} \quad \text{or}$$

$$= 0.04\times10^{-6} \text{ F/km} \quad \text{or}$$

$$f = 100 \text{ MHz}$$

$$= 10^{8} \text{ Hz}$$

Characteristic impedance

$$Z_o = \sqrt{\frac{L}{C}} \quad \text{or}$$

$$= \sqrt{\frac{96\times10^{-3}}{0.04\times10^{-6}}} \quad \text{or}$$

$$Z_o = 1549.19\ \Omega$$

Propagation constant $\gamma = \omega\sqrt{LC}$ or

$$= 2\pi\times100\times10^{6}\sqrt{96\times10^{-3}\times0.04\times10^{-6}} \quad \text{or}$$

$$= 2\pi\times10^{8}\times0.61967\times10^{-4} \quad \text{or}$$

$$\gamma = 38935.47\,(\text{m}^{-1})$$

9.9 What is the input impedance of 55 Ω lossless transmission line having length 0.2λ, if the load is a short circuit?

Solution:

It is given that $l = 0.2\ \lambda$

$$Z_o = 55\ \Omega \quad \text{and}$$

$$Z_l = 0 \qquad \text{(for a short-circuit line)}$$

Z_i (for a lossless line) is given as

$$Z_i = Z_o\left[\frac{Z_L + jZ_o \tan \beta l}{Z_o + jL_L \tan \beta l}\right]$$

Electric length

$$\beta l = \left(\frac{2\pi}{\lambda}\right)\times 0.2\lambda$$

$$= 0.4\pi$$

Thus,

$$Z_i = jZ_o \tan \beta l \quad \text{or}$$

$$= j\times 55\times \tan 0.4\pi \quad \text{or}$$

$$= j\times 55\times 0.0219 \quad \text{or}$$

$$Z_i = j\,1.20\ \Omega$$

9.10 A 55 Ω lossless transmission line has a length of 0.3λ. If it is terminated in an open circuit, find the input impedance.

Solution:

It is given that

$$l = 0.3\lambda \quad \text{and}$$

$$Z_o = 55\ \Omega$$

Electric length

$$(\beta l) = \frac{2\pi}{\lambda}\times 0\times 3\lambda \quad \text{or}$$

$$= 0.6\ \pi$$

Input impedance Z_{in} is given as

$$Z_{in} = Z_o\left[\frac{Z_L + jZ_o \tan \beta l}{Z_o + jZ_L \tan \beta l}\right]$$

but $Z_L = \infty$ (for a short-circuit line).

Thus,

$$Z_{in} = Z_o\left[\frac{1}{j\tan\beta l}\right] \quad \text{or}$$

$$Z_{in} = -jZ_o\cot\beta l \quad \text{or}$$

$$Z_{in} = -j\times 55\times\cot(0.6\pi) \quad \text{or}$$

$$Z_{in} = -1671.19\,j\,\Omega$$

9.11 A transmission line has a characteristic impedance of 40 Ω. It is terminated at a reactance of j 25 Ω. Find the input impedance of a section, which is 50 m long at a frequency of 150 MHz.

Solution:

It is given that $Z_o = 40\ \Omega$ and $Z_L = j\,25\ \Omega$. Also, frequency $(f) = 150$ MHz and

$$l = 50\text{ cm}$$

$$= \frac{1}{2}\text{ m}$$

Wavelength

$$\lambda = \left(\frac{v}{f}\right) \quad \text{or}$$

$$= \left(\frac{3\times10^8}{150\times10^6}\right) \quad \text{or}$$

$$= 2\text{ m}$$

Thus,

$$l = \lambda/4$$

It is a quarter-wave transmission line. Z_{in} in a quarter-wave transmission line is obtained as

$$Z_{in} = \frac{Z_o^2}{Z_L} \quad \text{or}$$

$$Z_{in} = \frac{(40)^2}{j25} \quad \text{or}$$

$$Z_{in} = -64\,j\,\Omega$$

9.12 A low loss transmission line has a characteristic impedance of 300 Ω. It is connected to a 500 Ω load. Find the reflection coefficient. Also, find the voltage standing wave ratio, in this case.

Solution:

It is given that $Z_o = 300\ \Omega$ and $Z_L = 500\ \Omega$. Reflection coefficient

$$\rho = \frac{(Z_L - Z_o)}{(Z_L + Z_o)} \quad \text{or}$$

$$= \left(\frac{500-300}{500+300}\right) \quad \text{or}$$

$$= \left(\frac{200}{800}\right) \quad \text{or}$$

$$\rho = \frac{1}{4}$$

In addition,

$$\text{VSWR} = \frac{(1+|\rho|)}{(1-|\rho|)} \quad \text{or}$$

$$= \frac{(1+(1/4))}{(1-(1/4))} \quad \text{or}$$

$$= \left(\frac{5/4}{3/4}\right) \quad \text{or}$$

$$= (5/3)$$

9.13 A transmission line has a characteristic impedance of 60 Ω. It is terminated in a purely resistance load. The minimum voltage upon it is 6 μV. The maximum voltage is 9 μV. What is the load impedance?

Solution:

It is given that $Z_o = 40\ \Omega$, $V_{\max} = 9\ \mu\text{V}$ and $V_{\min} = 6\ \mu\text{V}$.

Thus,

$$\text{VSWR} = \left(\frac{V_{\max}}{V_{\min}}\right) \quad \text{or}$$

$$= \frac{9}{6} \quad \text{or}$$

$$= 1.5$$

Reflection coefficient

$$\rho = \frac{(\text{VSWR}-1)}{(\text{VSWR}+1)}$$

$$= \frac{(1.5-1)}{(1.5+1)} \quad \text{or}$$

$$= \frac{1}{5}$$

However,

$$\rho = \frac{(Z_L - Z_o)}{(Z_L + Z_o)}$$

Putting ρ in the expression, we get

$$\frac{1}{5} = \frac{(Z_L - 60)}{(Z_L + 60)} \quad \text{or}$$

$$(Z_L + 60) = (5Z_L - 300) \quad \text{or}$$

$$360 = 4\,Z_L \quad \text{or}$$

$$Z_L = 90\ \Omega$$

9.14 In a transmission line, $Z(d) = (20 + j80)\ \Omega$ and $Z_o = 40\ \Omega$. Find $\Gamma(d)$.

Solution:

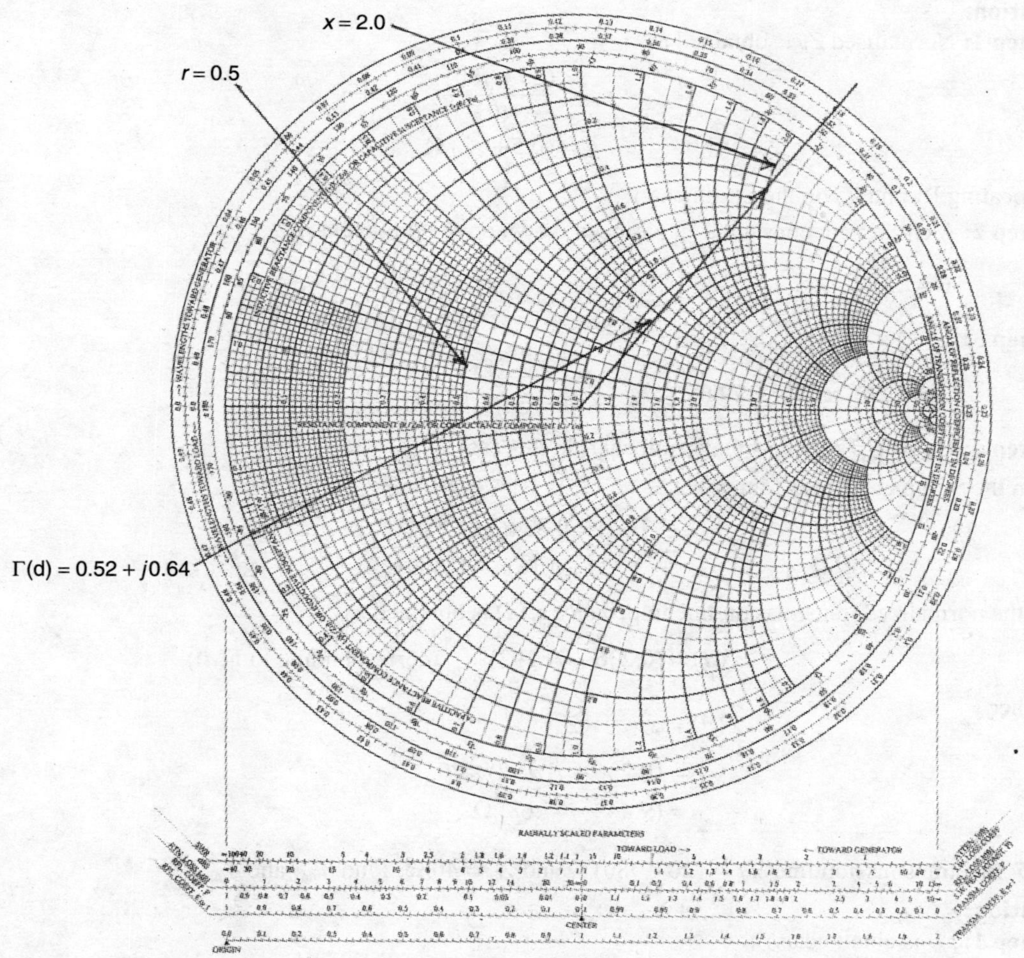

Figure 9.16 *Smith chart representing vector* $\Gamma(d) = 0.52 + j0.64$

Step 1: Normalizing the given impedance

$$z_n(d) = \frac{(20 + j30)}{40} = (0.5 + j2.0)$$

Step 2: Locating normalized resistance circle $r = 0.5$.

Step 3: Locating normalized reactance arc $x = 2.0$.

Step 4: Locating the intersection of the circle at $r = 0.5$ and arc at $x = 2.0$. The Smith chart in this case is shown in Figure 9.16.

The vector representing reflection coefficient $\Gamma(d)$ is given as

$$\Gamma(d) = 0.5 + j0.64$$
$$= 0.8246\angle 50.906°$$

9.15 In the case of a transmission line, find $Z(d)$ and $\Gamma(d)$, if $Z_R = (20 + j80)\ \Omega$, $Z_o = 40\ \Omega$ and $d = 0.18\lambda$.

Solution:

Step 1: Normalised Z_R is obtained as

$$z_{nr} = \frac{(20 + j80)}{40}$$
$$= (0.5 + j2)\ \Omega$$

Locating Γ_R and Z_R on the Smith chart.

Step 2: A circle with constant $|\Gamma_R|$ has been drawn on the Smith chart.

$$\theta = 2\beta d$$

Step 3: $$= 2 \times \frac{2\pi}{\lambda} \times 0.18\lambda$$
$$= 2.262 \text{ rad. or } 129.6°$$

Step 4: Figure 9.17 depicts the Smith chart in this case.

From the Smith chart, it is found that

$$\Gamma(d) = 0.8246\angle -78.7°$$
$$= (0.161 - j0.809)$$

and the normalized transmission line impedance $z_n(d)$ is obtained as

$$z_n(d) = (0.236 - j1.192) \qquad \text{(corresponding to } \Gamma(d))$$

Further,

$$Z(d) = z_n(d) \times Z_o$$
$$= (0.236 - j1.192) \times 40$$
$$= (9.44 - j47.68)\ \Omega$$

9.16 In a transmission line $Z_R = (20 + j80)\ \Omega$ and $Z_o = 40\ \Omega$. Find d_{max} and d_{min}.

Solution:

Step 1: z_{nR} is obtained as

$$z_{nR} = \frac{(20 + j80)}{40} = (0.5 + j2)\ \Omega$$

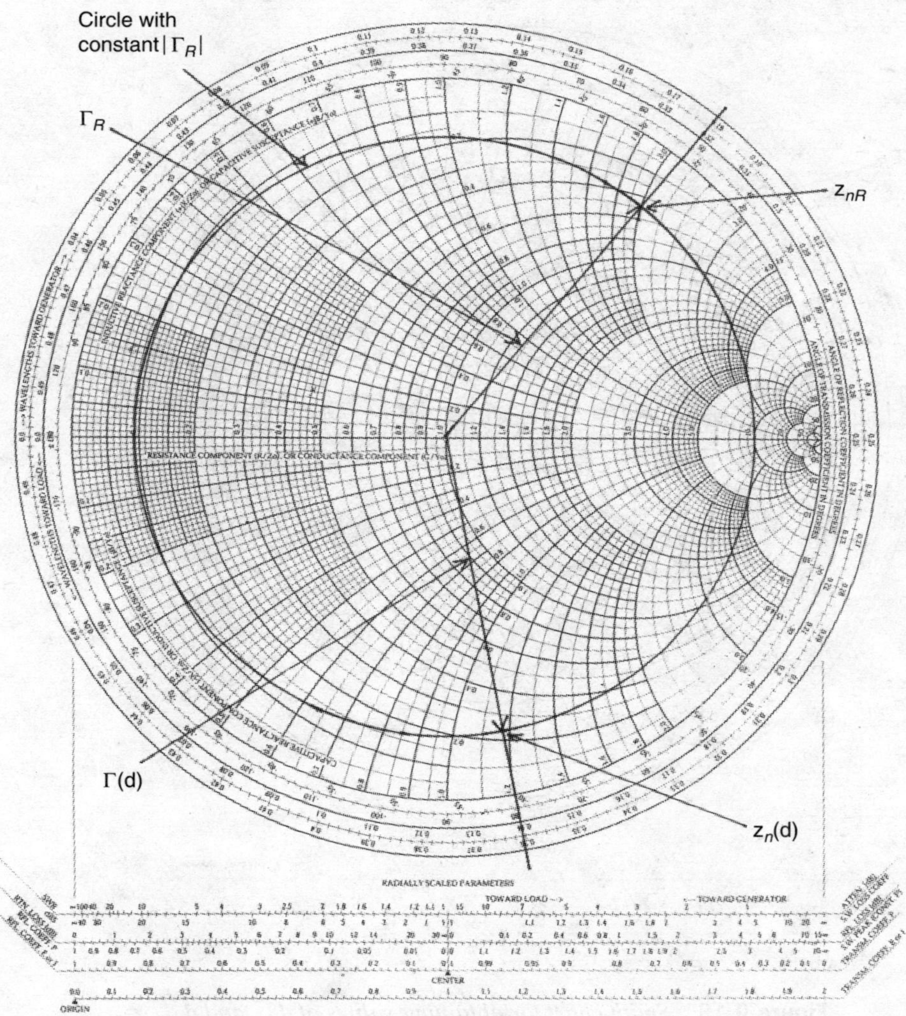

Figure 9.17 *Smith chart representing* $|\Gamma_R|, \Gamma_R$ *and* $\Gamma(d)$

Step 2: Drawing a circle of $|\Gamma_R|$, θ_s corresponding to $d_{\max}$ and $d_{\min}$ has also been shown. Please refer the Smith chart (Figure 9.18).

Step 3: From the outer graduation of the Smith chart, observing and recording the distances normalized to the wavelength,

$$2\beta d_{\max} = 50.9^\circ$$
$$\therefore\ d_{\max} = 0.0707\lambda \quad \text{and}$$

$$2\beta d_{\min} = 230.9^\circ$$
$$\therefore d_{\min} = 0.3207\lambda$$

Also, Im $(Z_R) > 0$.

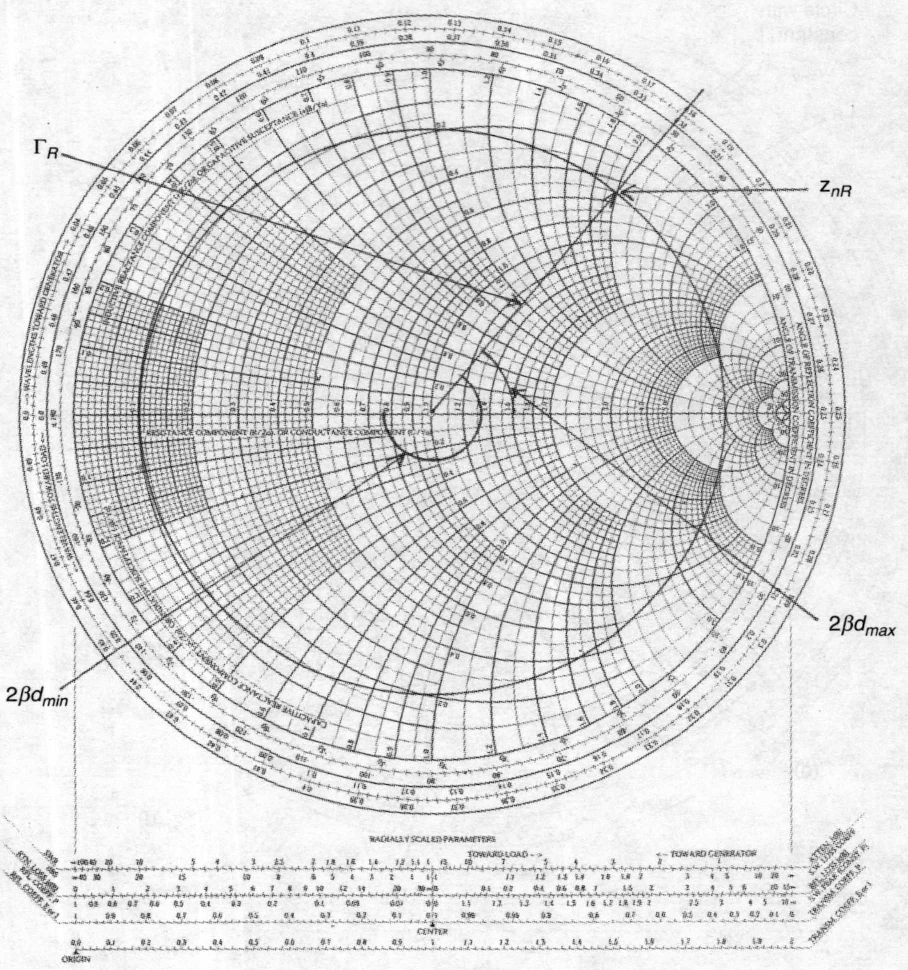

Figure 9.18 *Smith chart for obtaining values of d_{max} and d_{min}*

9.17 The thickness of a strip line is 0.6 mm and the dielectric constant is 2.25. Find out the required width b of the metal strip for the strip line to have a characteristic resistance of 100 Ω. Also, determine L, C and υ_p of the line. Neglect the losses and fringing effects.

Solution:

$$b = \frac{d}{z_0}\sqrt{\frac{\mu}{\varepsilon}} = \frac{0.6\times10^{-3}}{100}\times\frac{\eta_0}{\sqrt{\varepsilon_r}}$$

$$= \frac{0.6\times10^{-3}}{100\sqrt{2.25}}\times 377 = 1.5\times10^{-3}\,\text{m}$$

Also,

$$L = \mu \frac{d}{b} = 4\pi \times 10^{-7} \times \frac{0.6 \times 10^{-3}}{1.5 \times 10^{-3}} = 5.02 \times 10^{-7} \text{ H/m}$$

$$C = \varepsilon_0 \varepsilon_r \frac{b}{d} = \frac{10^{-9}}{36\pi} \times 2.25 \times \frac{1.5 \times 10^{-3}}{0.6 \times 10^{-3}} = 49.7 \times 10^{-12} \text{ F/m}$$

$$v_p = \frac{1}{\sqrt{\mu\varepsilon}} = \frac{C}{\sqrt{\varepsilon_r}} = \frac{C}{\sqrt{2.25}} = \frac{C}{1.5} = 2 \times 10^8 \text{ m/s}$$

9.18 We can measure the standing wave ratio S of a transmission line. Show how the value of terminating resistance of a lossless line of known characteristic impedance R_0 can be determined by S.

Solution:

As the terminating impedance is purely resistive,

$$Z_L = R_L$$

Transmission line wave is shown in Figure 9.19.

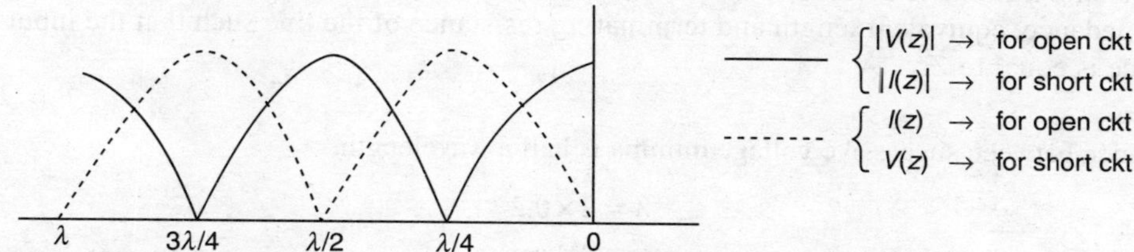

Figure 9.19 *Voltage and current standing wave on an open- and short-circuit lossless line*

Therefore,

$$R_2 > R_0 \qquad (\text{if } V_{max} \text{ is at } z = 0, \lambda/2, \lambda, \text{etc.})$$

Also,

$$R_2 < R_0 \qquad (\text{if } V_{min} \text{ is at } z = 0, \lambda/2, \lambda, \text{etc.})$$

For $R_2 > R_0$ $\theta_\Gamma = 0$, $|V_{max}|$ and $|I_{max}|$ occur at $\beta_z = 0$. In addition, $|V_{min}|$ and $|I_{min}|$ occur at $\beta_z = \pi/2$:

$$|V_{max}| = V_L, \quad |V_{min}| = I_L \frac{R_L}{R_0}$$

$$|I_{min}| = I_L, \quad |I_{max}| = I_L \frac{R_L}{R_0}$$

Thus,

$$\frac{|V_{max}|}{|V_{min}|} = \frac{|I_{max}|}{|I_{min}|} = S = \frac{R_L}{R_0} \quad \text{or}$$

$$R_L = SR_0$$

Now again for $R_L < R_0, \theta_\Gamma = -\pi$. Both $|V_{\min}|$ and $|I_{\min}|$ occur at $\beta z = 0$ and, $|V_{\max}|$ and $|I_{\max}|$ occur at $\beta_z = \pi/2$:

$$|V_{\min}| = V_L, \quad |V_{\max}| = V_L \frac{R_0}{R_L}$$

$$|I_{\max}| = I_L, \quad |I_{\min}| = I_L \frac{R_L}{R_0}$$

Therefore,

$$\frac{|V_{\max}|}{|V_{\min}|} = \frac{|I_{\max}|}{|I_{\min}|} = S = \frac{R_L}{R_0} \quad \text{or}$$

$$R_L = \frac{R_0}{S}$$

9.19 The standing wave ratio of a lossless 100 Ω transmission line terminated in unknown load impedance is found to be 5.0. The distance between successive voltage minima is 20 cm. The first minima is located at 6 cm from the load. Determine the reflection coefficient, load impedance, equivalent length and terminating resistance of the line such that the input impedance is equal to Z_L.

Solution:

The distance between successive voltage minima is half a wavelength.

$$\lambda = 2 \times 0.2$$
$$= 0.4 \text{ (m)}$$

$$\beta = \frac{2\pi}{\lambda} = 5\pi \text{ (rad/m)}$$

Now,

$$|\Gamma_L| = \frac{(S-1)}{(S+1)} \quad \text{or}$$

$$= \frac{(5-1)}{(5+1)} = 0.66 \quad \text{and}$$

$$\theta_{\Gamma_L} = (2\beta z - \pi)$$
$$= (2 \times 5\pi \times 0.06 - \pi)$$
$$= -1.25 \text{ rad}$$

$$\Gamma_2 = |\Gamma_2| e^{j\theta} \quad \text{or}$$

$$= 0.66\, e^{-j1.25}$$

The load impedance

$$Z_L = 100\frac{(1-j\,1.25)}{1+j\,1.25} = (-21.95 - j\,97.5)$$

Now for equivalent and terminating length resistance, we have

$$(-21.95 - j\,97.5) = 100\frac{(R_m + J\,100\tan\beta\,l_m)}{(100 + jR_m\tan\beta\,l_m)}$$

As

$$(Z_m + l_m) = \alpha/2 \quad \text{and} \quad R_m = R_0/S$$

$$l_m = \frac{\alpha}{2} - Z_m$$

$$= (0.2 - 0.06)$$

$$= 0.140 \text{ m} \quad \text{and}$$

$$R_m = \frac{100}{5} = 20\ \Omega$$

9.20 Given $Z_L = (195 + j200)$, find Y_L.

Solution:

$$Y_L = \frac{1}{Z_L} = \frac{(1+0j)}{(195 + j\,200)} \quad \text{or}$$

$$= (0.0024 - j\,0.0025)$$

9.21 The radial component of radiated power density $P_r = Am\sin\theta/r^2$. Find the total radiated power.

Solution:

Total radiated power

$$= \int_0^{2\pi}\int_0^{2\pi} \frac{Am\sin\theta}{r^2}\cdot r^2\sin\theta\,d\theta\,d\phi \quad \text{or}$$

$$= 2\pi\ Am\int_0^{\pi}\left(\frac{1-\cos 2\theta}{2}\right)d\theta \quad \text{or}$$

$$= 2\pi\ Am\left(\frac{\theta}{2} - \frac{\sin 2\theta}{4}\right)_0^{\pi} \cdot \frac{2\pi\ Am\times\pi}{2} \quad \text{or}$$

$$= \pi^2\,Am$$

9.22 For an electromagnetic wave $H_x = H_0e^{j(wt-\beta z)}$ in free space and $H_y = H_z = 0$. Find the value of E_y.

Solution:

$$E_y = H_0e^{j(wt-\beta z)} \quad \text{and}$$

$$E_x = E_z = 0$$

9.23 It is given that for a travelling electromagnetic wave, $\mu_r = 2\times10^{-6}$ and $f = 2$ MHz. Find the value of δ.

Solution:

$$\delta = \sqrt{\frac{1}{\pi f \mu\sigma}} = \frac{1}{\sqrt{\pi f \mu\sigma}} = 1.06\times10^{-4}\ \text{m}$$

9.24 For an electromagnetic wave, it is given that $\vec{E}_y = Ae^{j(w(t)-x/v)}$. $\vec{E}_x = \vec{E}_z = 0$. Find the different components of $\vec{H}$.

Solution:

$$H_x = \frac{E_y}{\eta}$$

As

$$\eta = \sqrt{\frac{\mu_0}{\varepsilon_0}}$$

$$H_x = \sqrt{\frac{\varepsilon_0}{\mu_0}} \cdot Ae^{j(wt-x/v)} \quad \text{and}$$

$$H_y = H_z = 0$$

9.25 Distilled water has $\sigma = 0$, $\varepsilon_r = 64$ and $\mu_r = 1$. Find the refractive index, phase velocity and group velocity.

Solution:

Refractive index $= \sqrt{\varepsilon_r} = \sqrt{64} = 8$

Phase velocity $= \dfrac{c}{\sqrt{\varepsilon_r}} = \dfrac{3\times10^8}{8} = 0.375\times10^8$ (m/s)

Group velocity $= v_p \times v_g = c^2$ or

$$v_g = \frac{(3\times10^8)^2}{(0.375\times10^8)} = \frac{3\times3\times10^8}{0.375} \quad \text{or}$$

$$v_g = 24\times10^8\ \text{m/s}$$

9.26 A 5 MHz plane wave travelling in a normally dispersive lossless medium has a phase velocity of 250×10^6 m/s. v_p (phase velocity) is a function of λ, given as $v_p = K\sqrt{\lambda}$ where K is a constant. Find the group velocity.

Solution:

$$v_g = v_p - \lambda\frac{d(v_p)}{d\lambda}$$

Group velocity $= K\sqrt{\lambda} - \lambda\frac{d}{d\lambda}\left(K\sqrt{\lambda}\right)$ or

$$= K\sqrt{\lambda} - \lambda.K\frac{1}{2}\left(\sqrt{\lambda}\right)^{-1/2} \quad \text{or}$$

$$= K\sqrt{\lambda} - \frac{K\sqrt{\lambda}}{2} \quad \text{or}$$

$$= v_\rho\left(1-\frac{1}{2}\right) \quad \text{or}$$

$$= \frac{v_\rho}{2}$$

$$v_g = \frac{250\times10^6}{2} = 125\times10^6\ \text{m/s}$$

9.27 Travelling electric and magnetic waves in free space (medium 1) are incident normally on the interface with a perfect dielectric medium (medium 2) of relative permittivity (5.0). Find E_r/E_i and E_t/E_i.

Solution:

Given that

$$\varepsilon_2 = 5\varepsilon_0, \quad \varepsilon_1 = \varepsilon_0, \quad \eta_1 = \sqrt{\frac{\mu_1}{\varepsilon_1}} \quad \text{and}$$

$$\eta_2 = \sqrt{\frac{\mu_2}{\varepsilon_2}}$$

$$\mu_1 = \mu_2 = \mu_0$$

$$\frac{\varepsilon_{t1}}{\varepsilon_i} = \frac{(\eta_2-\eta_1)}{(\eta_2+\eta_1)} = \frac{\sqrt{\frac{\mu_0}{5\varepsilon_0}} - \sqrt{\frac{\mu_0}{\varepsilon_0}}}{\sqrt{\frac{\mu_0}{5\varepsilon_0}} + \sqrt{\frac{\mu_0}{5\varepsilon_0}}} = \frac{\left(\frac{1}{\sqrt{5}}-1\right)}{\left(\frac{1}{\sqrt{5}}+1\right)} = -0.39$$

$$\frac{E_t}{E_i} = 1+\frac{E_r}{E_i} = \frac{2\eta_2}{\eta_2+\eta_1} = 2\times\sqrt{\frac{\mu_0}{5\varepsilon_0}} = 0.89\sqrt{\frac{\mu_0}{\varepsilon_0}}$$

9.28 A lossy cable has $R = 4.75\ \Omega/\text{m}$, $L = 2.0\ \mu\text{H/m}$ at $C = 2$ pF/m and $G = 0$. It operates at $f = 0.6$ GHz. What are the values of propagation content and phase constant of the line?

Solution:

Given $R = 4.75\ \Omega/\text{m}$, $L = 2.0\,\mu\text{H/m}$, $C = 2\times10^{-12}$ F/m, $G = 0$ and $f = 600\times10^6$ Hz.

Propagation constant

$$\gamma = \sqrt{(R+j\omega L)(G+j\omega C)} \quad \text{or}$$

$$\gamma = \sqrt{(4.75 + j2\pi \times 600 \times 10^{6} \times 2 \times 10^{-6})(0 + j2\pi \times 600 \times 10^{6} \times 2 \times 10^{-12})} \quad \text{or}$$

$$\gamma = \sqrt{(4.75 + j7536)(j7536 \times 10^{-6})}$$

$$\gamma = (2.617 \times 10^{-3} + 7.49\, j)$$

Also, phase constant $\beta = 7.49$.

9.29 An air line has a characteristic impedance of 80 Ω. Its phase constant is 8 rad/m at a frequency of 15 MHz. Find the inductance per meter. Also, find the capacitance per meter of this air line.

Solution:
An air line is a lossless line. Thus, $6 \simeq 0$. Also, $R = G = 0$ and $\alpha = 0$.

$$Z_0 = R_0 = \sqrt{\frac{L}{C}} \quad \text{and}$$

$$\beta = \omega\sqrt{LC}$$

From the above expressions, we get

$$\frac{R_0}{\beta} = \frac{1}{\omega c} \quad \text{or}$$

$$C = \left(\frac{\beta}{\omega R_0}\right) \quad \text{or}$$

$$= \frac{8}{2\pi \times 15 \times 10^{6} \times 80} = 1.06 \text{ nf}$$

$$C = 1.06 \text{ nf} \quad \text{and}$$

$$L = R_0^2 C$$

$$= (80)^2 \times 1.06 \times 10^{-9}$$

$$L = 6.79\ \mu\text{H/m}$$

9.30 A distortion line has $Z_0 = 80\ \Omega$, $\alpha = 20$ mNp/m and $\upsilon = 0.8c$, where $c = 3 \times 10^{8}$ m/s. Find the value of R, L, G, C and λ at 100 MHz.

Solution:
In the case of a distortion line, we know that

$$RC = GL \quad \text{or}$$

$$G = \left(\frac{RC}{L}\right)$$

Further,

$$Z_0 = \sqrt{\frac{L}{C}}$$

Thus,

$$\alpha = \sqrt{RG} \quad \text{or} \quad R = \sqrt{\frac{C}{L}} \quad \text{or}$$

$$= \frac{R}{Z_0} \quad \text{or}$$

$$R = \alpha Z_0$$

However, we know that

$$\upsilon = \frac{\omega}{\beta} \quad \text{or}$$

$$= \frac{1}{\sqrt{LC}}$$

Thus,

$$R = \alpha Z_0 = (20\times 10^{-3})\times 80$$

$$= 1.6\ \Omega/\text{m}$$

Also,

$$L = \left(\frac{Z_0}{\upsilon}\right) = \frac{80}{0.8\times 3\times 10^{8}} = 3.33\times 10^{-7} \quad \text{or}$$

$$= 333.3\ \text{nH/m} \quad \text{and}$$

$$G = \frac{\alpha^2}{R} = \frac{20\times 20\times 10^{-6}}{1.6} = 2.5\times 10^{-4}\ \text{S/m}$$

Now,

$$\upsilon Z_0 = \frac{1}{C} \Rightarrow C = \frac{1}{0.8\times 3\times 10^{8}\times 80} = 5.20\times 10^{-11}\ \text{F/m} \quad \text{and}$$

$$\lambda = \frac{\upsilon}{f} = \frac{0.8\times 3\times 10^{8}}{100\times 10^{6}} = 2.4\ \text{m}$$

9.31 An open transmission line has $R = 5\ \Omega/\text{m}$, $L = 5.2\times 10^{-8}$ H/m, $G = 7.2\times 10^{-3}$ mho/m, $C = 4\times 10^{-10}$ F/m and $f = 6$ GHz. Find Z_0 Y and υ.

Solution:

$$\upsilon = \frac{1}{\sqrt{LC}} = \frac{1}{\sqrt{5.2\times 10^{-8}\times 4\times 10^{-10}}}$$

$$= 2.19\times 10^{8}\ \text{m/sec}$$

Now,

$$\omega = 2\pi f = 2\times 3.14\times 6\times 10^{9} = 3.76\times 10^{10}\text{ rad}$$

$$Z = \sqrt{\frac{12 + jL\omega}{G + jC\omega}}$$

$$\Rightarrow (R + jL\omega) = (5 + j\times 5.2\times 10^{-\delta}\times 2\times 3.14\times 6\times 10^{3}) = (5 + j1955.2)$$

$$= 1955.20\angle 89.85°$$

$$\Rightarrow G + jC\omega = (7.2\times 10^{-3} + j\times 4\times 10^{-10}\times 3.76\times 10^{10}) = (7.2\times 10^{-3} + j15.04)$$

$$= 15.03\angle 89.97°$$

9.32 In the case of a transmission line, series inductance is 0.8 mH and capacitance 0.3 mF/km. Losses are negligible. What are the values of characteristic impedance Z_0 and phase velocity v_p?

Solution:
We know that

$$Z_0 = \sqrt{\frac{L}{C}} \quad \text{(in the case of a lossless line)}$$

$$Z_0 = \sqrt{\frac{0.8\times 10^{-3}}{0.3\times 10^{-3}}}$$

$$Z_0 = 1.63\ \Omega$$

Phase velocity

$$v_p = \frac{1}{\sqrt{LC}} = \frac{1}{\sqrt{(0.8)\times 10^{-3}\times (0.3)\times 10^{-3}}} \quad \text{or}$$

$$v_p = 2.041\times 10^{3}\text{ km/sec}$$

9.33 An ideal lossless transmission line of $Z_0 = 50\ \Omega$ is connected to unknown Z_L. If SWR = 4, find Z_L, the reflection coefficient and the transition coefficient.

Solution:

$$\Rightarrow Z_L = Z_0 \times \text{SWR} = 50\times 4 = 100\ \Omega$$

$$\Rightarrow \rho = \frac{(Z_L - Z_o)}{(Z_L + Z_o)} = \frac{(100-50)}{(100+50)} = \frac{50}{150} = 0.333$$

$$\Rightarrow \rho \text{ (refection coefficient)} = 0.333$$

Also, transmission coefficient

$$T = (1 + \rho)$$

$$T = (1 + 0.33) = 1.33$$

9.34 The transmission line of characteristic impedance of 50 Ω is terminated on a load of $(200 + j\,200)$ Ω. Find the reflection coefficient and SWR.

Solution:

Reflection coefficient $= \dfrac{(Z_L - Z_o)}{(Z_L + Z_o)} = \dfrac{(200 + j200 - 50)}{(200 + j200 + 50)}$

$$\Rightarrow \rho = \frac{(150 + j200)}{(250 + j200)} = \frac{25\angle 53.13^\circ}{320.15\angle 38.65^\circ}$$

$$\rho = 0.780\angle 14.48^\circ$$

$$\text{SWR} = \frac{1+|\rho|}{1-|\rho|}$$

$$= \frac{(1+0.780)}{(1-0.780)}$$

$$= \frac{1.780}{0.22} = 8.09$$

$$\text{SWR} = 8.09$$

9.35 A transmission line of $Z_0 = 100$ Ω is terminated at $R_L = Z_L = 150$ Ω. Find SWR, $Z_{\min}$ and $Z_{\max}$.

Solution:

Reflection coefficient

$$\rho = \frac{(Z_L - Z_o)}{(Z_L + Z_o)} = \frac{(150-100)}{(150+100)} = \frac{50}{250} = 0.2$$

$$\text{SWR} = \frac{1+|\rho|}{1-|\rho|} = \frac{1+0.2}{1-0.2} = \frac{1.2}{0.8} = 1.5$$

At the point of voltage minimum, we know that current is minimum.

$$Z_{\max} = Z_0\left(\frac{1+|\rho|}{1-|\rho|}\right) = Z_0 \times \text{SWR} = 100 \times 1.5 = 150\ \Omega$$

$$Z_{\max} = 150\ \Omega$$

At the point of voltage minimum, the current is maximum.

$$\therefore V_{\max} = V + e^{-\alpha z}\left(1+|\rho|\right) \quad \text{and}$$

$$V_{\min} = V + e^{-\alpha z}\left(1-|\rho|\right)$$

$$I_{\max} = \frac{V + e^{-\alpha z}}{Z_0}\left(1+|\rho|\right)$$

$$Z_{\min} = \frac{V_{\min}}{I_{\max}} = Z_0\left(\frac{1-|\rho|}{1+|\rho|}\right) = \frac{Z_0}{\text{SWR}} = \frac{100}{1.5} = 66.66$$

$$Z_{\min} = 66.66\ \Omega$$

9.36 A transmission line has the propagation constant $Y = (0.1 + j10)$ m and characteristic impedance $Z_0 = (50 + j5)\ \Omega$. The line is terminated in an impedance $(100 - j30)\ \Omega$. Find the impedance at a distance of 1.5 m from the load.

Solution:

We know that

$$Z(l) = Z_0\left[\frac{Z_L \cos hyl + Z_0 \sin hyl}{Z_L \sin hyl + Z_0 \cos hyl}\right] \quad (1)$$

For $l = 1.5$ m

$$\cos hyl = \cos h\{(0.1 + j10)1.5\} = (-0.7683 + 0.0979j)$$

$$\sin hyl = \sin h = \{(0.1 + j10)1.5\} = (-0.1144 + j0.6576)$$

and the impedance is given by

$$Z(l) = Z_0\left[\frac{Z_L \cos hyl + Z_o \sin hyl}{Z_L \sin hyl + Z_o \cos hyl}\right]$$

$$= (57.3175 + j38.6619)\ \Omega$$

9.37 It is given that $Y = (2.42 + j42)$ per meter. Calculate the attenuation constant in nepers/m and dB/m. If the voltage of a forward travelling wave at $t = 0$ and $x = 0$ is 6.2 V, find the voltage at $x = 2$ m at $t = 150$ nsec. What is the peak voltage at $x = 2$ m? Also, the wave has a 30° phase at $t = 0$ and $x = 0$. Consider $\beta = 28.2$ rad/m and $f = 1$ GHz.

Solution:

Given $Y = (2.42 + j42)$ per meter

$$\Rightarrow \alpha = 2.42 \text{ neper/m} = 8.68 \times 2.42 = 21.0056 \text{ dB/m}$$

The voltage for the forward travelling wave is given as

$$V(t) = \text{Re}\left\{V^+ e^{-\alpha x} e^{j(\omega t - \beta x)}\right\}$$

Taking $V^+ = |V^+| e_{j\phi}$, where φ is the initial phase, we get

$$V(t) = |V^+| e^{-\alpha x} \cos(\varphi + \omega t - \beta x)$$

At $x = 0$ and $t = 0$, it is given that $V(t) = 6.2$ V.

Therefore,
$$6.2 = |V^+| \cos 30°$$
$$\Rightarrow |V^+| = 7.159$$

The instantaneous voltage at $x = 2$ m and $t = 150$ nsec is obtained as
$$V(t) = 7.15e^{-2.42\times 2} \cos(30 + 2\pi\times 10^9 \times 150\times 10^{-3} - 28.2\times 2)$$
$$V(t) = 5.71\times e^{-(4.84)} = 0.045 \text{ V}$$

Peak voltage at $x = 2$/m is
$$V_p = 7.1e^{-2.42} = 0.631 \text{ V}$$

9.38 A 60 Ω lossless transmission line has its length of 0.6 λ. It is terminated in an open circuit. Find its input impedance. Given that $\lambda = 20$ cm.

Solution:
Given here
$$l = 0.6\lambda \text{ and}$$
$$Z_0 = 60\ \Omega$$

Electric length
$$(\beta l) = \frac{2\pi}{\lambda}\times 0.6\lambda = 1.2\pi$$

Input impedance Z_{in} is given as
$$Z_{in} = Z_o\left[\frac{Z_L + jZ_o \tan\beta l}{Z_o + jZ_L \tan\beta l}\right]$$

however, $Z_L = \infty$ (for a short-circuit line). Thus,
$$Z_{in} = Z_o\left[\frac{1}{j\tan\beta l}\right]$$
$$= -j Z_0 \cot\beta l$$
$$= -j\times 60\times \cot(1.2\pi)$$
$$= -j910.5\ \Omega$$
$$Z_{in} = -j910.5\ \Omega$$

9.39 Find the input impedance of a 40 Ω lossless transmission line having length 0.4λ if the load is a short circuit.

Solution:
It is given that $l = 0.4\lambda$, $Z_0 = 40\ \Omega$ and $Z_i = 0$ (for a short-circuit line). Z_i (for a lossless line) is given as
$$Z_i = Z_o\left[\frac{Z_L + jZ_o \tan\beta l}{Z_o + jZ_L \tan\beta l}\right]$$

Electric length

$$\beta l = \frac{2\pi}{\lambda} \times 0.4\lambda$$

$$= 0.8\pi$$

Thus,

$$Z_i = jZ_o \tan \beta l$$

$$= j \times 40 \tan(0.8\pi)$$

$$= j1.755\ \Omega$$

$$Z_i = j1.755\ \Omega$$

9.40 A transmission line has a characteristic impedance of 60 Ω. It is terminated at a resistance of $j40$ Ω. Find the input impedance of a section, which is 25 cm long at a frequency of 300 MHz.

Solution:

Given that $Z_0 = 60\ \Omega$ and $Z_L = j40\ \Omega$. Also, frequency $(f) = 200$ MHz.

$$l = \frac{25}{100}\ \text{m} = 0.25\ \text{m}$$

Further,

$$\lambda = \left(\frac{v}{f}\right) = \frac{3 \times 10^8}{300 \times 10^6} = 1\ \text{m}$$

Thus,

$$l = \frac{\lambda}{4} \quad \text{(from above)}$$

It is a quarter-wave transmission line. Z_{in} in a quarter-wave transmission line is obtained as

$$Z_{in} = \frac{Z_o^{\ 2}}{Z_L} \quad \text{or}$$

$$Z_{in} = \frac{(60)^2}{j40} = -90\,j\ \Omega$$

$$Z_{in} = -90\ j\ \Omega$$

9.41 A transmission line has a characteristic impedance of 80 Ω. It is terminated in a purely resistance load. The minimum voltage upon it is 3 μV. The maximum voltage is 5 μV. What will be the load impedance?

Solution:

Given here that $Z_o = 80\ \Omega$, $V_{\max} = 5\ \mu\text{V}$ and $V_{\min} = 3\ \mu\text{V}$

Thus,

$$\text{VSWR} = \left(\frac{V_{\max}}{V_{\min}}\right) \quad \text{or}$$

$$= \frac{5}{3} = 1.66$$

Reflection coefficient
$$\rho = \frac{(\text{VSWR} - 1)}{(\text{VSWR} + 1)}$$

$$= \frac{(1.66 - 1)}{(1.66 + 1)} = \frac{0.66}{2.66} = 0.248$$

However,

$$\rho = \frac{(Z_L - Z_o)}{(Z_L + Z_o)}$$

By putting the value of Z_o and ρ, we get

$$0.248 = \frac{(Z_L - 80)}{(Z_L + 80)}$$

$$\Rightarrow 0.248 Z_L + 19.84 = Z_L - 80$$

$$\Rightarrow -0.752\, Z_L = -99.84$$

$$\Rightarrow Z_L = 132.7\ \Omega$$

9.42 In the case of a transmission line, open- and short-circuit impedances are $40\angle -30°\ \Omega$ and $30\angle -20°\ \Omega$, respectively. It is operating at 3 kHz. Find Z_0 of this transmission line.

Solution:

It is given that $(Z_i)_{oc} = 40\angle -30°$ and $(Z_i)_{sc} = 30\angle -20°$. The characteristic impedance Z_0 of a transmission line is obtained as

$$Z_0 = \sqrt{(Z_i)_{oc} . (Z_i)_{sc}}$$

$$\Rightarrow Z_0 = \sqrt{40\angle -30° \times 30\angle -20°}$$

$$= \sqrt{40 \times 30}\, \frac{\angle -30° - \angle 20°}{2}$$

$$= 34.64\angle -25.5°\ \Omega$$

9.43 A transmission line has a characteristic impedance of $Z_0 = R_0 = 50\ \Omega$. The transmission line is terminated in an impedance $Z_R = (60 + j60)\ \Omega$. What is impedance Z_i at a point that is $\lambda/6$ m away from the load?

Solution:

Given that $l = \lambda/6$ and $\beta = 2\pi/\lambda$.

We know that
$$Z_i = Z_0 \left[\frac{Z_R + jZ_0 \tan \beta l}{Z_0 + jZ_R \tan \beta l} \right]$$

Electrical length
$$(\beta l) = \frac{2\pi}{\lambda} \times \lambda/6 = \frac{\pi}{3}$$

Thus,

$$Z_i = Z_0\left[\frac{Z_R + jZ_0 \tan\pi/3}{Z_0 + jZ_R \tan\pi/3}\right] \quad \text{or}$$

$$= 50\left[\frac{(60+j60)+j50\tan\dfrac{\pi}{3}}{50+j(60+j60)\tan\dfrac{\pi}{3}}\right] \quad \text{or}$$

$$= 50\left[\frac{(60+j60)+j0.91}{50+(60j-60).018}\right]$$

$$= 50\left[\frac{60+j60.91}{50+1.092j-1.092}\right]$$

$$= 50\left[\frac{60+j60.91}{48.90+j1.092}\right]$$

$$Z_i = \left(\frac{3000+j3045.5}{48.90+j1.092}\right)\Omega$$

$$Z_{in} = 44.72\left[\frac{-81+j37.11}{-24.14+j124.5}\right]\Omega$$

9.44 A 40 Ω lossless transmission line is connected to a load of (40 + *j* 40) Ω. The maximum voltage measured on the line is 60 V. Find the power delivered to the load and the peak voltage at the load-end of the line.

Solution:

The magnitude of the reflection coefficient is obtained as

$$|\Gamma_L| = \left|\frac{Z_L - Z_0}{Z_L + Z_0}\right| = \frac{40+j40-40}{40+j40+40}$$

$$= \frac{j40}{40(2+j)} = \frac{j}{2+j} = 0.33+0.66j$$

$$= 0.74$$

$$\text{VSWR} = \frac{1+|\Gamma_L|}{1-|\Gamma_L|} = \frac{(1+0.744)}{(1-0.744)} = \frac{1.744}{0.255} = 6.83$$

Since the line is lossless, the power loss at any point on the line is the same as the power delivered to the load. We can therefore conveniently choose a point on the line where the voltage is maximum (so current is minimum). The impedance at this location is real and its value is $\rho Z_0 = 273.2$. The power loss at this point (which is the same as power delivered to the load) is obtained as

$$P_L = \frac{1}{2}\frac{|V_{\max}|^2}{\rho Z_0} = \frac{1}{2} \times \frac{60 \times 60}{273.2} = 6.58$$

Now the, power loss in the load can also be written in term of the voltage across the load as

$$p_L = \frac{1}{2}\frac{|V_L|^2}{R_L}$$

$$V_L = \sqrt{2 \times 6.58 \times 40} = 22.94 \text{ V}$$

9.45 A lossless transmission line is 60 cm long and operates at a frequency of 500 MHz. The line parameters are $L = 0.2$ μH/m and $C = 100$ pF/m. Find the characteristic impedance, the phase constant, the velocity on the line and the input impedance for $Z_L = 150\ \Omega$.

Solution:

Since the line is lossless, both R and G are zero. The characteristic impedance is obtained as

$$Z_0 = \sqrt{\frac{L}{C}} = \sqrt{\frac{0.2 \times 10^{-6}}{100 \times 10^{-12}}} = 44.72\ \Omega$$

Since $$Y = (\alpha + j\beta) = \sqrt{(R + j\omega L)(G + j\omega C)} = j\omega\sqrt{LC}$$

$$\beta = \omega\sqrt{LC} = 2\pi \times 500 \times 10^6 \sqrt{0.2 \times 10^{-6} \times 100 \times 10^{-12}}$$

$$= 14.04 \text{ rad/m}$$

Also,

$$v_p = \frac{\omega}{\beta} = \frac{2\pi \times 500 \times 10^6}{14.04} = 2.2 \times 10^8 \text{ m/sec}$$

Z_{in} is found as

$$Z_{in} = Z_o\left[\frac{Z_L \cos\beta l + jZ_o \sin\beta l}{Z_o \cos\beta l + jZ_L \sin\beta l}\right]$$

$$= 44.72\left[\frac{150\cos(14.04 \times 0.6) + j44.72\sin(14.04 \times 0.6)}{44.72\cos(14.04 \times 0.6) + j150\sin(14.04 \times 0.6)}\right]$$

9.46 Two very long lossless cables of characteristic impedance 40 Ω and 80 Ω, respectively, are to be joined for a reflectionless transmission. Find the suitable matching transformer.

Solution:

The quarter-wave length transformer is appropriate for this case, as we have to match two real impedances. The characteristic impedance of the transformer section is

$$Z_o = \sqrt{40 \times 80} = 56.56\ \Omega$$

The length of the transformer should be an odd multiple of $\lambda/4$.

9.47 A 70 Ω lossless transmission line has a loss of 1.6 dB/m. The velocity of the voltage wave on the line is 2×10^8 m/sec. A section of the line is used to make a series resonant circuit at 2 GHz. Find the input impedance of the line, its quality factor and the 3 dB bandwidth of the resonant circuit.

Solution:

Here,

$$\lambda = \frac{v}{f} = \frac{2\times10^8}{2\times10^9} = 0.1 \text{ m}$$

The phase constant is obtained as

$$\beta = \frac{2\pi}{\lambda} = \frac{2\pi}{0.1} = 20\pi \text{ rad/m}$$

If we take an open-circuited section of the line for series resonance the length of the section would be

$$l_{oc} = \frac{\lambda}{4}, \frac{3\lambda}{4}, \ldots$$
$$= 0.25, 0.75\ldots \text{ (m)}$$

If we take a short-circuited section of the line for series resonance, the length of the section would be

$$l_{sc} = \frac{\lambda}{2}, \frac{2\lambda}{2}, \ldots$$
$$= 0.05, 0.1, \ldots \text{ (m)}$$

The loss of the line is $\alpha = 1.6$ dB/m $= 1.6/8.68 = 0.184$ nepers/m. For series resonance, the input impedance is

$$Z_{in} = Z_0 \alpha l = 70 \times 0.184 l = 12.88 l$$

where $l = l_{oc}$ or l_{sc}. The quality factor is obtained as

$$\varphi = \frac{\beta}{2\alpha} = \frac{20\pi}{2\times0.184} = 170.65$$

The 3 dB bandwidth is

$$\text{BW} = \frac{f_0}{\varphi} = \frac{2\times10^9}{170.65} = 11.71 \text{ MHz}$$

$$\text{BW} = 11.71 \text{ MHz}$$

9.48 A transmission line has $L = 0.25$ μH/m, $C = 100$ pF/m and $G = 0$. What should be the value of R for the line so that the line can be treated as a lossless line? The frequency of operation is 100 MHz.

Solution:

The phase constant of the lossless line is

$$\beta \approx \omega\sqrt{LC} = 2\pi\sqrt{0.25\times10^{-6}\times100\times10^{-12}}\times10^8$$
$$= \pi$$

The attenuation constant is given as

$$\alpha = \frac{R}{2}\sqrt{\frac{C}{L}} \qquad (G = 0)$$

$$= \frac{R}{2}\sqrt{\frac{10^{-10}}{0.25\times10^{-6}}} = 0.01\,R \text{ neper/m}$$

For a lossless line, $\beta >> \alpha$. Taking $\alpha < 1\%$ of β, we get

$$0.01R < \frac{\beta}{100}$$

$$\Rightarrow \qquad R < \pi\ (3.14)\ \Omega/\text{m}$$

9.49 A 45 Ω transmission line is connected to a parallel combination of a 90 Ω resistance and 1 nF capacitance. Find the VSWR on the line at a frequency of 3 MHz. Also, find the maximum and minimum resistance seen on the line.

Solution:

The load impedance Z_L is a parallel combination of $R = 90\ \Omega$ and $C = 1$ nF.

Therefore,

$$Z_L = \frac{R(1/j\omega C)}{R+(1/j\omega C)} = \frac{R}{1+\omega RC}$$

$$= \frac{90}{1+j2\pi\times3\times10^{6}\times90\times10^{-9}}$$

$$= \frac{90}{1+j0.69} = \frac{(90-j0.69)}{0.52}$$

$$Z_L = (171.78 - j1.32)$$

The reflection coefficient at the load-end of the line is

$$\Gamma_L = \frac{(Z_L - Z_0)}{(Z_L + Z_0)} = \frac{(171.78 - j2.32 - 45)}{(171.78 + j2.32 + 45)}$$

$$= \frac{(126.78 - j1.32)}{(216.78 + j1.32)}$$

$$|\Gamma_L| = 0.584$$

VSWR,

$$\rho = \frac{1+|\Gamma_L|}{1-|\Gamma_L|} = \frac{1.584}{0.415} = 3.80$$

The maximum resistance on the line is $R_{max} = \rho Z_0 = 171\ \Omega$.

The minimum resistance on the line is $R_{min} = Z_0/\rho = 11.842\ \Omega$.

9.50 A two-conductor transmission line is 20 cm long. A sinusoidal signal of 2 V peak amplitude is applied to one end of the line. If the signal travels on the line with a speed of 2×10^8 m/sec, find

(a) The transmit time on the line.
(b) The frequency at which the transmit time is 15% of the signal period.
(c) Signal voltage on the other end of the line at any instant when the input signal has a frequency of 600 MHz.

Solution:

(a) Transmit time

$$t_r = \frac{\text{Length of the line}}{\text{Velocity}} = \frac{0.2}{2\times10^8} = 1\,\text{n sec}$$

(b) The transmit time should be 15% of the period T, i.e.
Transmit time $t_r = 0.15T$

$$\Rightarrow T = \frac{t_r}{0.15} = \frac{1\times10^{-3}}{0.15} = 6.6\,\text{n sec}$$

$$\text{Frequency} = \frac{1}{T} = \frac{1}{6.6\times10^{-9}} = 0.151\times10^9\,\text{Hz} = 151\,\text{MHz}$$

(c) Let the signal voltage be given as

$$v(t) = A\cos(2\pi ft)V$$

Given $A = 2, f = 600$ MHz. Therefore,

$$v(t) = 2\cos(2\pi\times600\times10^6\times t)$$

$$= 2\cos(\pi\times12\times10^8 t)$$

Taking the instant as $t = 0$, when the input voltage $v(t) = 2V$, the signal at the other end of the line will correspond to $t = -t_r$. The voltage at the other end of the line therefore is

$$\cos(-\pi\times10^9 t_r) = \cos(-\pi\times10^9\times10^{-9}) = -1\,\text{V}$$

9.51 A voltage wave at 1 GHz is travelling on a transmission line in the $-x$ direction. The primary contents of the line are $R = 0.8\ \Omega$/m, $L = 0.1\ \mu$H/m, $G = 0.2\ \mho$/m and $C = 50$ pF/m. The wave has a 45° phase at $t = 0$ and $x = 0$. If the voltage of the forward travelling wave at $t = 0$ and $x = 0$ is 8.66 V, find the voltage at $x = 1$ m at $t = 100$ nsec. Given $Y = (2.23 + j28.2)$.

Solution:
The voltage for a wave travelling in the $-x$ direction is given as

$$v(t) = R_e\left\{V^- e^{\alpha x} e^{j(\omega t+\beta x)}\right\}$$

Taking $V^- = \left|V^-\right| e^{J\phi}$, we have

$$v(t) = \left|V^-\right| e^{\alpha x}\cos(\varphi + \omega t + \beta x)$$

At $x = 0$ and $t = 0$, $v(t) = 8.66$ V (given);

$$8.66 = \left|V^-\right| \cos 45°$$

therefore,

$$\Rightarrow \left|V^-\right| = 12.24 \text{ V}$$

The voltage at $x = 1$ m and $t = 100$ n sec is obtained as

$$v(t) = 12.24e^{(2.23\times1)} \cos(40° + 2\pi \times 10^9 \times 100 \times 10^{-9}(\text{rad}) + 28.2 \times 1(\text{rad}))$$

$$= 113.83\cos(40° + 2\pi \times 100 + 28.2)$$

$$v(t) = -91.43 \text{ V}$$

9.52 A voltage wave of 2 GHz is travelling on a transmission line. The primary constants of the line are $R = 0.8$ Ω/m, $L = 0.4$ μH/m, $G = 0.2$ ℧/m and $C = 200$ pF/m The wave has a 45° phase at $t = 0$ and $x = 0$. Find the phase of the wave at $x = 45$ cm and at $t = 2$ μsec.

Solution:

$$\omega = 2\pi \times 2 \times 10^9 \text{ rad/m}$$

$$= 4\pi \times 10^9 \text{ rad/m}$$

The propagation constant Y is obtained as

$$Y = \sqrt{(R + j\omega L)(G + j\omega C)}$$

$$= \sqrt{(0.8 + j \times 4\pi \times 10^9 \times 0.4 \times 10^{-6})(0.2 + j \times 4\pi \times 10^9 \times 200 \times 10^{-12})}$$

$$= \sqrt{(0.8 + j5024)(0.2 + j2.512)}$$

$$Y = (4.47 + 112.42j)$$

$$\beta = 11.242 \text{ rad/m}$$

Phase of the wave = Initial phase + $\omega t - \beta x$

$$= (45 + 2\pi \times 2 \times 10^{-6} - 112.42 \times 0.45)$$
$$= 992.965 \text{ rad}$$

9.53 The lossless transmission line has a characteristic impedance of 50 Ω and a phase constant of 4 rad/m at 120 MHz. Find the inductance and capacitance of the line per meter.

Solution:

Characteristic impedance is given as

$$Z_0 = \sqrt{\frac{L}{C}}$$

Also,

$$r = \beta = \omega\sqrt{LC}$$

Therefore,

$$\frac{Z_0}{\beta} = \frac{\sqrt{L/C}}{\omega\sqrt{LC}} = \frac{1}{C\omega} \quad \text{or}$$

$$\frac{Z_0\omega}{\beta} = \frac{1}{C} \quad \text{or}$$

$$\frac{50 \times 2\pi f}{4} = \frac{1}{C} \quad \text{or}$$

$$\frac{50 \times 2 \times \pi \times 120 \times 10^6}{4} = \frac{1}{C}$$

$$C = 1.06 \times 10^{-10} \text{ F/m}$$

As

$$L = Z_0{}^2 C$$

$$= (50)^2 \times 1.06 \times 10^{-10}$$

$$L = 2.852 \times 10^{-7} \text{ H/m}$$

9.54 In a lossless transmission line, the velocity of propagation of wave is 12.5×10^8 m/s. The capacitance of the line is found to be 40 pF/m. Find the following for the line:

(a) Inductance
(b) Phase constant at 120 MHz
(c) Characteristic impedance

Solution:
We are given here $\upsilon_p = 12.5 \times 10^8$ m/s and frequency $f = 120$ MHz.
Further,

$$\upsilon_p = \frac{\omega}{\beta} \quad \text{or}$$

$$\beta = \frac{2\pi \times 120 \times 10^6}{12.5 \times 10^8}$$

$$= 0.603 \text{ rad/m}$$

Therefore, the phase constant is 0.603 rad/m.

As

$$\upsilon_p = \frac{1}{\sqrt{LC}} \quad \text{or}$$

$$\upsilon_p^2 = \frac{1}{LC}$$

$$L = \frac{1}{Cv_\rho^{\,2}} \quad \text{or}$$

$$= \frac{1}{40\times10^{-12}(12.5\times10^{8})^2} \quad \text{or}$$

$$L = 1.6\pi10^{-8}\ \text{H/m}$$

Characteristic impedance

$$Z_0 = \sqrt{\frac{L}{C}} \quad \text{or}$$

$$= \sqrt{\frac{1.6\times10^{-8}}{40\times10^{-12}}}$$

$$= 20\ \Omega$$

9.55 Find the input impedance of a short-circuited transmission line of length $\lambda/8$ having $Z_0 = 500\ \Omega$.

Solution:

$$(Z_{in})_{sc} = jZ_0 \tan(\beta l) \quad \text{or}$$

$$= j500\times\tan\left(\frac{2\pi}{\lambda}\times\frac{\lambda}{8}\right) \quad \text{or}$$

$$= j500$$

$$(Z_{in})_{sc} = j500$$

9.56 A transmission line has a characteristic impedance of 75 Ω. It is terminated into a load of $(150 + j\,150)\ \Omega$. Find the reflection coefficient and standing wave ratio, SWR.

Solution:

Reflection coefficient

$$\Gamma_L = \frac{(Z_L - Z_0)}{(Z_L + Z_0)} = \frac{(150 + j\,150 - 75)}{(150 + j\,150 + 75)}$$

$$\Gamma_L = \frac{(75 + j\,150)}{(225 + j\,150)} \quad \text{or}$$

$$= \frac{(167.7\angle 63.43°)}{(270.4\angle 33.69°)}$$

$$= 0.6201\angle 29.74°$$

$$|\Gamma_L| = 0.62$$

Also, SWR is obtained as

$$\text{SWR} = \frac{(1+|\Gamma_L|)}{(1-|\Gamma_L|)}$$

$$= \frac{(1+0.62)}{(1-0.62)}$$

$$= 4.263$$

9.57 A transmission line has $Z_0 = 60\ \Omega$. It is terminated at $Z_L = R_L = 150\ \Omega$. Find the values of VSWR, Z_{min} and Z_{max}.

Solution:

As we know, reflection coefficient

$$\Gamma_L = \frac{(Z_L - Z_o)}{(Z_L + Z_o)}$$

Therefore,

$$\Gamma_L = \frac{90}{210} \quad \text{or}$$

$$= 0.428$$

As we know, at the point of maximum voltage, the current is minimum; therefore,

$$Z_{max} = Z_o\left(\frac{1+|\Gamma_L|}{1-|\Gamma_L|}\right) \quad \text{or}$$

$$= Z_o\,(\text{SWR}) \quad \text{or}$$

$$= Z_o \times \left(\frac{1+0.42}{1-0.42}\right) \quad \text{or}$$

$$= 60 \times 2.44 \quad \text{or}$$

$$Z_{max} = 146.8$$

Also, at the point of minimum voltage, current is maximum; therefore,

$$V_{max} = V+e^{-\alpha z}\left(1+|\Gamma_L|\right)$$

$$V_{min} = V+e^{-\alpha z}\left(1-|\Gamma_L|\right)$$

$$I_{max} = \frac{V+e^{-\alpha z}}{Z_o}\left(1+|\Gamma_L|\right) \quad \text{and}$$

$$Z_{min} = \frac{V_{min}}{I_{max}} \quad \text{or}$$

$$= Z_o \frac{(1-|\Gamma_L|)}{(1+|\Gamma_L|)} \quad \text{or}$$

$$= \frac{Z_o}{\left(\frac{1+|\Gamma_L|}{1-|\Gamma_L|}\right)} \quad \text{or}$$

$$= \frac{Z_o}{\text{SWR}}$$

$$= 24.59\ \Omega$$

9.58 An open-wire transmission line has $Z_o = 500\ \Omega$. It is terminated at $Z_L = 1000\ \Omega$. Find the reflection coefficient and SWR.

Solution:
Reflection coefficient is given by

$$\Gamma_L = \frac{(Z_L - Z_o)}{(Z_L + Z_o)} = \frac{(1000-500)}{(1000+500)} \quad \text{or}$$

$$= \frac{500}{1500} \quad \text{and}$$

$$= 0.33$$

$$\text{SWR} = \frac{(1+|\Gamma_L|)}{(1-|\Gamma_L|)}$$

$$= \frac{(1+0.33)}{(1-0.33)}$$

$$= 1.98$$

9.59 An ideal lossless transmission line has $Z_o = 120\ \Omega$. It is connected to an unknown load resistance Z_L. If the standing wave ratio SWR is found to be 6, find the values of Z_L, the reflection coefficient and the transmission coefficient.

Solution:
As it is given that

$$\text{SWR} = \frac{Z_L}{Z_o} \quad \text{or}$$

$$Z_L = Z_o \times \text{SWR}$$

$$= 120 \times 6 \quad \text{or}$$

$$Z_L = 720\ \Omega$$

$$\Gamma_L = \frac{(Z_L - Z_o)}{(Z_L + Z_o)} \quad \text{or}$$

$$= \frac{(720-120)}{(720+120)} \quad \text{or}$$

$$= \frac{600}{840}$$

$$= 0.71$$

Therefore, the transmission coefficient T is = $(1 + \Gamma_L)$

$$T = (1 + 0.71) \quad \text{or}$$

$$= 1.71$$

9.60 A radio frequency extension has Z_o = 500 Ω. It is terminated at $Z_L = (200 + j100)$ Ω. Find the value of SWR.

Solution:

Reflection coefficient

$$\Gamma_L = \frac{(Z_L - Z_o)}{(Z_L + Z_o)}$$

$$= \frac{(200 + j\,100 - 500)}{(700 + j\,100)}$$

$$= \frac{(-300 + j\,100)}{(700 + j\,100)}$$

$$= \frac{316.2\angle 161.5^\circ}{767.1\angle 8.1^\circ}$$

$$= 0.447\angle 153.4^\circ$$

Now,

$$\text{SWR} = \frac{(1+0.447)}{(1-0.447)} = 2.61$$

9.61 A lossless transmission line has Z_o = 25 Ω. It is terminated at an unknown load impedance Z_L. VSWR is 2. Find the first voltage minima from the load, which is at 20 cm. Find the value of load impedance if the supply frequency is 120 MHz.

Solution:

Given Z_o = 25 Ω and f = 120 MHz,

$$\lambda = \frac{c}{f} = \frac{3\times 10^8}{1.2\times 10^8} = 2.5 \text{ m}$$

The minima occurs at 20 cm:

$$(2\beta d - \theta) = \pi$$

$$(2 \times \frac{2\pi}{2.5} \times 20 \times 10^{-2} - \theta) = \pi$$

$$(0.32\pi - \theta) = \pi$$

$$\theta = (0.32\pi - \pi)$$

$$\theta = -0.68\pi$$

$$\theta = -2.13 \text{ rad} \quad \text{or}$$

$$\theta = -122.04°$$

Now, SWR

$$s = \frac{(1+|\Gamma_L|)}{(1-|\Gamma_L|)}$$

Therefore,

$$|\Gamma_L| = \frac{(s-1)}{(s+1)} \quad \text{or}$$

$$= \frac{(2-1)}{(2+1)} \quad \text{or}$$

$$= \frac{1}{3}$$

$$|\Gamma_L| = 0.33$$

$$\Gamma_L = |\Gamma_L| e^{j\theta} = \frac{(Z_L - Z_o)}{(Z_L + Z_o)} \quad \text{or}$$

$$\Gamma_L = 0.33 e^{-j122°} = \frac{(Z_L - Z_o)}{(Z_L + Z_o)} \quad \text{or}$$

$$0.33\,[\cos(122°) - j(\sin 122°)] = \frac{(Z_L - Z_o)}{(Z_L + Z_o)} \quad \text{or}$$

$$0.33\,(-0.529 - j0.848) = \frac{(Z_L - Z_o)}{(Z_L + Z_o)}$$

$$(-0.174 - j0.279) = \frac{(Z_L - Z_o)}{(Z_L + Z_o)}$$

$$-(0.174 + j\,0.279) = \frac{(Z_L - Z_o)}{(Z_L + Z_o)}$$

$$(0.174 + j\,0.279)(Z_L + 25) = (25 - Z_L)$$

$$(0.174\,Z_L + j0.279\,Z_L + 4.35 + j\,6.975 + Z_L - 25) = 0$$

$$Z_L(1.174 + j0.279) = (20.65 - j\,6.975)$$

$$Z_L = \frac{(20.65 - j6.975)}{(1.174 + j\,0.279)}$$

$$= \frac{21.7\angle -18.6^\circ}{1.20\angle 13.3^\circ}$$

$$= 18.08\angle -31.9^\circ$$

9.62 A lossless transmission line having Z_o = 400 Ω is terminated at unknown load impedance. The standing wave ratio is found to be 3. The first voltage minima is located at 10 cm from the load. Calculate the frequency and Z_L.

Solution:

$$|\Gamma_L| = \frac{(s-1)}{(s+1)} = \frac{(3-1)}{(3+1)} = 0.5$$

$$(2\beta\,d - \theta) = \pi$$

Let us assume that $\theta = 0$.

Therefore, $$|\Gamma_L| = \frac{(Z_L - Z_o)}{(Z_L + Z_o)} \quad \text{or}$$

$$0.5 = \frac{(Z_L - 400)}{(Z_L + 400)}$$

$$Z_L + 400 = 2Z_L - 800$$

Therefore, $$Z_L = 1200\ \Omega$$

$$2\,\beta\,d = \pi \quad \text{or}$$

$$2 \times \frac{2\pi}{\lambda} \times .10 = \pi \quad \text{or}$$

$$\frac{0.4\pi}{\lambda} = \pi$$

$$\lambda = 0.4\ \text{m}$$

As $$c/f = \lambda$$

$$f = \frac{3 \times 10^8}{0.4} = 7.5 \times 10^8\ \text{Hz}$$

9.63 The voltage minima occurs at 15 cm from the load in a transmission line. The distance between two consecutive minima is 20 cm. If SWR $s = 3$ and the characteristic output impedance $Z_o = 200\ \Omega$, find the value of unknown load impedance and frequency.

Solution:

The distance between two consecutive minima is given as $\lambda/2 = 20$ cm. Therefore, l = 40 cm or l = 0.4 m.

Frequency

$$f = \frac{c}{\lambda} = \frac{3\times10^8}{0.4}$$

$$f = 7.5\times10^8\ \text{Hz}$$

Now,

$$|\Gamma_L| = \frac{(s-1)}{(s+1)} = \frac{(3-1)}{(3+1)}$$

$$= 0.5$$

$$(2\beta\, d_{\min} - \theta) = \pi \quad \text{or}$$

$$\left(2\times\frac{2\pi}{0.4}\times.15 - \theta\right) = \pi \qquad \text{or}$$

$$\theta = 0.5\pi$$

$$\theta = 1.57\ \text{rad} \quad \text{or}$$

$$\theta = 89.9^\circ$$

Therefore,

$$|\Gamma_L| e^{j(89.9)} = \frac{(Z_L - Z_o)}{(Z_L + Z_o)}$$

$$\frac{|Z_L| - Z_o}{|Z_L| + Z_o} = \Gamma_L = 0.5 \quad \text{or}$$

$$\frac{(Z_L - 200)}{(Z_L + 200)} = 0.5e^{j(90^\circ)}$$

$$\frac{(Z_L - 200)}{(Z_L + 200)} = 0.5(\cos 90^\circ + J\sin 90^\circ)$$

$$\frac{(Z_L - 200)}{(Z_L + 200)} = 0.5(0 + j)$$

$$\frac{(Z_L - 200)}{(Z_L + 200)} = \frac{j}{2}$$

$$(2Z_L - 400) = (jZ_L + j200)$$

$$Z_L(2-j) = 400 + j200$$

$$Z_L = \frac{(400 + j\,200)}{(2-j)} = (120 + 160j)\,\Omega$$

9.64 An electric dipole consists of a positive and negative charge. The charge in the case of an electric dipole is 5 μC. The distance between them is 5 mm. Obtain the dipole moment.

Solution:

Given that

Charges $q = +\,5\ \mu\text{C}$ and $-5\ \mu\text{C}$

Distance $d = 5$ mm

Dipole moment $|\vec{p}|$ is obtained as

$$|\vec{p}| = qC\,d$$

$$= 5\times10^{-6}\times5\times10^{-3}\ \text{C-m}$$

$$= 25\times10^{-9}\ \text{C-m}$$

The direction of $\vec{P}$ will be from a –ve charge towards a +ve charge.

UNSOLVED QUESTIONS

9.1 A transmission line characteristic impedance is 50 Ω. Its phase constant is 2 rad/m at 120 MHz. Find the capacitance per meter and inductance per meter of this transmission line.

9.2 Attenuation on a 70 Ω distortionless transmission line is 0.05 dB/m. The line capacitance is 0.15 nF/m. What is the wave propagation velocity?

9.3 A distortionless line has $Z_o = 70\ \Omega$, $\alpha = 15$ mNp/m and $u = 0.5c$. c is the speed of light in vacuum. Find values of R, L, G, C and λ at 200 MHz.

9.4 A 120 Ω transmission line is to be matched to a 25 Ω antenna by using another piece of transmission line. The operating frequency is 200 MHz. Find the needed length of the transmission line. Also, find Z_o of this transmission line.

9.5 A 70 Ω transmission line is half wave long. It is terminated in a 200 load. What is its input impedance? It is operated at half frequency of the original operating frequency. Find the change in value of input impedance.

9.6 The maximum and minimum RMS voltages of a transmission line are 80 V and 10 V, respectively. Find the value of the reflection coefficient. Assume that this transmission line terminates at 200 Ω load.

9.7 A 70 Ω transmission line terminates at $Z_L = (80 + j20)\ \Omega$. Find the value of voltage reflection coefficient. Also, find the voltage standing wave ratio.

9.8 A transmission line operates at 10^6 (rad/s). Its $\alpha = 7$ dB/m, $\beta = 1.2$ rad/m and $Z_0 = (40 + j20)\ \Omega$. It is 1.5 m long. If the transmission line is connected to a source of 12 $\angle 0°$ Volt, $Z_g = 30$ W and terminates at a load of $(10 + j60)\ \Omega$, find the input impedance and current at the middle transmission line.

9.9 A dissipationless transmission line's characteristic impedance is 100 Ω. It is connected to a load of $(120 + j80)$ ohm. It operates at a frequency of 200 MHz. What is the input impedance of this line? The length of the line is 50 cm.

9.10 A 250 Ω transmission line is terminated in a load of $(200 + j300)$ Ω. Find the voltage reflection coefficient.

9.11 What will be needed, position and length of the short-circuited stub so as to match a 300 Ω load to a transmission line? The characteristic impedance of the transmission line is 500 Ω.

9.12 In the case of a distortionless line having $Z_o = 70\ \Omega$, $\alpha = 30$ mNp/m, $v_p = 4.5\times10^8$ m/s, Find line parameters R, L, C and G at 150 MHz.

9.13 What is the Smith chart? What are its advantages?

9.14 What is the procedure for obtaining VSWR from the Smith chart?

9.15 In a transmission line, $Z_R = (20 - j80)\ \Omega$ and $Z_o = 40\ \Omega$. Find d_{max} and d_{min}.
Answers: $d_{max} = 0.4293\lambda$ and $d_{min} = 0.1793\lambda$

9.16 Give various applications of transmission lines.

9.17 In a transmission line $Z_{R1} = (20 + j80)\ \Omega$ and $Z_{R2} = (20 - j80)\ \Omega$, $Z_0 = 40\ \Omega$. Find VSWR.
Answer: 10.4

9.18 In a transmission line $Z_R = (20 + j80)\ \Omega$ and $Z_o = 40\ \Omega$. Find the normalized admittance.
Answer: $(0.117 - j\,0.470)$

SUMMARY

1. Transmission lines facilitate guided wave propagation.
2. Transmission lines connect a source with a load.
3. Transmission line is modelled with the help of its line equations and it is analysed with the help of the Smith chart.
4. A transmission line is described in terms of resistance per unit length, conductance per unit length, inductance and capacitance per unit length.
5. Velocity of propagation in a lossless transmission line is given as $1/\sqrt{LC}$.
6. The various types of transmission lines allowing transverse electromagnetic (TEM) wave propagation are coaxial transmission lines, planar transmission lines, two were transmission lines and microstrip line.
7. In coaxial transmission lines,

$$C = \frac{2\pi\varepsilon}{\ln(b/a)}, \quad L = \frac{\mu_o}{2\pi}\ln(b/a)$$

$$R = \frac{1}{2\pi\sigma}\ln(b/a) \quad \text{and} \quad G = \frac{2\pi\sigma}{\ln(b/a)}$$

8. In planar transmission lines, $C = \varepsilon b/d$, $L = \mu d/b$, $R = 2/\sigma_c \delta b$ and $G = \sigma b/d$.

9. In two were transmission lines,

$$C = \frac{\pi\varepsilon}{\cos h^{-1}(d/2a)}, \quad L = \frac{\mu}{\pi}\cos h^{-1}(d/2a), \quad R = \frac{1}{\pi a \delta \sigma_c} \text{ (at high frequencies),}$$

$$\frac{2}{\sigma_c \pi a^2} \text{ (at low frequencies)} \quad \text{and} \quad G = \frac{\pi\sigma}{\cos h^{-1}(d/2a)}$$

10. For a TEM wave, $V = -\int \vec{E} \cdot d\vec{l}$ and $I = \oint \vec{H} \cdot d\vec{l}$.

11. Characteristic impedance Z_0 is given as

$$\sqrt{\frac{Z}{Y}} = \left(\sqrt{\frac{(R + j\omega L)}{(G + j\omega C)}}\right)$$

12. In a lossless transmission line, conductors of the line are perfect with $\sigma_c \approx \infty$ and the dielectric medium separating them is also lossless $(\sigma \simeq 0)$.

13. In a distortionless transmission line,

$$\frac{R}{L} = \frac{G}{C}$$

14. A distortionless transmission line has constant velocity and non-vanishing attenuation constant. Telephone lines must be distortionless.

15. In a shorted line, load impedance $Z_L \to 0$ and in the case of an open-circuited line load impedance $Z_L \to \infty$.

16. In a matched line, $Z_L = Z_o$.

17. In a quarter-wave transmission line, the line length "l" is equal to quarter of the wavelength $(\lambda/4)$.

18. In a half-wave transmission line, $l = \lambda/2$.

19. For maximum power transfer, load must be matched with the characteristic impedance of transmission line.

20. A single-stub tuner has an open or shorted section of the transmission line, which is connected in parallel to the main line.

21. The Smith chart is basically a polar plot of complex reflection coefficients (Γ). It provides information about transmission line behaviour for a wide range of impedances.

22. Normal complex load impedance z_n is obtained as Z_L/Z_o.

23. A Smith chart depicts reflection coefficient domains for normalized resistance varying between 0 and ∞, and normalized reactance varying between −∞ and ∞.

24. If the voltage standing wave ratio (VSWR) is high, then it refers to high mismatch.

25. Smith charts are used for obtaining reflection coefficient, line impedance, $d_{\max}$, $d_{\min}$, VSWR, characteristic admittance (Y_0), noise figure, stability, etc.

26. Transmission lines are used for the transmission of high-frequency signals as a filter, as a pulse generator, etc.

MULTIPLE-CHOICE QUESTIONS

1. The Smith chart is a __________ dimensional representation of complex impedances with respect to coordinates.
(a) Two (b) Three (c) Four (d) None of these

2. Z_L/Z_o is equal to __________.
(a) $\frac{\Gamma(d)}{1-\Gamma(d)}$ (b) $\frac{\Gamma(d)}{1+\Gamma(d)}$ (c) $\frac{1-\Gamma(d)}{1+\Gamma(d)}$ (d) $\frac{1+\Gamma(d)}{1-\Gamma(d)}$

3. In a __________ transmission line, the reflection coefficient domain is a circle of unitary radius.
(a) Lossy (b) Lossless (c) None of these (d) Cannot say

4. The Smith chart represents a family of circles obtained by varying normalized resistance in the range of __________.
(a) −1 to 0 (b) 0 to 1 (c) 0 to ∞ (d) −∞ to ∞

5. A family of arcs is obtained in the Smith chart by varying normalized reactance in a range of __________.
(a) −1 to 0 (b) 0 to 1 (c) 0 to ∞ (d) −∞ to ∞

6. Smith charts depict a wide impedance range of transmitter for __________ VSWR.
(a) Low (b) High (c) Medium (d) None of these

7. A cable or a wire is considered as a transmission line if its length is __________ of the wavelength of the signal travelling through it.
(a) Greater than 1/10 (b) Less than 1/10
(c) Greater than 1/5 (d) Less than 1/5

8. Transmission lines facilitate __________ propagation of energy.
(a) Guided (b) Unguided (c) None of these (d) Cannot say

9. A transmission line is specified in terms of __________.
(a) *R*, *G*, *L* (b) *G*, *L*, *C* (c) *R*, *G*, *L*, *C* (d) None of these

10. In the case of a lossless transmission line__________.

(a) $\frac{G}{C}=\frac{\sigma}{\varepsilon}$ (b) $LC = \mu\varepsilon$ (c) Both (a) and (b) (d) None of these

11. Velocity of propagation in a lossless transmission line is given as __________.

(a) $\sqrt{\frac{L}{C}}$ (b) $\sqrt{\frac{C}{L}}$ (c) $\sqrt{LC}$ (d) $\frac{1}{\sqrt{LC}}$

12. Conductance per unit length of a coaxial transmission line is given as

(a) $\frac{2\pi\sigma}{\ln(b/a)}$ (b) $\frac{2\pi\varepsilon}{\ln(b/a)}$ (c) $\frac{1}{2\pi\sigma}\ln(b/a)$ (d) $\frac{1}{2\pi\varepsilon}\ln(b/a)$

13. In planar transmission lines, $C =$ __________.

(a) $\sigma b/d$ (b) $\mu d/b$ (c) $\varepsilon b/d$ (d) None of these

14. In a microstrip line, there is __________ metal strip over the substrate.

(a) A thin (b) A thick (c) No (d) None of these

15. Characteristic impedance $Z_o =$ __________.

(a) $\sqrt{\frac{Y}{Z}}$ (b) $\sqrt{\frac{Z}{Y}}$ (c) $\sqrt{ZY}$ (d) None of these

16 In a fully lossless line, $\sigma =$ __________.

(a) 0 (b) 1 (c) ∞ (d) None of these

17. In a distortionless line, attenuation constant α is frequency __________.

(a) Dependent (b) Independent (c) None of these (d) Cannot say

18. A distortionless line has __________.

(a) Non-vanishing attenuation constant (b) Constant velocity
(c) Constant real characteristic impedance (d) All of these

19. Voltage reflection coefficient is the ratio of __________ wave to __________ wave.

(a) Incident, reflected (b) Reflected, incident
(c) Incident, absorbed (d) Absorbed, incident

20. In a quarter-wave transformer, input impedance is equal to __________.

(a) $\frac{Z_o}{Z_L}$ (b) $\frac{Z_L}{Z_o}$ (c) $\frac{Z_o^2}{Z_L}$ (d) $\frac{Z_L^2}{Z_o}$

ANSWERS TO MULTIPLE-CHOICE QUESTIONS

(1) c; (2) d; (3) b; (4) c; (5) d; (6) b; (7) a; (8) a; (9) c; (10) c;
(11) d; (12) a; (13) c; (14) a; (15) b; (16) c; (17) b; (17) d; (19). b; (20) c

Wave Guides

10

10.1 INTRODUCTION

A transmission line guides electromagnetic (EM) energy from one point (generator point) to another point (load point). A "wave-guide" has a similar function. However, wave guide fundamentally differs from a transmission line. Wave guides are special cases of transmission lines. Table 10.1 summarizes the distinction between transmission line and a wave guide.

Table 10.1 *Comparison between a transmission line and wave guide*

Sl No.	Transmission Line	Wave Guide
1.	It supports the transverse electromagnetic wave (TEM) only	It supports many possible configurations
2.	Due to the "skin effect" and higher dielectric losses, especially in frequency range 3–300 GHz, the transmission line becomes inefficient	Wave guides can be used efficiently at this frequency range. Also, these give a lower signal attenuation and a larger bandwidth in this frequency range (3–300 GHz)
3.	Transmission lines operate from DC to a very high frequency.	Wave guides can operate only above a specific frequency. This frequency is called the "cut-off" frequency. Thus, a wave guide is a kind of "high-pass-filter" also. A wave guide cannot be used in DC applications. Below microwave frequencies, a wave guide becomes too large.

Wave guides can be of any arbitrary shape but must have a uniform cross-section. Commonly used wave guides are either circular or rectangular. An analysis of circular wave guides is based on "Bessel functions". Lossless waveguides are characterized by $\sigma_c \simeq \infty$ and $\sigma \approx 0$. Maxwell's equations with appropriate boundary conditions give different wave propagation modes, and also their corresponding $\vec{E}$ and $\vec{H}$ fields.

10.2 RECTANGULAR WAVE GUIDE

A rectangular wave guide is shown in Figure 10.1. A wave guide is filled with a free source ($\rho_v = 0$ and $J = 0$) and lossless dielectric material ($\sigma \simeq 0$). Walls are perfectly conducting. i.e. $\sigma_c \simeq \infty$.

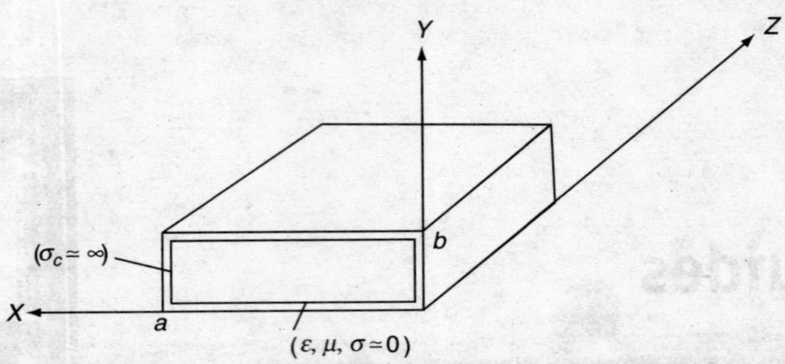

Figure 10.1 *Rectangular wave guide*

In the case of a lossless medium, Maxwell's equations are

$$\nabla^2 E + K^2 E = 0 \qquad \text{(in phasor form) and}$$

$$\nabla^2 H + K^2 H = 0 \qquad \text{(in phasor form)}$$

where $K = \omega\sqrt{\mu\varepsilon}$. Also,

$$\vec{E} = (E_x, E_y, E_z) \quad \text{and}$$

$$\vec{H} = (H_x, H_y, H_z)$$

$\vec{E}$ and $\vec{H}$ fields are obtained by solving these six scalar equations. For the z-component,

we get
$$\frac{\partial^2 E_z}{\partial x^2} + \frac{\partial^2 E_z}{\partial y^2} + \frac{\partial^2 E_z}{\partial z^2} + K^2 E_z = 0$$

Using Maxwell's equations so as to find the field components E_z and H_z,

we get
$$\nabla \times \vec{E} = -j\omega\,\mu\vec{H} \quad \text{and}$$

$$\nabla \times \vec{H} = -j\omega\varepsilon\vec{E}$$

Thus,

$$\frac{\partial E_z}{\partial y} - \frac{\partial E_y}{\partial z} = -j\omega\mu H_x$$

$$\frac{\partial H_z}{\partial y} - \frac{\partial H_y}{\partial z} = -j\omega\varepsilon E_x$$

$$\frac{\partial E_x}{\partial z} - \frac{\partial E_z}{\partial x} = -j\omega\mu H_y$$

$$\frac{\partial H_x}{\partial z} - \frac{\partial H_z}{\partial x} = -j\omega\varepsilon E_y$$

$$\frac{\partial E_y}{\partial x} - \frac{\partial E_x}{\partial y} = -j\omega\mu H_z \quad \text{and}$$

$$\frac{\partial H_y}{\partial x} - \frac{\partial H_x}{\partial y} = -j\omega\varepsilon E_z$$

E_x, E_y, H_x, H_y in terms of E_z and H_z are given as

$$E_x = -\frac{\gamma}{h^2}\frac{\partial E_z}{\partial x} - \frac{j\omega\mu}{h^2}\frac{\partial H_z}{\partial y} \quad \text{and}$$

$$E_y = -\frac{\gamma}{h^2}\frac{\partial E_z}{\partial y} - \frac{j\omega\mu}{h^2}\frac{\partial H_Z}{\partial x}$$

$$H_x = \frac{j\omega\varepsilon}{h^2}\frac{\partial E_z}{\partial y} - \frac{\gamma}{h^2}\frac{\partial H_z}{\partial x} \quad \text{and}$$

$$H_y = -\frac{j\omega\varepsilon}{h^2}\frac{\partial E_z}{\partial x} - \frac{\gamma}{h^2}\frac{\partial H_z}{\partial y}$$

where $h^2 = (\gamma^2 + K^2)$. There are several types of "field patterns" or "field configurations". These distinct field patterns are called "modes". Four different mode categories are given as follows:

(i) TEM mode ($E_z = 0$ and $H_z = 0$)
In the transverse electromagnetic (TEM) mode, $\vec{E}$ and $\vec{H}$ fields are transverse to the wave propagation direction. If $E_z = H_z = 0$, all field components vanish. Rectangular wave guides cannot support the TEM mode.

(ii) TE mode ($E_z = 0$ and $H_z \neq 0$)
In this mode, E_x and E_y of electric field are transverse to the propagation direction a_z. These fields are called "transverse electric (TE) modes". Figure 10.2 depicts fields in the TE mode.

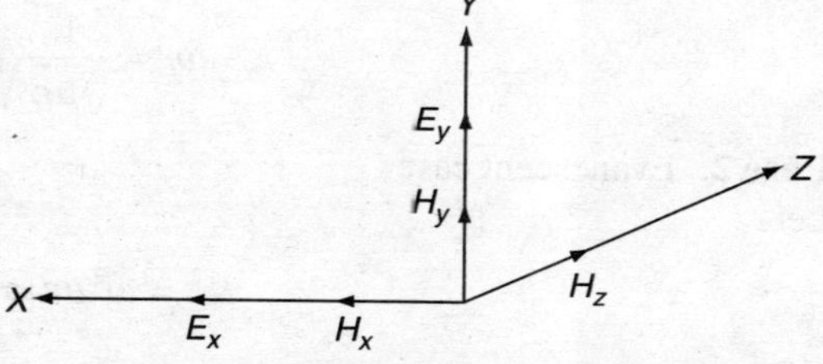

Figure 10.2 *Fields in the TE mode*

(iii) TM mode ($E_z \neq 0$ and $H_z = 0$)
The $\vec{H}$ field is transverse to the wave propagation direction. It is called "transverse magnetic (TM) modes". Figure 10.3 depicts fields in TM modes.

(iv) HE mode ($E_z \neq 0$ and $H_z \neq 0$)
Neither $\vec{E}$ nor $\vec{H}$ is transverse to the field. It is transverse to the wave propagation direction. It is also called the "hybrid mode".

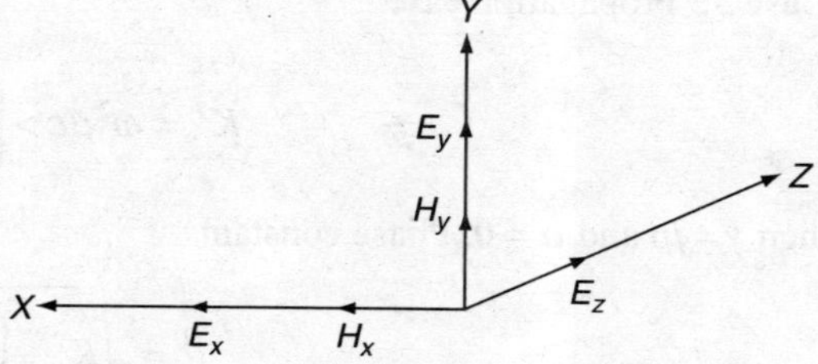

Figure 10.3 *Fields in the TM mode*

10.3 VARIOUS MODES

The various modes are as follows.

10.3.1 Transverse Magnetic (TM) Modes

In these modes, the magnetic field's component is transverse (or normal) to the wave propagation direction. Thus, $H_z = 0$. E_x, E_y, E_z, H_x and H_y can be determined using boundary conditions.

Here,

$$h^2 = (K_x^2 + K_y^2) \quad \text{or}$$

$$= \left(\frac{m\pi}{a}\right)^2 + \left(\frac{n\pi}{b}\right)^2$$

Integers m and n give different field patterns or modes. It is denoted as the TM_{mn} mode, where m refers to the number of half-cycle variations in the x-direction. n refers to the number of half-cycle variations in the y-direction.

For (m, n) equal to (0, 0) or $(0, n)$ or $(m, 0)$, all field components will vanish. Neither m nor n is zero. TM_{11} is the lowest order mode among all TM_{mn} modes.
Propagation constant γ is given as

$$\gamma = \sqrt{\left(\frac{m\pi}{a}\right)^2 + \left(\frac{n\pi}{b}\right)^2 - K^2}$$

where $K = \omega\sqrt{\mu\varepsilon}$ and $\gamma = \alpha + j\beta$.
There are three cases depending on K, m and n:

Case 1: Cut-off case
Let

$$K^2 = \omega^2\mu\varepsilon = \left(\frac{m\pi}{a}\right)^2 + \left(\frac{n\pi}{b}\right)^2$$

then $\gamma = 0$ and $\alpha = \beta = 0$. The cut-off angular frequency ω_c is given as

$$\omega_c = \frac{1}{\sqrt{\mu\varepsilon}}\sqrt{\left(\frac{m\pi}{a}\right)^2 + \left(\frac{n\pi}{b}\right)^2}$$

Case 2: Evanescent case
Let

$$K^2 = \omega^2\mu\varepsilon < \left(\left(\frac{m\pi}{a}\right)^2 + \left(\frac{n\pi}{b}\right)^2\right)$$

then $\gamma = \alpha$ and $\beta = 0$. There is no wave propagation.

Case 3: Propagation case
Let

$$K^2 = \omega^2\mu\varepsilon > \left(\left(\frac{m\pi}{a}\right)^2 + \left(\frac{n\pi}{b}\right)^2\right)$$

then $\gamma = j\beta$ and $\alpha = 0$. Phase constant

$$\beta = \sqrt{K^2 - \left(\frac{m\pi}{a}\right)^2 - \left(\frac{n\pi}{b}\right)^2}$$

Each mode is characterized by a set of integers m and n. In each mode, there is a cut-off frequency f_c also.
The cut-off frequency f_c is basically an operating frequency below which there is attenuation.
Wave guides operate like a high-pass filter with cut-off frequency f_c given as

$$f_c = \frac{\omega_c}{2\pi} = \frac{1}{2\pi\sqrt{\mu\varepsilon}}\sqrt{\left(\frac{m\pi}{a}\right)^2 + \left(\frac{n\pi}{b}\right)^2} \quad \text{or}$$

$$f_c = \frac{u'}{2}\sqrt{\left(\frac{m}{a}\right)^2 + \left(\frac{n}{b}\right)^2}$$

where $u' = (1/\sqrt{\mu\varepsilon})$ (known as "phase velocity" of uniform plane). Wave in a lossless dielectric medium is characterized by $\sigma = 0$, μ and ε. The cut-off wavelength λ_c is given as

$$\lambda_c = \left(\frac{u'}{f_c}\right) \quad \text{or}$$

$$\lambda_c = \frac{2}{\sqrt{\left(\frac{m}{a}\right)^2 + \left(\frac{n}{b}\right)^2}}$$

Phase constant

$$\beta = \omega\sqrt{\mu\varepsilon}\sqrt{1-\left(\frac{f_c}{f}\right)^2}$$

γ in the case of evanescent mode is obtained as

$$\gamma = \alpha = \omega\sqrt{\mu\varepsilon}\sqrt{\left(\frac{f_c}{f}\right)^2 - 1} \qquad (\gamma \text{ in terms of } f_c)$$

Phase velocity u_p and wavelength λ in wave guides are given as

$$u_p = \frac{\omega}{\beta} \quad \text{and}$$

$$\lambda = \frac{2\pi}{\beta}$$

Intrinsic wave impedance η of this mode is obtained as

$$\eta_{TM} = \left(\frac{E_x}{H_y}\right) \quad \text{or}$$

$$= \left(-\frac{E_y}{H_x}\right) \quad \text{or}$$

$$= \sqrt{\frac{\mu}{\varepsilon}}\sqrt{1-\left(\frac{f_c}{f}\right)^2}$$

Also, $\eta = (\sqrt{\mu/\varepsilon})$ (intrinsic impedance of uniform plane wave).

10.3.2 Transverse Electric (TE) Modes

In TE modes, electric field is transverse (or normal) to the wave propagation direction. Here, $E_z = 0$. E_x, E_y, H_x, H_y and H_z are determined from the wave equations and relevant boundary conditions. h and γ give TM modes. m and n denote the number of half-cycle variations in x–y cross-sections of the wave guide. Cut-off frequency f_c, cut-off wavelength λ_c, phase constant β, phase velocity u_p and wavelength λ in this case of TE modes are the same as that for TM modes. In TE modes (m, n) can be (0, 1) or (1, 0). In TE modes (m, n) can never be (0, 0). If both m and n are zero at the same time, this will cause other field components to vanish. Usually, $a > b$; thus, $(1/a^2) < (1/b^2)$. TE_{10} is the lowest mode given as

$$f_c(\text{in}\, TE_{10}) = \frac{u'}{2a} < f_c\ (\text{in}\, TE_{01})$$
$$= \frac{u'}{2b}$$

TE_{10} mode is called the "dominant mode" of the wave guide. The cut-off wavelength λ_C (for the TE_{10} mode) is

$$\lambda_C = 2a$$

The dominant mode has the lowest cut-off frequency f_c or the longest cut-off wavelength λ_c. An EM wave with frequency $f < f_c$ cannot get propagated through the wave guide. Intrinsic impedance in the case of the TE mode is obtained as

$$\eta_{\text{TE}} = \left(\frac{E_x}{H_y}\right) \quad \text{or}$$
$$= \left(-\frac{E_y}{H_x}\right) \quad \text{or}$$
$$= \left(\frac{\omega\mu}{\beta}\right) \quad \text{or}$$
$$= \sqrt{\frac{\mu}{\varepsilon}}\, \frac{1}{\sqrt{1-\left(\frac{f_C}{f}\right)^2}} \quad \text{or}$$
$$= \frac{\eta'}{\sqrt{1-\left(\frac{f_C}{f}\right)^2}}$$

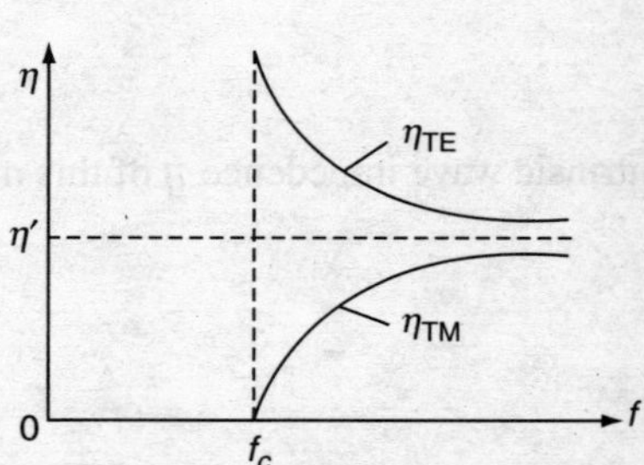

Figure 10.4 *η_{TE} and η_{TM} variation with frequency*

η_{TE} and η_{TM} are purely resistive. η_{TE} and η_{TM} vary with frequency. It is shown in Figure 10.4.

Also, $\eta_{\text{TE}}\ \eta_{\text{TM}} = (\eta')^2$ = (an important relationship).

10.4 WAVE PROPAGATION THROUGH THE GUIDE

Field components involve sine or cosine terms of $(m\pi/a)x$ or $(n\pi/b)y$ times $e^{-\gamma z}$. These are given as

$$\sin\theta = \frac{1}{2j}(e^{j\theta} - e^{-j\theta}) \quad \text{and}$$
$$\cos\theta = \frac{1}{2}(e^{j\theta} + e^{-j\theta})$$

A wave within the wave guide can be very easily resolved into combination of plane waves, which get reflected from wave guide walls. For the TE_{10} mode, we find

$$E_y = \frac{-j\omega\mu a}{\pi}\sin\left(\frac{\pi x}{a}\right)e^{-j\beta z} \quad \text{or}$$

$$= \frac{-\omega\mu a}{2\pi}(e^{j\pi x/a} - e^{-j\pi x/a})e^{-j\beta z} \quad \text{or}$$

$$= \frac{\omega\mu a}{2\pi}[e^{-j\beta(z+\pi x/\beta a)} - e^{-j\beta(z-\pi x/\beta a)}]$$

Here, the first term represents a wave travelling in the positive z-direction. Its angle with the z-axis is given as

$$\theta = \tan^{-1}\left(\frac{\pi}{\beta a}\right)$$

The second term represents a wave travelling in the positive z-direction at an angle $-\theta$. The field can be depicted as a sum of two plane TEM waves, which are propagating along zigzag paths within guide walls at $x = 0$ and $x = a$.

It is illustrated in Figure 10.5. Decomposition of the TE_{10} mode into two plane waves can be easily extended to TE and TM modes. If both n and m are different from zero, then four plane waves are obtained from decomposition.

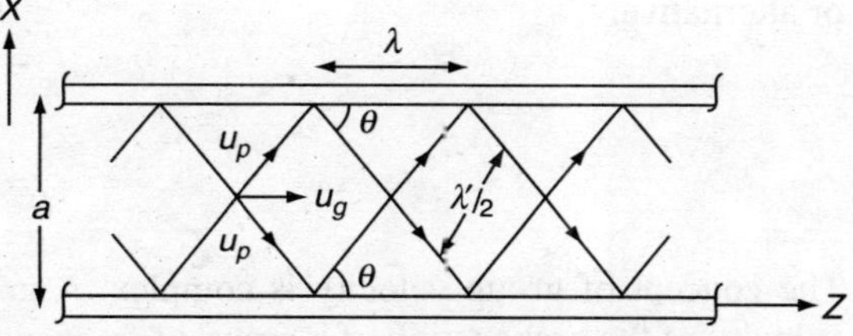

Figure 10.5 *Decomposition of TE_{10} mode into two plane waves*

The wave component in the z-direction has a different wavelength than that of plane waves. The wavelength along the axis of the guide is termed the "wave guide wavelength". It is given as

$$\lambda = \frac{\lambda'}{\sqrt{1-\left(\frac{fc}{f}\right)^2}}$$

where $\lambda' = \mu'/f$.

Due to zigzag propagation paths, there are three types of velocities, viz. medium velocity u', phase velocity u_p and group velocity u_g. Figure 10.6 demonstrates relationships between these three velocities. Medium velocity $\mu'(=1\sqrt{\mu\varepsilon})$ is the same as that explained in previous sections. Phase velocity u_p is defined as the velocity at which the loci of the constant phase propagate down the guide. It is given as

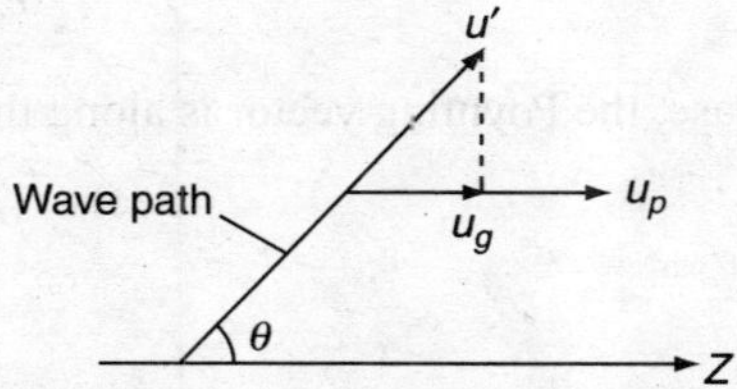

Figure 10.6 *Relationships between u', u_p and u_g*

$$u_p = \frac{\omega}{\beta}$$

Thus,

$$u_p = \frac{\mu'}{\cos\theta} = \frac{\mu'}{\sqrt{1-\left(\frac{fc}{f}\right)^2}}$$

where $u_p \geq u'$ as $\cos\theta \leq 1$.

If $\mu' = c$, then u_p is quite greater than the speed of light in vacuum. Einstein's relativity theory states that messages cannot travel faster than the speed of light. Information or energy in a wave guide generally does not travel at phase velocity. Information travels at group velocity. Group velocity is less than the speed of light. Group velocity u_g is defined as the velocity at which resultant repeated reflected waves travel down the guide. It is given as

$$u_g = \frac{1}{\partial\beta/\partial\omega}$$

or alternatively

$$u_g = u'\cos\theta = u'\sqrt{1-\left[\frac{fc}{f}\right]^2}$$

The concept of group velocity is complex. A group velocity is essentially the wave packet envelop velocity of the propagation of a group of frequencies. It is basically energy propagation velocity in the guide. It is always either less than or equal to u'. From the above relations, we get

$$u_p\, u_g = (u')^2$$

Variation of u_p and u_g with frequency is similar to that as shown in Figure 10.5 for η_{TE} and η_{TM}.

10.5 POWER TRANSMISSION AND ATTENUATION

The average Poynting vector is needed so as to determine power flow in wave guides. The average Poynting vector is given as

$$P_{ave} = \frac{1}{2}\text{Re}^1(\vec{E}\times\vec{H}^*)$$

In this case, the Poynting vector is along the z-direction. Thus,

$$P_{ave} = \frac{1}{2}\text{Re}\,(E_x H_y{}^* - Ey\,Hx^*)\,\vec{a}_z \quad \text{or}$$

$$= \frac{|Ex|^2 + |Ey|^2}{2\eta}\vec{a}_z$$

where $\eta = \eta_{TE}$ for TE modes or $\eta = \eta_{TM}$ for TM modes. The total average power transmitted through the cross-section of the wave guide is given as

$$P_{ave} = \int P_{ave}\cdot d\vec{s} \quad \text{or}$$

$$= \int_{x=0}^{a}\int_{y=0}^{b} \frac{|Ex|^2 + |Ey|^2}{2\eta}\, dy\, dx$$

Attenuation in a lossy wave guide is very important from a practical point of view. Here, for a lossless wave guide, it is assumed that $\sigma = 0$ and $\sigma_c \simeq \infty$ for $\alpha = 0$ and $\gamma = j\beta$. If the dielectric medium is lossy, i.e. $\sigma \neq 0$ and guide walls are not perfectly conducting, i.e. $\sigma_c \neq \infty$, there will be a continuous loss of power as the wave propagates (along the guide). Power flow in the guide is given as

$$P_{ave} = P_o e^{-2\alpha z}$$

If the rate of decrease in P_{ave} equals the time average power loss P_L per unit length, then energy is conserved. It is given as

$$P_L = \frac{-dP_{ave}}{dz}$$

$$= 2\alpha p_{ave} \quad \text{or}$$

$$\alpha = \frac{P_L}{2P_{ave}}$$

Generally,

$$\alpha = (\alpha_c + \alpha_d)$$

where α_c and α_d are attenuation constants due to ohmic (or conduction losses; $\sigma_c \neq \infty$) and dielectric losses ($\sigma \neq 0$), respectively.

In the case of a lossy dielectric medium, $\sigma \neq 0$. All equations except that for $\gamma = j\beta$ will still hold. The equation for $\gamma = j\beta$ needs to be modified. Thus, we find

$$\gamma = (\alpha_d + j\beta_d)$$

$$= \sqrt{\left(\frac{m\pi}{a}\right)^2 + \left(\frac{n\pi}{b}\right)^2 - \omega^2\mu\varepsilon_c}$$

Here,

$$\epsilon_c = (\epsilon' - j\epsilon'')$$

$$= \left(\epsilon - j\frac{\sigma}{\omega}\right)$$

Squaring above, we get

$$\gamma^2 = (\alpha_d^2 - \beta_d^2 + 2j\alpha_d\beta_d)$$

$$= \left[\left(\frac{m\pi}{a}\right)^2 + \left(\frac{n\pi}{b}\right)^2 - \omega^2\mu\epsilon + j\omega\mu\sigma\right]$$

Equating real and imaginary parts, we get

$$(\alpha_d^2 - \beta_d^2) = \left[\left(\frac{m\pi}{a}\right)^2 + \left(\frac{n\pi}{b}\right)^2 - \omega^2\mu\epsilon\right] \quad \text{and}$$

$$2\,\alpha_d\,\beta_d = \omega\mu\sigma$$

It yields

$$\alpha_d = \frac{\omega\mu\sigma}{2\beta_d}$$

Assume $\alpha_d^2 << \beta_d^2$ and $\alpha_d^2 - \beta_d^2 \simeq -\beta_d^2$. Thus, we get

$$\beta_d = \sqrt{\omega^2\mu \in -\left(\frac{m\pi}{a}\right)^2 - \left(\frac{n\pi}{b}\right)^2}$$

$$= \omega\sqrt{\mu \in}\sqrt{1-\left(\frac{fc}{f}\right)^2} \quad \text{and}$$

$$\alpha_d = \frac{\sigma\eta'}{2\sqrt{1-\left(\frac{fc}{f}\right)^2}}$$

where $\eta' = \sqrt{\mu/\in}$. The determination of α_c for TM_{mn} and TE_{mn} modes is quite cumbersome and time consuming. Here, finding α_c for TE_{10} mode is given. For TE_{10} mode, only E_y, H_x and H_z exist. Thus, we get

$$P_{ave} = \int_{x=0}^{a}\int_{y=0}^{b}\frac{|Ey|^2}{2n}\,dx\,dy$$

$$= \frac{\omega^2\mu^2 a^2 H_o^2}{2\pi^2\eta}\int_0^b dy\int_0^a \sin^2\frac{\pi x}{a}\,dx$$

Finally,

$$P_{ave} = \frac{\omega^2\mu^2 a^3 H_o^2 b}{4\pi^2\eta}$$

The total power loss per unit length P_L in walls is given as

$$P_L = P_L\big|_{y=0} + P_L\big|_{y=b} + P_L\big|_{x=0} + P_L\big|_{x=a} \quad \text{or}$$

$$= 2\left(P_L\big|_{y=0} + P_L\big|_{x=0}\right)$$

The same amount of power is dissipated in walls, i.e. $y = 0$ and $y = b$ or $x = 0$ and $x = a$. For wall $y = 0$, we find

$$P_L\big|_{y=0} = \frac{1}{2}\text{Re}\left[\eta_c\int\left(|H_x|^2 + |H_z|^2\right)dx\right]_{y=0} \quad \text{or}$$

$$= \frac{1}{2}R_s\left[\int_0^a \frac{\beta^2 a^2}{\pi^2}H_0^2\sin^2\frac{\pi x}{a}\,dx + \int_0^a H_0^2\cos^2\frac{\pi x}{a}\,dx\right] \quad \text{or}$$

$$= \frac{R_s\,aH_o^2}{4}\left(1+\frac{\beta^2 a^2}{\pi^2}\right)$$

where R_s is the real part of intrinsic impedance η_c of the conducting wall. It is given as

$$R_s = \frac{1}{\sigma_c \delta}$$

where δ is the skin depth. R_s is the skin resistance of wall. It is a resistance of 1 m by δ by 1 m of conducting material. For a wall $x = 0$, we find

$$P_L\big|_{x=0} = \frac{1}{2}\mathrm{Re}\left[\eta_c \int\left(|Hz|^2\right)dy\right]\Big|_{x=0}$$

$$= \frac{1}{2}R_s \int_0^b H_0^2\, dy \quad \text{or}$$

$$= \frac{R_s\, b H_0^2}{2}$$

Thus,

$$P_L = R_s\, H_0^2\left[b + \frac{a}{2}\left(1 + \frac{\beta^2 a^2}{\pi^2}\right)\right] \quad \text{and}$$

$$\alpha_c = \frac{R_s\, H_0{}^2\left[b + \frac{a}{2}\left(1 + \frac{\beta^2 a^2}{\pi^2}\right)\right]2\pi^2\eta}{\omega^2 \mu^2 a^3 H_0^2\, b}$$

α_c can be conveniently expressed in terms of f and f_c. For the TE$_{10}$ mode, we get

$$\alpha_c = \frac{2\,R_s}{b\eta'\sqrt{1 - \left[\frac{f_c}{f}\right]^2}}\left(\frac{1}{2} + \frac{b}{a}\left[\frac{f_c}{f}\right]^2\right)$$

Similarly, the attenuation constant for TE$_{mn}$ modes ($n \neq 0$) is obtained as

$$\alpha_c\big|_{TE} = \frac{2\,R_s}{b\eta'\sqrt{1 - \left[\frac{fc}{f}\right]^2}}\left[\left(1 + \frac{b}{a}\right)\left[\frac{fc}{f}\right]^2 + \frac{\frac{b}{a}\left(\frac{b}{a}m^2 + n^2\right)}{\left(\frac{b^2}{a^2}m^2 + n^2\right)}\left(1 - \left[\frac{fc}{f}\right]^2\right)\right]$$

For TM$_{mn}$ modes, we find

$$\alpha_c\big|_{TE} = \frac{2R_s}{b\eta'\sqrt{1 - \left[\frac{fc}{f}\right]^2}}\,\frac{((b/a)^3 m^2 + n^2)}{((b/a)^2 m^2 + n^2)}$$

10.6 WAVE GUIDE CURRENT AND MODE EXCITATION

For TM and TE modes, the surface current density K on wave guide walls is found by

$$\vec{K} = \vec{a}_n \times \vec{H}$$

where $\vec{a}_n$ is the unit outward normal to the wall. $\vec{H}$ is the field intensity evaluated on the wall. The current flow on the wave guide walls for TE_{10} mode propagation is depicted in Figure 10.7.

The surface charge density ρ_s on the wave guide walls is given by

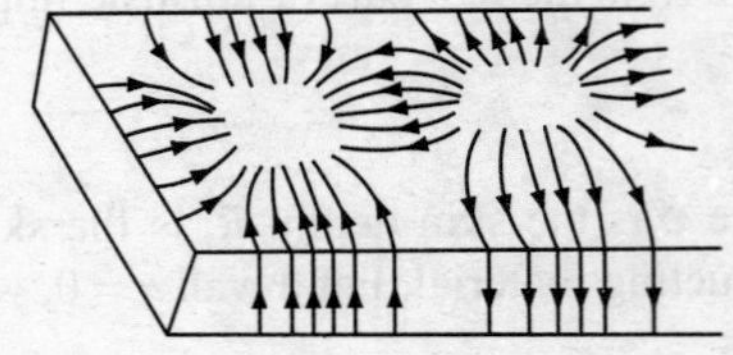

Figure 10.7 *Surface current on the wave guide walls for TE_{10} mode*

$$\rho_s = \vec{a}_n \cdot \vec{D}$$

$$= \vec{a}_n \cdot \in \vec{E}$$

where $\vec{E}$ is the electric field intensity. It is evaluated on the wave guide walls.

A wave guide is fed (or excited) by a coaxial line or alternatively by another wave guide. Often, a probe is basically a central conductor of a coaxial line, which establishes field intensities of the desired mode(s) and achieves maximum power transfer. The probe is located in such a way so as to produce $\vec{E}$ and $\vec{H}$ fields. It is roughly parallel to the lines of $\vec{E}$ and $\vec{H}$ fields of the desired mode(s). For the TE_{10} mode, E_y has a maximum value at $x = a/2$. Here, the probe is located at $x = a/2$ in order to excite the TE_{10} mode. It is shown in Figure 10.8(a). Field lines are similar to those as shown in Figure 10.9. The TM_{11} mode is launched by placing the probe along the z-direction. It is shown in Figure 10.8(b).

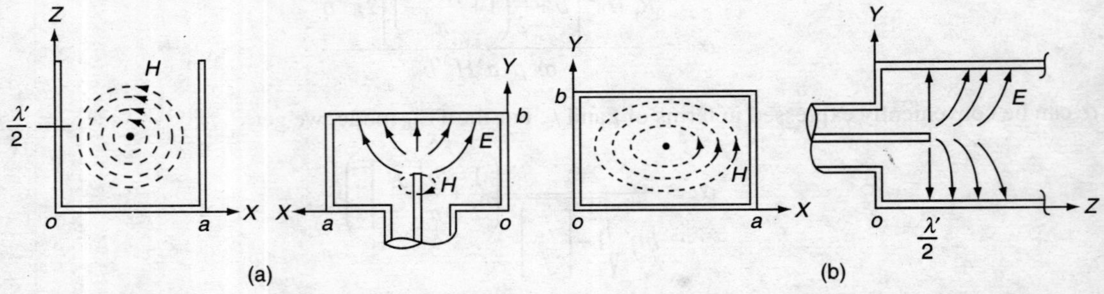

Figure 10.8 *Excitation of TE_{10} and TM_{11} modes in the case of a rectangular wave guide*

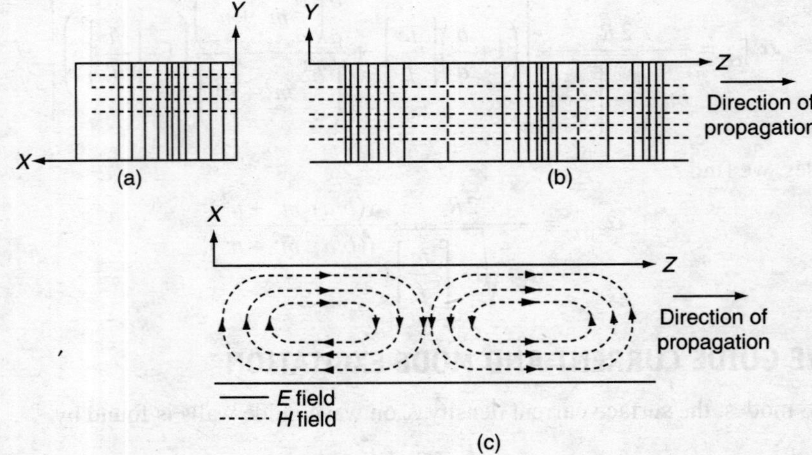

Figure 10.9 *Field lines for the TE_{10} mode (a) end view (b) side view and (c) top view*

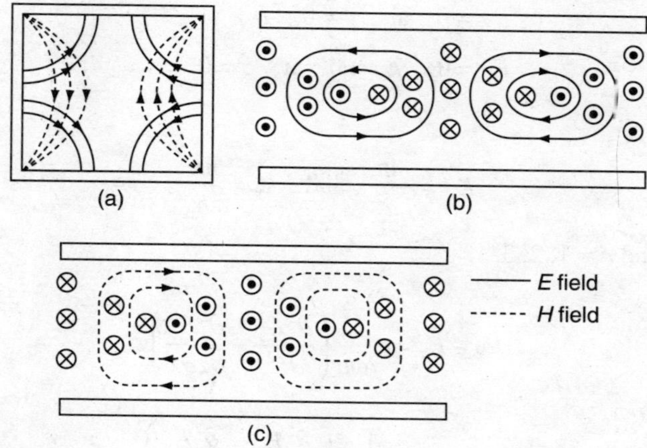

Figure 10.10 *TE_{11} mode views (a) end view (b) side view and (c) top view*

10.7 WAVE GUIDE RESONATORS

Resonators are used for the storage of energy. At high frequencies like 100 MHz and even above, RLC circuit elements become ineffectual when used as resonators. This is because circuit dimensions are comparable to operating wavelength, and therefore unwanted radiations take place. Thus, at high frequencies, RLC resonant circuits are usually replaced with electromagnetic cavity resonators. These resonator cavities are widely used in Klystron tubes, bandpass filters, wave meters, etc. A household microwave oven consists of a power supply, a wave guide feed and an oven cavity.

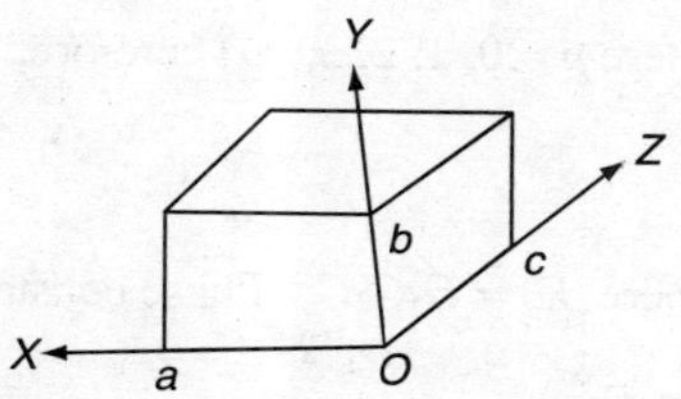

Figure 10.11 *Rectangular cavity*

Figure 10.11 depicts a rectangular cavity (a closed conducting box). The cavity is simply a rectangular wave guide that is shorted at both ends. Here, there will be a standing wave and TM and TE modes of wave propagation. Depending on how the cavity excitation wave will propagate in the x, y or z-directions, the $+z$-direction has been considered as "direction of wave propagation". As a matter of fact, there is no wave propagation; rather, there are "standing waves". A "standing wave" is a combination of two waves, which are travelling in opposite directions.

(i) TM mode to Z. Here, $H_z = 0$ and

$$E(x, y, z) = X(x)\, Y(y)\, Z(z)$$

$$X(x) = C_1 \cos K_x\, x + C_2 \sin K_x\, x$$

$$Y(y) = C_3 \cos K_y\, y + C_4 \sin K_y\, y$$

$$Z(z) = C_5 \cos K_z\, z + C_6 \sin K_z\, z$$

Here,

$$K^2 = K_x^2 + K_y^2 + K_Z^2 = \omega^2 \mu \in$$

Boundary conditions are given as

$$E_Z = 0 \quad \text{at} \quad x = 0, \;\; a$$

$$E_Z = 0 \quad \text{at} \quad y = 0, \quad b$$

$$E_y = 0, \quad E_x = 0 \quad \text{at} \quad z = 0, \quad c$$

If $C_1 = 0 = C_3$ then

$$K_x = \frac{m\pi}{a} \quad \text{and} \quad k_y = \frac{n\pi}{b}$$

where $m = 1, 2, 3, \ldots$ and $n = 1, 2, 3, \ldots$
If $H_z = 0$, then

$$j\omega \in E_x = \frac{1}{j\omega\mu}\left(\frac{\partial^2 E_x}{\partial z^2} - \frac{\partial^2 E_z}{\partial z \partial x}\right)$$

Further, if $Hz = 0$, then

$$j\omega\varepsilon E_y = \frac{1}{-j\omega\mu}\left(\frac{\partial^2 E_z}{\partial y\, \partial z} - \frac{\partial^2 E_y}{\partial z^2}\right)$$

Thus,

$$\frac{\partial E_z}{\partial z} = 0 \text{ at } z = 0, c$$

We get $C_6 = 0$ and consequently $\sin kz\ C = 0 = \sin p\pi$.

Thus,

$$Kz = \frac{p\pi}{c}$$

where $p = 0, 1, 2, 3, \ldots$ Therefore,

$$E_z = E_o \sin\left(\frac{m\pi x}{a}\right)\sin\left(\frac{n\pi y}{b}\right)\cos\left(\frac{p\pi z}{c}\right)$$

where $E_o = C_2C_4C_5$. Phase constant β is obtained as

$$\beta^2 = K^2 = \left[\left[\frac{m\pi}{a}\right]^2 + \left[\frac{n\pi}{b}\right]^2 + \left[\frac{p\pi}{c}\right]^2\right]$$

As $\beta^2 = \omega^2 \mu \in$, resonant frequency f_r is obtained as

$$2\pi f_r = \omega_r = \frac{\beta}{\sqrt{\mu \in}} = \beta u'$$

It yields

$$f_r = \frac{u'}{2}\sqrt{\left[\frac{m}{a}\right]^2 + \left[\frac{n}{b}\right]^2 + \left[\frac{p}{c}\right]^2}$$

The corresponding resonant wavelength λ_r is obtained as

$$\lambda_r = \frac{u'}{f_r} = \frac{2}{\sqrt{\left[\frac{m}{a}\right]^2 + \left[\frac{n}{b}\right]^2 + \left[\frac{p}{c}\right]^2}}$$

The lowest-order TM mode is TM_{110}.

(ii) TE Mode to Z. Here, $E_z = 0$ and

$$H_Z = (b_1 \cos K_x x + b_2 \sin K_x x)(b_3 \cos K_y y + b_y \sin K_y y)$$
$$(b_5 \cos K_z z + \sin K_z z)$$

Here, $H_z = 0$ at $z = 0, c$.

Thus,

$$\frac{\partial H_z}{\partial x} = 0 \text{ at } x = 0, a \text{ and}$$

$$\frac{\partial H_z}{\partial y} = 0 \text{ at } y = 0, b$$

It yields

$$H_z = H_o \cos\left(\frac{m\pi x}{a}\right)\cos\left(\frac{n\pi y}{b}\right)\sin\left(\frac{p\pi z}{c}\right)$$

where $m = 0, 1, 2, 3,\ldots$, $n = 0, 1, 2, 3, \ldots$ and $p = 1, 2, 3,\ldots$. m or n (but not both) at the same time can be zero in the case of TE modes. This is because field components will be zero, if both m and n are zero. The mode having the lowest resonant frequency for a given cavity size (a, b, c) is called the "dominant mode". If $a > b < c$, then $1/a < 1/b > 1/c$. Thus, the dominant mode is TE_{101}. For $a > b < c$, the resonant frequency of the TM_{110} mode is higher than that in the case of the TE_{101} mode. Therefore, TE_{101} is dominant. If different modes possess the same resonant frequency, then modes are said to be degenerate modes. One mode dominates the other mode depending upon how the cavity is being excited.

A practical resonant cavity has walls with finite conductivity σ_c. Therefore, the cavity is capable of losing stored energy. Quality factor Q is a measure of determining this loss of energy. Q is defined as

$$Q = 2\pi \times \left[\frac{\text{Time of averge energy stored}}{\text{Energy loss per cycle of oscillation}}\right]$$

$$= 2\pi \times \left[\frac{W}{p_L T}\right]$$

$$= \omega\left[\frac{W}{P_L}\right]$$

where $T = 1/f$ (period of oscillation). P_L is called the time average power loss of the cavity, W is called the total time average energy stored in electric and magnetic fields in the cavity. Q is usually very high in the case of a cavity resonator as compared to the Q for an RLC resonant circuit. Quality factor Q for the dominant TE_{101} mode is given as

$$Q_{TE_{101}} = \frac{(a^2 + c^2)abc}{\delta[2b(a^3 + c^3) + ac(a^2 + c^2)]}$$

where

$$\delta = \left(\frac{1}{\sqrt{\pi f_{101} \mu_0 \sigma_c}}\right)$$

is called as the skip depth of the cavity walls.

SOLVED QUESTIONS

10.1 Find the magnetic field intensity for a TEM wave with an electric field intensity of 2 μV/m in air and a lossless dielectric constant of $\varepsilon_r = 7$.

Solution:

$$\frac{\vec{E}}{\vec{H}} = \eta = 377\ \Omega \quad \text{in free space}$$

$$\eta = \sqrt{\frac{\mu}{\varepsilon}}$$

In air,

$$\vec{H} = \frac{\vec{E}}{\eta} = \frac{\vec{E}}{377} \qquad \text{(In air, } \mu = \mu_0,\ \varepsilon = \varepsilon_0 \text{ and } \varepsilon_r = 1)$$

$$= 5.30 \times 10^{-9}\ \text{A/m}$$

Lossless dielectric medium with $\varepsilon_r = 7$,

$$\eta = \sqrt{\frac{\mu}{\varepsilon}} = \sqrt{\frac{\mu_0}{\varepsilon_0 \varepsilon_r}} = \frac{377}{\sqrt{\varepsilon_r}} = \frac{377}{\sqrt{7}}$$

$$= 142.4$$

$$\vec{H} = \frac{\vec{E}}{\eta} = 1.40 \times 10^{-8}\ \text{A/m}$$

10.2 Prove that in a rectangular wave guide TM_{01} and TM_{10} do not exist.

Solution:

From the field equation for TM waves, we find

$$E_z = E_{0z} \sin\left(\frac{m\pi}{a}x\right) \sin\left(\frac{n\pi}{b}y\right) e^{-j\beta z}$$

If m or n is equal to 0, then E_z will be 0. Therefore, it implies that it will result in TEM. Hence, TM_{01} and TM_{10} will not exist.

10.3 A dominant mode TE_{10} is propagated in a rectangular guide of dimension 8 cm $\times$ 6 cm. The distance between the maximum and minimum is 5 cm. Find the signal frequency of the dominant mode.

Solution:

Signal frequency f_c is given as

$$f_c = \frac{c}{2}\sqrt{\left(\frac{m}{a}\right)^2 + \left(\frac{n}{b}\right)^2}$$

$$(f_c)_{10} = \frac{3 \times 10^8}{2} \sqrt{\frac{1}{(8 \times 10^{-2})^2}}$$

$$= 1.875\ \text{GHz}$$

$$\beta = \omega^2 \mu\varepsilon \quad \text{and}$$

$$\frac{\lambda_g}{2} = 0\cdot05 \qquad \text{(given)} \quad \text{or}$$

$$\lambda_g = 100 \text{ mm}$$

$$\lambda_g = 0.1 \text{ m}$$

$$\lambda_g = \frac{\lambda_o}{\sqrt{1-(f_c/f)^2}}$$

as

$$\lambda_o = \frac{c}{f_c} = 0.16 \text{ m} \quad \text{or}$$

$$1-(f_c/f)^2 = \left(\frac{\lambda_o}{\lambda_g}\right)^2 = 1.6$$

$$\left(\frac{f_c}{f}\right)^2 = 0.6$$

$$f^2 = \frac{f_c^{\,2}}{(0\cdot6)^2}$$

$$f = 3.125 \text{ GHz}$$

10.4 For the TE_{11} mode, it is given that f = 10 GHz, $a = b = \sqrt{5}\,\text{cm}$, $\beta_g = 2\,\text{rad/cm}$, $H_z = H_o \cos((\pi/\sqrt{b})x)\cos(\pi/\sqrt{b})y)$ A/m and $E_z = 0$.

Find the values f_c, H_y, λ_g and v_p.

Solution:

$$H_y = H_{oy}\cos\left(\frac{\pi}{\sqrt{b}}x\right)\sin\left(\frac{\pi}{\sqrt{b}}y\right)$$

$$H_{oy} = -j\frac{\beta g}{K^2}$$

where $K^2 = \omega^2\mu\in -\beta g^2$, $\omega = 2\pi f$ and

$$f_c = \frac{c}{2}\sqrt{\left(\frac{1}{a}\right)^2+\left(\frac{1}{b}\right)^2}$$

$$= \frac{c}{2}+\sqrt{\frac{1}{5}+\frac{1}{5}}$$

$$= \frac{c}{2}\times\sqrt{2/5}$$

$$= \frac{c}{\sqrt{10}} = 9.48 \text{ GHz}$$

$$\lambda_g = \frac{\lambda_0}{\sqrt{1-(f_c/f)^2}} = \frac{c/f_c}{\sqrt{1-(f_c/f)^2}}$$

$$= \frac{3\times 10^{10} \div 9.48\times 10^{-12}}{\sqrt{1-\left(\frac{9.48}{10}\right)^2}} = 31.2 \text{ cm}$$

Also,

$$v_p = \frac{c}{\sqrt{1-(f_c/f)^2}}$$

$$v_p = 29.6\times 10^8 \text{ m/s}$$

10.5 It is given that for TE_{10}, $f = 50$ Hz, $f_c = 2$ GHz and $b = a = 4$. The time average power flowing through the guide is 2 kW. Find the magnitude of E_y and H_x.

Solution:

Figure 10.12 *Guide*

$$E_x = E_z = 0 = H_y$$

$$E_y = E_{oy} \sin\left(\frac{\pi}{a}x\right)e^{-j\beta z}$$

where

$$E_{oy} = j\frac{\omega\mu}{2}$$

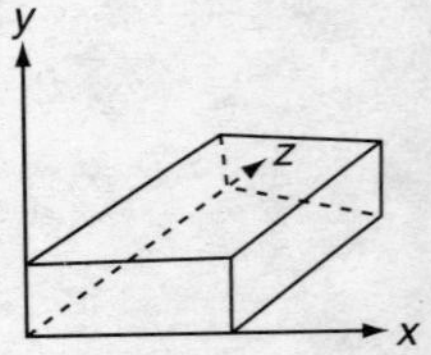

$$H_x = \frac{E_{oy}}{z_g}\sin\left(\frac{\pi x}{a}\right)e^{-j\beta z}$$

$$H_z = H_{oz}\cos\left(\frac{\pi x}{a}\right)e^{-j\beta z}$$

Here,

$$H_{oz} = c$$

Further,

$$z_g = \frac{\omega\mu_o}{\beta_g} \quad \text{and} \quad \beta_g = \omega^2\mu_0\varepsilon_0 - K^2$$

Therefore,

$$\beta_g = \sqrt{\omega^2\mu_0\varepsilon_0 - \left(\left(\frac{\pi}{a}\right)^2 + \left(\frac{o}{b}\right)^2\right)} \quad \text{or}$$

$$= \sqrt{\frac{\omega^2}{c^2} - \left(\frac{\pi}{a}\right)^2} \quad \text{or}$$

$$= \sqrt{\frac{4\pi^2 f^2}{c^2} - \frac{\pi^2}{a^2}} \quad \text{or}$$

$$\beta_g = \pi\sqrt{\frac{4f^2}{c^2} - \frac{1}{a^2}}$$

$$\beta_g = 0.695 \text{ rad/m}$$

$$(f_c)_{10} = \frac{c}{2}\sqrt{\left(\frac{1}{a}\right)^2}$$

Therefore,

$$\frac{c}{2a} = f_c \quad \text{or}$$

$$\frac{c}{2a} = 2\times10^9$$

$$a = \frac{c}{4\times10^9}$$

$$a = 0.075\,\text{m}$$

10.6 It is given that $a = 8$ cm and $b = 6$ cm. The TE_{10} mode power density = 1 HP = 746 Joules and $f = 10$ GHz. Find the value of the electric field intensity.

Solution:

$$E_x = E_z = 0$$

$$E_y = E_{yo}\sin\left(\frac{\pi}{a}x\right)e^{-j\beta z}$$

$$H_x = \frac{E_{yo}}{z_g}\sin\frac{VV^*}{2W}$$

$$H_z = H_{oz}\cos\left(\frac{\pi x}{a}\right)e^{-j\beta z}$$

where $z_g = \omega\mu_o/\beta_g$

$$\beta_g = \sqrt{\omega^2\mu_o\,\varepsilon_0 - \left(\frac{\pi}{a}\right)^2}$$

$$= \sqrt{\frac{4\pi^2 f^2}{c^2} - \left(\frac{\pi^2}{64\times10^{-4}}\right)^2}$$

$$= \pi\sqrt{4444.4 - 156.2}$$

$$\beta_g = 205.72 \text{ rad/m}$$

Further,

$$P = \frac{1}{2}\text{Re}\int_0^a\int_o^b (E_y \times H_y^*)\,d\vec{s} \quad \text{or}$$

$$= \frac{1}{2}\frac{E_{yo}^2}{z_g}\int_o^a\int_o^b \sin^2\left(\frac{\pi x}{a}\right)dy\,dx \quad \text{or}$$

$$= \frac{1}{2}\frac{E_{yo}^2\,\beta_g}{\omega\mu_0}\frac{ab}{2} \quad \text{or}$$

$$= \frac{E_{yo}^2}{\omega\mu_0}\frac{ab}{4} \times 205.72$$

$$= 746$$

Therefore,

$$E_{yo}^2 = \frac{746 \times 4 \times 2 \times 3.14 \times 10 \times 10^9 \times 4\pi \times 10^{-7}}{205.72 \times 48 \times 10^{-4}}$$

$$E_{yo}^2 = 238.6 \times 10^6$$

$$E_{yo} = 15.44 \text{ kV/m}$$

10.7 It is given that $a = 5$ cm, $b = 1$ cm and $f = 12$ GHz for a TE_{10} with $E_{oy} = 100$ V/m. Find Z_g, V and I.

Solution:

$$Z_g = \frac{\omega\mu_o}{\beta_g} \quad \text{and} \quad \beta = \sqrt{\omega^2 \mu_0 \varepsilon_0 - \left(\frac{\pi}{a}\right)^2}$$

$$\beta_g = \sqrt{\frac{4\pi^2 f^2}{c^2} - \frac{\pi^2}{a^2}}$$

$$= \pi\sqrt{\frac{4f^2}{c^2} - \frac{1}{a^2}}$$

$$= 77.4\pi$$

$$Z_g = \frac{2\pi \times 12 \times 10^9 \times 4\pi \times 10^{-7}}{77.42\pi}$$

$$= 389.6$$

$$H_{ox} = \frac{E_{oy}}{Zg} = \frac{100}{389.6}$$

$$= 0.256 \text{ mA/m}$$

Current = 0.0256 A and Voltage = 1 V.

10.8 Show that a linearly polarized wave can be resolved into a right-hand circularly polarized wave and a left-hand circularly polarized wave of equal amplitude.

Solution:

Let us assume a linearly polarized wave propagating in a positive z-direction. Considering that E is polarized in the x-direction, we have

$$\vec{E}(z) = E_0 e^{-j\beta z} \vec{a}_x$$

The above expression may be written as

$$E(z) = \text{Re}(E(z)) + \text{Im}(E(z))$$

where

$$\text{Re}(E(z)) = \frac{E_0}{2} e^{-j\beta z} (\vec{a}_x - j\vec{a}_y) \quad \text{and}$$

$$\text{Im}(E(z)) = \frac{E_0}{2} e^{-j\beta z}(\vec{a}_x + j\vec{a}_y)$$

The above two expressions represent right-hand and left-hand circularly polarized waves, respectively, each having an amplitude of $E_o/2$. Hence, proved.

10.9 Find the wave impedance and wave guide length at a frequency equal to twice the cut-off frequency in the wave guide for TM and TE modes.

Solution:

At $f = 2f_c$, this frequency is above the cut-off frequency.

So,

$$f = 2f_c, \quad (f_c/f)^2 = \frac{1}{4}$$

$$\sqrt{1-(fc/f)^2} = \sqrt{3/2} = 0.866$$

Thus,

$$Z_{\text{TM}} = 0.866\,\eta < \eta \quad \Rightarrow \lambda_{\text{TM}} = 1.55\lambda > \lambda \quad \text{and}$$

$$Z_{\text{TM}} = 1.155\,\eta > \eta \quad \Rightarrow \lambda_{\text{TM}} = 1.1555\lambda\alpha > \lambda$$

10.10 Determine the energy transport velocity of the TM_n mode in a lossless parallel plate wave guide.

Solution:

$$\vec{\rho}_{avg} = \frac{1}{2}\text{Re}\left(\vec{E} \times \vec{H}\right)$$

$$= \frac{1}{2}\text{Re}\left(E_y^0\ H_x^0\ \vec{a}_x + E_z^0\ H_x^0\ \vec{a}_y\right)$$

Therefore,

$$\vec{\rho}_{\text{avg}} \cdot \vec{a}_z = \frac{1}{2}\text{Re}\left(E_y^0\ H_x^0\right)$$

$$= \frac{\omega\varepsilon\beta}{2h^2} A_n^2 \cos^2\left(\frac{\eta\pi y}{b}\right)$$

Electric energy density $= \frac{1}{2}\varepsilon E^2 = \frac{1}{2}\varepsilon \vec{E} \cdot \vec{E}^*$

Magnetic energy density $= \frac{1}{2}\mu H^2 = \frac{1}{2}\mu \vec{H} \cdot \vec{H}^*$

Average electric energy density $= \frac{\varepsilon}{4}\text{Re}\left(\vec{E} \cdot \vec{E}^*\right)$

Average magnetic energy density $= \frac{\mu}{4}\text{Re}\left(\vec{H} \cdot \vec{H}^*\right)$

Also,

$$(P_z)_{avg} = \int_0^b \vec{\rho}_{avg} \cdot \vec{a}_z\ dy$$

$$= \frac{\omega \varepsilon \beta b}{4h^2} An^2$$

$$E_z^0(y) = A_n \sin\left(\frac{\eta \pi y}{b}\right) \quad \text{and}$$

$$E_y^0(y) = \frac{-r}{h} A_n \cos\left(\frac{\eta \pi y}{b}\right)$$

Substituting these above, we get

$$\text{Average electric energy density} = \frac{\varepsilon}{4} A_n^2 \left[\sin^2\left(\frac{\eta \pi y}{b}\right) + \frac{\beta^2}{h^2} \cos^2\left(\frac{\eta \pi y}{b}\right)\right]$$

Also,

$$\int_0^b (\text{Average electric energy density}) dy = \frac{\varepsilon}{8} b A_n^2 \left[1 + \frac{\beta^2}{h^2}\right]$$

$$= \frac{\varepsilon b}{8h^2} K^2 A_n^2$$

Here,

$$\beta^2 + h^2 = K^2$$

Similarly, we can also find the expression for the average magnetic energy density as

$$\text{Average magnetic energy density} = \frac{\mu}{4}\left(\frac{\omega^2 \varepsilon^2}{h^2}\right) A_n^2 \cos^2\left(\frac{\eta \pi y}{b}\right)$$

also,

$$\int_0^b (\text{Average magnetic energy density})\, dy = \frac{\mu b}{8h^2} (\omega^2 \varepsilon^2) A_n^2$$

$$= \frac{\varepsilon b}{8h^2} K^2 A_n^2$$

The above expressions are time average stored energy per unit wave guide width.

Therefore,

$$v_{en} = \frac{\omega \beta}{K^2} = \frac{\omega}{K}\left(\frac{\beta}{K}\right)$$

$$= v \sqrt{1 - \left(\frac{fc}{f}\right)^2}$$

Therefore, energy transport velocity = group velocity.

10.11 A sinusoidal plane wave is transmitted through a medium whose breakdown voltage is 32 kV (rms). $\varepsilon_r = 4$. Find average power density and peak value of magnetic field intensity.

Solution:

$$P_{\text{avg}} = \frac{E_0^2}{2\eta} \quad \text{or}$$

$$P_{\text{avg}}(\text{rms}) = \frac{E_r^2 \text{ms}}{\eta}$$

As

$$E_{\text{rms}} = \frac{E_0}{\sqrt{2}}$$

$$\eta = \sqrt{\frac{\mu}{\varepsilon}} = 120\pi\sqrt{\frac{1}{4}} = 60\pi$$

$$P_{\text{avg}} = \frac{(32\times10^3)^2}{60\pi} = 5.43\times10^6\ \text{Watt/m}^2$$

$$H_{\text{rms}} = \frac{E_{\text{rms}}}{\eta} = \frac{32\times10^3}{60\pi} = 169.7\ \text{A/m}$$

10.12 A 400 m long line has the following constants:

$R = 5.5$ KΩ, $L = 1.15$ mH, $G = 50$ m mho and $C = 10$ nF. The operating frequency is 12 MHz. Find the propagation constant, characteristic impedance and velocity of propagation.

Solution:

Propagation velocity is given as

$$v_p = \frac{1}{\sqrt{LC}} \quad \text{or}$$

$$v_p = \sqrt{\frac{1}{\frac{1.15\times10^{-3}}{300}\,\frac{10\times10^{-9}}{300}}} \quad \text{or}$$

$$v_p = \frac{300}{\sqrt{1.15\times10\times10^{-12}}} \quad \text{or}$$

$$v_p = 8.84\times10^7\ \text{m/s}$$

Further,

$$R = \frac{5.5\times10^3}{300} = 18.33\ \Omega/\text{m}$$

$$L = \frac{1.15\times10^{-3}}{300} = 3.8\times10^{-6}\ \text{H/m}$$

$$C = \frac{10\times10^{-9}}{300} = 33.3\ \text{pF/m} \quad \text{and}$$

$$G = \frac{50\times10^{-3}}{300} = 166.6\times10^{-6}\ \text{mho/m}$$

$$= 0.16\ \text{m mho/m}$$

As

$$\omega = 2\pi f \quad \text{or}$$

$$= 2\times\pi\times12\times10^6$$

$$= 75.39\times10^6\ \text{rad}$$

$$Z_0 = \sqrt{\frac{R + j\omega L}{G + j\omega C}}$$

Now,

$$R + j\omega L = (18.33 + j \times 75.39 \times 10^6 \times 3.8 \times 10^{-6}) \text{ or}$$

$$= (18.33 + j286.51) \text{ or}$$

$$= 287.09\angle 86.33° \text{ and}$$

$$G + j\omega C = (0.16 \times 10^{-3} + j75.39 \times 10^6 \times 33.3 \times 10^{-12})$$

$$= 2.5 \times 10^{-3} \angle 86.35°$$

$$Z_0 = \sqrt{\frac{R + j\omega L}{G + j\omega C}} = \sqrt{\frac{277.09\angle 86.33°}{2.5 \times 10^{-3} \angle 86.35°}}$$

$$= 107.16\angle 0.02°$$

10.13 An air-filled wave guide has dimensions of $a = 8$ cm and $b = 6$ cm. The signal frequency is 30 GHz. Calculate the following for the TE_{10} mode: (a) f_c, (b) λ_g and (c) β_g.

Solution:

$$f_c = \frac{1}{2\sqrt{\mu\varepsilon}} \cdot \sqrt{\left(\frac{m}{a}\right)^2 + \left(\frac{n}{b}\right)^2}$$

It is airy media; therefore,

$$\left.\begin{aligned} \mu = \mu_o, \quad \mu_r = \mu_o \\ \varepsilon = \varepsilon_o, \quad \varepsilon_r = \varepsilon_o \end{aligned}\right\}$$

$$(f_c)_{10} = \frac{3 \times 10^8}{2} \sqrt{\frac{1}{(8 \times 10^{-2})^2} + 0} \text{ or}$$

$$= 1.875 \text{ GHz}$$

$$\beta_g = \omega^2 \mu\varepsilon \text{ or}$$

$$= (2\pi f)^2 \times \mu_0 \varepsilon_0 \text{ or}$$

$$= 0.39 \text{ rad/m}$$

$$\lambda_g = \frac{2\pi}{\beta_g} \text{ or}$$

$$\lambda_g = 16.1 \text{ m}$$

10.14 A position charge Q to 208 pC is kept in air. Draw equipotential surfaces at radii 100 mm and 20 mm. Find the potential rise between these two radii.

Solution:

The absolute potential V_r due to a point charge at distance r is given as

$$V_r = \frac{Q}{4\pi\varepsilon_0 r} \quad \text{(Here, } r \text{ is the radius)}$$

For $r = a = 100$ mm, the potential curve is shown in Figure 10.13.
The absolute potential will be obtained as

$$V_a = \frac{Q}{4\pi\varepsilon_0} \frac{1}{a}$$

$$= \frac{208\times10^{-12}\times9\times10^9}{100\times10^{-3}} = 18.72\ V$$

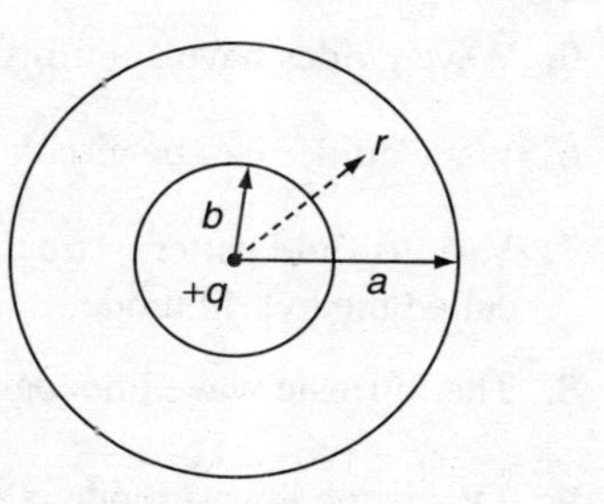

Figure 10.13 *Equipotential surface at radii a and b*

Similarly, for $r = b = 20$ mm,

$$V_b = \frac{Q}{4\pi\varepsilon_0} \frac{1}{b}$$

$$V_b = \frac{208\times10^{-12}\times9\times10^9}{20\times10^{-3}} = 93.6\text{ V}$$

Potential rise V_{ab} is obtained as

$$V_{ab} = V_b - V_a$$

$$= 74.88\text{ V}$$

UNSOLVED QUESTIONS

10.1 Make a detailed comparison of transmission lines and wave guides.

10.2 What are rectangular wave guides? Explain in detail. List the applications of wave guides.

10.3 Explain four different modes with suitable diagrams and mathematical derivations.

10.4 Define the Poynting Vector. Explain it.

10.5 Derive an expression for the total average power transmitted through wave guides.

10.6 Discuss attenuation in a lossy wave guide. What are attenuation constants?

10.7 Write technical notes on the following:
(a) Wave guide current and mode excitation and
(b) Wave guide resonators

SUMMARY

1. A wave guide is similar to a transmission line for its function. However, the wave guide is a special case of a transmission line.
2. A transmission line propagates TEM waves only, whereas a wave guide is more versatile.
3. A transmission line is an inefficient option for wave propagation in frequency range 3–300 GHz, whereas a wave guide is quite efficient in this range.
4. Transmission lines operate from DC (zero frequency) to a very high frequency. A wave guide is a kind of high-pass filter. It can only be used above a cut-off frequency. It cannot be used in DC applications.

5. Wave guides have a uniform cross-section.

6. Wave guides can be circular or rectangular.

7. Various field patterns are the TEM mode, TE mode, TM mode and HE mode. The HE mode is also called the hybrid mode.

8. The intrinsic wave impedance of a uniform plane wave η is given as $\sqrt{\frac{\mu}{\varepsilon}}\sqrt{1-\left(\frac{f_c}{f}\right)^2}$.

9. TE_{10} is the lowest mode, also called the dominant mode of the wave guide.

10. A wave within the wave guide is resolved into a combination of plane waves, which get reflected from the wave guide walls.

11. Group velocity is the velocity at which resultant repeated reflected waves travel down the guide.

12. Wave guide resonators are used for energy storage.

MULTIPLE-CHOICE QUESTIONS

1. __________ are very efficient in a frequency range of 3–300 GHz.
(a) Transmission line (b) Wave guides (c) Both (a) and (b) (d) None of these

2. Wave guides __________ in DC applications.
(a) Can also be used (b) Cannot be used (c) None of these (d) Cannot say

3. Wave guides are a type of __________ filter.
(a) Low pass (b) All pass (c) Band pass (d) High pass

4. For lossless wave guides, $\sigma_c \approx$ __________ and $\sigma \approx$ __________.
(a) ∞, 0 (b) 0, ∞ (c) 0, 0 (d) ∞, ∞

5. In a lossless medium, Maxwell's equation in the phasor form is given as __________ (Here, $K = \omega\sqrt{\mu\varepsilon}$).
(a) $\nabla^2 H = 0$ (b) $K^2 H = 0$ (c) $\nabla^2 H + k^2 H = 0$ (d) None of these

6. In the TM mode, $E_z =$ __________ and $H_z =$ __________.
(a) 0, 0 (b) 0, ∞ (c) ∞, 0 (d) None of these

7. In the TEM mode, $E_z =$ __________ and $H_z =$ __________.
(a) 0, 0 (b) 0, ∞ (c) ∞, 0 (d) None of these

8. In the HE mode __________ is transverse to the field.
(a) $\vec{E}$ (b) $\vec{H}$ (c) Both $\vec{E}$ and $\vec{H}$ (d) Neither $\vec{E}$ nor $\vec{H}$

9. Below the cut-off frequency, there is __________.
(a) Attenuation (b) No attenuation (c) Amplification (d) None of these

10. Cut-off wavelength λ_c in the TE_{10} mode is ______________.
(a) a (b) $2a$ (c) $4a$ (d) $8a$

11. __________ velocity is the one at which loss of constant phase propagates down the guide.
(a) Medium (b) Phase (c) Group (d) None of these

12. Information or energy through the wave guide travels at __________velocity.
(a) Medium (b) Phase (c) Group (d) None of these

13. __________ velocity is that whose resultant repeated reflected waves travel down the guide.
(a) Medium (b) Phase (c) Group (d) None of these

14. __________ vector is needed to determine power flow in wave guides.
(a) Poynting (b) Average Poynting (c) Power (d) Phase

15. Surface current density K on wave guide walls for __________ is given as $\vec{a}_n \times \vec{H}$.
(a) TM mode (b) TE mode (c) Both (a) and (b) (d) None of these

16. At __________ frequencies, RLC resonant circuits are replaced by electromagnetic cavity resonators.
(a) Low (b) Medium (c) High (d) None of these

17. Resonator cavities are used in __________.
(a) Klystron tubes (b) Band-pass filters (c) Wave meters (d) All of these

18. Cavity is a rectangular wave guide __________.
(a) Open at both ends (b) Shorted at both ends
(c) With one end open and the other shorted (d) None of these

19. Quality factor θ for a cavity is expensed as __________.
(a) $\dfrac{\omega \times W}{P_L}$ (b) $\dfrac{\omega}{WP_L}$ (c) $\dfrac{WP_L}{\omega}$ (d) $\dfrac{P_L \times \omega}{W}$

20. The skin depth for the dominant TE_{101} mode is given as __________.
(a) $\dfrac{1}{\sqrt{\pi f_{101} \mu_o \sigma_c}}$ (b) $\dfrac{1}{\sqrt{\pi f_{101} \mu_o}}$ (c) $\dfrac{1}{\sqrt{\pi f_{101}}}$ (d) $\dfrac{1}{\sqrt{f_{101}}}$

ANSWERS TO MULTIPLE-CHOICE QUESTIONS

(1) b; (2) b; (3) d; (4) a; (5) c; (6) d; (7) a; (8) d; (9) a; (10) b;
(11) b; (12) c; (13) c; (14) b; (15) c; (16) c; (17) d; (18) b; (19) a; (20) a

Radiation and Antennas

11

11.1 INTRODUCTION

The antenna is the most important device in any transmitting station. Antennas are very commonly used for radio and television broadcasting. Also, antennas are used for dedicated point-to-point radio communication, radar, wireless LAN, space exploration, etc. An antenna is a logical arrangement of several conductors also called "elements". An antenna also called an aerial is basically a transducer. The antenna is the device that sends a radio frequency (RF) signal through the space to the receiver. This antenna is also called the transmitting antenna. An alternating current is produced by applying a voltage to the transmitting antenna elements, which causes the elements to radiate the electromagnetic field. The receiving antenna picks up the RF signal (basically a travelling electromagnetic field or wave through the space) that is being sent by the transmitting antenna. When an RF signal is received by the receiving antenna, it induces a voltage into it. Thus, it can be easily said that an antenna converts electromagnetic waves into electrical currents and vice versa. In the receiving antenna, this electrical current produced due to the induced voltage is converted back into the information that was transmitted in the form of the RF signal. Thus, a receiving antenna does exactly the reverse of a transmitting antenna. Due to reciprocity, all antenna parameters applicable to a transmission antenna are also applicable to a receiving antenna.

The size (dimensions) and design of an antenna should be such that it is efficiently able to radiate the generated RF signal. In no case should the power supplied by the transmitter to it go waste. The transmitting antenna size should be exact, which is determined by the transmitting frequencies. The receiving antenna size is not an important issue especially in relatively low RFs. The receiving antenna size, design and installation become an important issue as the frequency of the signal being received increases. It happens in the case of a household television-receiving antenna. When we raise it from the ground and/or give it a turn (change in direction), then a snowy or blurred picture can be changed into a vivid picture. Some receiving antennas like the parabolic and the horn incorporate shaped reflective surfaces for collecting electromagnetic waves from free space. It is then directed or focused onto the actual conductive elements. Some special-purpose short-distance antennas can even be placed in the water or soil or rock. Antenna tuning is basically altering an electrical resonance antenna. An antenna's electrical resonance is changed by adjusting its length. Antenna tuning is done by adjusting the values of inductance or capacitance, which are externally combined to the active antenna. The inductive and capacitive reactances combine with the active antenna's internal reactance and establish a resonance in the circuit, which also includes an active antenna. Such resonance frequency is different from the active antenna's electrical resonant frequency. Various antennas are the Yagi-Uda beam antenna, rooftop TV antenna, shortwave antenna, broadband antenna, multi-band rotary directional antenna, terrestrial microwave radio antenna, rotatable log-periodic array antenna, etc. Antenna shields are used for noise rejection.

11.2 BASIC ANTENNA PARAMETERS

An antenna's performance depends upon its parameters. Antenna parameters depend upon its design. Transmit antennas have a maximum power rating, whereas receiving antennas differ in their noise rejection properties. Figure 11.1 depicts a VHF antenna. Figure 11.2 depicts a satellite dish antenna.

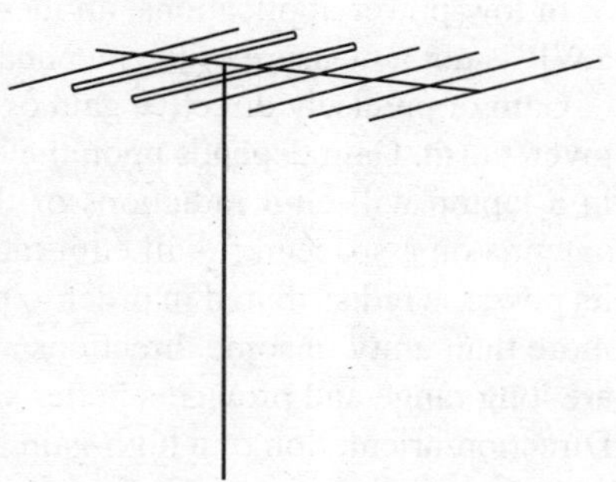

Figure 11.1 *A typical VHF antenna*

All these antenna parameters can be measured through various means and methods. The important antenna parameters are given as:

1. Resonant frequency
2. Impedance
3. Gain
4. Aperture (radiation pattern)
5. Polarization
6. Bandwidth
7. Efficiency

Figure 11.2 *A typical dish antenna*

Every antenna is tuned for a specific frequency. The range of frequencies for an antenna is centered on the resonant frequency. Antenna properties especially radiation pattern and impedance are badly affected by the change with the frequency. Some antennas have multiple resonant frequencies. Antennas can be made resonant on harmonic frequencies by having their lengths as a fraction of the target wavelength. The resonant frequency of an antenna depends upon its electrical length. Electrical resonance is also dependent on the antenna's electrical length. The electrical length is given as

$$\text{Electrical length} = \{(\text{Physical length of the wire})/(\text{Velocity factor})\}$$

Radiation in an antenna is affected by radiation resistance. Radiation resistance is a part of total resistance, which also includes loss resistance. Loss resistance lowers the antenna efficiency. Loss resistance is seen in terms of heat generation instead of radiations. The impedance of an antenna depends on its electrical length at a given wavelength. The impedance of an antenna must be matched to the feed line and radio. It can be done by adjusting the impedance of the feed line with the help of an impedance transformer. As an electromagnetic wave travels through the different parts of the antenna system like radio, feed line, antenna, free space then encounters differences in impedance. At each interface due to the impedance mismatch, some fraction of the wave's energy is reflected back to the source. Thus, it forms a standing wave in the feed line. Minimizing impedance mismatches at each interface though proper impedance matching reduces SWR but maximizes the flow of power through the various parts of the antenna system. The standing wave ratio (SWR) is given as:

Standing wave radio (SWR) = {(Maximum measurable power of the wave)/
(Minimum measurable power of the wave)}

In low-power applications, an ideal SWR is 1:1. The SWR of 1.5:1 is marginally acceptable. Alone, SWR is not a measure of the antenna's efficiency.

Gain or precisely directive gain or power gain of an antenna measures its efficiency with respect to a given norm. Gain depends upon the antenna's directionality. A low-gain antenna like the Wi-Fi antenna in a laptop will emit radiations of the same power in all directions. A high-gain antenna like a dish antenna on a spacecraft will emit radiations in particular directions. Antenna energy is conserved, and its power is redistributed in order to provide more radiated power in a certain direction. Antenna gain is more than unity in some directions and less than unity in other directions. Antennas with a higher gain are long range and provide a better signal quality, whereas antennas with low gain are of a short range. Direction/orientation of a high-gain antenna is a very crucial issue, whereas the direction of a low-gain antenna is a trivial issue. Gain of an antenna is given as

Gain = {(Radiation intensity power per unit surface in a given direction at an arbitrary distance)/
(Radiation intensity of a hypothetical isotropic antenna at the same distance)}

Gain is expressed in the units of "decibels (dB) over dipole". Half-wave dipole is taken as a reference. The most common wide band antenna/aerial is logarithmic or log periodic. Its gain is much lower than the gain of a specific or narrower band antenna/aerial.

The aperture or radiation pattern of an antenna is represented by a three-dimensional graph. It may even be polar plots having horizontal and vertical cross-sections. The aperture or radiation pattern is the geometric pattern of the relative field strengths. This field strength pertains to the fields that are emitted by the antenna. The geometric radiation pattern is a sphere in the case of an ideal isotropic antenna. The geometric radiation pattern is a toroid in the case of a typical dipole antenna. The antenna gain is minimum at the radiation pattern sidelobes and it is maximum at the radiation pattern backlobes.

Polarization is determined by the physical structure and orientation of an antenna. Polarization is affected by the reflections. Polarization in the case of an antenna is defined as the orientation of the electric field (E-plane) due to radio waves with respect to the Earth's surface. A simple straight wire antenna when mounted vertically will have one polarization and if mounted horizontally will have a different polarization. A vertical omni-directional Wi-Fi antenna has vertical polarization. In the case of directional antennas, the polarization of side lobes is different from the polarization of a main propagation lobe. In the case of elongated wave guide antennas, vertically placed antennas have horizontal polarization. It is an exception. Many vendors mark their antennas according to their emitted signal polarization.

Polarization is calculated as the sum of E-plane orientations over time. These E-plane orientations are projected onto an imaginary plane, which is perpendicular to the radio wave motion direction. An elliptical polarization (oblong projection) means that the antenna polarization varies over time. Polarization is categorized as follows:

(a) Linear polarization and
(b) Circular polarization

Linear polarization can be either horizontal or vertical depending upon antenna orientation. In the case of circular polarization, the antenna continuously varies radio wave electric field through all possible orientation values with regard to the Earth's surface. Further, circular polarizations like elliptical polarizations, on the basis of the direction of the propagation, are classified as follows:

(a) Right-hand polarization
(b) Left-hand polarization

Linearly, polarized antennas should be matched for good signal strength like horizontal should be used with horizontal and vertical with vertical. That is why transmitters on vehicles use circularly polarized antennas having good motional freedom. Electromagnetic wave polarization filters are used for filtering out the wave energy of an undesired polarization and for passing the wave energy of a desired polarization.

The bandwidth of an antenna is defined as the range of frequencies over which it is most effective. Usually, the antenna bandwidth is centered on the resonant frequency. The antenna bandwidth can be conveniently increased by using tapering antenna components (a feed horn) or thicker wires or replacing wires with cages, thus, simulating a thicker wire or just by combining multiple antennas into a single assembly (multiple antennas-single assembly). In the case of "multiple antennas-single assembly", the natural impedance selects the best antenna. However, there is a limit for antenna bandwidth and efficiency, which depends upon the antenna size. Transmission antennas possess a maximum power rating (typically in microwatt range), and any transgression of it may lead to heating or arcing or sparking of various antenna components.

The efficiency of an antenna is defined as follows:

Antenna efficiency = {(Actually radiated power)/(Power put into the antenna terminals)}

Also, antenna efficiency is defined as

Antenna efficiency = {(Radiation resistance)/(Total resistance)}

11.3 ARRAYS

An antenna array is a combination of several antenna elements. In certain cases, single-element antennas do not give the desired gain or radiation pattern. Therefore, an antenna array is required. Antenna array can be one or two-dimensional. Every type of antenna exhibits a unique radiation pattern. If an antenna array (several antennas) is used, then the overall radiation pattern also changes. An antenna array is characterized by a factor known as "array factor". The array factor in the case of an antenna array depends on several things like spacing among the elements, number of elements, amplitude and phase of signal applied to every antenna array element. An array factor quantifies the effect of combined radiating elements. The array factor does not take into the account the element-specific radiation pattern. The overall radiation pattern of an antenna array is determined by its array factor along with the radiation pattern of the antenna element. The overall radiation pattern results in a particular directivity. If the efficiency of the antenna is 100%, then the antenna directivity and antenna gain must be equal. Array directivity increases as the number of antenna array elements is increased in number. Antenna array elements can be isotropic. An isotropic radiating antenna array element will radiate an equal amount of power in all directions. An isotropic radiating antenna array element having an efficiency of 100% will have a directivity of 1 (i.e. 0 dB) and a gain of 1 (i.e. 0 dB). Antenna array elemental spacing greatly influences the array factor. If the antenna array elemental spacing is large, then directivity will be high. On the basis of the radiation pattern, the antenna array is classified as follows:

(a) Broadside array and
(b) End fire array

In the case of a broadside array, radiation is perpendicular to array orientation. It is shown in Figure 11.3.

In the case of a end fire array, radiation is in the same direction as the array orientation. It is shown in Figure 11.4.

The number of elements and elemental spacing determine the aperture. An aperture in the case of an antenna array is the total surface area of the radiating structure. An antenna array will have a large gain if it has a larger aperture. An aperture is characterized on the basis of aperture efficiency.

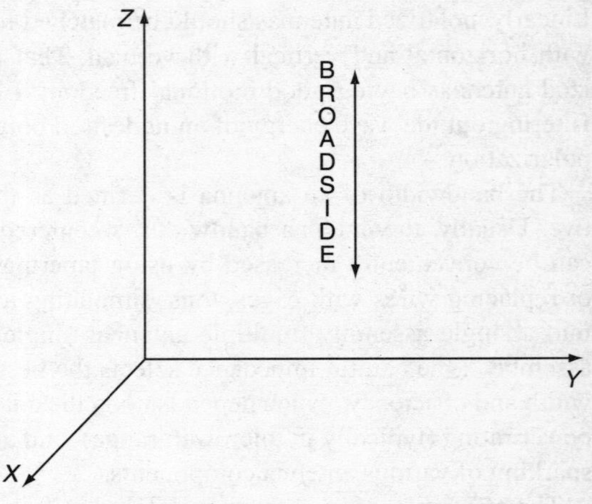

Figure 11.3 *Broadside array*

11.4 RETARDED POTENTIALS

Retarded potentials are the delayed potentials that get established because of the time-varying field. This delay is precisely called propagation time. The propagation time depends upon the distance and the propagation velocity. Delay is given by r/v_0. Retarded potential is given as

$$V(r, t) = (1/4\pi\varepsilon) \int_v ((\rho(r, t - r/v_0))/r)dv$$

The vector magnetic potential is given as

$$A(r, t) = (\mu / 4\pi) \int_v ((\mathbf{J}(r, t - r/v_0))/r)dv$$

Without considering the propagation time (delay), the potential for the time-varying field is given as

$$V(r, t) = (1/4\pi\varepsilon) \int_v ((\rho(r, t))/r)dv$$

and the vector magnetic potential is given as

$$A(r, t) = (\mu/4\pi) \int_v ((\mathbf{J}(r, t - r/v_0))/r)dv$$

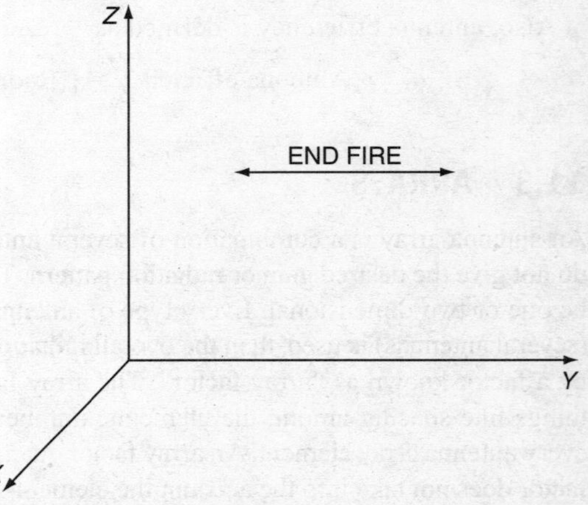

Figure 11.4 *End fire array*

where J is the conduction current density. Its units are A/m^2. If phase variation is also considered, then retarded potential is given as

$$V(r, t) = (1/4\pi\varepsilon) \int_v ((\rho(r, t - r/v_0)) / r)e^{-j\beta r}dv$$

The vector magnetic potential is given as

$$A(r, t) = (\mu/4\pi) \int_v ((\mathbf{J}(r, t - r/v_0))/r)e^{-j\beta r}dv$$

11.5 SHORT DIPOLE ANTENNA

A short dipole antenna is made by a simple wire. A short dipole antenna has a center-fed driven element for transmitting and also for the purpose of receiving RF energy. In a short dipole antenna, current amplitude I_0 is maximum at the center (where the dipole is fed) and zero at the ends. Short dipole antennas are used in televisions, FM bands, etc. A dipole is characterized by its length, wavelength, frequency, radiation pattern, gain and the feeder line.

The EM radiations E_θ from the dipole are expressed as

$$E_\theta = \frac{-iI_o \sin\theta}{4\varepsilon_o cr}\frac{L}{\lambda}e^{i(\omega t-kr)}$$

where L is the total length of the dipole. Figure 11.5 shows a short dipole antenna.

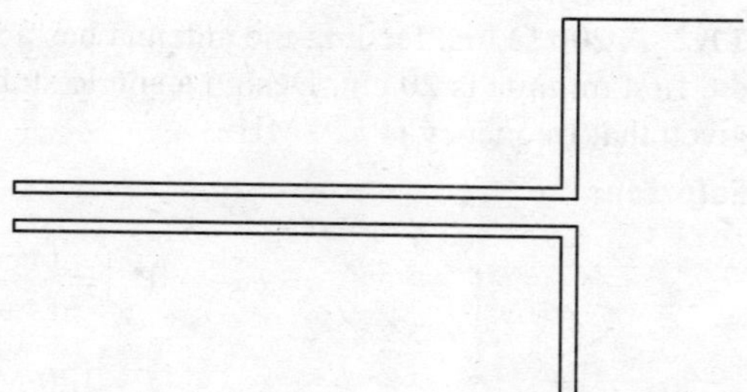

Figure 11.5 *A short dipole antenna*

11.6 RADIATION RESISTANCE

Radiation resistance is the resistive component of series impedance of dipole. The EM radiations emitted from a dipole cause this series impedance. From the EM fields, EM power can be very easily obtained. Radiation resistance for the case $L \ll \lambda$ is given as

$$R_{series} = \frac{\pi}{6}Z_0\left(\frac{L}{\lambda}\right)^2$$

11.7 BROADSIDE ARRAY

It is basically an antenna array whose direction of maximum radiation is always perpendicular to the line or the plane of the array according to the elements that lie on a line or a plane. A broadside array looks like a ladder. A uniform broadside array is a linear array. The elements of a uniform broadside array contribute EM fields of equal amplitude and phase. Figure 11.6 shows a broadside array.

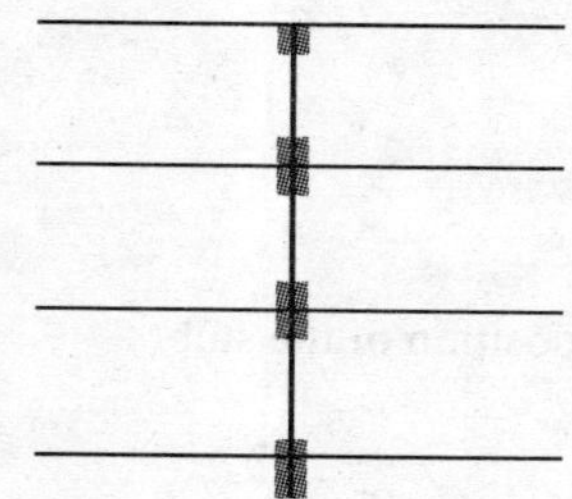

Figure 11.6 *A broadside array*

SOLVED QUESTIONS

11.1 A 100 Ω line feeding the antenna has SWR = 4, and the distance from the load to the first minima is 15 cm. Design a single stub matching to make SWR = 1, given that $f = 200$ MHz.

Solution:

$$|p| = \frac{\text{SWR}-1}{\text{SWR}+1} = \frac{3}{5} = 0.6$$

$$f = 200 \text{ MHz}$$

$$\therefore \lambda = \frac{c}{f} = \frac{3\times10^8}{200\times10^6} = \frac{3}{2} = 1.5 \text{ m}$$

We know that

$$2\beta d_{\min} - \varphi = \pi$$
$$\Rightarrow 2\beta d_{\min} = \pi + \varphi$$

It gives sufficient parameters to be computed by the reader to design a stub matching.

11.2 A 200 Ω line feeding the antenna has a standing wave ratio of 2.5. The distance from the load to the first minima is 20 cm. Design a single stub matching to make the standing wave ratio of $s = 1$. It is given that frequency is 120 MHz.

Solution:

$$|\Gamma_L| = \frac{(s-1)}{(s+1)} = \frac{(2.5-1)}{(2.5+1)} = 0.42$$

$$f = 120 \text{ MHz}$$

$$\lambda = \frac{f}{c} = \frac{3\times 10^8}{120\times 10^6}$$
$$\lambda = 2.5 \text{ m}$$

As we know that

$$(2\beta\ d_{\min} - \theta) = \pi$$
$$2\beta d_{\min} = \pi + \theta$$

$$= 2\times\frac{2\pi}{2.5}\times 0.2$$
$$= 0.32$$

the position of the stub

$$l_s = \frac{\lambda}{4\pi}(2\beta d_{\min} - \cos^{-1}(\Gamma_L))$$

$$|l_s| = \frac{\lambda}{4\pi}(0.32\pi - 1.13)$$

$$= \frac{0.2}{4\pi}(1.00 - 1.13)$$
$$(l_s) \cong 1.98 \text{ mm}$$

Therefore, l_t the length of the stub

$$l_t = \frac{\lambda}{2\pi}\tan^{-1}\sqrt{\frac{1-|\Gamma_L|^2}{2|\Gamma_L|}}$$

$$= \frac{2.5}{2\pi}\tan^{-1}\sqrt{\frac{0.823}{0.84}}$$
$$\cong 17.7 \text{ m}$$

UNSOLVED QUESTIONS

11.1 What is the effect of spacing?

11.2 How are the elements arranged in a broadside array?

11.3 Why do broadside arrays, if not being operated at the designed frequency, lose efficiency?

11.4 If broadside arrays are not operated at their designed frequency, why do they loose efficiency?

11.5 How are the elements arranged when two elements are used in a broadside array?

11.6 What happens to the major lobes as spacing between elements in a broadside array is increased?

SUMMARY

1. An antenna is a logical arrangement of several conductors. It is a transducer. It is also commonly termed an aerial.
2. The transmitting antenna transmits electromagnetic field waves.
3. The receiving antenna picks up this electromagnetic field (RF) signal.
4. An antenna converts electrical currents in the RF signal at the transmission side and then converts back this RF signal into an electrical current at the reception side.
5. Antenna designs are reciprocal in the sense that all transmitter antenna parameters are also applicable to the receiving antenna (reciprocity theorem).
6. Antenna tuning refers to charging the electrical resonance in an antenna, which is done by adjusting the length of the antenna.
7. Antenna shields are used for noise rejection.
8. The transmitting antenna should have a high power rating.
9. The receiving antenna should have good noise rejection properties.
10. Important antenna parameters are resonant frequency, impedance, gain, aperture (radiation pattern), polarization, bandwidth and efficiency.
11. Antennas are made resonant on harmonic frequencies by having their lengths a fraction of the desired wavelength.
12. The electrical length of an antenna is the ratio of the physical length of the wire to the velocity factor.
13. The standing wave ratio (SWR) is the ratio of maximum to minimum measurable power of the wave.

14. Gain of an antenna is the ratio of radiation intensity power per unit surface in a given direction at an arbitrary distance to radiation intensity of a hypothetical isotropic antenna at the same distance.

15. Aperture or radiation pattern of an antenna is a 3-D graph.

16. In an antenna, polarization is the orientation of an electric field (E-plane) due to the radio wave with respect to the Earth's surface.

17. Polarization can be linear or circular.

18. On the basis of the direction of propagation, elliptical polarization can be right handed or left handed.

19. The bandwidth of an antenna is the range of frequencies over which it is most effective.

20. The efficiency of an antenna is defined as the ratio of radiation resistance to total resistance.

21. Alternatively, antenna efficiency is also defined as the ratio of actually radiated power to the amount of power put into antenna terminals.

22. An antenna array is a combination of several antenna elements. It can be of broadside or end fire types.

23. Retarded potentials are delayed potentials that get established because of time-varying field.

24. A short dipole antenna is a simple antenna, which is fed at the centre for transmitting and receiving RF signals. These antennas are used in TV, FM bands, etc.

25. Radiation resistance is given as

$$\frac{\pi}{\sigma} z_o \left(\frac{L}{\lambda}\right)^2$$

MULTIPLE-CHOICE QUESTIONS

1. Antenna tuning is done by changing its __________.
   (a) Inductive reactances (b) Capacitive reactances (c) Both (a) and (b) (d) None of these

2. Resonant frequency of an antenna depends upon its __________.
   (a) Physical length (b) Electrical length (c) None of these (d) Cannot say

3. Impedance of an antenna is matched to feed __________.
   (a) Line (b) Radio (c) Both (a) and (b) (d) None of these

4. In low-power applications, ideal SWR is __________.
   (a) 1:1 (b) 1:2 (c) 2:1 (d) 2:2

5. Which of the following is a low-gain antenna?
   (a) Dish antenna on a space craft (b) Wi-Fi antenna
   (c) Both (a) and (b) (d) None of these

6. In the case of an ideal isotropic antenna, the geometric radiation pattern is __________.
   (a) Circle (b) Sphere (c) Parabola (d) Hyperbola

7. Antenna gain is minimum at radiation pattern __________.
   (a) Side lobes (b) Back lobes
   (c) None of these (d) Gain is independent of the radiation pattern

8. Elliptical polarization can be of __________.
   (a) Left hand (b) Right hand (c) Either (a) or (b) (d) None of these

9. Transmitters used on vehicles are__________ polarized.
   (a) Linearly (b) Circularly (c) Both (a) and (b) (d) None of these

10. Antenna bandwidth can be increased by using __________.
    (a) A feed horn (b) Thicker wires (c) Cages (d) All of these

11. Antenna array is __________.
    (a) 1-D (b) 2-D (c) Either (a) or (b) (d) None of these

12. Array factor depends upon __________.
    (a) Number of elements (b) Spacing among elements
    (c) Phase of the applied signal (d) All of these

13. If antenna directivity and antenna gain are equal, then antenna efficiency is __________%.
    (a) 20 (b) 50 (c) 75 (d) 100

14. If antenna array elemental spacing is large, then directivity will be __________.
    (a) Small (b) More (c) None of these (d) Cannot say

15. On the basis of the radiation pattern, an antenna array is __________.
    (a) Broadside (b) End fire (c) Either (a) or (b) (d) None of these

16. In __________ array, radiation is perpendicular to array orientation.
    (a) Broadside (b) End fire (c) None of these (d) Cannot say

17. In __________ array, radiation is in the same direction as the array orientation.
    (a) Broadside (b) End fire (c) None of these (d) Cannot say

18. A large aperture means antenna gain will be ________.
(a) Large (b) Small (c) None of these (d) Cannot say

19. A short dipole antenna is ___________.
(a) Left end-fed (b) Right end-fed (c) Centre-fed (d) None of these

20. Radiation resistance for the case $L << \lambda$ is given as__________.
(a) $\frac{\pi}{6} Z_o \left(\frac{L}{\lambda}\right)$ (b) $\frac{\pi}{6} Z_o \left(\frac{L}{\lambda}\right)^2$ (c) $\frac{\pi}{3} Z_o \left(\frac{L}{\lambda}\right)$ (d) $\frac{\pi}{3} Z_o \left(\frac{L}{\lambda}\right)^2$

ANSWERS TO MULTIPLE-CHOICE QUESTIONS

(1) c; (2) b; (3) c; (4) a; (5) b; (6) b; (7) a; (8) c; (9) b; (10) d;
(11) c; (12) d; (13) d; (14) b; (15) c; (16) a; (17) b; (18) a; (19) c; (20) b.

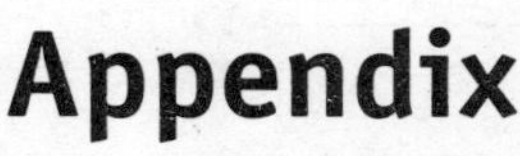

Appendix

Program 1: Addition of Two Vectors

```
clc
clear all
disp('----------------------------------------------------------');
disp('           ADDITION OF TWO VECTORS P & Q (P + Q)           ');
disp('----------------------------------------------------------');
disp('Enter values for vector P ex [1 -1 0]');
P = input('Enter P:');
disp('Enter values for vector Q ex [2 3 -1]');
Q = input('Enter Q:');
disp('Sum of two vectors P and Q is');
C = P + Q                  % Sum of two vectors
C_mag = sqrt(dot(C,C));  % magnitude of vector C
fprintf('Magnitude of the sum of two vectors is %f\n',C_mag);
```

Simulation Results of Program 1 Implementation (Addition of Two Vectors)

```
----------------------------------------------------
         ADDITION OF TWO VECTORS P & Q (P + Q)
----------------------------------------------------
Enter values for vector P ex [1 -1 0]
Enter P:[0 2 4]
Enter values for vector Q ex [2 3 -1]
Enter Q:[2 1 -1]
Sum of two vectors P and Q is

C =

     2     3     3

Magnitude of the sum of two vectors is 4.690416
```

Program 2: Subtraction of Two Vectors

```
clc
clear all
disp('-----------------------------------------------------------');
disp('          SUBTRACTION OF TWO VECTORS Q & R (Q - R)          ');
disp('-----------------------------------------------------------');
disp('Enter values for vector Q ex [1 -1 0]');
Q = input('Enter Q:');
disp('Enter values for vector R ex [2 3 -1]');
R = input('Enter R:');
disp('Difference of two vectors Q and R is');
C = Q - R                   % difference of two vectors
C_mag = sqrt(dot(C,C));  % magnitude of vector C
fprintf('Magnitude of the difference of two vectors is %f\n',C_mag);
```

Simulation Results of Program 2 Implementation (Subtraction of Two Vectors)

```
-----------------------------------------------------
       SUBTRACTION OF TWO VECTORS Q & R (Q - R)
-----------------------------------------------------
Enter values for vector Q ex [1 -1 0]
Enter Q:[2 1 -1]
Enter values for vector R ex [2 3 -1]
Enter R:[2 1 0]
Difference of two vectors Q and R is

C =

   0     0    -1

Magnitude of the difference of two vectors is 1.000000
```

Program 3: Finding the Magnitude of a Vector

```
clc
clear all
disp('-----------------------------------------------------------');
disp('              MAGNITUDE OF VECTOR Q                       ');
disp('-----------------------------------------------------------');
disp('Enter values for vector Q ex [1 -1 0]');
Q = input('Enter Q:');
Q_mag = norm(Q);  % magnitude of vector Q
fprintf('Magnitude of the difference of two vectors is %f\n',Q_mag);
```

Simulation Results of Program 3 Implementation (Finding the Magnitude of a Vector)

```
-----------------------------------------------------
          MAGNITUDE OF VECTOR Q
-----------------------------------------------------
Enter values for vector Q ex [1 -1 0]
Enter Q:[2 1 -1]
Magnitude of the difference of two vectors is 2.449490
```

Program 4: Finding the Unit Vector

```
clc
clear all
disp('-------------------------------------------------------------');
disp('             UNIT VECTOR R                                   ');
disp('-------------------------------------------------------------');
disp('Enter values for vector R ex [1 -1 0]');
R = input('Enter R:');
R_mag = norm(R);  % magnitude of vector R
R_unit = R/R_mag
```

Simulation Results of Program 4 Implementation (Finding the Unit Vector)

```
-----------------------------------------------------
        UNIT VECTOR R
-----------------------------------------------------
Enter values for vector R ex [1 -1 0]
Enter R:[2 1 0]

R_unit =

    0.8944    0.4472         0
```

Program 5: Finding the Dot Product of Two Vectors

```
clc
clear all
disp('---------------------------------------------------------------');
disp('          DOT PRODUCT OF VECTORS Q & R (Q . R)                 ');
disp('---------------------------------------------------------------');
disp('Enter values for vector Q ex [1 -1 0]');
Q = input('Enter Q:');                % user inputs vector Q
disp('Enter values for vector R ex [2 3 -1]');
R = input('Enter R:');                % user inputs vector R
c = dot(Q,R);                        %dot is a keyword in matlab
fprintf('The dot product of vector Q and vector R is: %d\n',c)
```

Simulation Results of Program 5 Implementation (Finding the Dot Product of Two Vectors)

```
-----------------------------------------------------
          DOT PRODUCT OF VECTORS Q & R (Q . R)
-----------------------------------------------------
Enter values for vector Q ex [1 -1 0]
Enter Q:[2 1 -1]
Enter values for vector R ex [2 3 -1]
Enter R:[2 1 0]
The dot product of vector Q and vector R is: 5
```

Program 6: Finding the Angle Between Two Vectors

```
clc
clear all
disp('--------------------------------------------------------');
disp('          ANGLE BETWEEN TWO VECTORS P & Q               ');
disp('--------------------------------------------------------');
disp('Enter values for vector P ex [1 -1 0]');
P = input('Enter P:');
P_mag = norm(P);
disp('Enter values for vector Q ex [2 3 -1]');
Q = input('Enter Q:');
Q_mag = norm(Q);
C = dot(P,Q);                          % Dot product of two vectors
theta =acos(C/(P_mag*Q_mag))*(180/pi);
fprintf('Angle between two vectors is %f\n',theta);
```

Simulation Results of Program 6 Implementation (Finding the Angle Between Two Vectors)

```
--------------------------------------------------------
          ANGLE BETWEEN TWO VECTORS P & Q
--------------------------------------------------------
Enter values for vector P ex [1 -1 0]
Enter P:[0 2 4]
Enter values for vector Q ex [2 3 -1]
Enter Q:[2 1 -1]
Angle between two vectors is 100.519735
```

Program 7: Finding the Cross Product of Two Vectors

```
clc
clear all
disp('-------------------------------------------------------------');
disp('          CROSS PRODUCT OF VECTORS P & Q (P x Q)          ');
disp('-------------------------------------------------------------');
disp('Enter values for vector P ex [1 -1 0]');
P = input('Enter P:');                  % user inputs vector P
P_mag = sqrt(dot(P,P));                  % magnitude of vector P
fprintf('Magnitude of vector P: %f\n', P_mag);
disp('Enter values for vector Q ex [2 3 -1]');
Q = input('Enter Q:');                  % user inputs vector Q
Q_mag = sqrt(dot(Q,Q));                  % magnitude of vector Q
fprintf('Magnitude of vector Q: %f\n', Q_mag);S
disp('Cross product (P x Q) is ');
c = cross(P,Q)                          %cross is a keyword in matlab
c_mag = norm(c);                         % norm is a keyword in matlab
fprintf('Magnitude of vector C: %f\n', c_mag);
disp('The unit vector normal to P and Q is')
Unit_V = c/c_mag
```

Simulation Results of Program 7 Implementation (Finding the Cross Product of Two Vectors)

```
-------------------------------------------------------------
          CROSS PRODUCT OF VECTORS P & Q (P x Q)
-------------------------------------------------------------
Enter values for vector P ex [1 -1 0]
Enter P:[0 2 4]
Magnitude of vector P: 4.472136
Enter values for vector Q ex [2 3 -1]
Enter Q:[2 1 -1]
Magnitude of vector Q: 2.449490
Cross product (P x Q) is

c =

    -6     8    -4

Magnitude of vector C: 10.770330
The unit vector normal to P and Q is

Unit_V =

   -0.5571    0.7428   -0.3714
```

Program 8: Finding the Cross Product of Two Vectors (Another Program)

```
clc
clear all
disp('--------------------------------------------------------');
disp('         CROSS PRODUCT OF VECTORS P & Q (P x Q)       ');
disp('--------------------------------------------------------');
disp('Enter values for vector P ex [1 -1 0]');
P = input('Enter P:');              % user inputs vector A
P_mag = sqrt(dot(P,P));              % magnitude of vector A
fprintf('Magnitude of vector P: %f\n', P_mag);
disp('Enter values for vector Q ex [2 3 -1]');
Q = input('Enter Q:');              % user inputs vector B
Q_mag = sqrt(dot(Q,Q));              % magnitude of vector B
fprintf('Magnitude of vector Q: %f\n', Q_mag);
disp('Cross product (P x Q) is ');
c = cross(P,Q)                      %cross is a keyword in matlab
c_mag = norm(c);                     % norm is a keyword in matlab
fprintf('Magnitude of vector C: %f\n', c_mag);
disp('The unit vector normal to P and Q is')
Unit_V = c/c_mag
phi = 180 - (asin(norm(cross(P,Q))/(norm(P)*norm(Q))))*180/pi;
fprintf('Angle between two vectors is: %f\n', phi);
```

Simulation Results of Program 8 Implementation
(Finding the Cross Product of Two Vectors (Another Program))

```
--------------------------------------------------------
         CROSS PRODUCT OF VECTORS P & Q (P × Q)
--------------------------------------------------------
Enter values for vector P ex [1 -1 0]
Enter P:[0 2 4]
Magnitude of vector P: 4.472136
Enter values for vector Q ex [2 3 -1]
Enter Q:[2 1 -1]
Magnitude of vector Q: 2.449490
Cross product (P × Q) is

c =

    -6     8    -4

Magnitude of vector C: 10.770330
```

```
The unit vector normal to P and Q is

Unit_V =

   -0.5571    0.7428   -0.3714

Angle between two vectors is: 100.519735
```

Program 9: Finding P. (Q x R)

```
clc
clear all
disp('--------------------------------------------------------------');
disp('                           P.(Q x R)                          ');
disp('--------------------------------------------------------------');
disp('Enter values for vector P ex [1 -1 0]');
P = input('Enter P:');                 % user inputs vector P
disp('Enter values for vector Q ex [2 3 -1]');
Q = input('Enter Q:');                 % user inputs vector Q
disp('Enter values for vector R ex [1 0 2]');
R = input('Enter R:');                 % user inputs vector R
disp('Cross product (Q x R) is ');
c = cross(Q,R)                         %cross is a keyword in matlab
D = dot(P,c);
fprintf('The value of the expression P.(Q x R) is %d\n',D)
```

Simulation Results of Program 9 Implementation (Finding P. (Q x R))

```
-----------------------------------------------------
                  P.(Q × R)
-----------------------------------------------------
Enter values for vector P ex [1 -1 0]
Enter P:[0 2 4]
Enter values for vector Q ex [2 3 -1]
Enter Q:[2 1 -1]
Enter values for vector R ex [1 0 2]
Enter R:[2 1 0]
Cross product (Q × R) is

c =

     1    -2     0

The value of the expression P.(Q × R) is -4
```

Program 10: Finding (P x Q) x R

```
clc
clear all
disp('---------------------------------------------------------------');
disp('                    (P x Q) x R                                ');
disp('---------------------------------------------------------------');
disp('Enter values for vector P ex [1 -1 0]');
P = input('Enter P:');                   % user inputs vector P
disp('Enter values for vector Q ex [2 3 -1]');
Q = input('Enter Q:');                   % user inputs vector Q
disp('Enter values for vector R ex [1 0 2]');
R = input('Enter R:');                   % user inputs vector R
disp('Cross product (P x Q) is ');
c = cross(P,Q)                           %cross is a keyword in matlab
disp('The value of final expression (P x Q) x R')
D = cross(c,R)
```

Simulation Results of Program 10 Implementation (Finding (P x Q) x R)

```
----------------------------------------------------
                    (P x Q) x R
----------------------------------------------------
Enter values for vector P ex [1 -1 0]
Enter P:[0 2 4]
Enter values for vector Q ex [2 3 -1]
Enter Q:[2 1 -1]
Enter values for vector R ex [1 0 2]
Enter R:[2 1 0]
Cross product (P x Q) is

c =

    -6     8    -4

The value of final expression (P x Q) x R

D =

     4    -8   -22
```

Program 11: Finding the Dot and Cross Products of Vectors

```
clc
clear all
disp('--------------------------------------------------------');
disp('     DOT AND CROSS PRODUCT OF VECTORS                   ');
disp('--------------------------------------------------------');
syms t
P = [t^2 cos(t) 4];
Q = [exp(t) t sin(t)];
disp('Dot product of P and Q is');
dot(P,Q)
disp('Cross product of P and Q is');
cross(P,Q)
```

Simulation Results of Program 11 Implementation (Finding the Dot and Cross Products of Vectors)

```
--------------------------------------------------------
     DOT AND CROSS PRODUCT OF VECTORS
--------------------------------------------------------
Dot product of P and Q is

ans =

conj(t)^2*exp(t)+cos(conj(t))*t+4*sin(t)

Cross product of P and Q is

ans =

[ cos(t)*sin(t)-4*t, 4*exp(t)-t^2*sin(t),   t^3-cos(t)*exp(t)]
```

Program 12: Coordinate System and Transformation Cartesian to Cylindrical

```
clc
clear all
disp('-----------------------------------------------------------');
disp('          COORDINATE SYSTEM & TRANSFORMATION                ');
disp('-----------------------------------------------------------');
disp('-----------------------------------------------------------');
disp('          CARTESIAN TO CYLINDERICAL CONVERSION              ');
disp('-----------------------------------------------------------');
x = input('Enter value of x co ordinate:');
y = input('Enter value of y co ordinate:');
z = input('Enter value of z co ordinate:');
rho = sqrt(x^2+y^2);
phi = (atan(y/x))*180/pi;
disp('The values for cylindrical co ordinate systems are');
disp(rho)
disp(phi)
disp(z)
```

Simulation Results of Program 12 Implementation (Coordinate System and Transformation Cartesian to Cylindrical)

```
----------------------------------------------------
         COORDINATE SYSTEM & TRANSFORMATION
----------------------------------------------------
----------------------------------------------------
         CARTESIAN TO CYLINDERICAL CONVERSION
----------------------------------------------------
Enter value of x co ordinate:1
Enter value of y co ordinate:2
Enter value of z co ordinate:4
The values for cylindrical co ordinate systems are
    2.2361

   63.4349

     4
```

Program 13: Coordinate System and Transformation Cartesian to Spherical

```
clc
clear all
disp('-----------------------------------------------------------');
disp('          COORDINATE SYSTEM & TRANSFORMATION                ');
disp('-----------------------------------------------------------');
disp('-----------------------------------------------------------');
disp('          CARTESIAN TO SPHERICAL CONVERSION                 ');
disp('-----------------------------------------------------------');
x = input('Enter value of x co ordinate:');
y = input('Enter value of y co ordinate:');
z = input('Enter value of z co ordinate:');
r = sqrt(x^2+y^2+z^2);
theta1 = sqrt(x^2+y^2)/z;
theta = atan(theta1)*180/pi;
phi = atan(y/x)*180/pi;
disp('The values for spherical co ordinate systems are');
disp(r)
disp(theta)
disp(phi)
```

Simulation Results of Program 13 Implementation (Coordinate System and Transformation Cartesian to Spherical)

```
---------------------------------------------------------
         COORDINATE SYSTEM & TRANSFORMATION
---------------------------------------------------------
---------------------------------------------------------
         CARTESIAN TO SPHERICAL CONVERSION
---------------------------------------------------------
Enter value of x co ordinate:2
Enter value of y co ordinate:1
Enter value of z co ordinate:3

The values for spherical co ordinate systems are
    3.7417

   36.6992

   26.5651
```

Program 14: Coulomb's Law

```
clc
clear all
disp('------------------------------------------------------------------');
disp('                          COULOMB LAW                              ');
disp('------------------------------------------------------------------');
e0 = 8.8541e-12;
k = 1/(4*pi*e0);
Q1 = input('Enter value of 1st charge Q1 (in coulomb):');
Q2 = input('Enter value of 2nd charge Q2 (in coulomb):');
r =  input('Enter value of distance R (in meter):');
F1 = Q1*Q2/r^2;
F = k*F1;
fprintf('Force between Q1 and Q2 is (in Newton)%f\n',F);
```

Simulation Results of Program 14 Implementation (Coulomb's Law)

```
-------------------------------------------------------
                  COULOMB'S LAW
-------------------------------------------------------
Enter value of 1st charge Q1 (in coulomb):-2e-6
Enter value of 2nd charge Q2 (in coulomb):-20
Enter value of distance R (in meter):0.5
Force between Q1 and Q2 is (in Newton)1438022.548577
```

Program 15: Coulomb's Law (Another Program)

```
clc
clear all
disp('-------------------------------------------------------');
disp('                     COULOMB LAW                       ');
disp('-------------------------------------------------------');
e0 = 8.8541e-12;
k = 1/(4*pi*e0);
Q1 = input('Enter value of 1st charge Q1 (in coulomb):');
Q2 = input('Enter value of 2nd charge Q2 (in coulomb):');
P1x = input('Enter X- coordinate of point P1:');
P1y = input('Enter Y- coordinate of point P1:');
P1z = input('Enter Z- coordinate of point P1:');
P2x = input('Enter X- coordinate of point P2:');
P2y = input('Enter Y- coordinate of point P2:');
P2z = input('Enter Z- coordinate of point P2:');
P = [P2x-P1x P2y-P1y P2z-P1z]
r = sqrt((P2x-P1x)^2+(P2y-P1y)^2+(P2z-P1z)^2);
fprintf('The distance between two charges is %f\n',r);
F1 = Q1*Q2/r^2;
F = k*F1;
fprintf('Force between Q1 and Q2 is (in Newton)%f\n',F);
```

Simulation Results of Program 15 Implementation (Coulomb's Law (Another Program)

```
-------------------------------------------------------
                COULOMB LAW
-------------------------------------------------------
Enter value of 1st charge Q1 (in coulomb):242e-9
Enter value of 2nd charge Q2 (in coulomb):50e-6
Enter X- coordinate of point P1:0.03
Enter Y- coordinate of point P1:0.01
Enter Z- coordinate of point P1:-0.04
Enter X- coordinate of point P2:0.03
Enter Y- coordinate of point P2:0.08
Enter Z- coordinate of point P2:0.02

P =

        0    0.0700    0.0600

The distance between two charges is 0.092195
Force between Q1 and Q2 is (in Newton)12.794171
```

Program 16: Finding the Electric Field Intensity

```
clc
clear all
disp('-----------------------------------------------------------');
disp('                    ELECTRIC FIELD INTENSITY              ');
disp('-----------------------------------------------------------');
B = input('Enter magnetic flux density B ex[0.4 0.5 0]:');
V = input('Enter velocity V ex [1 2 3]:');
disp('The electric field intensity is ');
E = -cross(V,B)
```

Simulation Results of Program 16 Implementation (Finding the Electric Field Intensity)

```
------------------------------------------------------
          ELECTRIC FIELD INTENSITY
------------------------------------------------------
Enter magnetic flux density B ex[0.4 0.5 0]:[0.4 -0.6 1]
Enter velocity V ex [1 2 3]:[1e4 0 0]
The electric field intensity is

E =

         0       10000        6000
```

Program 17: Finding the Electric Field Intensity (Another Program)

```
clc
clear all
disp('-----------------------------------------------------------');
disp('                ELECTRIC FIELD INTENSITY                   ');
disp('-----------------------------------------------------------');
B = input('Enter magnetic flux density B ex [0.4 0.5 0]:');
V = input('Enter velocity V ex [1 2 3]:');
E = input('Enter electric field intensity ex [2 0 1]:');
disp('Force on electron due to E and B is ');
F = (cross(V,B)+E);
F1= F*1.6e-19
```

Simulation Results of Program 17 Implementation (Finding the Electric Field Intensity (Another Program))

```
-----------------------------------------------------------
                 ELECTRIC FIELD INTENSITY
-----------------------------------------------------------
Enter magnetic flux density B ex [0.4 0.5 0]:[0.4 -0.6 1]
Enter velocity V ex [1 2 3]:[1e4 0 0]
Enter electric field intensity ex [2 0 1]:[2 1 3]
Force on electron due to E and B is

F1 =

  1.0e-014 *

    0.0000   -0.1600   -0.0960
```

Program 18: Finding the Force on a Wire

```
clc
clear all
disp('------------------------------------------------------------');
disp('                 FORCE ON A WIRE                            ');
disp('------------------------------------------------------------');
L = input('Enter length of wire L (in meter):');
theta = input('Enter angle created by wire on the solenoid (in degree):');
theta1 = theta*(pi/180);
tht = sin(theta1);
I = input('Enter current (ampere):');
B = input('Enter magnetic field due to solenoid (in tesla):');
F = L*I*B*tht;
fprintf('The force on the wire (in newton) is %f\n',F);
```

Simulation Results of Program 18 Implementation (Finding the Force on a Wire)

```
---------------------------------------------------------
             FORCE ON A WIRE
---------------------------------------------------------
Enter length of wire L (in meter):0.04
Enter angle created by wire on the solenoid (in degree):60
Enter current (ampere):3
Enter magnetic field due to solenoid (in tesla):0.25
The force on the wire (in newton) is 0.025981
```

Program 19: Finding the Magnetic Field of a Coil

```
clc
clear all
disp('----------------------------------------------------');
disp('       MAGNETIC FIELD OF A COIL                     ');
disp('----------------------------------------------------');
N = input('Enter number of turns of the coil:');
R = input('Enter radius of coil (in meter):');
I = input('Enter current in coil (in ampere):');
meu0 = 4*pi*1e-7;
B = (meu0*N*I)/(4*pi*2*R);
fprintf('Magnitude of the magnetic field is (in tesla)%f\n', B);
```

Simulation Results of Program 19 Implementation (Finding the Magnetic Field of a Coil)

```
----------------------------------------------------
       MAGNETIC FIELD OF A COIL
----------------------------------------------------
Enter number of turns of the coil:60
Enter radius of coil (in meter):0.045
Enter current in coil (in ampere):6
Magnitude of the magnetic field is (in tesla)0.000400
```

Program 20: Finding the Magnetic Field of a Coil (Another Program)

```
clc
clear all
disp('-------------------------------------------------------');
disp('          MAGNETIC FIELD OF A COIL                     ');
disp('-------------------------------------------------------');
N = input('Enter number of turns of the coil:');
R = input('Enter radius of coil (in meter):');
I = input('Enter current in coil (in ampere):');
meu0 = 4*pi*1e-7;
B = (meu0*N*I*R^2)/(4*pi*(R^2+R^2)^1.5);
fprintf('Magnitude of the magnetic field is (in tesla)%f\n', B);
```

Simulation Results of Program 20 Implementation (Finding the Magnetic Field of a Coil (Another Program)

```
-----------------------------------------------------
      MAGNETIC FIELD OF A COIL
-----------------------------------------------------
Enter number of turns of the coil:60
Enter radius of coil (in meter):0.045
Enter current in coil (in ampere):6
Magnitude of the magnetic field is (in tesla)0.000283
```

Program 21: Plotting Polar Plot

```
clc
clear all
close all
disp('-----------------------------------------------------');
disp('                    POLAR PLOT                       ');
disp('-----------------------------------------------------');
theta = 0:0.01:2*pi;
r = sin(2*theta);
subplot(121)
polar(theta,r)
title('Polar plot for sin(2*theta)');
r1 = cos(2*theta);
subplot(122)
polar(theta,r1)
title('Polar plot for cos(2*theta)');
```

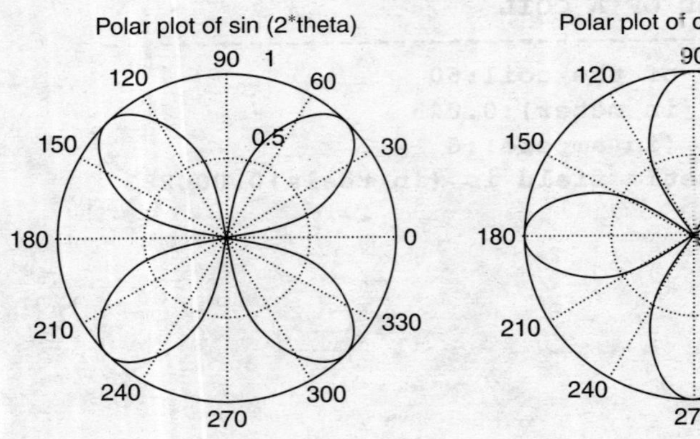

Program 22: Polar to Cartesian Conversion

```
clc
clear all
close all
disp('-------------------------------------------------');
disp('      POLAR TO CARTESIAN CONVERSION              ');
disp('-------------------------------------------------');
theta = 0:0.01:2*pi;
r = sin(2*theta);
x = r.*cos(theta);
y = r.*sin(theta);
plot(x,y)
title('Plot of x and y where x=r*cos(theta) and y = r*sin(theta)')
```

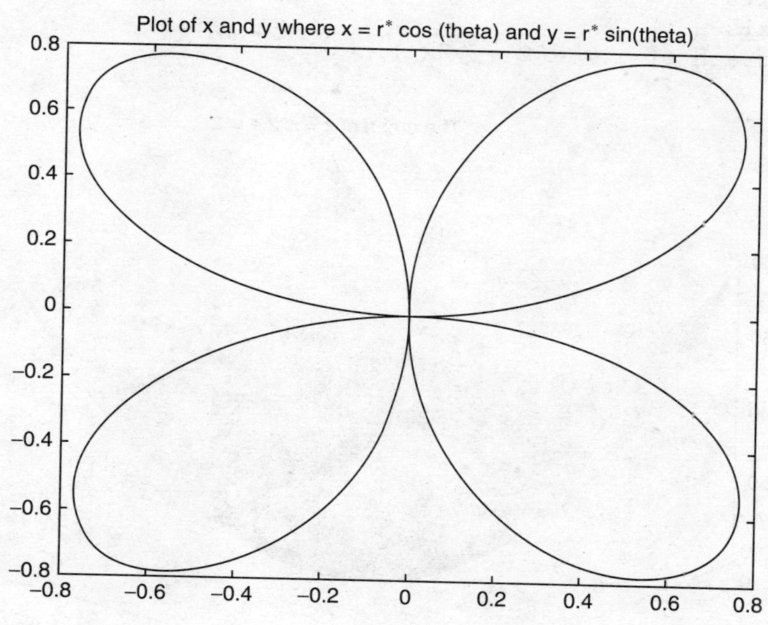

Program 23: Plotting the Mesh Plot in a Rectangular Region

```
clc
clear all
close all
disp('---------------------------------------------------');
disp('MESH PLOT OF z = x^2+y^2 IN RECTANGULAR REGION  ');
disp('---------------------------------------------------');
x=-10:0.01:10;          % define range of x
y=-10:0.01:10;          % define range of y
[x1,y1]=meshgrid(x,y);  % meshgrid produces 3-d array
z=x1.^2+y1.^2;
mesh(x1,y1,z);
xlabel('x-axis-->')
ylabel('y-axis-->')
zlabel('z-axis-->')
title('The graph of z = x^2 + y^2')
```

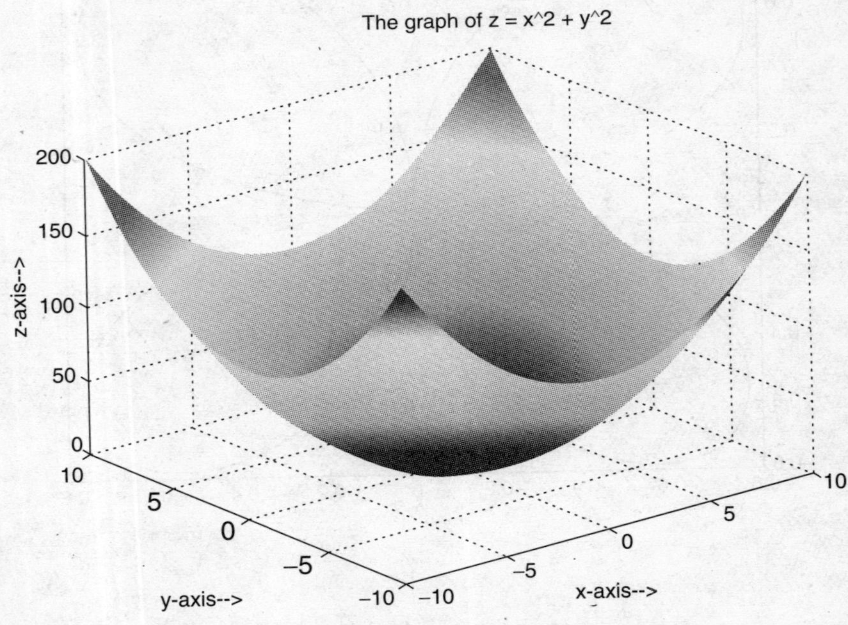

Program 24: Plotting the Mesh Plot in Cylindrical Coordinates

```
clc
clear all
close all
disp('-------------------------------------------------');
disp('         MESH PLOT IN CYLINDRICAL COORDINATES    ');
disp('-------------------------------------------------');
R=0:0.01:10;
theta=0:pi/12:10*pi;                          % define theta
[r1,theta1]=meshgrid(R,theta);
x=r1.*cos(theta1);                             % x = rcos(theta)
y=r1.*sin(theta1);                             % y = rsin(theta)
z = sqrt(x.^2+y.^2);
mesh(x,y,z);
xlabel('x-axis')
ylabel('y-axis')
zlabel('z-axis')
title('Mesh plot in cylindrical coordinates')
```

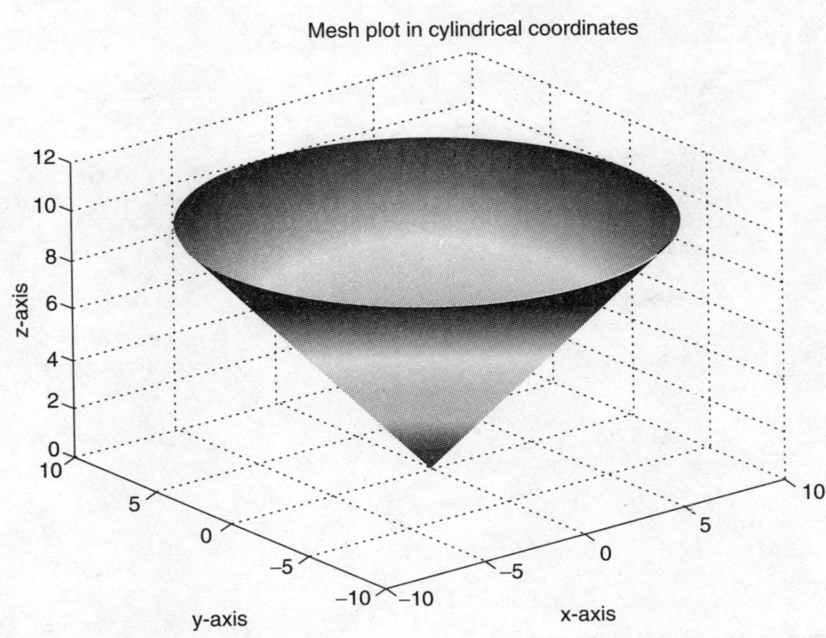

MATLAB has different commands for coordinate conversion. The summary of the commands are given below.

Coordinate Conversion	Command	Implementation
Cartesian to Spherical	cart2sph	`[th,phi,r] = cart2sph(x,y,z)`
Cartesian to Polar	cart2pol	`[th,r] = cart2pol(x,y)`
Spherical to Cartesian	sph2cart	`[x,y,z] = sph2cart(th,phi,r)`
Polar to Cartesian	pol2cart	`[x,y] = pol2cart(th,r)`

Index